RCAF

*To Dad
With our love and best wishes
Patti, Gord, Kate and Scott
April 1999*

Historical Publication 14.
**Canadian War Museum, National Museum of Man,
National Museums of Canada**
John Swettenham, Editor

Publications in this series by the Canadian War Museum, National Museum of Man, may be obtained from Marketing Services Division, National Museums of Canada, Ottawa, Ontario.

[1] *Canada and the First World War*, by John Swettenham. Canadian War Museum, Ottawa, 1968, Bilingual.

[2] *D-Day*, by John Swettenham. Canadian War Museum, Ottawa, 1969, Bilingual.

[3] *Canada and the First World War*, by John Swettenham. Based on the Fiftieth Anniversary Armistice Display at the Canadian War Museum, Ryerson, Toronto, 1969. Illustrated. Republished in paperback by McGraw-Hill Ryerson, 1973.

[4] *Canadian Military Aircraft*, by J.A. Griffin. Queen's Printer, Ottawa, 1969, Bilingual.

5. *The Last War Drum: The North West Campaign of 1885*, by Desmond Morton. Hakkert, Toronto, 1972.

6. *The Evening of Chivalry*, by John Swettenham. National Museums of Canada, Ottawa, 1972. Edition française, *Le crépuscule de la chevalerie*.

7. *Valiant Men: Canada's Victoria Cross and George Cross Winners*, edited by John Swettenham. Hakkert, Toronto, 1973.

8. *Canada Invaded 1775-1776*, by George F.G. Stanley. Hakkert, Toronto, 1973.

9. *The Canadian General, Sir William Otter*, by Desmond Morton. Hakkert, Toronto, 1974.

10. *Silent Witnesses*, by John Swettenham and Herbert F. Wood. (*Témoins Silencieux*, adapté par Jacques Gouin).

11. *Broadcast From the Front: Canadian Radio Overseas in The Second World War*, by A.E. Powley. Hakkert, Toronto, 1975.

12. *Canada's Fighting Ships*, by K.R. Macpherson. Samuel Stevens Hakkert & Co., Toronto, 1975.

13. *Canada's Nursing Sisters*, by G.W.L. Nicholson. Samuel Stevens Hakkert & Co., Toronto 1975.

NATIONAL MUSEUM OF MAN
NATIONAL MUSEUMS OF CANADA

RCAF

SQUADRON HISTORIES AND AIRCRAFT
1924-1968

SAMUEL KOSTENUK
JOHN GRIFFIN

SAMUEL STEVENS
HAKKERT & COMPANY
TORONTO & SARASOTA
1977

This book has been written and published with the aid of funds provided by the National Museum of Man, National Museums of Canada. In the writing of this book the inferences drawn and the opinions expressed are those of the author himself, and the National Museums of Canada are in no way responsible for his presentation of the facts as stated.

© Crown Copyright 1977
All Rights Reserved
It is illegal to reproduce this book, or any portion thereof, without written permission from the publishers. Reproduction of this material without authorization by any duplication process whatsoever is a violation of copyright.

Design by Helmut Rath

International Standard Book Number: 0-88866-577-6
International Standard Serial Number 0317-3860
Library of Congress Catalogue Number 76-50945

Published by A.M. Hakkert Ltd.,
554 Spadina Crescent,
Toronto, Canada M5S 2J9 and
Samuel Stevens & Co.,
3807 Bond Place,
Sarasota, Florida 33580 U.S.A.

This book was set in Palladium by Attic Typesetting.

Printed and bound in Canada by The Hunter Rose Co.

Canadian Cataloguing in Publication Data

Kostenuk, Samuel.
  R. C. A. F. squadron histories and aircraft, 1924-1968

(Historical publications ; 14 ISSN 0317-3860)

Includes index.
ISBN 0-88866-577-6

1. Canada. Royal Canadian Air Force — History. 2. Airplanes, Military — History. 3. Aeronautics, Military — Canada — History. I. Griffin, John. II. Title. III. Series: Historical publications (Ottawa, Ont.) ; 14.

UG635.C2R62   358.4'00971   C77-001191-8

CONTENTS

FOREWORD
   by Air Marshal C.R. Dunlap, C.B.E., C.D. IX
PREFACE XI
ACKNOWLEDGEMENTS XIII
ABOUT THIS BOOK XV

PART ONE: THE HERITAGE: 1909-1920
The Origins, 1909 1
The First World War, 1914-1918 1
   Canadian Aviation Corps, 1914-1915 1
   Royal Flying Corps/Royal Air Force
   in Canada, 1917-1918 1
   Canadian Air Force (England), 1918-1920 2
   Royal Canadian Naval Air Service, 1918 4
   Canadians in the British air services 5
Unit Histories: Canadian Air Force (England) 7

PART TWO: CANADIAN AIR FORCE, 1920-1923 ... 9

PART THREE:
ROYAL CANADIAN AIR FORCE 1924-1968 11
Organization — A Civil Air Force 11
   Squadrons Formed 11
   Formation of the Directorate of
   Civil Government Air Operations 11
   Unit Histories : 1925-1927 13
RCAF Reorganized as a Military Force 17
   Squadrons Re-formed 17
   Air Commands Formed 17
   RCAF — An Independent Force 17
Non-Permanent (later, Auxiliary) Active Air Force
1932-1939 18

The Second World War, 1939-1945
Wartime Expansion 18
British Commonwealth Air Training Plan 18
   Training Commands 1940-1945 19
Home War Establishment
   Organization 19
   Eastern Air Command 20
   Western Air Command 20
Unit Histories: Home Squadrons 21
The RCAF Overseas
   Personnel in England 75
   Proposal to Form Overseas Units 75
   First Squadrons Overseas 75
   The "400-Block" Squadrons 75

RCAF Units in RAF Commands
   Army Co-operation Command 75
   Fighter Command
      Day Fighter Squadrons 76
      Night Fighter and Intruder Squadrons 76
   Bomber Command
   No. 6 (RCAF) Group 76
   No. 331 Wing in North Africa 77
   Coastal Command 77
   South East Asia Command 77
   Second Tactical Air Force 77
   Transport Command 77
   British Air Forces of Occupation (Germany) 78
   "Tiger Force" (Pacific) 78
Summary 78
Unit Histories: "400 Block" Overseas Squadrons 80

The Post-War Years, 1945-1968
   Establishment and Organization 144
   Air Transport Command 144
   Maritime Air Command 145
   Air Defence Command 145
   North American Air Defence Command 145
   North Atlantic Treaty Organization 146
   No. 1 Air Division Europe 146
   The Auxiliary 147
   The Unification of the Forces, 1964-1968 148
Unit Histories: "400-Block" Post-War Squadrons 150

PART FOUR: HIGHER FORMATIONS
AND ANCILLARY UNITS, 1939-1968
Headquarters 207
Commands 208
Air Divisions 211
Groups 211
Sectors and Bases 213
Wings 215
Ancillary Units, ... 226

PART FIVE: APPENDICES
1. British Commonwealth Air Training Plan 229
2. RAF "Tiger" Force (Pacific) 231
3. RCAF Fighter Pilots in Korea 232
4. Unit Codes 233
5. Orders of Battle 235

Indexes 241

FOREWORD

In historical terms the life span of the Royal Canadian Air Force can hardly be described as anything but brief, a mere 44 years from creation to unification with Canada's other two armed services. Yet, from its modest beginning of fewer than two thousand men, it reached a wartime peak that exceeded two hundred thousand. The true measure of its greatness, however, is to be found in its growth and accomplishments, here recorded, which, both in peace and war, were truly magnificent — hardly a surprising fact, for Canadians have always demonstrated a remarkable aptitude for flying.

Those who served in the RCAF, as well as those who carry on its traditions in the present-day unified force, will, I feel sure, derive satisfaction and inspiration from this book. The record of the squadrons has been strengthened by the inclusion of a photographic record of the squadron aircraft used through more than 50 years. Thus we have a single reference work that will prove invaluable both to students and to modellers, as well as to the general reader.

Air Marshal C.R. Dunlap, C.B.E., C.D.

PREFACE

Whether through modesty or simply complacency Canadians have tended not to record in detail their own history and when at last some sign of interest whets their curiosity they must rely on what others have recorded on their behalf. So it has been with the history of the Royal Canadian Air Force. Though there have been several good narrative accounts of the exploits of the force published over the years, there have been few meaty details for the serious researcher. In keeping with the Canadian tradition, the material already published, relating to the overseas squadrons, exceeds that which is available on the home defence squadrons because the Royal Air Force historians have been of service.

The achievements of the Royal Canadian Air Force from 1924 to 1968 have been worthy of broader public understanding and appreciation. What the RCAF has accomplished would take many volumes to recount, but the scoreboard after all is to be found in the squadron orderly room. It is here that the history of an air force is written because the squadron is the means of implementing in part, the grand strategy, or even the minor task. The squadron was the force in the field, whether in peace or in war. How well the whole force performed had meaning only in the ability of the squadrons to perform in part their role.

Initially it was my desire to gain only an understanding of these units, their organization, their equipment and the essence of their achievements. As the project gained momentum and the privilege of access to official records became a reality, the simple search for some of the remote units expanded to a desire to compile a comprehensive summary of the total squadron strength of the Royal Canadian Air Force throughout its forty-four years of service to Canada. There was no thought that this might become the stuff from which a book might emerge.

One of those strokes of good fortune which bring two researchers together, in any field of endeavour, was arranged by John Swettenham of the Canadian War Museum when a project in hand by John Griffin was matched with my undertaking and soon a plan was cast with the publisher to create this book. My record of squadron research would be illustrated with John Griffin's photographic record of the aircraft of the RCAF squadrons and published as a single reference work.

I hope that the reader will gain as much from our efforts as I did in researching and assembling the written material. This work was a labour of love which brought about many new friendships and provided the opportunity to appreciate the contribution made by the men and women of the Royal Canadian Air Force whose struggles and sacrifices are summarized in these pages.

S.K.

ACKNOWLEDGEMENTS

It would have been impossible to collect and compile the information for this book without the active assistance and participation of a number of organizations and individuals.

The authors wish primarily to acknowledge the help of the Canadian Forces Directorate of History which provided guidance and encouragement for compilation and gave full access to unit diaries and historical reports. We wish to thank those Directors who gave support and approval while research was in progress: C.P. Stacey, S.F. Wise and W.A.B. Douglas; and the following staff members: R.V. Dodds, Hugh Halliday, P.A.C. Chaplin and Robert Stokeley.

To Mr. Henry Hrushoway and his staff at the Records Branch, Department of Veterans Affairs, our appreciation for the provision of unit Daily Routine Orders which yielded much valuable information.

All the photographs of unit badges were supplied by the Directorate of Ceremonial of the Canadian Forces, the acknowledged source. The sources of the aircraft photographs which illustrate this book are varied, and are listed in a separate section.

The search for the illustrations was greatly assisted by the following individuals of the Department of National Defence who willingly made contributions and whom the authors wish to thank: Lieutenant-Commander J.A. Young and Mel Lundy of the Photographic Unit; Mr. N.A. Buckingham and Captain J.A. MacNaughton of the Directorate of Ceremonial; and Major R.J. Tracy, Managing Editor of *SENTINEL* magazine. Much assistance was received from the Public Archives of Canada, in particular Messrs. Peter Robertson and Carl Vincent. Mr. Robert Bradford of the National Museum of Science and Technology was of great help. All these services are gratefully acknowledged.

The Canadian War Museum of the National Museum of Man was particularly helpful in discovering new sources of photographs and the authors are grateful to R.V. Manning, P.E. Butler, A.A. Azar, and Fred Gaffen. The contribution of Mr. D. Mayne of the Imperial War Museum helped to expand the photographic coverage of the RCAF's overseas activities and his assistance is acknowledged with thanks.

The following individuals generously made the contents of their private collections available to the authors and their contribution is gratefully acknowledged.

D.E. Anderson	George Neal
Bert Arnold	D.F. Parrott
J.S. Beilby	W.R. Richardson
Douglas Binns	Geoff. Rowe
William Breadman	W. Sawchuck
W. Brown	M.J. Smedley
Charles Catalano	T. Stachiw
George Chivers	G.E. Stewart
D.R. Dunsmore	Harvey Stone
J.R. Ellis	J.M. Taillefer
George Fuller	C.R. Vincent
Brian Gibbens	A. Walker
Stephen Harper	W.J. Wheeler
D.A. Kennedy	Ingwald Wikene
Jack McNulty	Gerald Wohl
L. Milberry	Graham Wragg
K.M. Molson	R.W. Wylie

Two persons, who are not recorded elsewhere in this book, have made a major contribution. Mr. Robert Baglow and Master Corporal R.H. Smith have devoted much personal time to uncovering many rare photographs from official sources and private collections. These illustrations bear the acknowledgement to the owner of the collection, and to thus fail to recognize the behind-the-scenes research which the authors would like to recognize with special thanks.

The authors express their gratitude to the Canadian War Museum for undertaking the sponsorship of this book, in particular to the Chief Curator, Lee Murray for his encouragement and approval; John Swettenham for his editorial skill, co-ordination and liaison with the publisher; Frank McGuire for much help in preparation of the material; Fred Gaffen for his meticulous checking of statements and fact; and A.A. Azar for the compilation of the index.

To each person and organization mentioned, we express our deepest appreciation.

Finally, to Air Marshal Dunlap, who reviewed the manuscript and wrote the Foreword, we owe much. This distinguished airman, to whom the preservation of Air Force history is a way of life, has done far more than can be stated here to ensure the accuracy of the facts presented.

ABOUT THIS BOOK...

The format of this illustrated history of Royal Canadian Air Force squadrons was developed to serve the needs of two specific groups — students of Canadian military aviation history, and aircraft modellers. The following notes will describe and explain the format.

Squadron histories have been grouped into three sections:

Home squadrons that served exclusively in Canada before and during the Second World War;

"400-Block" Overseas squadrons that served in conjunction with the Royal Air Force during the Second World War;

"400-Block" post-war squadrons that served in Canada and in Europe up to 1968, when the RCAF was unified with the Royal Canadian Navy and the Canadian Army as the Canadian Forces.

Because the history and service of the squadrons within each grouping are very similar and of a routine nature, and to avoid repeating long individual squadron narratives, each of the groupings is preceded by a short narrative describing the organization and main roles of the RCAF and its squadrons as a whole in that period.

The individual squadron histories consist of two parts, a very brief narrative and a point-form summary of its history.

Squadron Number All squadrons are listed only once within each grouping in numerical sequence and are officially considered as a single unit when calculating the 25 years of cumulative service required to qualify for a Squadron Standard. For example, as of 1 February 1968, No. 413 Squadron had accumulated 20¼ years of service during three separate periods of service — as No. 413(GR) during the Second World War and the two post- war periods — and No. 414 Squadron 19½ years during four periods. When a squadron was directly renumbered, service under its previous number also counted towards a Squadron Standard. For example, when No. 1 (Fighter) Squadron was renumbered No. 401 (Fighter) Squadron on 1 March 1941, the latter assumed the previous service, history and battle honours of No. 1 (F) Squadron. This was also the case when, during the Second World War, six "100-block" Home squadrons were transferred overseas complete with air and ground personnel and were renumbered in the "400-block" overseas squadrons: No. 14 (Fighter) to No. 442 (Fighter); 111 (F) to 440 (Fighter Bomber); 118 (F) to 438 (FB); 123 (Army Co-operation Training) to 439 (FB); 125 (F) to 441 (F); 127 (F) to 443 (F). It is because of this direct relationship that both Nos. 440 and 442 Squadrons carry the battle honour "ALEUTIANS", earned by Nos. 111 and 14 respectively for offensive operations against Japanese forces on Kiska and Attu islands in 1942-43.

Unit Badge The details of unit badges — motto, authority and description — are from official sources, chiefly Air Force Administrative Order 62.00/02, Appendix "C".

Brief Chronology A statement of the squadron's formation, designation, redesignations and disbandment. Each entry shows the event; where it occurred — if not at the same place as the preceding event; and the date (by day, month, year).

Title or Nickname The question of titles and nicknames was referred to the Directorate of History, Canadian Forces Headquarters, as late as 22 April 1975, and the reply is quoted:

"In answer to your questions, it is true that in some instances so-called squadron 'titles' were part of the unit's official designation, an example being No. 110 (later No. 400) (City of Toronto) (Army Co-operation) Squadron. It is also true that during the Second World War a number of squadrons came to be known by unofficial titles, such as No. 421 Squadron, known as the Red Indian Squadron.

"The use of either official or unofficial titles for security reasons is a somewhat cloudy question. The late Dr. Fred Hitchins informed Ron Dodds of this Directorate that when The R.C.A.F. Overseas was being written, the historians received instructions not to refer to squadrons or wings by number, but rather by title, so that information about the order of battle would not be published or broadcast while the war was still on.

"I have discussed your suggestion for differentiating between titles and nicknames with Mr. H.A. Diceman who designed many of the squadron badges. We agreed that it would be difficult to say whether a badge really indicated an official title, because the name sometimes suggested the design for the badge and the badge sometimes suggested the name. For example, 414 Squadron's Black Knight was derived from the name of a squadron commander, and Mr. Diceman consequently designed a badge to reflect this. But 439 squadron derived its title from the sabre toothed tiger that Mr. Diceman used for the badge. It is probably safe to say therefore, that almost all titles are 'unofficial.' Thus I would suggest differentiating between squadrons with city or county affiliations (mostly arising out of places adopting squadrons during the Second World War), and squadrons with titles derived from various other origins, whether or not these titles are reflected in the badge." It will be seen from the above that the matter must be approached with caution.

Certain pre-war, and all post-war Auxiliary squadrons were officially named after the city in which they were located. Such a title is included in the squadron heading (under unit histories) and is also shown as a separate entry under "Title."

The names of the "400-block" Overseas squadrons, because they received a measure of official recognition, are not included in the heading but are shown as a separate entry under "Title or Nickname."

Unofficial names acquired during the war by "100-block" Home squadrons, and those of the "400-block" post-war squadrons, are shown as a separate entry under "Nickname".

Adoption During the Second World War, service clubs and city councils adopted overseas squadrons and supplied the members of those squadrons with such amenities as cigarettes, books, sweaters and stockings.

Ancestry This indicates the direct inheritance by the squadron named of the history of another, and both histories should be consulted. For example, No. 401 (Fighter) Squadron of the "400-block" Overseas squadrons traces its ancestry directly to both No. 1 (Fighter) Squadron of the Regular Force and No. 115 Squadron of the Auxiliary, both listed in the Home squadrons grouping.

Commanders Although the title commander was never used by the RCAF, it is employed here for simplicity.

The person in command of a squadron could either be a "commanding officer" or an "officer commanding," depending on the status of the commander of the station on which the squadron was located. Each title denoted specific powers which could be exercised by the person as prescribed in regulations.

Up to 1962, commanders of single-engine fighter squadrons were of squadron leader rank, while those of multi-engine squadrons were of wing commander rank; both then became wing commanders. Higher appointments included commanding officers of wings; group commanders; and air-officer-commanding groups, air divisions or commands.

Wartime appointment dates were obtained from unit diaries while peacetime dates came from station and unit daily routine orders, or from press releases, and may be in error as to the exact date on which changes of command took place. The occasional gap between relinquishment of command by one officer and assumption by his successor indicates that one or more acting commanders were appointed for a time.

A notation in brackets after a person's name means that he was not a member of the RCAF. For example, (Can/RAF) means that the individual was a Canadian serving in the Royal Air Force.

Each entry gives the rank, initials and name, decoration(s) held or received while commanding the unit, period of command and, in certain cases, an explanatory remark. If the officer was merely posted elsewhere, this column remains blank, while reasons other than posting are indicated as follows in italics:

KIA killed in action.
KIFA killed in a flying (non-operational) accident.
MIA missing in action
OTE operational tour expired, either in number of operational hours or sorties flown. A numeral preceding signifies more than one tour, 2 OTE (second), 3 OTE (third) operational tour expired.
POW prisoner of war.
repat repatriated.
ret retired from the RCAF, or transferred to the non-active supplementary reserve list.

For ranks and decorations, the following abbreviations are used:

RCAF		Army Equivalent	
A/C/M	Air Chief Marshal	General	Gen
A/M	Air Marshal	Lieutenant-General	Lt-Gen
A/V/M	Air Vice-Marshal	Major-General	Maj-Gen
A/C	Air Commodore	Brigadier	Brig
G/C	Group Captain	Colonel	Col
W/C	Wing Commander	Lieutenant-Colonel	Lt-Col
S/L	Squadron Leader	Major	Maj
F/L	Flight Lieutenant	Captain	Capt
F/O	Flying Officer	Lieutenant	Lt
P/O	Pilot Officer	Second Lieutenant	2nd Lt
WO1	Warrant Officer Class I	Warrant Officer Class I	WO1
WO2	Warrant Officer Class II	Warrant Officer Class II	WO2
FS	Flight Sergeant	Staff Sergeant	S/Sgt
Sgt	Sergeant	Sergeant	Sgt
Cpl	Corporal	Corporal	Cpl
LAC	Leading Aircraftman	Lance Corporal	L/Cpl

(Prior to 27 August 1919, when air force ranks were introduced in the RAF, army ranks had been used. On 1 February 1968, when the Canadian armed services were united, army ranks were again adopted.)

VC	Victoria Cross
GC	George Cross
GCB	Knight of the Grand Cross of the Most Honourable Order of the Bath
GCMG	Knight of the Grand Cross of the Most Distinguished Order of St. Michael and St. George
GBE	Knight of the Grand Cross of the Most Excellent Order of the British Empire
KCB	Knight Commander of the Most Honourable Order of the Bath.
KCMG	Knight Commander of the Most Distinguished Order of St. Michael and St. George
KBE	Knight Commander of the Most Excellent Order of the British Empire
CB	Companion of the Most Honourable Order of the Bath
CMG	Companion of the Most Distinguished Order of St. Michael and St. George
CBE	Commander of the Most Excellent Order of the British Empire.
DSO	Companion of the Distinguished Service Order

OBE	Officer of the Most Excellent Order of the British Empire
MBE	Member of the Most Excellent Order of the British Empire
DSC	Distinguished Service Cross
MC	Military Cross
DFC	Distinguished Flying Cross
AFC	Air Force Cross
DCM	Distinguished Conduct Medal
CGM	Conspicuous Gallantry Medal
GM	George Medal
DSM	Distinguished Service Medal
MM	Military Medal
DFM	Distinguished Flying Medal
AFM	Air Force Medal
BEM	British Empire Medal
ED	Efficiency Decoration
CD	Canadian Forces Decoration
MiD	Mentioned in Despatches

Higher Formations and (Unit) Locations The RCAF adopted the Royal Air Force command and control structure of command-group-wing-squadron. The United States Air Force, on the other hand, reversed the group-wing relationship and established the wing as being higher than the group formation.

Where control was shared by two or more commands, both are shown with the nature of the sharing noted in brackets, e.g., (for adm) denotes administrative control, (for ops) operational control. The date of passing of control from one formation to another is given in brackets.

All higher formations controlling the home squadrons are RCAF unless otherwise noted by a prefix — RAF for Royal Air Force, and US for United States (RCAF units serving in Alaska).

Those listed in the "400-block" overseas squadron histories are RAF unless noted as "Allied", indicating an integrated command.

Post-war squadrons all served under RCAF higher formations except those in the integrated Canada/USA North American Air Defence (NORAD) Command, and the North Atlantic Treaty Organization (NATO) military command in Europe (listed as Allied Command Europe).

The date on which a squadron moved from station to station is based on the movement of the squadron's aircraft and crews. For higher formations, the date shown refers to the movement of the main headquarters staff. If, during the stay at a location, the squadron had a detachment located elsewhere, or was itself temporarily relocated for training, e.g. at an Armament Practice Camp for air-to-air or air-to-ground firing, this fact is indicated by italics.

Counties of the United Kingdom in which squadrons were stationed are abbreviated as follows:

In England

Beds.	Bedfordshire	Lincs.	Lincolnshire
Bucks.	Buckinghamshire	Northumb.	Northumberland
Cambs.	Cambridgeshire	Notts.	Nottinghamshire
Glos.	Gloucestershire	Oxon.	Oxfordshire
Hants.	Hampshire	Salop.	Shropshire
Herts.	Hertfordshire	Som.	Somersetshire
Hunts.	Huntingdonshire	War.	Warwickshire
Lancs.	Lancashire	Wilts.	Wiltshire
Leics.	Leicestershire	Yorks.	Yorkshire

In Scotland

Aber.	Aberdeen	Caith.	Caithness
Ayr.	Ayrshire	E. Loth	East Lothian

In Northern Ireland

Ferm.	Fermanagh

Representative Aircraft Interwoven with the colourful exploits of the Royal Canadian Air Force over the years, the aircraft which it flew emerge as the means by which the many squadrons were able to leave their mark on the record for posterity. The very nature of the squadron organization and its role makes its aircraft an inseparable element in its triumphs and defeats.

The extent of the detailed information which could be devoted to the subject of the aircraft, their individual identification and description, received much consideration and it was concluded that such detail was beyond the scope of this account. Perhaps a future publication, dealing solely with the aircraft of the RCAF, may be devoted to this aspect of the history of the RCAF, but for the present we have provided sufficient reference material to enable the serious modeller and artist to "see for himself" the nature of things.

For all of those who have more than a passing interest in the aircraft a listing appears with each squadron account, headed "Representative Aircraft." These are specific aircraft known to have served the squadron in the time period indicated. These are not all of the aircraft of each type which served the unit; the list provides only a representative sampling. The squadron identification codes are provided where they apply and, where the squadron record had clearly related serial numbers and the aircraft identification letter, this information is also provided.

Of general interest to all, the photographs have been selected to illustrate each variant of each type of aircraft flown operationally by the squadron. On occasion, we have also included aircraft attached to the units for training or communications duties. No apology is offered for photographs which have appeared in other publications or for those of inferior quality. One of the aims of this book has been to consolidate the best but, frequently, the illustrations are the only photographs of these aircraft known to exist. We have not resorted to the use of photographs which merely indicate an aircraft of a type similar to that in use. In the few instances where transfers have caused the aircraft to be marked other than appropriate to the period illustrated, this fact is noted in the caption.

The description of aircraft used in this book requires some amplification to explain what might appear to be inconsistencies in terminology. In the section dealing with pre-war and home defence squadrons, the type names are those in use by the RCAF as are those in use in the post-war section. All overseas aircraft between 1940 and 1945 are identified using Royal Air Force terminology. Where a squadron used an aircraft type in a time period, but in two or more sub-variants simultaneously, the heading indicates the sub-types in use, but the serial numbers listed are not individually amplified by Mark number. Those wishing to investigate this detail further are advised to consult **Canadian Military Aircraft: Serials and Photographs** by J.A. Griffin and **British Military Aircraft Serials** by Bruce Robertson.

The Royal Air Force used popular names to identify aircraft in service; and Mark numbers to identify significant variants of the basic type. Mark numbers were generally indicated by the abbreviation Mk. followed by a Roman numeral identifying the variant. On occasion a suffix letter appeared to denote armament characteristics or other special features. In 1942 the prefix role identifier was introduced to the Mark number and appeared as L.F. Mk. IXe with the "L.F." denoting, in this example, a Spitfire whose role was a low altitude fighter and the suffix "e" denoting a wing mounting two 20-mm cannon and two .50-calibre machine guns. Other role designators were B. (bomber), F. (fighter), H.F. (high altitude fighter), P.R. (photographic reconnaissance), etc.

The use of Roman numerals in Mark numbers was discontinued in 1950 and replaced by Arabic numerals. Aircraft built prior to 1950 continued to use the Roman designation but when extensive modifications were carried out, it was common practice to change to Arabic Mark numbers. The RCAF Expeditors were modified to Mk. 3 standard after 1950 and were designated in Arabic form.

None of the aircraft flown by Canadian squadrons permanently based overseas and under the operational control of the Royal Air Force during the Second World War were on charge to the Canadian government. They were issued, withdrawn, overhauled, modified and marked by the Royal Air Force for use by the RCAF during the 1939-45 period. The first squadrons to go overseas brought aircraft from Canada but these were withdrawn by the RAF as being below acceptable operational standards. Operational aircraft manufactured in Canada for the Ministry of Aircraft Production were delivered to the Royal Air Force and some of these found their way to Canadian units as a result of RAF equipment assignments. The Canadian-built Lancaster B. Mk. X's were delivered to Canadian squadrons in this manner.

Post-war RCAF contributions to NATO were fully supplied from the resources of the RCAF with the attendant freedom to select aircraft and equipment and establish marking standards.

Unit Code The unit code (shown in parentheses following "Representative Aircraft") was normally a two-letter combination assigned to the squadron and applied to the fuselage sides of the aircraft. Where more than one set of unit codes was assigned, then the period of use for each set is indicated.

The use of unit codes, plus the individual aircraft letter, is explained in detail in Appendix 4. Briefly, however, the following rule has been applied in listing either the full three-letter or partial code after the individual aircraft serial number.

For those units in the home squadrons, where it could be confirmed either by photographic evidence or by explicit mention in the unit diary that the full code was applied to the aircraft fuselage sides — the two-letter unit code plus the individual aircraft code letter — that code is listed. Listings of individual aircraft letters only are based on information from the diaries and do not necessarily mean that such letters did in fact appear on the aircraft, especially after early 1943.

In the "400-block" overseas squadrons, all aircraft carried the full set of three code letters although only the individual aircraft letter is listed after the aircraft serial number. The exception to this rule was the three photographic reconnaissance squadrons — Nos. 400, 414 and 430 of Second Tactical Air Force — whose aircraft carried only individual letters. Also, during certain periods, some of the coastal squadrons — Nos. 404, 422 and 423 — used a single numeral in place of two letters.

For post-war squadrons, three-letter codes were carried for a brief period between 1947 and 1951 as part of the "VC-" registration, and such cases are listed where known. Between 1951 and 1958 the individual aircraft letter was replaced by the last three digits of its serial number; and in 1958 the unit code was replaced by "RCAF."

Operational History This portion applies only to squadrons that were operationally employed during the Second World War from 10 September 1939, when Canada declared war on Germany, to either 8 May 1945, when Germany surrendered, or on 15 August when Japan capitulated.

Operational is defined as war flights listed in the squadron's "Operational Record Book" (ORB). Non-operational flights are all others, such as training and test flights.

First and Last Mission First and last entries in the squadron ORB. For day fighter and fighter bomber squadrons, the First Offensive Mission is also listed, and represents the first mission over enemy-occupied territory by a majority of the squadron's aircraft (eight or more) operating either on a fighter sweep or as fighter escorts.

First Victory For fighter squadrons, this refers to the first recorded engagement of an enemy aircraft, with results. Where the initial claim is shown either as probably destroyed or as damaged, it is immediately followed by a further entry covering the first instance of an enemy aircraft destroyed.

Triple Victory For fighter squadrons, three or more enemy aircraft destroyed, in the air, by one pilot during a single sortie — not the total claimed in a day's air fighting.

Enemy Aircraft — Abbreviations German aircraft opposing the RCAF during the Second World War were popularly known by a system of two letters and two or three numerals. The letters represented the manufacturer's name and the numerals identified an aircraft type. Suffix letters identified subsequent models of a design much as the Mark numbering system served the British.

The abbreviations which appear in the book are amplified in the following list of two-letter German aircraft manufacturers' identifications:

Ar — Arado Flugzeugwerke G.m.b.H.
Bf — Bayerisch Flugzeugwerke
Bv — Blohm & Voss, Hamburger Flugzeugbau
Do — Dornierwerke G.m.b.H.
Fw — Focke-Wulf Flugzeugbau
He — Ernst Heinkel Flugzeugwerke G.m.b.H.
Ju — Junkers Flugzeug and Motorenwerke A.G.
Me — Messerschmitt A.G.

Victories, U-boat For coastal squadrons, a list of all submarines sunk or damaged.

Summary A statement of the squadron's wartime activities, compiled mainly from unit diaries and ORB's. These sources are not always complete or accurate, but it is estimated that the compilations are approximately 90 percent accurate. **Missions** Number of operational (combat) squadron assignments. **Sorties:** Number of individual operational (combat) fights. (Where, for example, a squadron despatches eight aircraft for a specific purpose, this is listed as 1 mission and 8 sorties.)

Operational/Non-operational Flying Hours Figures for the respective categories are shown, and divided by an oblique stroke.

Victories: Aircraft Expressed in terms of the numbers of enemy aircraft *claimed by the squadron* as destroyed, probably destroyed, and damaged. **U-boat** The number sunk or damaged is given, plus the number of recorded sightings and attacks; and, where known, the number and type of depth charges dropped. **Shipping** The number of enemy ships, and total tonnage, sunk or damaged. **Ground** Here the total weight of bombs dropped is stated (usually 500- and 1,000-pound bombs), and the number and type of ground targets destroyed and/or damaged. For close support aircraft, one of the key credits was the number of road and rail cuts effected, and these are stated.

Casualties Both operational and non-operational casualties are shown — aircraft lost or damaged beyond repair, and human casualties by type.

Squadron Aces This term, though unofficial, is used for fighter pilots credited with five or more victories while serving with the squadron. The score given does not include victories previously or subsequently gained in another squadron.

Top Scores Where there were no squadron "aces", a list of squadron pilots credited with three or more destroyed, or of the three leading scorers, is given. The figures given, e.g., 4½-1-3, represent the numbers credited as destroyed — probably destroyed — damaged.

Battle Honours Squadron battle honours are as listed in Air Force Administrative Order 62.00/04, Appendix "B." Lesser battle honours are shown in italics. Additional battle honours earned but not awarded, as the units had since been disbanded and never reactivated, were "Atlantic 1939-1945" (bomber reconnaissance squadrons of Eastern Air Command) and "Pacific Coast 1941-1945" (bomber reconnaissance squadrons of Western Air Command).

Part One
THE HERITAGE
1909-1920

The Origins — 1909

On 23 February 1909 J.A.D. McCurdy made the first successful aeroplane flight in Canada, piloting the *Silver Dart* for a distance of half-a-mile over the ice-covered surface of Baddeck Bay in Nova Scotia.[1] Next day he made a longer flight, covering four-and-a-half miles in a complete circle back to his starting point. Both flights were recognized by the Royal Aero Club of the United Kingdom as the first successful powered heavier-than-air flights by a British subject anywhere in the British Empire.[2]

On 2 August 1909, during the annual militia training camp held at Petawawa, Ontario, McCurdy made four demonstration flights with the *Silver Dart* in an attempt to interest the Department of Militia and Defence in the aeroplane as a weapon of war. On the last flight of the day, the machine was wrecked in a heavy landing and a second machine, *Baddeck No. 1*, crashed a few day's later.[3] Officials who witnessed some of these flights were not impressed, and it was decided to await the outcome of similar tests being made in Britain before considering what should be Canada's policy on military aviation matters. In the next few years, repeated efforts to have the department form an aviation section were frustrated on the grounds that "no funds are available."

The First World War — 1914-1918

When Canada found herself at war with Germany on 4 August 1914, several European nations were employing the aeroplane as a military weapon.[4] In Canada, on the other hand, there were neither pilots nor aeroplanes in the armed forces.

Colonel Sam Hughes, Minister of Militia and Defence, who was responsible for assembling and despatching the Canadian Expeditionary Force (CEF) for service overseas, asked the British Secretary of War about the need for aviators. He was advised that Britain could accept six expert aviators immediately and more later. Surveying his resources, Hughes could find no trained aviators to meet the British needs. He did, however, approve the formation of a small Canadian aviation unit to accompany the CEF to England.

Canadian Aviation Corps — 1914-1915
On 16 September 1914, Hughes personally authorized a Canadian Aviation Corps to consist of two officers and one mechanic. He appointed E.L. Janney as the "Provisional Commander of the CAC" with the rank of captain[5] and approved the expenditure of a sum not to exceed five thousand dollars for the purchase of a suitable aeroplane. Janney was able to locate and purchase a biplane from the Burgess-Dunne Company in Massachusetts for that amount, and arranged for its delivery to Quebec City.[6]

While he was engaged in this transaction, the other two members of the CAC — Lieutenant W.F.N. Sharpe[7] and Staff Sergeant H.A. Farr[8] — were waiting at Camp Valcartier.

When the aeroplane arrived on 1 October, it was immediately loaded aboard one of the 30 ships in which the CEF was to sail next day. The convoy docked at Plymouth on 17 October 1914 and the biplane was unloaded and trucked to Salisbury Plain where the troops were encamped. It was never to fly in England, however, as not one of the three members of the CAC was a qualified pilot. Staked out in the damp English climate, the Burgess-Dunne quickly deteriorated and was eventually written off.

By 7 May 1915 the Canadian Aviation Corps had ceased to exist. Thus ended the short life of Canada's first military aviation force.[9]

Royal Flying Corps/Royal Air Force in Canada — 1917-1918
From the beginning of the war, both of the British air services, the War Office's Royal Flying Corps (RFC) and the Admiralty's Royal Naval Air Service (RNAS), which were amalgamated on 1 April 1918 to form the independent Royal Air Force (RAF), had turned to Canada as a source of recruits. At first they would accept only trained pilots holding *Fédération Aeronautique Internationale* (FAI) licences. There were few such men, so that the hundreds of young Canadians who sought to volunteer for the RFC or the RNAS were first required to attend a civilian flying school at their own expense.[10] Civilian schools were inadequate to meet the demands of the rapidly expanding air services and, as sufficient facilities were unavailable in England, the RFC decided in early 1917 to establish its own training organization in Canada.[11]

The RFC plan called for the formation of four training stations with one or more aerodromes at each station, a station to have five training squadrons with Curtiss JN-4 "Canuck" aircraft (manufactured in Canada).[12] Following consultation with Canadian officials in January and February, plans were completed for a revised organization of three stations.

1 The Silver Dart *flown by J.A.D. McCurdy at Baddeck Bay, N.S. on 23 February 1909. This qualified as the first flight by a British subject in the British Empire.*

2 The Silver Dart *at Petawawa, Ont. during the demonstration flights made before officials of the Department of Militia and Defence on 2 August 1909.*

RFC Station Camp Borden, encompassing 850 acres, would be the main training site. Work was begun on 4 February 1917 and although units were in operation by mid-March, it was not until 2 May that the RFC officially took over. By this time the station was a going concern with all five of its squadrons, plus the School of Aerial Gunnery, formed by mid-April.

RFC Station Deseronto, consisting of aerodromes at Mohawk and Rathbun, was formed on 1 May. By the end of the month it had all five squadrons in operation.

RFC Station North Toronto, formed on 15 June, consisted of aerodromes at Long Beach, Leaside and Armour Heights. At the end of the month three of its five squadrons were in operation.

During October 1917, in keeping with the training organization in England, the RFC in Canada was reorganized into a Training Brigade and the stations became numbered wings: No. 42 (Camp Borden), No. 43 (Deseronto) and No. 44 (North Toronto).[13]

The United States entered the war in April 1917 and a reciprocal training scheme was then negotiated between the RFC and the US Army's Signal Corps. American aero squadrons and personnel were to be trained in Canada in return for training facilities for the RFC in snow-free Texas during the winter of 1917-18.[14]

The advance headquarters group of the RFC arrived at Fort Worth on 26 September and was stationed at Camp Taliferro. The School of Aerial Gunnery was the first RFC unit to move to Texas, having completed one course at Camp Borden on 30 October. It started the next course in Texas on 5 November. Two RFC wings, Nos. 42 and 43, ceased flying in Canada on 14 November and, after a 1600-mile rail trip, commenced training operations in Texas three days later. No. 44 Wing remained at North Toronto to test the feasibility of training under Canadian winter conditions. This experiment proved so successful that it was decided to train all airmen in Canada during the winter of 1918-19.

In April 1918 the RFC returned to its Canadian aerodromes and resumed its training schedule. In addition to providing preliminary pilot and observer training, it was decided to form several advanced training units with emphasis on pilot training in aerial marksmanship and tactics. The School of Aerial Gunnery was renamed the School of Aerial Fighting and given its own training complex. In addition, the School of Armament came into being to handle the ground instruction of pilots and thus free the School of Aerial Fighting to concentrate on airborne work.

When the Armistice was signed on 11 November 1918, the Royal Air Force (formerly Royal Flying Corps) establishment in Canada had a total strength of 11,928 all ranks. It was staffed by 993 officers and 6158 other ranks and had 4333 cadet pilots and 444 other officers under training. In its twenty and one-half months in Canada the RFC/RAF training organization had enlisted a total of 16,663 personnel — 9200 flight cadets, 7463 mechanics. It graduated 3135 pilots (excluding Americans) of whom 2539 went overseas and 356 remained in Canada as instructors; and 137 observers, of whom 85 were sent overseas. At the time of the Armistice it had an additional 240 pilots and 52 observers ready for overseas service. The RFC/RAF in Canada suffered 130 fatal casualties in aircraft accidents.

Canadian Air Force (England) — 1918-1920

In 1915 the British Army Council suggested that the dominions raise complete air units for service with the RFC.[15] It was not until the spring of 1918, however, that any action was taken by Canada on this proposal.

In a memorandum dated 30 April 1918, the Canadian High Commissioner in London[16] suggested that the government consider forming a Canadian Air Force (CAF) in England. His proposal was based on the fact that many Canadians were serving with the Royal Air Force and making a valuable contribution, and had expressed a desire to serve in Canadian squadrons. In considering it, the authorities made a study in July, and this showed some 13,000 Canadians in the RAF, of whom 850 were on secondment from the Overseas Military Forces of Canada. In view of these figures, the Canadian Privy Council agreed to discuss with the British Air Ministry the question of forming Canadian squadrons within the RAF — the eventual aim being the formation of the CAF.

A proposal was then put forward to form a wing of eight squadrons for duty with the Canadian Corps in France, the cost of equipping and maintaining these units to be borne by the Canadian government. To raise the eight squadrons, it was suggested that RAF squadrons in France be surveyed to determine how many had between 60 and 80 per cent

3   The Baddeck No. 1 *being erected at Petawawa, Ont. for the demonstration flights which took place in mid-August 1909.*

4   *Following its delivery flight from Marblehead, Massachusetts, the* Burgess-Dunne *is taken aboard the S.S.* Athenia *on 1 October 1914 for shipment overseas with the Canadian Expeditionary Force.*

5   *The Curtiss School of Aviation at Hanlan's Point, on Toronto Island, had facilities to train pilots in Canada in 1915. Pilots wishing to join the RFC and the RNAS had first to qualify before proceeding overseas.*

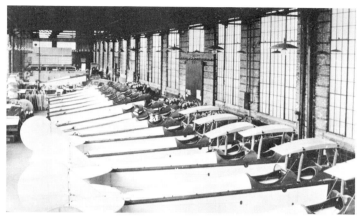

6   *Curtiss JN-4 fuselages in the manufacturing shop of Canadian Aeroplanes Limited at Toronto, Ontario. Between 1917 and 1918, 1288 of these uniquely Canadian trainers were built for use by the RFC/RAF flying training establishment in Canada. These JN-4's had control sticks instead of control wheels, ailerons on the lower wings and a completely redesigned tail assembly of fabric covered metal tubing. To differentiate the type from its American counterpart it was named the Canuck.*

7   *A typical line-up of Canadian-built Curtiss JN-4 training aircraft of No. 85 Training Squadron at Camp Mohawk, Ont. in the summer of 1918.*

8   *Armour Heights, located five miles north of Toronto, Ont., was the centre of advanced flight training, including instructor courses.*

Canadian aircrew. From these, eight would be selected as "Canadian" and would have their non-Canadian aircrew replaced by Canadians from other units. But the British Air Ministry and the RAF in the field felt that this procedure would unduly impair the RAF's fighting ability. Another serious factor was an almost total lack of trained Canadian groundcrew. It was then decided to train the required groundcrew first, and only then to form new "Canadian" squadrons.[17]

On 5 August 1918 the Air Ministry authorized the formation of two Canadian squadrons, one to be a fighter and the other a two-seater day bombing squadron.[18] The Canadian Privy Council, on 19 September, approved the formation of the Canadian Air Force in England comprising these two squadrons.[19] A Canadian Air Force Section (which later became the CAF Directorate of Air Services) was formed as a branch of the General Staff of the Overseas Military Forces of Canada. Lieutenant-Colonel W.A. Bishop, Canada's first airman to be awarded the Victoria Cross and the British Empire's leading "ace" (72 victories), became the first commander of the CAF in England.[20]

A firm proposal was put forward on 26 October by the CAF Section to increase the CAF to eight squadrons — the two already authorized, two more to be formed later in 1918, and four early in 1919. The signing of the Armistice on 11 November 1918 postponed any decision by the Canadian government on this proposal, pending the final outcome of discussions in parliament as to whether to maintain a permanent peacetime air force. A possible nucleus for such a force, for which planning continued, would be a wing consisting of the two existing squadrons.

On 20 November 1918, nine days after the signing of the Armistice, No. 1 Squadron (fighter) was formed at Upper Heyford, Oxfordshire; it was followed on the 25th by No. 2 Squadron (day bombing).[21] Four months later, on 25 March 1919, No. 1 Wing was also formed at Upper Heyford to administer the two squadrons, but the headquarters did not commence its duties until 1 April when the wing and the two squadrons had moved to Shoreham-by-Sea, Sussex.[22]

The Canadian government decided not to form a permanent peace-time air force. Orders directing all flying to cease immediately, and all aircraft and equipment belonging to the Canadian government to be dismantled and packed ready for shipment to Canada, reached England on 14 June 1919. No. 1 Squadron was disbanded on 28 January 1920, and No. 2 Squadron and the wing headquarters on 5 February. A Canadian Air Force Packing Section was formed to complete the packing.[23] The Directorate of Air Services was disbanded on 9 August,[24] ending the history of Canada's second military aviation force, composed for a time of her most outstanding and most decorated airmen.[25]

Royal Canadian Naval Air Service — 1918
The threat posed by German submarines in the First World War led the British Admiralty, in 1917, to adopt the convoy system for Atlantic shipping. Because of its natural and safe harbour, Bedford Basin at Halifax was chosen as the assembly area for convoys bound for Britain. To protect this harbour, and the early stages of the eastbound convoys, the Admiralty suggested the establishing of stations and air units at Eastern Passage (Dartmouth) and Sydney, Nova Scotia. At the same time, they expressed regret that

9 Gunnery training was an essential part of the flight programme in Canada, as illustrated by this Curtiss JN-4 with a wing-mounted camera gun.

10 Home of No. 1 Canadian Wing and its two squadrons, Shoreham-by-Sea, Sussex in the summer of 1919.

11 National pride displayed at Shoreham-by-Sea with "No. 1 Fighter Squadron, C.A.F." carefully detailed in coloured stones.

12 *One of twelve Curtiss HS-2L flying boats operated by the United States Navy from bases at Halifax and Sydney, N.S. during 1918.*

13 *No. 1 Squadron, Canadian Air Force was initially equipped with Sopwith Dolphins but several fatal crashes led to their replacement by the more reliable Royal Aircraft Factory S.E.5a.*

14 *Captain W.R. Kenny in one of No. 1 Squadron's S.E.5a's. Aft of the roundel the presentation inscription has been applied to identify "No. 168 Presented by Addis Ababa Branch Overseas Club and Patriotic League."*

very little assistance could be expected from the British government and suggested that Canada form her own air force. As the United States was sending large numbers of its troops overseas through Halifax, the Americans too were very desirous of having aerial protection for their troopships and so offered their assistance to Canada in bringing the two stations into operation, and in providing the necessary air units until Canadians could carry out patrols. On 16 August 1918, under the command of Lieutenant Richard E. Byrd, who later gained fame for his flights over the Polar icecaps, two HS-2L flying boats of the United States Naval Flying Corps arrived at Dartmouth. They carried out the first patrols on the 25th.[26]

Regulations governing the Royal Canadian Naval Air Service (RCNAS) were drawn up and approved by the Canadian government on 5 September 1918. These stated that personnel were to be trained in lighter-than-air and heavier-than-air aeronautics in both the United States and the United Kingdom. Up to the Armistice, the RCNAS had recruited 81 cadets, of whom 60 were under training in the United States, 13 in the United Kingdom and eight in Canada awaiting instructions. Six coxswains had also enlisted and were in the United Kingdom serving as airship coxswains.

On 5 December 1918 the RCNAS was disbanded and the cadets under training were demobilized, thus ending the life of Canada's third air force formation.

Canadians in the British Air Services
During the First World War, as members of the Royal Flying Corps, the Royal Naval Air Service and, finally, the Royal Air Force, Canadians flew on almost every front. Of 27 British fighter pilots credited with 30 or more victories, ten were Canadians, including the Empire's two leading surviving aces — Bishop with his 72 victories, and Major Raymond Collishaw with 60. Canadians manned day and night bombers,[27] flying boats, observation balloons and army co-operation aircraft. They flew over the Western Approaches, the English Channel and the North Sea;[28] over France and Belgium; over Italy,[29] the Adriatic and the Mediterranean; over Macedonia and Thrace, the Aegean Sea and the Dardanelles; over Egypt, Palestine and Mesopotamia; over the Red Sea and the Indian Ocean and even German East Africa.

The total number of Canadians who flew with the British has been estimated at more than 23,000.[30] When the war ended, some 20,500 — mainly pilots, observers or cadets — were on the RAF's strength of 281,165. The Book of Remembrance in the Peace Tower's Memorial Chamber in Ottawa records the names of 1563 who gave their lives. Well over half that number were decorated, including three with the Victoria Cross — Bishop, Second Lieutenant A.A. McLeod, and Major W.G. Barker.

Canadian air achievements in the First World War are not only part of the rich tradition of the RAF; they are the heritage of the Royal Canadian Air Force.

---

[1] The *Silver Dart* was the fourth production aeroplane of the Aerial Experimental Association which had been formed in Baddeck in September 1907 under the leadership of Dr. Alexander Graham Bell, the inventor of the telephone. Associated with him were J.A.D. McCurdy and R.W. Baldwin, Canadian engineers; Glen Curtiss, an American motorcycle racer and engine manufacturer, and Lt. Thomas Selfridge, an American army

officer. When the AEA expired on 31 March 1909, McCurdy and Baldwin formed the Canadian Aerodrome Company and continued their experiments in aviation.

[2] The first successful powered heavier-than-air flight had been made by Wilbur and Orville Wright of Dayton, Ohio, on 17 December 1903.

[3] This Canadian machine (sent to Petawawa for further flights) crashed on Friday the thirteenth and the militia camp was closed before repairs could be completed.

[4] The British War Office had formed an Air Battalion on 1 April 1911 consisting of 14 officers and 150 other ranks, organized into a headquarters and two self-contained companies — one of kite balloons and the other aeroplanes. On 13 April 1913 the Air Battalion became the Royal Flying Corps. The Admiralty's Royal Naval Air Service was formed on 1 July 1914.

[5] Janney's appointment was quite extraordinary in that he had little or no training or experience in aeronautics, and no prior military service. The appointment was never confirmed, nor was Janney ever commissioned in the CEF.

[6] To avoid violating American neutrality, the biplane was shipped by rail from the factory at Marblehead, Massachusetts to East Alburg, Vermont, where it was reassembled and flown to Quebec City by Clifford Webster of the Burgess-Dunne Company with Janney as his passenger — Janney's first known flight.

[7] Sharpe had attended the Curtiss Flying School at San Diego, California before the war. Although not in possession of an Aero Club of America licence, he had a testimonial from the school stating that he was well advanced as a pupil. He appears to have offered his services to the CEF as an aviator on or about 7 September, and to have been assigned to the 1st Battalion (infantry), and thence to the Canadian Aviation Corps, without ever being duly attested.

[8] Farr, born in England, had resided in the United States. At the outbreak of war he was chief motor mechanic at the Christofferson Aircraft Company, San Francisco. He attempted to enlist in Victoria, British Columbia in August 1914 but according to the available documents it was at Valcartier, on 23 September, that his service began. Officially, he was on the strength of Divisional Headquarters.

[9] Janney had offered to resign on 23 December; his resignation was accepted on the 31st, and he returned to Canada in January 1915. Sharpe, according to a British Air Ministry source, was transferred to the Royal Flying Corps on 3 February 1915 and assigned to No. 3 Reserve Aeroplane Squadron at Shoreham, Sussex, for training. On 4 February, flying a Maurice Farman (No. 731), he was killed and thus became Canada's first military aviation casualty. Farr was discharged at Shorncliffe on 7 May 1915 "owing to the Canadian Flying (sic) Corps being disbanded" (letter from Officer-in-Charge, Canadian Records, London) and applied for a commission in the RFC, but was advised that a pilot's licence was necessary. He returned to Canada in the summer of 1916, completed a pilot's course in the United States, and was commissioned in the RFC on 2 February 1917. He was discharged on 15 October 1918 owing to ill health.

[10] The cost was approximately $400 for 400 minutes of flying training. The Curtiss School of Aviation at Toronto, under the management of J.A.D. McCurdy, trained 129 pilots in 1915-16 and gave partial training to hundreds of others.

[11] The advance party of 14 officers and 76 other ranks, along with 15 motor vehicles, departed from England aboard four ships between 9 and 18 January 1917 under the command of Lt-Col (later Brig-Gen) C.G. Hoare. It arrived in Toronto on 25 January and established headquarters there.

[12] The British Air Ministry assigned Nos. 78 to 97 (Canadian) Reserve Squadrons, redesignated in May 1917 as Canadian Training Squadrons.

[13] Each wing was organized as a composite training school with three of the squadrons providing elementary flying training and two advanced flying training. The latter two squadrons each had three flights, one of which handled cross-country and photographic training; the second handled wireless telegraphy training; the third provided bomber training.

[14] Negotiations for this reciprocal scheme were begun on 9 July and concluded on 29 August 1917. It called for the training of ten American squadrons consisting of 300 pilots, 144 other flying officers, 20 administrative and equipment officers, and 2000 groundcrew.

Only eight of the ten squadrons were formed and trained by the RFC in both Canada and Texas:

17th Aero Squadron was organized as the 29th Aero Squadron at Kelly Field, Texas on 16 June 1917; was redesignated 17th Aero Squadron on 20 July; trained in Canada from 4 August to 13 October, then in Texas until 23 December 1917, when it proceeded overseas.

22nd Aero Squadron was organized as the 17th Aero Squadron at Kelly Field, Texas on 16 July 1917; was redesignated 22nd Aero Squadron on 20 June; trained in Canada from 12 August to 21 October, then in Texas until 21 January 1918, when it proceeded overseas.

27th Aero Squadron was organized as the 21st Aero Squadron at Kelly Field, Texas on 15 June 1917; was redesignated 27th Aero Squadron on 23 June; trained in Canada from 18 August to 28 October, then in Texas until 25 January 1918, when it proceeded overseas.

28th Aero Squadron was organized at Kelly Field, Texas on 22 June 1917; trained in Canada from 25 August to 4 November, then in Texas until 25 January 1918, when it proceeded overseas.

147th Aero Squadron was organized at Kelly Field, Texas on 11 November 1917; trained with the RFC in Texas from 12 November 1917 until 18 February 1918, when it proceeded overseas.

77th Aero Squadron was organized at Waco, Texas on 20 February 1918 and trained with the RFC in Texas until April 1918.

78th Aero Squadron was organized at Waco, Texas on 28 February 1918 and trained with the RFC in Texas during March and April 1918.

79th Aero Squadron was organized at Waco, Texas on 22 February 1918 and trained with the RFC in Texas during March and April 1918.

[15] Australia was the only dominion of the British Empire to adopt the suggestion, and formed No. 1 Squadron, Australian Flying Corps, which left on 16 March 1916 for Egypt. By the end of the war, the AFC had four operational squadrons serving in conjunction with the Royal Air Force — No. 1 Squadron in Palestine and Nos. 2, 3 and 4 in France. In addition, there was a wing of four training squadrons (Nos. 5, 6, 7 and 8) in England and the Central Flying School (formed in February 1914) in Australia.

[16] Sir George Perley.

[17] On 22 August 1918 a Canadian Air Force Detachment was formed at the RAF School of Technical Training at Halton to train 199 Canadian army personnel as riggers and mechanics. There were also 26 armament trainees at Uxbridge and 12 non-commissioned officers at Eastchurch training as observers.

[18] Air Ministry Memorandum A.O./967 dated 5 August 1918 designated the two squadrons as No. 93 and 123 (Canadian); A.O./969 dated 5 August 1918 designated No. 93 as fighter and No. 123 a two-seater day bombing squadron; A.O./1117 dated 19 October 1918 changed the designation of the fighter squadron from No. 93 to No. 81 owing to a change in aircraft to have been issued to the unit; Air Ministry Instruction No. 11930 dated 19 October 1918, referred to the two squadrons as No. 81 Sopwith Dolphin Squadron (Canadian) and No. 123 D.H.9a Squadron (Canadian) — No. 93 was to have received Sopwith Snipes.

[19] Authority: PC 1984, dated 19 September 1918.

[20] Lt-Col Bishop was first posted to Headquarters, Overseas Military Forces of Canada on 5 August 1918 to supervise the screening of RAF squadrons for Canadian aircrew and the CEF for suitable men to train as groundcrew to man the CAF squadrons.

[21] No. 2 Squadron was also to have been formed on the 20th but was delayed while waiting for an RAF squadron to vacate the quarters assigned to No. 2 Squadron.

[22] No. 1 Squadron did not move until 1 May when the exchange of Dolphins for S.E.5a's was complete.

[23] Authority: Southern Area Letter S.A./799, Air, dated 29 November 1919.

[24] At this time, a Canadian Air Force liaison officer position was created at the British Air Ministry.

[25] At a CAF dinner held at Shoreham in June 1919, Maj A.E. McKeever, commander of No. 1 Squadron, announced that the total number of enemy aircraft credited to those attending was 219. Lt-Col Bishop's total of 72 was not included as he had returned to Canada some time earlier.

[26] The United States Naval Flying Corps eventually had two detachments, each with six HS-2L flying boats and two kite balloons, stationed at Eastern Passage (near Dartmouth) and at North Sydney. When the war ended, this equipment was given to Canada by the US government.

[27] When No. 3 (Naval) Wing was formed in the autumn of 1916 as the first long-range strategic bomber force, 44 of its 74 pilots were Canadians.

[28] Canadians were involved in the shooting down of six of 12 Zeppelins destroyed, and in the sinking of one German submarine.

[29] Fourteen Canadian fighter pilots were credited with destroying more than 130 aircraft in the Italian theatre. Leading the list were Maj W.G. Barker with 27 victories — his final score was 50 — and Lt C.M. McEwen with 20.

[30] British records frequently fail to differentiate between airmen from Canada and those from other parts of the Empire. The problem of defining "Canadian" for statistical purposes is further complicated by the enlistment, before the United States entered the war, of many Americans professing to be Canadians.

UNIT HISTORIES
Canadian Air Force (England)

Directorate of Air Services

Formed as the Canadian Air Force Section of the General Staff, Headquarters Overseas Military Forces of Canada, London, England on 19 September 1918; reorganized as the Directorate of Air Services on 28 February 1919; disbanded on 9 August 1920.

Commanders
CAF Section:
Lt-Col W.A. Bishop, VC, DSO and Bar, MC, DFC 19 Sep 18 - 2 Dec 18.
Col R.H. Mulock, DSO 3 Dec 18 - 27 Feb 19.
Directorate of Air Services:
Col G.C. St. P. de Dombasle, OBE 28 Feb 19 - 31 Aug 19.
Maj D.R. MacLaren, DSO, MC and Bar, DFC 1 Sep 19 - 9 Aug 20.

No. 1 WING
(No. 1 Canadian Wing RAF)

Formed at Upper Heyford, Oxfordshire, England on 25 March 1919 by amalgamating the CAF Headquarters Detachment at Upper Heyford and the CAF Training Detachment at Halton, Buckinghamshire; commenced its duties 1 April 1919 when it moved to Shoreham-by-Sea; was disbanded on 5 February 1920.

Commanders
Lt-Col R. Leckie, DSO, DSC, DFC 25 Mar 19 - 19 Nov 19.[1]
Capt J.A. Glen, DSC 20 Nov 19 - 2 Dec 19.
Capt W.I. Bailey 3 Dec 19 - 5 Feb 20.[2]

Higher Formations and Wing Locations
RAF Home Command:
No. 2 Group,
Upper Heyford, Oxon. 15 Mar 19 - 31 Mar 19.
No. 1 Group,
Shoreham-by-Sea, Sussex 1 Apr 19 - 5 Feb 20.

---

[1] Accepted a permanent commission in the RAF. Soon afterwards, Lt-Col Leckie returned to Canada on loan to the Air Board.
[2] Simultaneously commanded both No. 1 Wing and No. 2 Squadron for this period.

15 Bristol F.2b F4336 with the distinctive leaf marking of No. 1 Squadron and the inscription "Huddersfield — Canada" which appeared under the exhaust manifold.

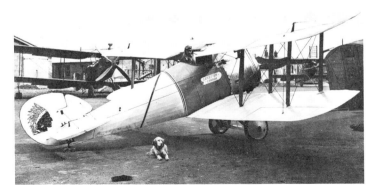

16 Photographed at Shoreham-by-Sea, Sopwith Snipe E8213 with the inscription "Leicester — Canada" is shown in front of Bristol F.2b F4336 and a Fokker D.VII. Both the Bristol and Snipe were later sent to Canada when No. 1 Squadron was disbanded.

No. 1 SQUADRON
(No. 81 Squadron (Canadian), RAF)

Formed as a scout (fighter) unit at Upper Heyford, Oxfordshire, England on 20 November 1918 the squadron flew Sopwith Dolphin and S.E.5a aircraft on training until disbanded at Shoreham-by-Sea, Sussex on 28 January 1920.

Commanders
Maj A.E. McKeever, DSO, MC and Bar 26 Nov 18 - 14 Aug 19.
Capt W.B. Lawson, DFC 15 Aug 19 - 12 Jan 20.

Higher Formations and Squadron Locations
RAF Home Command:
No. 2 Group,
Upper Heyford, Oxon. 20 Nov 18 - 30 Apr 19.
No. 1 Group, No. 1 Wing, CAF,
Shoreham-by-Sea, Sussex 1 May 19 - 28 Jan 20.

Representative Aircraft
Sopwith Dolphin (Nov 18 - Apr 19)[1]
E4764   F4767   F7076   F7085
S.E.5a (May 19 - Jan 20) E5747   E5755   F7982   F9020
The squadron also had the following aircraft on strength for varying periods:
Sopwith Pup: B4158, B4338, B5333,
Avro 504K: E4207
Bristol F.2B Fighter: E4336 "Huddersfield"
Sopwith Snipe: E8213 "Leicester"
Fokker D.VII: 6823/18, 8493/18.

---

[1]On 13 February 1919 the Dolphin was grounded because of several fatal crashes. Next day the order was amended to permit straight and level flight only; aerobatics were forbidden.

17 *Fokker D.VII in the foreground, still wearing German markings while serving with No. 1 Squadron. Three of the squadron's S.E.5a aircraft complete the line-up.*

18 *De Havilland D.H.9a bombers of No. 2 Squadron, Canadian Air Force lined up at Shoreham-by-Sea in 1919.*

No. 2 SQUADRON
(No. 123 Squadron (Canadian), RAF)

Formed as a day-bombing unit at Upper Heyford, Oxfordshire, England on 28 November 1918, the squadron flew two-seater D.H.9a's on training until disbanded at Shoreham-by-Sea, Sussex on 5 February 1920.

Commanders
Capt. W.B. Lawson, DFC 26 Nov 18 - 22 May 19.
Capt J.O. Leach, MC, AFC 29 May 19 - 19 Aug 19.
Capt J.A. Glen, DSC and Bar 20 Aug 19 - 19 Nov 19.
Capt W.I. Bailey[1], 20 Nov 19 - 5 Feb 20.

Higher Formations and Squadron Locations
RAF Home Command:
No. 2 Group,
Upper Heyford, Oxon. 15 Nov 18 - 31 Mar 19.
No. 1 Group, No. 1 Wing, CAF,
Shoreham-by-Sea, Sussex 1 Apr 19 - 5 Feb 20.

Representative Aircraft
de Havilland D.H.9a (Nov 18 - Jan 20)[2]
E731   E732   E8495   F1611

---

[1]Also commanded No. 1 Wing from 3 December 1919 to 5 February 1920.
[2]The squadron also had two Fokker D. VII's between May 1919 and January 1920: 6849/18 and 8482/18. Capt A.D. Carter (19 victories) was killed when 8482/18 crashed at Shoreham on 22 May 1919.

19 *Captain W.B. Lawson, the commanding officer of No. 2 Squadron, in one of the squadron's de Havilland D.H.9a's.*

Part Two
CANADIAN AIR FORCE 1920-1923

After the First World War there was bitter debate in parliament as to the value of a permanent air force; to resolve the question, the government passed the Air Board Act on 6 June 1919, authorizing a seven-member board to regulate and control all aeronautics in Canada. The Air Board was to form three divisions — a Civil Aviation Branch for the control of commercial and civil flying,[1] a Civil Operations Branch in charge of all non-military flying operations,[2] and a Canadian Air Force primarily responsible for training rather than defence.[3] The first Air Board was constituted on 23 June 1919, and on 22 December it submitted to the Privy Council a memorandum outlining recommendations for the organization of the Canadian Air Force.

This second, but home-based, CAF was authorized on 18 February 1920, and on 23 April approval was granted to appoint six officers and men with temporary rank. The CAF was a non-permanent organization to provide biennial 28-day refresher training to former officers and airmen of the wartime Royal Air Force. On 31 August a CAF Association was established, with branches in all provinces, to maintain a roster and select personnel for training. The programme began at Camp Borden using the hangars and other installations that had been erected by the RAF in Canada during the war, and the aircraft and other equipment that had been donated by the British and American governments. By the end of 1922, when refresher training was suspended, 550 officers and 1271 airmen had completed the course.

While the CAF had neither a permanent establishment nor embodied units, and was involved only in providing service training to its part-time personnel, it was seconding trained personnel to the Air Board for its Civil Operations Branch.[4]

One operation carried out by the Air Board deserves special mention, the first Trans-Canada flight, in 1920. Starting from Halifax on 7 October and using relays of Civil Operations Branch and CAF aircraft and crews, it ended at Vancouver ten days later. The flying time over the 3341-mile route was 49 hours and seven minutes, for an average speed of 68 miles per hour.

It is also noteworthy that the Air Board took an early interest in Arctic flying. In the summer of 1922 it detailed Squadron Leader R.A. Logan to accompany a Department of the Interior expedition into the Arctic. He visited the islands of Baffin, Bylot, Ellesmere and North Devon, and upon his return to Ottawa submitted a valuable report on the possibilities of flying in northern Canada.

By the spring of 1922 it had become obvious that a com-

20 Camp Borden, Ont., photographed in 1921, was the largest airport in Canada. A product of the RFC/RAF training programme of 1917-18, Camp Borden was the central training ground of the Air Board and the Canadian Air Force. For many years all RCAF flying training was based here.

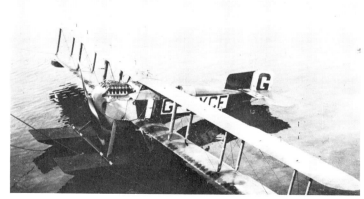

21 Ready to participate in the trans-Canada flight, the Fairey Transatlantic was to have flown the first leg of the programme from Halifax to Winnipeg non-stop. The flight ended at Whelpley's Point, on the St. John River, where a forced landing was made under difficult conditions. The Transatlantic was never flown again by the Canadian Air Force.

plete reorganization of the CAF, the "service" arm of the Air Board, was necessary. Eighteen months experience had demonstrated that biennial training by a non-permanent, non-professional force[5] was not requalifying many pilots, and no new pilots were being trained. It was therefore decided to reorganize the CAF on a permanent basis, the transitional period being from April 1922 to March 1924. The first stage, begun by the Air Board on 29 June 1922, was the consolidation of the Civil Operations Branch and the Canadian Air Force into a single military organization.[6]

---

[1] Lt-Col J.S. Scott was appointed Superintendent, Certificate Branch (later known as Controller Civil Aviation) on 3 November 1919.
[2] Lt-Col R. Leckie, formerly commanding No. 1 Wing CAF (England) and now on loan from the RAF, was appointed Superintendent, (later Director) Flying Operations on 15 December 1919. Although the RAF had adopted distinctive air force ranks on 27 August of that year, Leckie continued to sign himself "Lt-Col" rather than as "W/C" (Wing Commander).
[3] Maj-Gen Sir Willoughby Gwatkin, a British officer who had served as Canada's Chief of the General Staff throughout the war, was appointed Inspector-General of the new Canadian Air Force with the rank of Air Vice-Marshal on 23 April 1920. (Like Lt-Col Leckie and, indeed, many CAF officers until late November 1922, he was still better known by army rank.) On 17 May 1920 Lt-Col A.K. Tylee was appointed Air Officer Commanding, CAF, with the rank of Air Commodore.
[4] During the 1920-21 seasons the Civil Operations Branch, under Lt-Col Leckie, operated from six stations located at Vancouver, High River, Winnipeg, Ottawa, Roberval and Dartmouth. With a staff of 110, of whom 32 were licensed pilots, it flew more than 2200 hours for nine federal departments and three provincial governments (British Columbia, Ontario and Quebec; at this time the natural resources of the three Prairie provinces were under Federal control). Four F3 flying boats, 18 HS-2L flying boats, 11 D.H.4 landplanes and three Avro Viper seaplanes were employed on forest fire patrols, anti-smuggling patrols, "treaty money" flights to Indians, and general communications and transport work.
[5] Up to now the CAF had had no permanent staff. The personnel of the Civil Operations Branch were civilians appointed by the Civil Service Commission and when necessary they were granted leave for service training with the Canadian Air Force.
[6] On 28 June 1922 the National Defence Act was passed to incorporate into one department, under a single minister, Militia and Defence, the Naval Service and the Air Board. When the Department of National Defence was formed, on 1 January 1923, the Air Board ceased to exist as a separate department.

23 Felixstowe F-3 N4016, later registered G-CYBT, flew the leg from Rivière du Loup, Que., to Winnipeg, Man., with stops at Ottawa, Sault Ste Marie, Kenora and Selkirk.

24 The western legs of the trans-Canada flight were flown by de Havilland D.H.9a aircraft. G-CYAN developed engine trouble on the flight from Winnipeg to Regina, and G-CYAJ was pressed into service for the leg from Regina to Calgary, Alta.

22 The Curtiss HS-2L G-CYAG was pressed into service to replace the Fairey Transatlantic and successfully flew the first leg from Halifax to Rivière du Loup, Que., with a refuelling stop at Fredericton, N.B.

25 The final leg from Calgary to Vancouver, B.C., with stops at the Crowle Ranch, near Revelstoke, and Merritt was flown in the third D.H.9a, G-CYBF.

Part Three
ROYAL CANADIAN AIR FORCE 1924-1968

The reorganization of the Canadian Air Force was completed, and the prefix "Royal" officially adopted, on 1 April 1924.[1] That date, on which Canada's fifth air force organization came into being as a permanent component of her defence forces, marked the birthday of the Royal Canadian Air Force.

Organization — A Civil Air Force

Under the new organization the RCAF was still administered by a director responsible to the Chief of Staff but it was now composed of three branches: a Permanent Active Air Force, a Non-Permanent (later, Auxiliary) Active Air Force, and a Reserve Air Force. The authorized establishment of the Permanent Force was set at 68 officers and 307 airmen but the actual strength was 62 officers[2] and 262 airmen. Like its predecessor (the Canadian Air Force) the RCAF was to be unique among the air forces of the world in that the greater part of its work was essentially non-military in character. Operating from the Air Board's original six stations, which were staffed by specialized detachments as required, the RCAF performed many valuable services as the government's civil aviation company. It photographed vast areas of Canada, helped open new sections of the interior, transported officials into inaccessible regions, blazed air routes, experimented in air mail service, carried "treaty money" to the Indians, patrolled forests and fishing grounds, assisted in the suppression of smuggling and illegal immigration and flew sick and injured traders, trappers, farmers and Indians from remote outposts to places where medical attention was available.

Squadrons Formed

On 19 May 1925 the Privy Council authorized an establishment for the RCAF that provided for service squadrons to fulfil the operational requirements of various government departments and agencies: RCAF Headquarters, Ottawa; No. 1 Flying Training Station, Camp Borden; No. 1 (Operations) Wing, Winnipeg; No. 1 (Operations) Squadron, Vancouver; No. 2 (Operations) Squadron, High River; No. 3 (Operations) Squadron, Ottawa; No. 4 (Operations) Squadron, Dartmouth.[3]

Formation of the Directorate of Civil Government Air Operations

In 1927 came strong opposition to the performance by a military organization, the RCAF, of civil operations. As a result, the Directorate of Civil Government Air Operations (DCGAO) was formed on 1 July 1927 to administer and control all air operations carried out by state aircraft (other than those of a military nature) and to administer and control such units, detachments and formations of the RCAF as would be placed under its control.[4]

Since DCGAO was supposedly a civilian organization — in reality it was commanded, administered and staffed by RCAF personnel who were either seconded or attached to the new directorate — it was organized into air stations with attached specialized detachments. All operational flying units of the RCAF were transferred to DCGAO and their military service designations lapsed.

For the 1927-28 fiscal year, the RCAF establishment was reduced to a headquarters, two training stations and five training squadrons: RCAF Headquarters, Ottawa; RCAF Station Camp Borden (landplane training); No. 1 (Ab Initio Training) Squadron; No. 2 (Advanced Training) Squadron; No. 3 (Service) Squadron, "A" Flight — Fighter with Siskin aircraft, "B" Flight — Army Co-operation with Atlas aircraft, "C" Flight — Communications with Fairchild 71 and Bellanca aircraft; RCAF Station Vancouver (seaplane training); No. 4 (Training) Squadron; No. 5 (Service) Squadron.

Impressive as this may have looked, it soon became ap-

26 Jericho Beach, B.C. served the Vancouver area and the British Columbia coast as a base of operations for the Air Board's flying boats in the summer of 1921. The following year the canvas hangars were replaced with permanent structures.

parent that the RCAF lacked sufficient funds, personnel and equipment to maintain such an organization and, at the same time, provide trained personnel for DCGAO. Except for the headquarters in Ottawa and small staffs at Camp Borden and Vancouver, it was only a paper force and never functioned to any degree in a military sense.

---

[1] Formal application to use the prefix "Royal" was made on 5 January 1923. Royal approval was granted by King George V on 15 February 1923, and the prefix first appeared in Air Force Orders on 13 March 1923. It was used in all correspondence beginning on the 14th, but was not authorized by the Canadian government until 1 April 1924, when **King's Regulations and Orders for the Royal Canadian Air Force 1924** (approved on 4 March) was promulgated. Also on 1 April a scale of pay and allowances, approved on 15 January, came into effect.
[2] Of the 62 officers, 43 were pilots (ten were given administrative work); four were technical; five photographic officers; and ten were in charge of stores.
[3] Authority: PC 774, retroactive to 1 April 1925. The new service designations "wing" and "squadron" were published in GO (General Order) 69/25, retroactive to 1 April, but did not come into use until July 1925. A No. 5 (Operations) Squadron was authorized as a unit of No. 1 Wing; apparently, however, it existed only on paper.
[4] Up to this date, Civil Government Air Operations and Civil Aviation had been under the control of the RCAF through assistant directors. The new organization had four branches within the Department of National Defence:
— the RCAF, which continued as the military branch of the aviation service, responsible to the Chief of Staff;
— Civil Government Air Operations, divorced from the RCAF and placed under a separate director, W/C J.L. Gordon;
— a Controller of Civil Aviation, responsible also for the organization of airways and related matters. The Controller himself was Mr. J.A. Wilson.
— an Aeronautical Engineering Division, created under W/C E.W. Stedman, to service the three directorates.
The three new branches were responsible to the Deputy Minister of National Defence.

27 *The Canadian Air Force air station at High River, Alta. photographed from the air on 10 July 1922.*

28 *Victoria Beach, Man. was the Canadian Air Force's first base of operations in the Winnipeg area. A few frame buildings surround the canvas hangar erected on a heavy plank platform. In the open field behind the beach two Felixtowe F-3 flying boats, G-CYEO and G-CYDH, are being erected for forestry operations.*

29 *Ottawa Air Station at Rockcliffe, Ont. had facilities for both wheel and float equipped aircraft. By the early 1930s it had become well established with permanent buildings and slipway. In 1921 the undeveloped site served as a base for the CAF in eastern Ontario.*

30 *Dartmouth, N.S. was the first CAF base on Canada's east coast and served the Maritime provinces as the centre of flying boat operations for photographic flights and fishery patrols.*

UNIT HISTORIES
1925-1927

No. 1 (OPERATIONS) WING

Formed at Winnipeg, Manitoba on 1 April 1925, the wing was employed on civil government air operations. It flew forestry patrols over Manitoba, Saskatchewan and western Ontario until transferred to the non-military Directorate of Civil Government Air Operations on 1 July 1927; its service designation then lapsed.

Commanders
S/L G.O. Johnson, MC 1 Apr 25 - 1 Jul 27.

Higher Formation and Wing Location
RCAF Headquarters:
Winnipeg, Man. 1 Apr 25 - 1 Jul 27.
Sub-bases at Victoria Beach, Norway House, Cormorant Lake, and Lac du Bonnet.

Representative Aircraft
Vickers Viking (Apr 25 - Jun 27)
G-CYET   G-CYEV   G-CYEX   G-GYEZ
Avro 552A (May 25 - Jun 27) G-CYGB   G-CYGG
G-CYGI   G-CYGS
Canadian Vickers Vedette (Oct 26 - Jun 27) G-CYGW
G-CYZO   G-CYZN   G-CYZS
Canadian Vickers Varuna (May - Jun 27) G-CYZT
G-CYZU

31 *Vickers Viking Mk.IV   G-CYEX and Avro 552A G-CYGG on beaching gear at Cormorant Lake Sub-Base in 1925. The tents were soon replaced by more permanent structures as the base developed.*

32 *Canadian Vickers Vedette Mk.II   G-CYZN on photographic operations out of Winnipeg, Man. in 1927. The terrain is typical of the forest and lake country in central Manitoba over which crews flew with minimal navigational aids.*

No. 1 (OPERATIONS) SQUADRON

Formed at Jericho Beach (Vancouver), British Columbia on 1 April 1925, the squadron was employed on civil government air operations. It flew West Coast forestry and fishery patrols until transferred to the non-military Directorate of Civil Government Air Operations on 1 July 1927; its service designation then lapsed.

Commanders
S/L A.E. Godfrey, MC, AFC 1 Apr 25 - 18 May 25.
S/L J.H. Tudhope, MC 19 May 25 - 1 Jul 27.

Higher Formation and Squadron Location
RCAF Headquarters:
Jericho Beach, B.C. 1 Apr 25 - 1 Jul 27.
Sub-base at Casey Cove, B.C.

Representative Aircraft
Curtiss HS-2L (Apr 25 - Jul 27)
G-CYDU   G-CYEB   G-CYGA   G-CYGN

34   *Curtiss HS-2L flying boat G-CYGA on the slipway at Jericho Beach, B.C., photographed on 6 January 1925 prior to joining No. 1 (Operations) Squadron.*

33   *Canadian Vickers Varuna Mk.II G-CYZT on the ramp at Cormorant Lake, Man., in 1927. The Varuna was used to transport fire fighters and their equipment to the site of forest fires reported by patrolling Avro 552A's and Vedettes.*

35   *De Havilland D.H.4 G-CYDB at High River, Alta., used by No. 2 (Operations) Squadron.*

36   *Avro 552A G-CYFU served with No. 2 (Operations) Squadron at High River, Alta. until February 1927. This photo was taken at Camp Borden, Ont. soon after all of the 552A's were reduced to training duties.*

No. 2 (OPERATIONS) SQUADRON

Formed at High River, Alberta on 1 April 1925, the squadron was employed on civil government air operations. It flew forestry patrols over Alberta until transferred to the non-military Directorate of Civil Government Air Operations on 1 July 1927; its service designation then lapsed.

Commanders
S/L A.A.L. Cuffe 1 Apr 25 - 18 Oct 26.
F/L R. Collis 19 Oct 26 - 1 Jul 27.

Higher Formation and Squadron Location
RCAF Headquarters:
High River, Alta. 1 Apr 25 - 1 Jul 27.
Sub-bases at Pincher Creek and Eckville, Alta.

Representative Aircraft
de Havilland D.H.4 (May 25 - June 27) G-CYDB
G-CYDL G-CYDN G-CYEC
Avro 552A (Feb 26 - Feb 27) G-CYFT G-CYFU
G-CYFV G-CYFX
Avro 504N (Feb - Jun 27) G-CYAV G-CYBL
G-CYBM G-CYCD
The squadron also had 2 Siskins (22 June - 14 July 1927) for testing purposes. J-7758 crashed on 28 June, killing P/O C.M. Anderson; J-7759 was returned to Camp Borden.

No. 3 (OPERATIONS) SQUADRON

Formed at Rockcliffe (Ottawa), Ontario on 1 April 1925, the squadron was employed on civil government air operations. It flew forestry patrols over Ontario and Quebec, and also operated a test and development centre for new aircraft and photographic equipment, until transferred to the non-military Directorate of Civil Government Air Operations on 1 July 1927; its service designation then lapsed.

Commanders
S/L A.B. Shearer 1 Apr 25 - 11 Jan 26.
F/L R.S. Grandy 12 Jan 26 - 1 Jul 27.

Higher Formation and Squadron Location
RCAF Headquarters:
Rockcliffe, Ont. 1 Apr 25 - 30 Apr 25.
Shirleys Bay (Ottawa) Ont. 1 May 25 - 1 Jul 27.

Representative Aircraft
Curtiss HS-2L (Apr 25 - Jun 27)
G-CYGL G-CYGP
Vickers Viking (Apr 25 - Jun 27) G-CYES
Canadian Vickers Varuna (Apr 25 - Jun 27) G-CYGV
Canadian Vickers Vedette (Apr 25 - Jun 27) G-CYFS
G-CYGW G-CYFZ
Avro 552A (Apr 25 - Jun 27) G-CYGB G-CYGK
G-CYGG

37 Armstrong Whitworth Siskin III J7759 was flown on trials by No. 2 (Operations) Squadron at High River in the summer of 1927.

39 Vickers Viking Mk.IV G-CYES in Canadian Vickers' Basin Montreal, Que. Delivered to Station Ottawa in 1924, it continued in use with No. 3 (Operations) Squadron until July 1925.

38 Curtiss HS-2L G-CYGL tied down on her beaching dolly ready for the 1925 season with No. 3 (Operations) Squadron.

No. 4 (OPERATIONS) SQUADRON

Formed at Dartmouth, Nova Scotia on 1 April 1925, the squadron was employed on civil government air operations. It flew customs preventive patrols on the East Coast until transferred to the non-military Directorate of Civil Government Air Operations on 1 July 1927; its service designation then lapsed.

Commanders
F/L T.A. Lawrence 7 May 25 - 28 Nov 25.
Squadron inactive.
F/L B.N. Harrop 7 Apr 27 - 1 Jul 27.

Higher Formation and Squadron Location
RCAF Headquarters:
Dartmouth, N.S. 1 Apr. 25 - 1 Jul 27 (Squadron and station were inactive during 1926).

Representative Aircraft
Curtiss HS-2L (Apr 25 - Jun 27) G-CYEL G-CYGO G-CYGT
Canadian Vickers Varuna (Mar - Jun 27) G-CYZV

42 Wright powered Avro 552A G-CYGK with a typical hitch on a sandy beach. One of a kind, it served with No. 3 (Operations) Squadron from 1925 to 1927.

40 Canadian Vickers Varuna Mk.I G-CYGV, assigned to No. 3 (Operations) Squadron following the type-test flights at Station Ottawa in 1927.

43 Curtiss HS-2L flying boats G-CYGO and GT at No. 4 (Operations) Squadron, Dartmouth, N.S. in the summer of 1925.

41 Canadian Vickers Vedette Mk.I G-CYFS was the first of the type to enter service. It joined No. 3 (Operations) Squadron on 17 July 1925 and remained until lost at Torrance, Ont. in August 1927.

44 Canadian Vickers Varuna Mk.II G-CYZV joined No. 4 (Operations) Squadron at Dartmouth, N.S. on 12 March 1927 and remained there until it crashed at Fredericton, N.B. in November 1927.

The RCAF Reorganized as a Military Force

During the eight years from 1924 to 1932, the RCAF had expanded slowly but steadily. More than half of its efforts and funds, however, had been expended on civil aviation.

The world-wide depression brought, in 1932, a drastic curtailment of all service and civil aviation activities. On 1 April the strength of the RCAF was slashed by one-fifth, with 78 officers and 100 airmen released, leaving a strength of 103 officers and 591 airmen. On 1 November, to achieve maximum economy and efficiency with the funds available, the RCAF and the Directorate of Civil Government Air Operations were consolidated and, together with the Aeronautical Engineering Division, placed under the command of a Senior Air Officer responsible to the Chief of the General Staff.[1]

For three years the RCAF was barely able to survive. Then, in 1935, the situation began to improve as the annual appropriations were increased, and the RCAF started to grow again. At the same time, its character underwent a final major change.

Since its inception in 1924, the RCAF had been engrossed with the government's civil flying. The reorganization of 1927 resulted only in the creation of a quasi-military force to perform this work, and since 1932 operations had been almost at a standstill. In 1936 it was decided that the RCAF should be reorganized as a purely military organization, and on 1 November the Department of Transport was formed to establish and implement civil aviation policy. The Civil Aviation Branch of the Department of National Defence — less RCAF personnel, equipment and units —was transferred to the new department. Thereafter the RCAF's only involvement in civil operations was aerial photography, a task that would increase in importance as time passed.

Squadrons Re-formed
Largely freed of civil responsibilities, the RCAF was reorganized along service lines and developed into a military air force. Service units, which had seldom been used, were reactivated; and efforts were made to obtain up- to-date operational types of service aircraft.[2]

The first service squadrons had begun to appear in 1933 with the formation of No. 4 (Flying Boat) Squadron at Vancouver on 17 February, followed by No. 5 (Flying Boat) Squadron at Dartmouth on 16 April 1934. Early in 1936 the various specialized detachments employed on civil aerial photography and general purpose work (transportation and communications) were consolidated to form two more squadrons: No. 7 (General Purpose) at Ottawa on 29 January,and No. 8 (General Purpose) at Winnipeg on 14 February.[3]

Up to now the four squadrons formed were still closely associated with the RCAF's earlier civil aviation work, but with the creation of the Department of Transport to handle civil matters, the RCAF was authorized to form three purely military squadrons during the fiscal year 1936-37. No. 2 (Army Co-operation) and No. 6 (Torpedo Bomber) Squadron were to consist of two flights each, and No. 3 (Bomber) Squadron of one bomber and one fighter flight. All three units were located at Trenton, Ontario,[4] which had opened in 1931 as the RCAF's main training base. In May 1937 No. 3 Squadron was reorganized as a purely bomber unit, and its fighter flight became the nucleus of No. 1 (Fighter) Squadron — the eighth and last Permanent Force squadron to be formed before the Second World War.[5]

Air Commands Formed
During the period 1924 to 1935 it was possible for the RCAF to exercise control of its units under the Senior Air Officer through various headquarters directorates in Ottawa. Although these units were spread across the country — and communications were far from adequate or reliable — the relatively small size of the RCAF, and its work as the government's civil air service, made central control workable. By 1936, however, its growth and reorganization into a military air force made it necessary to decentralize; and the RCAF now requested authority to form four air commands.

Three of these commands (Eastern, Central and Western Air Command) were to be geographical operational commands in charge of all matters pertaining to operational training and the control of air and/or coastal defence, army co-operation, air transport and communications; the fourth (Air Training Command) would be a functional command. Their responsibilities, as visualized in 1936, would be:

> Eastern Air Command (with headquarters in Halifax, Nova Scotia), operational control of all units in Nova Scotia, New Brunswick and Prince Edward Island;
>
> Central Air Command (Winnipeg, Manitoba), operational control of all units in Manitoba, Saskatchewan and northwestern Ontario;
>
> Western Air Command (Vancouver, British Columbia), operational control of units in Alberta and British Columbia;
>
> Air Training Command (Toronto, Ontario) to control all basic aircrew and groundcrew training for the RCAF, and also to be responsible for the training organizations at Trenton and Camp Borden.

RCAF Headquarters in Ottawa would exercise operational control of all units in Ontario (excluding the northwestern portion) and Quebec.

When funds became available in 1937 as a result of the growing threat of conflict in Europe and the need to expand Canada's military forces against that threat, the RCAF began to implement the planned command organization. Western Air Command was formed on 1 March 1938, followed by Eastern Air Command and Air Training Command on 15 November. The formation of Central Air Command, however — although sanctioned — was not proceeded with; its responsibility was transferred to Western Air Command, which now controlled all units west of the Ontario-Manitoba border. The setting up of these new air commands reflected a growing concern with the problems of air defence. As a result, the commands were operational at the outbreak of the Second World War and were able to handle the rapid expansion, both of the RCAF's operational commitments and of its forces.

The RCAF — An Independent Force
On 19 December 1938 the RCAF, which had been under the Chief of Staff and later the army Chief of the General

Staff, became an independent arm directly under the Minister of National Defence. With that, the head of the RCAF became Chief of the Air Staff.

---

[1] The Controller of Civil Aviaton remained under the Deputy Minister.
[2] The last service type aircraft to be introduced into the RCAF had been the Siskin biplane fighter in April 1930.
[3] In February 1937 these two squadrons were consolidated and reorganized in Ottawa. No. 8 Squadron was assigned exclusively the role of aerial photography (but with no change in designation) while No. 7 Squadron assumed all responsibility for aerial communications and transport.
[4] In view of insufficient trained personnel and delay in obtaining aircraft from England — Wapitis for No. 3 Squadron, Sharks for No. 6 — the station improvised a series of incomplete squadron organizations: 15 April 1936: Nos. 2 and 3 Squadrons as flights under the command of the senior officer present (usually a flight commander). (Station orders of this period do not even mention No. 6 Squadron). 13 July 1936: No. 2 Squadron — two flights under the senior officer present; No. 3 Squadron, under the command of S/L A.H. Hull — two flights, of which the Bomber Flight was to include the nucleus of No. 6 (Torpedo Bomber) Squadron. 10 August 1936: Nos. 2 and 3 Squadrons and the nucleus of No. 6 to be reorganized as a single, composite squadron of four flights under S/L Hull with "A" and "B" Flights — Army Co-operation, from No. 2 Squadron, "C" Flight — the Bomber Flight of No. 3 Squadron, still including the nucleus of No. 6 Squadron, and "D" Flight — the Fighter Flight of No. 3 Squadron.
This third temporary organization was introduced primarily for continuity of command, as officers were frequently away on course or taking part in an exercise. Between November 1936 and April 1937, as more personnel became available, the composite squadron was gradually broken up and the three authorized squadrons formed or re-formed. 23 November 1936: No. 6 (TB) Squadron activated under F/L C.L. Trecarten. 1 December 1936: No. 2 (AC) Squadron reactivated under S/L T.A. Lawrence. 7 April 1937: No. 3 (B) Squadron, still with one bomber and one fighter flight, again became a separate unit (temporarily commanded by F/L B.G. Carr-Harris). On 17 May, S/L Hull assumed command of the newly- formed Flying Training School at Trenton.
[5] Three more squadrons had been authorized but were not formed prior to the outbreak of the Second World War: No. 9 (General Reconnaissance) in Western Air Command, and Nos. 10 (Torpedo Bomber) and 11 (General Reconnaissance) in Eastern Air Command.

The Non-Permanent (later Auxiliary) Active Air Force* 1932-1939

Although provision had been made in 1924 for the formation of the Non-Permanent Active Air Force, it was not until 1932 that it actually came into being. On 5 October the first of its three squadrons, all designated Army Co-operation, were authorized — No. 10 at Toronto, 11 at Vancouver and 12 at Winnipeg. While they could begin recruiting immediately, their organization was not sufficiently advanced for aircraft to be issued, and initial flying training started, until 1934.[1]

Two squadrons were formed in 1934 — No. 15 (Fighter) and No. 18 (Bomber) at Montreal, the only location to have more than one squadron — followed in 1935 by two bomber squadrons, No. 19 at Hamilton and No. 20 at Regina. In 1937 two fighter units, Nos. 13 and 21, were formed at Calgary and Quebec City respectively.

On 15 November 1937, to allow for the expansion of the Permanent Force, the Non-Permanent units were renumbered in the "100-block," No. 10 becoming No. 110, No. 11, 111, and so on.

The last three NPAAF squadrons came into being on 1 April 1938: No. 114 (Bomber) at London, No. 116 (Coast Artillery Co-operation) at Halifax, and No. 117 (Fighter) at Saint John. On 1 December of that year the NPAAF was redesignated the Auxiliary Active Air Force; and to its authorized peak strength in squadrons (twelve) were now added three wing headquarters: No. 100 in Vancouver, No. 101 in Toronto, and No. 102 in Montreal.

In September 1939, when units were mobilized, the Auxiliary represented one-third of the RCAF's total strength, and it supplied two of the first three complete squadrons to go to England in 1940.

---

[1] Each squadron was allotted five de Havilland Moths, and a Permanent Force detachment of two officers and five airmen to provide the initial flying training and groundcrew training and assist in maintaining the aircraft.

The Second World War 1939-1945

Wartime Expansion
The RCAF peace establishment called for a total of twenty-three squadrons, of which eight of eleven authorized Permanent Force squadrons and twelve Auxiliary had been formed by the eve of war. During the first month, however, it was found that only fifteen squadrons could be readily brought up to strength and mobilized — twelve for home defence, and three for service overseas. Of aircraft, there were twenty different types totalling 270 machines. Over half (146) were training or transport types and only 124 could be termed operational service types. First-line equipment was limited to nineteen Hurricane fighters and ten Battle day-bombers. Other aircraft were obsolete types such as the Atlas, Wapiti, Shark and Siskin. From this small nucleus, both in personnel and equipment, the RCAF expanded during the war to become the fourth largest air force of the Allied powers.

The wartime RCAF consisted of three main parts of which two were in Canada. One was a vast training establishment — the British Commonwealth Air Training Plan — while the other was an operational organization — the Home War Establishment — which was to deploy thirty-seven squadrons for coastal defence, protection of shipping, air defence and other duties in the Western Hemisphere. The third, with headquarters in London, England, was also an operational organization — the Overseas War Establishment. At the end of the war it had forty-eight squadrons serving in conjunction with the Royal Air Force in the Western European, Mediterranean and Far Eastern theatres.

British Commonwealth Air Training Plan
On 10 October 1939 it was announced that Australia, Canada, the United Kingdom and New Zealand had agreed in principle that Canada should provide a training ground for Commonwealth aircrew where, as in 1917-18, instruction could be carried out away from the actual battle zone. On 17 December the British Commonwealth Air Training Plan (BCATP) agreement was signed, converting Canada into what President Roosevelt of the United States later termed the "airdrome of democracy." When the terms were

---

*The official designation was NPAAF until Dec. 1, 1938. It was then AAAF (Auxiliary Active Air Force) until mobilization for the Second World War. The term "Auxiliary," however, was commonly used throughout these periods; and for convenience we have used "Auxiliary" in squadron titles — under "UNIT HISTORIES."

known, both the BCATP concept and Canada's responsibility under it were no less than staggering.

In the fiscal year which had ended on 31 March 1939, less than six months before the outbreak of war, the RCAF with a total strength of 4061 officers and airmen (including the Auxiliary) had produced only 45 pilots. Under the BCATP agreement the RCAF was to administer 40,000 trained personnel, and to instruct (and provide groundcrew for) 20,000 aircrew annually in 74 training schools or other installations. The first schools were to be opened by 29 April 1940 ("Z" Day), some four months after the signing of the agreement, with the first graduates to arrive in England by early 1941 and the full scheme to be in operation by the end of April 1942.

To meet the demands of the plan the RCAF called on the seventeen civilian flying clubs then in Canada to be responsible for elementary flying training; and a group of commercial operators from bush flying and charter flight operations undertook the training of observers. The Department of Transport assumed responsibility for locating suitable sites for airfields and schools; and for contracts for their construction.

As planned, the first schools were opened on "Z" Day and training began. By the end of September 1941, seven months ahead of schedule, all but three of the projected schools were in operation. Though the BCATP had not expected to graduate students until early 1941, accelerated progress became evident. As early as 30 September 1940 the first 39 pilots passed out of Camp Borden, followed by the first observers from Trenton on 24 October and the first 50 air gunners from Jarvis on the 28th.

In June 1942, under an extension of the plan, the number of training schools was increased to 67 (including 21 double schools); and ten specialist schools were added, in the main for operational and flying instructor training. Not only that, the administration of 27 RAF schools in Canada was to become a responsibility of the RCAF. At the close of 1943 the plan reached its peak strength, with an establishment of 104,000 personnel in four Training Commands (Nos. 1 to 4) operating 97 schools and 184 ancillary units on 231 sites (including the RAF's 40,000 personnel at 27 locations); and it was producing an average of 3000 graduates per month.

The signatories agreed on 16 February 1944, because of the large reserve of aircrew already trained or under instruction, to begin a gradual reduction of trainees and staff. In June, recruiting of aircrew and ground personnel for the RCAF was suspended; by October the closing down of schools was stepped up; and on 31 March 1945 the British Commonwealth Air Training Plan came to an end. Canada's responsibilities had been fulfilled, and the plan itself had proved to be a major factor in Allied mastery of the air.

Training commands 1940-1945

No. 1 Training Command

Formed at Toronto, Ontario on 1 January 1940; moved to Trenton Ontario on 14 January 1944; merged with No. 3 Training Command to form No. 1 Air Command on 14 January 1945.
Commanders
A/C A.A.L. Cuffe 1 Jan 40 - 10 Jun 40.
A/V/M G.E. Brookes, OBE 11 Jun 40 - 21 Jun 42.
A/V/M G.O. Johnson, MC 22 Jun 42 - 6 Jan 43.
A/V/M F.S. McGill 7 Jan 43 - 29 Nov 43.
A/V/M A.T.N. Cowley 30 Nov 43 - 14 Jan 45.

No. 2 Training Command

Formed at Winnipeg, Manitoba on 15 April 1940; merged with No. 4 Training Command to form No. 2 Air Command on 30 November 1944.
Commanders
A/V/M A.B. Shearer 15 Apr 40 - 5 Jan 43.
A/V/M T.A. Lawrence 6 Jan 43 - 30 May 44.
A/V/M K.M. Guthrie CB, CBE, 31 May 44 - 30 Nov 44.

No. 3 Training Command

Formed as No. 2 Training Group at Montreal, Quebec on 18 March 1940; redesignated No. 3 Training Command on 29 April 1940; merged with No. 1 Training Command to form No. 1 Air Command on 15 January 1945.
Commanders
A/C C.M. McEwen, MC, DFC 19 Mar 40 - 26 Mar 41.
A/C G.V. Walsh, MBE 27 Mar 41 - 1 Nov 41.
A/V/M J.L.E.A. de Niverville 2 Nov 41 - 19 Nov 43.
A/V/M A. Raymond, CBE 20 Nov 43 - 15 Jan 45.

No. 4 Training Command

Formed at Regina, Saskatchewan on 29 April 1940; moved to Calgary, Alberta on 1 October 1941; merged with No. 2 Training Command to form No. 2 Air Command on 30 November 1944.
Commanders
A/C L.F. Stevenson 7 May 40 - 28 Sep 40.
A/C A.T.N. Cowley 29 Sep 40 - 20 Mar 42.
A/V/M G.R. Howsam, MC 21 Mar 42 - 30 Nov 44.

Home War Establishment

Organization

When the war began, in September 1939, the RCAF's Home War Establishment (HWE) had two operational air commands (Eastern and Western Air Command) and seven understrength squadrons equipped with a variety of obsolescent aircraft with which to fulfil its role of defending Canada's two seaboards. Because the greatest threat to Allied shipping was posed by German surface and undersea (U-boat) raiders in the North Atlantic, top priority was given to expanding the facilities and capability of Eastern Air Command, and to re-equipping its operational squadrons with aircraft of more modern type. When Japan entered the war in December 1941 and occupied islands in the Aleutian chain — thus threatening the West Coast of Canada — priorities were reversed. Western Air Command was then provided with reinforcements and more modern aircraft.

From late 1941 until the spring of 1942, the HWE experienced its maximum growth. With Eastern Air Command moving squadrons into Newfoundland to extend its air coverage of the North Atlantic, and with Western Air Command providing air reinforcements to United States forces in Alaska, serious problems were experienced in exercising operational control; squadrons were widely spread and communications inadequate and unreliable. To overcome these difficulties, both air commands received authority to form operational sub-headquarters (called Groups) as required. Odd-numbered group headquarters were assigned to Eastern Air Command and even-numbered to Western Air Command. Throughout the war, however, the headquarters of both commands exercised control at both command and group levels, with the result that each had only one group headquarters under its command for any length of time. Eastern Air Command had No. 1 Group (with headquarters at St. John's, Newfoundland) and for a short time No. 5 (Gulf) Group at Gaspé, Quebec. Western Air Command had No. 4 Group at Prince Rupert, British Columbia — which covered northern British Columbia and Alaska — and for a very short period, No. 2 Group at Victoria which was formed to cover the move of Western Air Command's headquarters from Victoria to Vancouver.

With rumours prevalent concerning a German aircraft carrier operating off Halifax (which proved to be unfounded) and the threat posed by the Japanese along the West Coast, the Home War Establishment was authorized in early 1942 to increase its strength by ten fighter and six bomber reconnaissance squadrons.[1] Based on the experience of air fighting over Europe and China, it was decided to equip the East Coast fighter squadrons with Hurricane aircraft and those on the West Coast with Kittyhawks.

In November 1943 the HWE reached its peak strength with a total of 37 squadrons — 19 in Eastern Air Command and 18 in Western Air Command — plus a network of air stations on both coasts. Many air stations and posts were virtually inaccessible, except by air, and flying conditions were most uncertain because of sudden fog. Isolation, boredom and the weather were the chief enemies at these lonely outposts.

Eastern Air Command

Throughout the whole of the war the main concern of Canada's home forces was with the threat posed by German submarines to Allied shipping, save for a temporary preoccupation with Japan, as has been mentioned. The British Admiralty had immediately reintroduced the First World War convoy system, and selected Bedford Basin in Halifax as the western terminus and assembly point for the convoys. Eastern Air Command provided and directed the work of air protection for shipping passing through the Northwest Atlantic. To meet this responsibility, the Command had priority in the re-equipping of its bomber reconnaisance squadrons with Hudson, Bolingbroke, Catalina and, later, Liberator aircraft.[2] With that the command was able to extend its convoy escort and anti-submarine sweeps hundreds of miles out over the Atlantic and, from early 1944, right across.

The first 18 months of the war were relatively quiet, but from the spring of 1941 onwards the resources of the command were taxed to their utmost in the grim Battle of the Atlantic. Enemy submarines were sighted and attacked in Canadian coastal waters; they even ascended the St. Lawrence to sink vessels. The most critical period was from early 1942 to mid-1943 when submarine activity in the North Atlantic reached its peak. Then the tide turned, and although the introduction of acoustic torpedoes and the Schnorkel "breathing" device presented new and serious problems, these were overcome by more scientific detection apparatus and especially by extended air cover. The sea and air forces of Britain, the United States and Canada retained the upper hand until the U-boats surrendered with the end of the war in Europe.

Aircraft of Eastern Air Command sank six German submarines. This figure, however, is no full measure of the command's contribution nor would the total number of sightings and attacks express it. A better indication is to be found in the 180,132 hours flown through weather that was often appalling. Carefully searching the grey expanse of water, aircraft forced enemy submarines to crash-dive or remain submerged and thus kept them away from convoys. It was wearisome and unglamorous work but its importance cannot be too much stressed. The battle lines in Europe were fed by Atlantic ships.

Western Air Command

The first fifteen months of the war found Western Air Command with no opportunity to engage the enemy. Patrols were flown daily to identify and track all shipping along the West Coast, and to survey new sites for bases required to extend the patrol lines.

The sudden entry of Japan into the war, with its crippling attack on the United States naval and air forces at Pearl Harbor on 7 December 1941, quickly changed the tempo. Early in 1942, as a result of the serious losses suffered by the United States in the Pacific and the lack of immediate reinforcements for its troops in Alaska, an agreement was signed under which the RCAF was to assist in the defence of Alaskan bases.

Two squadrons were sent to Annette Island in May 1942 for the defence of Prince Rupert, a major Canadian seaport for supplies bound for Alaska. Known as "Y" Wing, these squadrons served on Annette Island until November 1943 when they were then redeployed to Terrace, a new airfield 50 miles east of Prince Rupert, and the wing ceased to exist.

A second formation, "X" Wing, was set up at Anchorage in June 1942. For months its squadrons, flying side by side with Americans in the "worst weather in the world", carried out reconnaissance and defensive fighter patrols. In September 1942 they commenced offensive fighter operations against Japanese forces on the island of Kiska in the Aleutians. On one of these missions S/L K.A. Boomer, commanding No. 111 (Fighter) Squadron, became the only member of a home unit of the RCAF to score a victory over an enemy aircraft. With the total withdrawal of Japanese forces from the Aleutians in the summer of 1943, the wing was disbanded and its squadrons were redeployed to stations in southern British Columbia.

---

[1] On 31 October 1939 seven of the home squadrons whose duty was primarily reconnaissance, but which could also be massed as bombers, were redesignated Bomber Reconnaissance (BR). Those affected were Nos. 4, 5, 8 and 11 (General Reconnaissance), 6 (Torpedo Bomber), and 119 and 120 (Bomber) Squadrons.

[2] Hudson aircraft were delivered to No. 11 (BR) Squadron in October 1939, Bolingbrokes to No. 119 in July 1940, Catalina flying boats to Nos. 5 and 116 in August 1941, and Liberators to No. 10 in April 1943.

When No. 1 (F) Squadron went overseas with its Hurricanes in May 1940, Canada's air defence capability consisted of a handful of obsolescent Grumman Goblin biplanes operated by No. 118 (F) Squadron at Halifax. It was not until November 1941 that Curtiss Kittyhawks became available.

UNIT HISTORIES
HOME SQUADRONS

No. 1 Squadron

Formed as a Fighter unit at Trenton, Ontario on 21 September 1937 with Siskin aircraft — the nucleus had come from the Fighter Flight of No. 3 (Bomber) Squadron on 17 May[1] — the squadron moved to Calgary, Alberta in August 1938, and was re-equipped with Hurricane aircraft in February 1939.[2] It was mobilized at St Hubert, Quebec on 10 September, and on 5 November it moved to Dartmouth, Nova Scotia.[3] On 28 May 1940, before going overseas, it absorbed No. 115 (Fighter) Squadron of the Auxiliary from Montreal. On 26 August 1940, the unit had its first encounter with German aircraft, and was the first squadron of the RCAF to engage the enemy, to score victories,[4] to suffer combat casualties, and to win gallantry awards.[5] The squadron was renumbered No. 401 (Fighter) Squadron at Driffield, Yorkshire, England on 1 March 1941.

Brief Chronology Formed at Trenton, Ont. 21 Sep 37. Mobilized at St Hubert, Que. 10 Sep 39. Absorbed No. 115 (F) Sqn (Aux), Dartmouth, N.S. 28 May 40. Renumbered No. 401 (F) Sqn, Driffield, Yorks., Eng. 1 Mar 41.

Ancestry Fighter Flight, No. 3 (B) Squadron. No. 115 (F) Squadron

Commanders
F/L B.G. Carr-Harris 21 Sep 37 - 28 Aug 38.
S/L E.G. Fullerton 29 Aug 38 - 31 Oct 39.
S/L E.A. McNab, DFC 1 Nov 39 - 1 Nov 40.
S/L G.R. McGregor, DFC 2 Nov 40 - 12 Dec 40.
S/L P.B. Pitcher 13 Dec 40 - 28 Feb 41.

Higher Formations and Squadron Locations
RCAF Headquarters:
Trenton, Ont. 21 Sep 37 - 29 Aug 38.
Western Air Command:
Calgary, Alta. 30 Aug 38 - 2 Sep 39.
Eastern Air Command:
St Hubert, Que. 10 Sep 39 - 5 Nov 39.
Dartmouth, N.S. 6 Nov 39 - 8 Jun 40.
En route overseas 9 Nov 40 - 19 Jun 40.
RAF Fighter Command:
No. 11 Group,
Middle Wallop, Hants 21 Jun 40 - 3 Jul 40.
Croydon, Surrey 4 Jul 40 - 16 Aug 40.
Northolt, Middlesex 17 Aug 40 - 10 Oct 40.
No. 13 Group,
Prestwick, Ayr., Scot. 11 Oct 40 - 8 Dec 40.
No. 14 Group,
Castletown, Caith., Scot. 9 Dec 40 - 10 Feb 41.
No. 12 Group,
Driffield, Yorks. 11 Feb 41 - 28 Feb 41.

Representative Aircraft
Armstrong Whitworth Siskin Mk. IIIA (Sep 37 - Jun 38)
302 303 304 309
Hawker Hurricane Mk.I (Canada Feb 39 - Jun 40, Unit Code NA) 310 to 319 324 327 to 329
(England Jun 40 - Feb 41, Unit Code YO) P3069 YO-A P3670 YO-E P3672 YO-F P3859 YO-B P3873 YO-H R4171 YO-L V7287 YO-F
(These are seven of ten aircraft known to have been used on 26 August 1940 in the squadron's first encounter with the *Luftwaffe* in the Battle of Britain.)

Operational History: First Mission - Canada 20 November 1939, Hurricane 324 with F/O E.M. Reyno from Dartmouth — naval co-operation (diving practice on naval vessels) in Bedford Basin (Halifax). **Last Mission - Canada** 24 April 1940, 2 Hurricanes from Dartmouth — 50-mile radius local reconnaissance. **First Mission - England** 18 August 1940, 1 Hurricane from Northolt — uneventful scramble. **First Victories** 26 August 1940, 10 Hurricanes from Northolt, operating from North Weald for the day, scrambled and intercepted an enemy bomber force of 25-30 Do. 215's. F/L G.R. McGregor (P3863 YO-H) credited with 1 destroyed. F/O T.B. Little credited with 1 damaged, later upgraded to probably destroyed. Squadron as a whole credited with 2 Do.215's destroyed and 2 damaged. Lost 1 Hurricane destroyed and 2 damaged; 1 pilot killed in action (F/O R.L. Edwards); 2 wounded, not seriously. **Last Mission** 26 February 1941, 2 Hurricanes from Driffield — patrol at 10,000 feet. **Summary** Canada (Sep 39 - Feb 40), Sorties: 28. Operational/Non-operational Flying Hours: 12/139. Casualties: Operational: nil. Non-operational: 2 aircraft. England (Jun 40 - Feb 41), Sorties: 1694. Operational/Non-operational Flying Hours: 1569/1201. Victories: Aircraft: 30 destroyed, 8 probably destroyed, 34 damaged. Casualties: Operational: 10 aircraft: 13 pilots, of whom 3 were killed, 10 wounded or injured. Non-operational: 2 personnel killed. **Top Scores** S/L E.A. McNab, DFC 4⅓ - 1 - 3. F/L G.R. McGregor, DFC 4 - 3 - 5. F/O B.D. Russel, DFC 3 - 2 - 3. F/O J.W. Kerwin 3 - 0 - 0. **Honours and Awards** 3 DFC's. **Battle Honours** Battle of Britain 1940. Defence of Britain 1940-1944. (see No. 401 Squadron)

---

[1] The Fighter Flight had been formed at Camp Borden on 1 April 1930 with Siskin aircraft, moved to Trenton in September 1931, and became the Fighter Flight of No. 3 (B) Squadron on 1 September 1935.
[2] The Hurricanes had been shipped in crates from England to Vancouver where they were uncrated, reassembled, test flown, and then ferried to the squadron at Calgary.
[3] At that time the squadron had seven Hurricanes: 311, 315, 316, 324, 327, 328 and 329.
[4] For the record, the first member of the RCAF to be credited with a victory was S/L F.M. Gobeil, who was on exchange duty with the RAF and commanding No. 242 (Canadian) Squadron. On 23 May 1940 he was credited with a Bf.109 probably destroyed and on the 25th with a Bf.109 destroyed, both victories gained while the squadron was in France. The first victory by a member of No. 1 (F) Squadron was scored by S/L McNab on 11 August 1940, when he destroyed a Do.215 over Westgate-on-Sea. At the time, he was attached to No.111 (F) Squadron for experience.
[5] S/L McNab was awarded the DFC on 22 October 1940, and F/L G.R. McGregor and F/O B.D. Russel received theirs on the 25th.

Formed as the Conversion Training Squadron at Picton, Ontario on 29 November 1940 to supply pilots for staff flying duties at bombing and gunnery schools, the squadron moved to Rockcliffe (Ottawa), Ontario in November 1941 and on 15 March 1943 was redesignated No. 1 (Refresher) Squadron. Its role was to provide additional training for student pilots of the service flying training schools who were above average in flying ability but had failed ground school training, and refresher training for pilots returning from overseas for flying duties in Canada. The squadron was disbanded on 1 November 1943.
Brief Chronology Formed as Conversion Training Sqn, Picton, Ont. 29 Nov 40. Redesignated No. 1 (Ref) Sqn, Rockcliffe, Ont. 15 Mar 43. Disbanded 1 Nov 43.

Commanders
S/L W.J. McFarlane 1 Dec 40 - 25 Jan 42.
S/L J.C. Huggard 21 Dec 42 - 16 Oct 42.
S/L D.I. Macklin, AFC 17 Oct 42 - 9 Aug 43.
F/L J.P. Howard 10 Aug 43 - 1 Nov 43.
Higher Formations and Squadron Locations
No. 1. Training Command, BCATP:
Picton, Ont. 29 Nov 40 - 27 Nov 41.
No. 3 Training Command, BCATP:
Rockcliffe, Ont. 28 Nov 41 - 1 Nov 43.
Representative Aircraft Various training types: Fairey Battle, North American Yale and Harvard, Airspeed Oxford, Bristol Bolingbroke, Avro Anson.

45 *Armstrong Whitworth Siskin IIIa 302 of No. 1 (Fighter) Squadron being prepared for flight at Trenton, Ont. in the summer of 1938.*

46 *Northrop Delta Mk.II 675, used by No. 1 (Fighter) Squadron to convert its pilots to Hurricanes at Vancouver in the autumn of 1939.*

47 *Hawker Hurricane Mk.I 313 as received by No. 1 (Fighter) Squadron in March 1939.*

No. 2 Squadron

Formed as an Army Co-operation unit at Trenton, Ontario on 1 April 1935 with Atlas aircraft,[1] the squadron moved to Rockcliffe (Ottawa), Ontario in June 1937. It returned to Trenton on 1 April 1939 to absorb the School of Army Co-operation, thus becoming both a training and an operational unit. Alerted for hostilities on 26 August 1939, the squadron immediately moved to Halifax, Nova Scotia. Mobilized at Saint John, New Brunswick on 10 September and selected for overseas duty with the Canadian Active Service Force, it moved to Rockcliffe in September for training and re-equipping with Lysander aircraft. The need to bring two other army co-operation units (Nos. 110 and 112 Squadrons) up to full strength resulted in the squadron being disbanded on 16 December 1939.

Brief Chronology Formed at Trenton, Ont. 1 Apr 35. Mobilized at Saint John, N.B. 10 Sep 39. Disbanded at Rockcliffe, Ont 16 Dec 39.

Commanders
F/L W.D. Van Vliet 1 Apr 35 - 1 Jul 36.
F/L F.G. Wait 2 Jul 36 - 9 Aug 36.
Squadron inactive.[2]
S/L T.A. Lawrence 1 Dec 36 - 4 Feb 37.
S/L G.R. Howsam, MC 9 Aug 37 - 31 Mar 38.
S/L T.A. Lawrence 1 Apr 38 - 30 Nov 38.
S/L W.D. Van Vliet 1 Dec 38 - 16 Dec 39.

Higher Formations and Squadron Locations
RCAF Headquarters:*
Trenton, Ont. 1 Apr 35 - 16 Jun 37.
Rockcliffe, Ont. 17 Jun 37 - 20 Mar 39.
Trenton, Ont. 21 Mar 39 - 26 Aug 39.
Eastern Air Command:
Halifax, N.S. 27 Aug 39 - 31 Aug 39.
Saint John, N.B. 1 Sep 39 - 30 Oct 39
"A" Flight remained at Halifax.
Air Force Headquarters:* Canadian Active Service Force, Rockcliffe, Ont. 1 Nov 39 - 16 Dec 39.

Representative Aircraft (Unit Code 1939 KO)
Armstrong Whitworth Atlas Mk.I (Apr 35 - Nov 39)
401 to 413 415
Westland Lysander Mk.II (Nov - Dec 39) 418 419 422 425

Operational History: First Mission 7 September 1939, 2 Atlas's from Halifax — reconnaissance patrol. **Last Mission** 30 October 1939, 1 Atlas from Saint John — reconnaisance patrol over the Bay of Fundy. **Summary** Sorties: 74. Operational/Non-operational Flying Hours: 168/212.

*From 31 August 1939, "Air Force Headquarters"

48 Armstrong Whitworth Atlas Mk.IAC 409 of No. 2 (AC) Squadron at Rockcliffe, Ont. in the summer of 1938.

49 Westland Lysander Mk.II aircraft at Rockcliffe, Ont. on 17 November 1939, used by No. 2 (AC) Squadron prior to its disbandment.

---

[1] Nucleus from the Army Co-operation Flight which had formed at Camp Borden on 1 April 1930, and moved to Trenton in September 1931.
[2] Its two flights were incorporated in a temporary, composite unit, under S/L A.H. Hull, the commander of No. 3 (B) Squadron.

No. 3 Squadron

Formed as a Bomber unit at Trenton, Ontario on 1 September 1935, the squadron was to have had one flight with bomber aircraft and one with fighter, but only the Fighter Flight was formed with Siskin aircraft; the Bomber Flight was waiting for the delivery of Wapiti aircraft from England. On 1 June 1937, following the arrival of the Wapitis, the unit was reorganized into a pure bomber squadron[1], and the Fighter Flight became the nucleus of No. 1 (Fighter) Squadron. Three weeks later, it moved to Rockcliffe (Ottawa), Ontario and, in October 1938, to Calgary, Alberta.[2] Alerted for hostilities on 26 August 1939, seven Wapiti aircraft immediately departed from Calgary for the squadron's war station at Halifax, Nova Scotia; the last aircraft reached Halifax eleven days later.[3] On 31 August, while in transit, the squadron was redesignated Fighter but was never converted to that role; instead, it was disbanded on 5 September 1939.[4]

Brief Chronology Formed as No. 3 (B) Sqn, Trenton, Ont. (Fighter Flight only) 1 Sep 35. Reorganized as a pure bomber sqn 1 Jun 37. Redesignated No. 3 (F) Sqn, Halifax, N.S. 31 Aug 39. Disbanded 5 Sep 39.

Commanders
S/L A.H. Hull 13 Jul 36 - 6 Apr 37.
F/L B.G. Carr-Harris 7 Apr 37 - 16 May 37.
S/L A. Lewis 17 May 37 - 5 Sep 39.

Higher Formations and Squadron Locations
RCAF Headquarters:
Trenton, Ont. 1 Sep 35 - 16 Jun 37.
Rockcliffe, Ont. 17 Jun 37 - 18 Oct 38.
Western Air Command:
Calgary, Alta. 21 Oct 38 - 26 Aug 39.
Eastern Air Command:
Halifax, N.S. 1 Sep 39 - 5 Sep 39.

Representative Aircraft (Unit Code 1939 OP)
Armstrong Whitworth Siskin Mk.IIIA (Sep 35 - May 37)
See No. 1 (F) Squadron.
Westland Wapiti (Jun 37 - Sep 39) 508   513   530   543

---

[1]With four Wapiti aircraft, five pilots and five air gunners.
[2]The squadron flew its Wapiti aircraft on the 2300-mile move from Rockcliffe to Calgary following the route established by Trans-Canada Airlines (now Air Canada). This was the first time that a squadron of the RCAF had completed a move by air over such a long distance. When No. 1 (F) Squadron had moved to Calgary in August 1938, its Siskin aircraft had been shipped by rail.
[3]The squadron's role was to have been air strike, either independent or in cooperation with the Royal Canadian Navy against any enemy naval forces that might appear between Port Mouton and Cape Canso.
[4]Personnel and aircraft went to No. 10 (Bomber) Squadron. An organization order for a No. 3 (Bomber Reconnaissance) Squadron was issued on 1 June 1943, but was cancelled shortly afterwards.

50   Westland Wapiti Mk.IIA   554, 541, and 538 of No. 3 (Bomber) Squadron at Rockcliffe, Ont. on 30 August 1939, en route to their war station at Halifax, N.S. from their former base at Calgary, Alta.

No. 4 Squadron

Formed as a Flying Boat unit at Jericho Beach (Vancouver), British Columbia on 17 February 1933,[1] the squadron was employed on civil government operations — preventive patrols against illegal immigration, fishery and forestry patrols, and some aerial photography. On 1 January 1938 it was redesignated General Reconnaissance and relieved of its civil work to begin service training. Mobilized on 10 September 1939, and redesignated Bomber Reconnaissance on 31 October, the squadron was engaged in West Coast anti-submarine duty until disbanded at Tofino, British Columbia on 7 August 1945.

Brief Chronology Formed as No. 4 (FB) Sqn, Jericho, Beach, B.C. 17 Feb 33. Redesignated No. 4 (GR) Sqn 1 Jan 38. Mobilized 10 Sep 39. Redesignated No. 4 (BR) Sqn 31 Oct 39. Disbanded at Tofino, B.C. 7 Aug 45.

Commanders
W/C A.B. Shearer 17 Feb 33 - 4 Nov 36.
W/C A.A.L. Cuffe 5 Nov 36 - 22 Feb 38.
W/C E.L. MacLeod 23 Feb 38 - 7 Nov 38.
S/L W.I. Riddell 8 Nov 38 - 2 Jan 39.
S/L A.J. Ashton 3 Jan 39 - 20 Oct 39.
S/L F.J. Mawdesley 21 Oct 39 - 11 Mar 40.
S/L R.C. Mair 12 Mar 40 - 27 Aug 40.
S/L C.M.G. Farrell, DFC 28 Aug 40 - 11 Feb 42.
S/L K.F. Macdonald 12 Feb 42 - 27 Jul 42.
S/L E.W. Beardmore 28 Jul 42 - 29 Sep 43.
S/L R.H. Lowry 30 Sep 43 - 31 Oct 44.
S/L R.W. McRae 1 Nov 44 - 7 Aug 45.

Higher Formations and Squadron Locations
RCAF Headquarters,
Western Air Command (1 Mar 38):
Jericho Beach, B.C. 17 Feb 33 - 2 May 40.
Ucluelet, B.C. 3 May 40 - 26 Aug 44
Movement exercise, Coal Harbour, B.C. 26 Jan - 8 Feb 43.
Tofino, B.C. 27 Aug 44 - 7 Aug 45.

Representative Aircraft (Unit Code 1939-42 FY, 1942 BO)
Canadian Vickers Vancouver Mk.II (Feb 33 - Jul 39)
902   to 906
Fairchild 71 (Sep 34 - Jul 39) 619   633   641
Canadian Vickers Vedette (Aug 36 - May 40)
809   812   813   816
Supermarine Stranraer (Jul 39 - Sep 43) 910
915 FY-B   934   954
Blackburn Shark Mk.III (May 40-Jan. 42)[2]
545 FY-C   546 FY-D   548 FY-F
Consolidated Canso A (Dec 42 - Aug 45)
9771   9802   11006   11044
Consolidated Catalina Mk.IV & IB (Apr -Aug 44)
JX211   FP291   FP294

Operational History: First Mission 12 September 1939, Vancouver 906 from Jericho Beach with FS J.W. McNee and crew — coastal patrol; returned early, trouble with port engine. **Last Mission** 31 July 1945, Canso A 11016 from Tofino — patrol; returned early, poor weather. **Summary** Sorties: 1762. Operational/Non-operational Flying Hours: 13,269/8408. Victories: nil. Casualties: Operational: nil. Non-operational: 2 aircraft; 11 aircrew killed. **Honours and Awards** 1 AFC. **Battle Honours** see page xix

---

[1]The history of No. 4 Squadron can be traced back to No. 1 (Operations) Squadron formed at Jericho Beach on 1 April 1925. Although its service designation lapsed on transfer to the non-military Directorate of Civil Government Air Operations on 1 July 1927, it continued to operate as the

RCAF's seaplane training unit until it became No. 4 (FB) Squadron.
²Majority of Shark aircraft were received from No. 6 (BR) Squadron, five being received on 1 May 1940. (serials 501, 503, 506, 545 & 546).

51   Looking like an Air Force Day display, the aircraft of No. 4 (BR) Squadron collected on the pier in May 1940 include Fairchild 71, 633; Canadian Vickers Vedette Mk.VA 816; Canadian Vickers Vancouver Mk.II 902; Supermarine Stranraer 912, and three unidentified Blackburn Sharks.

52   Blackburn Shark Mk.III   548 displaying the markings of No. 4 (BR) Squadron.

53   Supermarine Stranraer 915 in the markings of No. 4 (BR) Squadron at Vancouver, B.C. 1940.

54   Consolidated Canso A   9771 of No. 4 (BR) Squadron on patrol out of Ucluelet, B.C. in 1943.

No. 5 Squadron

Badge: A gannet in flight
Motto: *Volando vincimus* (By flying we conquer)
Authority: King George VI, June 1941

The area of this squadron's activities coincides with one of the few areas in which the gannet is found. Its habits in locating and attacking fish are symbolical of the functions of a general reconnaissance squadron in locating and attacking enemy shipping.

Formed as a Flying Boat unit at Dartmouth, Nova Scotia on 16 April 1934¹ by the amalgamation of the five detachments then in the Maritimes (Nos. 8 to 12, formed at Ottawa in 1932), the squadron flew preventive (anti-smuggling and illegal immigration) patrols for the Royal Canadian Mounted Police.² When the RCMP formed its own air division at the end of 1936, the unit concentrated on service training and duties. In 1937 it was redesignated Coastal Reconnaissance, and later the same year General Reconnaissance. Mobilized on 10 September 1939, and redesignated Bomber Reconnaissance on 31 October, the squadron flew Stranraer and Canso flying boats on East Coast antisubmarine duty, sank one U-boat, and delivered Eastern Air Command's last attack on a submarine.³ The squadron was disbanded at Gaspé, Quebec on 15 July 1945.

Brief Chronology Formed as No. 5 (FB) Sqn, Dartmouth, N.S. 16 Apr 34. Redesignated No. 5 (CR) Sqn 1 Apr 37. Redesignated No. 5 (GR) Sqn 1 Dec 37. Mobilized 10 Sep 39. Redesignated No. 5 (BR) Sqn 31 Oct 39. Disbanded at Gaspé, Que. 15 Jun 45.

Commanders
S/L F.C. Higgins 16 Apr 34 - 4 Nov 34.
W/C H. Edwards 5 Nov 34 - 27 Feb 38.
W/C G.E. Brookes, OBE 28 Feb 38 - 1 Jun 39.
S/L A.D. Ross 3 Jul 39 - 26 Jul 40.
W/C F.A. Sampson 27 Jul 40 - 6 Sep 40.
S/L R.C. Mair 7 Sep 40 - 2 Nov 40.
W/C H.M. Carscallen 3 Nov 40 - 24 Apr 41.
W/C S.W. Coleman 3 Jun 41 - 13 Sep 41.
W/C H.M. Carscallen 14 Sep 41 - 28 Jul 42.
S/L A. Fleming 29 Jul 42 - 27 Jan 43.
W/C F.J. Ewart 28 Jan 43 - 27 Oct 43.
S/L K.E. Krug 28 Oct 43 - 15 Apr 44.
S/L J.M. Viau 16 Apr 44 - 26 Jan 45.
W/C W.M. Doherty 27 Jan 45 - 15 Jun 45.

Higher Formations and Squadron Locations
RCAF Headquarters,
Eastern Air Command (17 Apr 39):
Dartmouth, N.S. 16 Apr 34 - 1 Nov 42.
3 aircraft, Gaspé, Que. 17 Jun - 21 Oct 40.
Sydney, N.S. 30 Jul - 18 Dec 40 and 26 May - 24 Sep 41.
No. 1 Group,
Gander, Nfld. 2 Nov 42 - 9 May 43.
4 aircraft, Torbay, Nfld. 24 Apr - 9 May 43.
Torbay, Nfld. 10 May 43 - 1 Aug 44.
6 aircraft, Goose Bay, Lab. 28 Jul - 9 Aug 43.
Eastern Air Command:
Yarmouth, N.S. 2 Aug 44 - 7 May 45.
Gaspé, Que. 8 May 45 - 15 Jun 45.

3 aircraft, Mont-Joli, Que. 19 Apr - 21 May 45; Sydney, N.S. 22 May - 15 Jun 45.
Representative Aircraft (Unit Code 1939-42 QN, 1942 DE)
Fairchild 71 (Nov 34 - Dec 38) 630   636   645   647
Supermarine Stranraer (Nov 38 - Sep 41) 913 QN-B   914 QN-O   916 QN-P
Consolidated Catalina Mk.I (Jun - Jul 41)[4]
W8430   W8431   Z2137   Z2139 Consolidated Canso A (Oct 41 - Jun 45) 9737 E   9738 E   9739 K   9740 L   9741 B   9742 X   9743 N   9744 P   9745 O   9753 C   9756 M   9764 D   9773 G   9807 F   9858 S   9875 R   9879 Q   11004   11052   11055
Operational History: First Mission 10 September 1939, Stranraer 908 from Dartmouth with F/L Price and crew — parallel track search; sighted five vessels but no enemy activity. **Victory: U-Boat** 4 May 1943, Canso A 9747 "W" from Gander with S/L B.H. Moffit and crew — close convoy support; sank U-630 at 5638N 4232W, Eastern Air Command's fourth kill. **Last Mission** 1 June 1945, Canso A 11054 from Sydney with F/O L.G. Empringham and crew — escort to convoy HGI/31. **Summary Sorties:** 3848. Operational/Non-operational Flying Hours: 27,117/8197. Victories: U-boat: 1 sunk; 17 attacks on 25 sightings. Casualties: Operational: nil. Non-operational: 3 aircraft; 11 aircrew killed. **Honours and Awards** 14 DFC's, 2 AFC's, 2 DFM's, 2 AFM's, 1 BEM, 7 MiD's.
Battle Honours see page xix

---

[1] The numeral 5 had been used on paper to designate an operations squadron at Winnipeg, Manitoba in 1925 under No. 1 Wing, and an advanced training squadron at Vancouver, British Columbia in 1927, but the effective history began with this formation.
[2] FS Gibb in Fairchild 633 flew the squadron's first RCMP patrol on 17 April 1934.
[3] On 3 May 1945, Canso A 9743 "N" from Yarmouth with F/L S. MacPherson and crew — convoy escort.
[4] Catalinas to No. 116 (BR) Squadron.

56   *Supermarine Stranraer 916 in the markings of No. 5 (BR) Squadron, in June 1940, at Dartmouth N.S.*

57   *Bristol Bolingbroke Mk.III   717, the sole example of the type to be fitted with floats. No. 5 (BR) Squadron operated this aircraft from 30 September 1940 until February 1941 for operational evaluation.*

58   *Consolidated Canso A   9737 of No. 5 (BR) Squadron photographed at Mont-Joli, Que., in late 1942.*

55   *Fairchild 71   630, which served with No. 5 (FB) Squadron in 1936.*

59   *Consolidated Catalina Mk.I   Z2137, which served with No. 5 (BR) Squadron.*

No. 6 Squadron

Authorized as a Torpedo Bomber unit at Trenton, Ontario on 4 March 1936, the squadron commenced service training in November of that year. Starting with Vedette flying boats, it received Shark aircraft from England in January 1937, moved to Jericho Beach (Vancouver), British Columbia in November 1938, and concentrated on torpedo-dropping training. Mobilized on 10 September 1939, and redesignated Bomber Reconnaissance on 31 October, the squadron flew Shark, Stranraer, Catalina and Canso A aircraft on West Coast anti-submarine duty until disbanded at Coal Harbour, British Columbia on 7 August 1945.

Brief Chronology
Authorized as No. 6 (TB) Sqn, Trenton, Ont. 4 Mar 36. Mobilized 10 Sep 39. Redesignated No. 6 (BR) Sqn, Jericho Beach, B.C. 31 Oct 39. Disbanded at Coal Harbour, B.C. 7 Aug 45.

Commanders
F/L C.L. Trecarten 23 Nov 36 - 17 Feb 38.
F/L E.A. Springall (RAF) 18 Feb 38 - 12 Jun 38.
W/C A.H. Hull 13 Jun 38 - 3 Feb 40.
S/L L.E. Wray 4 Feb 40 - 5 Nov 40.
S/L M.G. Doyle 6 Nov 40 - 24 Aug 41.
S/L B.N. Harrop 25 Aug 41 - 31 Mar 42.
S/L H.J. Winny, OBE 1 Apr 42 - 24 Aug 42.
S/L G.C. Upson 25 Aug 42 - 13 Dec 42.
S/L V.A. Margetts 14 Dec 42 - 21 Sep 43.
S/L L.A. Harling 22 Sep 43 - 19 Sep 44.
W/C A.C. Neale, AFC 20 Sep 44 - 7 Aug 45.

Higher Formations and Squadron Locations
RCAF Headquarters:
Trenton, Ont. 4 Mar 36 - 1 Nov 38.
Western Air Command:
Jericho Beach, B.C. 5 Nov 38 - 13 May 40.
2 aircraft, Ucluelet, B.C. 12 Sep 39 - 1 May 40.
No. 4 Group (16 Jun 42 - 1 Apr 44),
Alliford Bay, B.C. 15 May 40 - 21 Apr 44.
Movement exercise, Bella Bella, B.C. 19 Nov - 3 Dec 42.
Coal Harbour, B.C. 23 Apr 44 - 7 Aug 45.

Representative Aircraft (Unit Code 1939-42 XE, 1942 AF)
Canadian Vickers Vedette (Nov 36 - Jan 37) 806
Blackburn Shark Mk.II & III (Jan 37 - Dec 41) 501 503 XE-B 525 545
Supermarine Stranraer (Nov 41 - May 43)
907 922 930 948
Consolidated Canso A (Apr - Nov 43) 9762 9787 9788
Consolidated Catalina Mk.IB & IIIA (Sep 43 - Aug 45)
FP202 FP290 FP296 JX572
Consolidated Canso A (Mar 44 - Aug 45)
9762 9790 11007 11044
Noorduyn Norseman (1940 - Dec 41) 695 696 XE-Z 697

Operational History: First Mission 11 September 1939, Shark 501 from Jericho Beach — patrol of the Strait of Georgia area from Gabriola Island south to the Pender Islands. (Patrols had commenced on 2 September for the purpose of identifying and reporting all shipping in the Vancouver area.) **Last Mission** 1 August 1945, Catalina FP290 from Coal Harbour with F/O Erickson and crew — patrol. **Summary** Sorties: 2506. Operational/Non-operational Flying Hours: 11,716/10,565. Victories: nil.[1] Casualties: Operational: nil. Non-operational: 2 aircraft; 9 aircrew killed. **Honours and Awards** nil. **Battle Honours** see page xix

---

[1] The squadron had only one encounter with enemy weaponry — a Japanese fire balloon. On 12 March 1945, when F/L Moodie and crew (Canso 9702) were returning from a patrol, they sighted a partially deflated balloon at an altitude of 500 feet and drifting in an easterly direction over Rupert Inlet. By flying above the balloon, F/L Moodie was able to force it down on the south side of Rupert Arm, where it was later recovered by a ground party and shipped to Western Air Command headquarters for examination.

60 *Blackburn Shark Mk.II 503 in the markings of No. 6 (BR) Squadron.*

61 *Supermarine Stranraer 948 used by No. 6 (BR) Squadron from its base at Alliford Bay, B.C.*

62 *Noorduyn Norseman Mk.IV 696, marked XE-Z, with No. 6 (BR) Squadron at Patricia Bay, B.C. on 26 December 1941.*

No. 7 Squadron

Formed as a General Purpose unit at Rockcliffe (Ottawa), Ontario on 29 January 1936 by amalgamating the Test Flight, General Purpose Flight and two photographic detachments then based at Rockcliffe, the squadron was reorganized on 1 February 1937 to consist of two flights: the Test Flight which service-tested new aircraft types for military operations and developed new techniques in aerial photography, and the General Purpose Flight for aerial communications work.[1] On the outbreak of war, the squadron's personnel were required to bring other, more needed operational squadrons up to strength, and it was disbanded on 10 September 1939.[2]

Brief Chronology Formed at Rockcliffe, Ont. 29 Jan 36. Disbanded 10 Sep 39.

Commanders
S/L E.G. Fullerton 29 Jan 36 - 14 Aug 36.
S/L G.R. Howsam, MC 15 Aug 36 - 7 Aug 37.
S/L D.A. Harding 8 Aug 37 - 4 Dec 38.
F/L L.E. Wray 5 Dec 38 - 31 Jan 39.
F/L R.C. Davis 1 Feb 39 - 10 Sep 39.

Higher Formation and Squadron Location
RCAF/Air Force Headquarters:
Rockcliffe, Ont. 29 Jan 36 - 10 Sep 39.

Representative Aircraft
Fairchild 71 (Jan 36 - Sep 39) 634 638 641 644
Bellanca Pacemaker (Jan 36 - Sep 39) 601 603 610 612

---

[1] The Photographic Detachments went to No. 8 (GP) Squadron also at Rockcliffe.
[2] The General Purpose Flight became the Air Force Headquarters Communications Flight on 10 September 1939, and subsequently the nucleus of No. 12, later 412, (Communications) Squadron.

Re-formed as a Bomber Reconnaissance unit at Prince Rupert, British Columbia on 8 December 1941, the squadron flew Shark, Canso A and Catalina aircraft on West Coast anti-submarine duty until disbanded at Alliford Bay, British Columbia on 25 July 1945.

Brief Chronology
Formed at Prince Rupert, B.C. 8 Dec 41. Disbanded at Alliford Bay, B.C. 25 Jul 45.

Commanders
F/L R.H. Morris 12 Dec 41 - 31 Mar 42.
S/L J.E. Jellison 1 Apr 42 - 15 Jun 42.
S/L R.H. Morris 16 Jun 42 - 8 Dec 42.
S/L P.G. Grant 21 Dec 42 - 25 Apr 43.
S/L R. Dobson 26 Apr 43 - 26 Jun 44.
S/L A.C. Neale 27 Jun 44 - 11 Sep 44.
S/L T. Benson 12 Sep 44 - 25 Jul 45.

Higher Formations and Squadron Locations
Western Air Command:
No. 4 Group (16 Jun 42 - 1 Apr 44),
Prince Rupert, B.C. 8 Dec 41 - 22 Apr 44.
Alliford Bay, B.C. 23 Apr 44 - 25 Jul 45.

Representative Aircraft (Unit Code 1941-42 LT, 1942 FG)
Blackburn Shark Mk.III (Dec 41 - Sep 43)
518 523 536 545
Supermarine Stranraer (Feb 43 - Mar 44) 920 954 956
Consolidated Canso A (Apr 43 - Jul 45) 9788 9803 11007 11099
Consolidated Catalina Mk.IV (Jan 44 - Jul 45) JX206 JX209 JX212 JX213

Operational History: First Mission 15 December 1941, 2 Sharks (523 and 545) from Prince Rupert — anti-submarine patrol; returned early, adverse weather. **Last Mission** 14 July 1945, Canso A 11070 from Alliford Bay with F/O Craddock and crew — anti-submarine patrol. **Summary** Sorties: 2614. Operational/Non-operational Flying Hours: 11,554/6754. Victories: nil. Casualties: Operational: nil. Non-operational: 5 aircraft; 14 personnel, of whom 8 were killed, 4 injured, 2 died of natural causes. **Honours and Awards** 1 DFC, 1 AFC, 4 MiD's. **Battle Honours** see page xix

63 Bellanca Pacemaker 610 as used by No. 7 (GP) Squadron.

64 Blackburn Shark Mk.III 546 based at Prince Rupert, B.C. in the summer of 1942 while serving No. 7 (BR) Squadron.

65 *Consolidated Canso A with some of the personnel of No. 7 (BR) Squadron accentuating the wing span of this popular amphibian.*

66 *Consolidated Catalina Mk.IVA JX212 on patrol with No. 7 (BR) Squadron.*

No. 8 Squadron

Badge: The head of a musk ox affronte
Motto: Determined to defend
Authority: King George VI, June 1941

The activities of this unit covered a wide area in the north west of Canada, frequented by the musk ox. The musk ox is capable, as is a bomber reconnaissance squadron, of covering great distances and is also very formidable in attack.

Formed as a General Purpose unit at Winnipeg, Manitoba on 14 February 1936 by amalgamation of the General Purpose Flight and four General Purpose and Forestry Flights then operating in Manitoba and Saskatchewan, the squadron moved to Rockcliffe (Ottawa), Ontario where, on 1 February 1937, it was reorganized as a photographic unit although its service designation remained unchanged.[1] Alerted for hostilities on 26 August 1939, the unit recalled its photographic detachments from the field and proceeded to its war station at Sydney, Nova Scotia.[2] Mobilized on 10 September as a General Reconnaissance unit, and redesignated Bomber Reconnaissance on 31 October, it flew Delta and Bolingbroke aircraft on East Coast anti-submarine duty. In December 1941, after Japan had entered the war, it was transferred to the West Coast and, from June 1942 to March 1943, served in Alaska as part of the RCAF reinforcement to the United States Army Air Forces. The squadron was disbanded at Patricia Bay (Vancouver), British Columbia on 25 May 1945.

Brief Chronology Formed as No. 8 (GP) Sqn, Winnipeg, Man. 14 Feb 36. Reorganized (though not redesignated) as a photographic unit, Rockcliffe, Ont. 1 Feb 37. Mobilized as No. 8 (GR) Sqn, Sydney, N. S. 10 Sep 39. Redesignated No. 8 (BR) Sqn 31 Oct 39. Disbanded at Patricia Bay, B.C. 25 May 45.

Commanders
S/L R.S. Grandy, OBE 14 Feb 36 - 31 Jan 37.
S/L C.R. Slemon 1 Feb 37 - 16 Dec 37.
W/C W.W. Brown 17 Dec 37 - 18 Sep 40.
S/L H.H.C. Rutledge 19 Sep 40 - 7 Nov 40.
S/L S.S. Blanchard 8 Nov 40 - 22 Nov 40.
S/L R.B. Wylie 23 Nov 40 - 2 Jun 41.
W/C C.A. Willis 3 Jun 41 - 11 May 43.
S/L H.M. Lay 12 May 43 - 30 Oct 44.
W/C C.W. McNeill 31 Oct 44 - 25 May 45.

Higher Formations and Squadron Locations
RCAF Headquarters:
Winnipeg, Man. 14 Feb 36 - 31 Jan 37.
Rockcliffe, Ont. 1 Feb 37 - 26 Aug 39.
Eastern Air Command:
Sydney River Base, North Sydney, N.S. 27 Aug 39 - 14 Dec 39. *2 aircraft, St. John's Nfld. 4-13 Sep 39.*
Kelly Beach, North Sydney, N.S. 15 Dec 39 - 8 Feb 41.
Bolingbroke Detachment:
Uplands, Ont. 1 Jun - 15 Apr 40.
St Hubert, Que. 16 Apr - 26 May 40.
Moncton, N.B. 27 May - 19 Jul 40.
Yarmouth, N.S. (transferred to No. 119 (BR) Sqn) 20 Jul 40.

Sydney, N.S. 9 Feb 41 - 26 Dec 41.
Western Air Command:
Sea Island, B.C. 1 Jan 42 - 2 Jun 42.
No. 4 Group (for adm),
US Alaskan Command (for ops).
RCAF "X" Wing,
Anchorage, Alaska 7 Jun 42 - 26 Feb 43.
3 aircraft, Kodiak 27 Jun 42 - 12 Feb 43.
Nome 18 Jul 42 - 5 Dec 43.
Western Air Command:
Sea Island, B.C. 4 Mar 43 - 9 Dec 43.
6 aircraft, Tofino, B.C. 10-25 Oct 43.
4 aircraft, Port Hardy, B.C. 19 Oct - 9 Dec 43.
Port Hardy B.C. 10 Dec 43 - 19 Mar 44.
Patricia Bay, B.C. 20 Mar 44 -24 Mar 45.
Movement exercise, Terrace, B.C. 22-27 Jun 44.
Representative Aircraft (Unit Code 1939-42 YO, 1942 GA)
Fairchild 71 (Feb 36 - Aug 39) 629   631   643   644
Bellanca Pacemaker (Apr 36 - Aug 39) 601   to   611
Canadian Vickers Vedette (May 36 - Aug 39)
805   807   811
Northrop Delta Mk.II (Feb 37 - Nov 41)
666   to   674   677   682
Bristol Bolingbroke Mk.I & IV (Dec 40 - Aug 43)
702 YO-A   9025 YO-X   9048 YO-T
Lockheed-Vega Ventura G.R.Mk.V (May 43 - May 45)
2176 A   2177 H   2178 O   2186 N   2189 Q   2190 C
2194 B   2218 B   2222 J   2244 D   2261 B   2271 D
Operational History: First Mission 12 September 1939, Delta 674 from Sydney with Sgt Mitchell and crew — anti-submarine patrol. **Last Mission** 16 May 1945, Ventura 2185 from Patricia Bay with F/O Riley and crew — anti- submarine patrol. **Summary** Sorties: 2444. Operational/Non-operational Flying Hours: 9958/11,320. Victories: nil. Casualties: Operational: nil. Non- operational: 4 aircraft; 6 aircrew killed. **Honours and Awards** 2 AFC's, 1 AFM, 1 MiD. **Battle Honours** see page xix

¹The squadron absorbed the two Mobile Photographic Detachments of No. 7 (GP) Squadron and gave up the General Purpose Flight.
²On the way from Halifax to Sydney on 14 September 1939, FS J.E. Doan and LAC D.A. Rennie became the first RCAF casualties of the war, when Delta 673 was lost. The wreckage of the aircraft was not located until July 1958, north of Fredericton, N.B.

68   Bellanca Pacemaker 610 at Camp Borden, Ont. on 14 June 1939 while serving No. 8 (GP) Squadron.

69   Northrop Delta Mk.II 672, in the markings of No. 8 (BR) Squadron, is rolled out of the hangar for winter operations. During the summer months the squadron operated its Deltas on floats.

70   Bristol Bolingbroke Mk.IV 9048 of No. 8 (BR) Squadron during the unit's period of operations in Alaska.

67   Fairchild 71 643 of No. 8 (GP) Squadron.

71   Lockheed Vega Ventura G.R.Mk.V 2244, aircraft letter "D", of No. 8 (BR) Squadron in white camouflage.

No. 9 Squadron

Formed as a Bomber Reconnaissance unit at Bella Bella, British Columbia on 9 December 1941,[1] the squadron flew Stranraer, Canso A and Catalina aircraft on West Coast anti-submarine duty until disbanded on 1 September 1944.
Brief Chronology Formed at Bella Bella, B.C. 9 Dec 41. Disbanded 1 Sep 44.
Commanders
S/L F.S. Carpenter 9 Dec 41 - 19 May 42.
S/L J.W. McNee 18 Jun 42 - 15 Dec 42.
S/L P.E. Sorenson 16 Dec 42 - 18 May 43.
S/L A.W. Mitchell, AFC 19 May 43 - 18 Jun 44.
S/L R.W. McRae 19 Jun 44 - 1 Sep 44.
Higher Formations and Squadron Location
Western Air Command:
No. 4 Group (16 Jun 42 - 1 Apr 44),
Bella Bella, B.C. 9 Dec 41 - 1 Sep 44.
Movement exercise, Alliford Bay, B.C. 19 Nov - 2 Dec 42.
Representative Aircraft (Unit Code 1939-42 KA *1942* HJ)
Supermarine Stranraer (Dec 41 - Apr 44) 915 920 936 956
Consolidated Canso A (Apr 43 - Aug 44) 9761 9790 9800 11005
Consolidated Catalina Mk.I, IB & IV (Feb -Aug 44) W8432 FP293 FP297 JX207
Operational History: First Mission 9 December 1941, Stranraer 949 from Bella Bella with S/L Carpenter and crew — anti-submarine patrol. **Last Mission** 21 August 1944, Canso A 11005 from Bella Bella with F/O Asher and crew — patrol to a depth of 200 miles. **Summary** Sorties: 1314. Operational/Non-operational Flying Hours 8863/4559. Victories: nil. Casualties: Operational: nil. Non-operational: 1 aircraft; 1 killed, 3 seriously injured. **Honours and Awards** 1 AFC. **Battle Honours** see page xix

---

[1] A No. 9 (General Reconnaissance) Squadron was authorized in Western Air Command on 1 April 1938, but was not formed owing to a shortage of aircraft and personnel.

72 *Consolidated Catalina Mk.IB FP293 preparing to leave on a patrol with No. 9 (BR) Squadron.*

No. 10 Squadron

Formed as a Bomber unit at Halifax, Nova Scotia on 5 September 1939,[1] mobilized on the 10th, and redesignated Bomber Reconnaissance on 31 October, the squadron flew Wapiti, Digby and Liberator aircraft on East Coast anti-submarine duty. It established a record with attacks on 22 U-boats,[2] including 3 sinkings, and won the proud but unofficial, title "North Atlantic Squadron." The squadron was disbanded at Torbay, Newfoundland on 15 August 1945.
Brief Chronology Formed as No. 10 (B) Sqn, Halifax, N.S. 5 Sep 39. Mobilized 10 Sep 39. Redesignated No. 10 (BR) Sqn 31 Oct 39. Disbanded at Torbay, Nfld. 15 Aug 45.
Nicknames; "North Atlantic Squadron"; "Dumbo"
Ancestry No. 3 (B) Squadron
Commanders
S/L A. Lewis 5 Sep 39 - 16 Sep 39.
W/C R.C. Gordon 13 Nov 39 - 14 Apr 41.
W/C H.M. Carscallen 15 Apr 41 - 11 Sep 41.
S/L A. Laut 12 Sep 41 - 5 Mar 42.
W/C C.L. Annis 6 Mar 42 - 31 Jul 42.
S/L J.M. Young 1 Aug 42 - 6 Jan 43.
S/L A.M. Cameron, AFC 7 Jan 43 - 31 Mar 43.
W/C C.L. Annis, OBE 1 Apr 43 - 5 Aug 43.
W/C J.M. Young 6 Aug 43 - 4 Sep 43 *KIFA*.
W/C M.P. Martyn 4 Oct 43 - 4 Jul 44.
W/C A.M. Cameron, AFC 5 July 44 - 10 Dec 44.
S/L C.W. Bradley, DFC 11 Dec 44 - 15 Aug 45.
Higher Formations and Squadron Locations
Eastern Air Command:
Halifax, N.S. 5 Sep 39 - 14 Jun 40.
Dartmouth, N.S. 15 Jun 40 - 10 Apr 41.
"A" Flight, Gander, Nfld. 16 Jun 40 - 10 Apr 41.
No. 1 Group (formed 10 July 41),
Gander, Nfld. 11 Apr 41 - 10 Nov 42.
4 aircraft, Yarmouth, N.S. 20 Mar - 9 Jun 42.[3]
Eastern Air Command:
Dartmouth, N.S. 15 Nov 42 - 7 May 43.
No. 1 Group (disbanded 30 Jun 45),
Gander, Nfld. 8 May 43 - 10 Jun 45.
Torbay, Nfld. 11 Jun 45 - 15 Aug 45.
3 aircraft, Yarmouth, N.S. 17-31 Jul 45.[4]
Representative Aircraft (Unit Code 1939-42 PB *1942* JK)
Westland Wapiti Mk.IIA (Sep 39 - May 40) 509 513 535 544
Douglas Digby (Apr 40 - Apr 43) 739 P 740 R 745 W 748 PB-V 751 Y 755 J 756 M
Consolidated Liberator Mk.III,V and G.R. Mk.VI (Apr 43 -Aug 45)[5] 586 A 587 B 588 C 589 D 590 G 591 K 592 F 593 J 594 P 595 X 596 Y 597 L 598 Q 599 R 600 N 3704 Z 3706 W 3707 M 3713 E 3726 G 3732 P 3734 H 3735 L 3736 D 3738 T 3739 S 3742 A
Operational History: First Mission 17 June 1940, Digby 744 from the Gander Detachment with S/L Carscallen and crew — anti-submarine patrol: returned early, failing light. **First Completed Mission** 3 July 1940, Digby 757 PB-K from Dartmouth with F/O A. Laut and crew — harbour entrance patrol over Bedford Basin (Halifax). **Victories: U-Boat** 30 October 1942, Digby 747 "X" from Gander with F/L D.F. Raymes and crew — returning from patrol of convoy ON140, sank U-520 with four 250-pound depth charges at 4747N 4950W. This was the squadron's seventh attack

and Eastern Air Command's third kill. 19 September 1943, Liberator 586 "A" from Gander with F/L R.F. Fisher and crew — returning to Gander from Iceland after escorting Prime Minister Churchill (returning in HMS *Renown* from the Quebec Conference ONS18) sank U-341 at 5840N 2530W, Eastern Air Command's fifth kill. 26 October 1943, Liberator 586 "A" from Gander with F/L R.M. Aldwinkle and crew — convoy escort: sank U-420 at 5049N 4101W after an hour-long engagement, Eastern Air Command's sixth and last kill.[6] **Last Mission** 26 May 1945, Liberator 599 "R" from Torbay with F/O W.W. Adshead and crew — escort of convoy ONS50 (33 merchantmen, 4 escorts). **Summary** Sorties: 3414. Operational/Non-operational Flying Hours: 30,331/7976. Victories: U-boats: 3 sunk. Casualties: Operational: 7 aircraft; 25 aircrew, of whom 24 were killed or missing, 1 wounded. Non-operational: 27 fatal (including 3 drowned), 6 non-fatal. **Honours and Awards** 24 DFC's, 6 AFC's, 1 GM, 1 AFM, 3 BEM's, 33 MiD's. **Battle Honours** see page xix

---

[1] The formation of No. 10 (Torpedo Bomber) Squadron at Halifax was authorized on 7 April 1939, but the order stated it was "not being formed this year"; and the intended role was changed to Bomber on 31 August 1939. When formed, the squadron took over the aircraft and personnel of No. 3 (B) Squadron, which had been redesignated Fighter but was never re-formed in that role.
[2] Almost one-quarter of the total recorded in Eastern Air Command. The squadron also recorded Eastern Air Command's first U-boat sighting and attack on 25 October 1941. S/L C.L. Annis (command armament officer) and crew in Digby 740 dropped two 600-pound bombs; they failed to explode.
[3] Absorbed by No. 162 (BR) Squadron.
[4] The detachment consisted of 3 Liberators (3742 "A", 3835 "L" and 3732 "P"), 4 crews (28 personnel) and 42 ground-crew, and operated as a meteorological flight.
[5] First Liberator received on 22 April 1943, and by the end of May the squadron had its full complement of 15 aircraft.
[6] U-420 had previously been damaged by Liberator 587 "B" from Gander (P/O R.R. Stevenson and crew) on 3 July 1943, and forced to return to Germany for repairs. It had just returned to service when sunk.

74 Douglas Digby line-up including 748 in the foreground and all carrying the No. 10 (BR) Squadron PB markings.

75 Consolidated Liberator Mk.V 600, aircraft "N" of No. 10 (BR) Squadron, in standard Coastal Command finish, prepares to taxi on the inboard engines from an east coast dispersal in the summer of 1943.

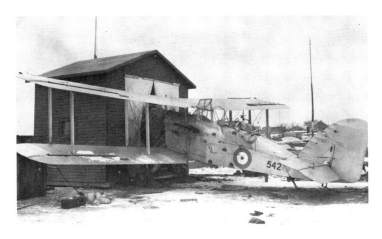

73 Westland Wapiti Mk.IIA 542 protected by a nose hangar against the raw east coast winter weather. Photographed at Halifax, N.S. with No. 10 (BR) Squadron on 13 March 1940.

76 Consolidated Liberator G.R.Mk.VI aircraft "W" of No. 10 (BR) Squadron on 9 December 1944 following the complete stripping of the camouflage finish, including serial numbers.

No. 11 Squadron

Formed as a Bomber Reconnaissance unit at Rockcliffe (Ottawa), Ontario on 3 October 1939, the squadron flew Hudson[1] and Liberator aircraft on East Coast anti-submarine duty until May 1945, when it was transferred to the West Coast. It was disbanded at Patricia Bay (Vancouver), British Columbia on 15 September 1945.

Brief Chronology Formed at Rockcliffe, Ont. 3 Oct 39. Disbanded at Patricia Bay, B.C. 15 Sep 45.
Nickname "The Joe Squadron"
Commanders
W/C A. Lewis 3 Nov 39 - 4 Jan 41.
S/L P.G. Baskerville 5 Jan 41 - 12 Jan 42.
S/L M.P. Martyn 13 Jan 42 - 15 Apr 42.
W/C W.C. Van Camp 16 Apr 42 - 26 May 43.
S/L K.C. Wilson 27 May 43 - 1 Oct 43.
W/C G.M. Cook, AFC 2 Oct 43 - 30 Nov 44.
S/L P. Wilkinson 1 Dec 44 - 20 Feb 45.
W/C W.H. Swetman, DSO, DFC 21 Feb 45 - 12 Sep 45.
Higher Formations and Squadron Locations
No. 3 Training Command:
Rockcliffe, Ont. 3 Oct 39 - 20 Oct 39.
Uplands, Ont. 21 Oct 39 - 2 Nov 39.
Eastern Air Command:
Dartmouth, N.S. 3 Nov 39 - 25 Oct 43
No. 1 Group, 4 aircraft, Torbay, Nfld. 26 Nov 41 - 31 May 42.[2]
2 aircraft, Mont-Joli, Que. 12 May - 11 Jun 42.
No. 1 Group,
Torbay, Nfld. 26 Oct 43 - 17 Jun 44.
Eastern Air Command:
Dartmouth, N.S. 18 Jun 44 - 23 May 45.
Western Air Command:
Patricia Bay, B.C. 31 May 45 - 15 Sep 45.
Representative Aircraft (Unit Code 1939-42 OY, 1942 KL)
Lockheed Hudson Mk. I (Oct. 39 - Jul 42)
759   763 OY-F   773 OY-O   785 OY-U
Lockheed Hudson Mk.III (Mar 42 - Sep 44) BW403 S
BW453 L   BW454 K   BW618 Q   BW621 Z   BW622 A
BW623 K   BW627 X   BW632 Y   BW644 H   BW646 F
BW649 D   BW658 M   BW660 A   BW685 T   BW712 E
BW716 W   BW717 L   BW718 O   BW719 N   BW724 P
Consolidated Liberator Mk.III, V and G.R. Mk. VI (Jul 43 - Sep 45) 3704 Z[3]   3709 E   3711 B   3712 P   3714 A
3715 F 3716 D   3717 I   3718 J   3719 N   3720 Q   3721 S
3722 C   3723 R   3724 T   3725 H   3727 T   3728 J
3730 S   3731 F   3733 L   3737 Z   11120 U   11121 X
Operational History: First Mission 10 November 1939, Hudson 761 from Dartmouth with S/L A. Lewis and crew — naval co-operation height finding and sighting practice for the anti-aircraft guns of HM Ships *Repulse* and *Furious*, and shore batteries at Halifax. **Last Mission** 11 August 1945, Liberator 3715 "H" from Patricia Bay with F/O D.L. Craig and crew — anti-submarine sweep. (The squadron flew an additional 23 meteorological sorties, the last on 12 September 1945 by F/O J.M. Clark and crew in Liberator 3730 "S".) **Summary** Sorties: 5098. Operational/Non-operational Flying Hours: 25,386/13,377. Victories: nil: 10 U-boat sightings, 8 attacks. Casualties: Operational: 4 aircraft; 12 aircrew missing. Non-operational: 13 aircraft; 25 aircrew, 7 passengers killed. **Honours and Awards** 8 DFC's, 1 AFC, 1 DFM, 6 MiD's **Battle Honours** see page xix

---

[1] The squadron was the first to operate Hudson aircraft.
[2] The Torbay detachment formed the nucleus of No. 145 (BR) Squadron.
[3] 3704 "Z" (christened "Basterpiece") was remodelled as a transport aircraft and used on the Dartmouth - Goose Bay - Reykjavik run in support of No. 162 (BR) Squadron.

77   Lockheed Hudson Mk.I 764 at Dartmouth, N.S. on 17 February 1941 in the markings of No. 11 (BR) Squadron.

78   Consolidated Liberator G.R.Mk.VI 3727, aircraft "T" of No. 11 (BR) Squadron at Summerside, P.E.I. on 19 February 1945.

79   Consolidated Liberator Mk.V 3704 converted to Transport configuration, and christened "Basterpiece," in the service of No. 11 (BR) Squadron at Dartmouth, N.S.

No. 12 Squadron

Formed as a Communications unit at Rockcliffe (Ottawa), Ontario on 30 August 1940,[1] the squadron flew a variety of aircraft on general air communication and light transport work, ferrying and testing new aircraft, and examining prospective RCAF pilots from the United States. It also maintained a practice flight for the benefit of Air Force Headquarters personnel who wanted to keep their hand in as pilots, navigators and wireless operators. Retained in the post-war force as a Composite unit, the squadron was renumbered No. 412 on 1 April 1947.

Brief Chronology Formed as No. 12 (Comm) Sqn, Rockcliffe, Ont. 30 Aug 40. Redesignated No. 412 (K*) Sqn 1 Apr 47.
Ancestry: No. 7 (GP) Squadron
Commanders
W/C J.L. Plant 30 Aug 40 - 2 Feb 41.
W/C H.M. Kennedy, AFC 3 Feb 41 - 22 Oct 43.
W/C G.G. Diamond, AFC 23 Oct 43 - 7 Nov 45.
W/C E.B. Hale, DFC 8 Nov 45 - 31 Jul 46.
W/C G.G. Diamond, AFC 1 Aug 46 - 21 Mar 47.
W/C W.H. Swetman, DSO, DFC 22 Mar 47 - 31 Mar 47.
Higher Formations and Squadron Location
Air Force Headquarters:
No. 9 (Transport) Group (5 Feb 45),
Rockcliffe, Ont. 30 Aug 40 - 31 Mar 47.
Representative Aircraft (Unit Code QE) Fairchild 71 and 51, Fleet Fawn, Hawker Tomtit, Grumman Goose, Northrop Delta, Barkley Grow, Lockheed Hudson, Boeing 247D, Noorduyn Norseman, North American Harvard, Lockheed 10-A, 12-A and 212, Avro Anson, Lockheed Lodestar, Douglas Dakota, Beechcraft Expeditor.
Flying Summary Jan 40 - Mar 47: 39,248 hours.

*The abbreviation for "Composite"

---

[1] From the AFHQ Communications Flight, which had formerly been the Communications Flight of No. 7 (General Purpose) Squadron.

80 *Northrop Delta Mk.II 675 in service with No. 12 (Comm) Squadron.*

81 *Stinson 105 3486 was one of two light communications aircraft used by No. 12 (Comm) Squadron in 1940.*

82 *Fairchild 51 625 utility transport on skis, serving No. 12 (Comm) Squadron.*

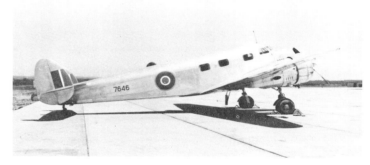

83 *Lockheed 12-A 7646 in the service of No. 12 (Comm) Squadron at Rockcliffe, Ont. on 22 June 1943.*

84 *Lockheed 10-A 1526 of No. 12 (Comm) Squadron visiting Camp Borden, Ont. on 10 January 1940.*

85 Barkley Grow 758, one of a kind in the RCAF and used by No. 12 (Comm) Squadron.

88 Lockheed Lodestar 567 flown by No. 12 (Comm) Squadron and wearing the Air Marshal's pennant on the nose.

86 Beechcraft Expeditor Mk.II 1388 in the markings of No. 12 (Comm) Squadron.

89 Grumman Goose Mk.II 917 at Rockcliffe, Ont. with No. 12 (Comm) Squadron.

90 North American Harvard Mk.IIB 3101, one of several which served with No. 12 (Comm) Squadron.

87 Douglas Dakota Mk.III 663 at Rockcliffe, Ont. with No. 12 (Comm) Squadron.

91 Fleet Fawn Mk.II 213 at Rockcliffe, Ont. with No. 12 (Comm) Squadron on 24 October 1939.

92 *Noorduyn Norseman Mk.VI 792 on skis while serving No. 12 (Comm) Squadron. One of the squadron's Avro Anson Mk.V's is visible in the background.*

93 *Hawker Tomtit 139 was flown by No. 12 (Comm) Squadron in 1940 from its base at Rockcliffe, Ont.*

94 *Lockheed 212 7642, one of a kind in RCAF service and operated by No. 12 (Comm) Squadron.*

No. 13 Squadron

Formed as the Seaplane and Bomber Reconnaissance Training School at Sea Island (Vancouver), British Columbia on 1 May 1940. When it was decided to include landplane training in its functions the school was redesignated the Operational Training Squadron on 13 July, and No. 13 (Operational Training) Squadron on 30 July. On 29 June 1942, landplane training was discontinued, and the squadron concentrated on the training of flying boat crews. It was disbanded at Patricia Bay (Vancouver), British Columbia on 9 November 1942.[1]

Brief Chronology Formed as S&BRT School, Sea Island, B.C. 1 May 40. Redesignated OT Sqn 13 Jul 40. Redesignated No. 13 (OT) Sqn 30 Jul 40. Disbanded at Patricia Bay, B.C. 9 Nov 42.

Commanders
W/C R.G. Briese 1 May 40 - 14 Jul 41.
S/L Z.L. Leigh 15 Jul 41 - 3 Jul 42.
S/L C.C. Austin 26 Jul 42 - 9 Nov 42.

Higher Formation and Squadron Locations
Western Air Command:
Sea Island, B.C. 1 May 40 - 29 Oct 40.
Patricia Bay, B.C. 1 Nov 40 - 9 Nov 42.

Representative Aircraft (Unit Code 1939-42 AN, 1942 MK)
Canadian Vickers Vancouver (May - Nov 40) 902 903
Canadian Vickers Vedette (May 40 - May 41)
809 812 816 817
Fairchild 71 (May 40 - Apr 42) 633 638 647
Noorduyn Norseman Mk.IV (May 40 - Nov 42)
694 697 2480 2481
Lockheed Hudson Mk.I (Aug 40 - Jun 42)[2]
Grumman Goose (Nov 40 - Oct 42) 917 924 MK-G
940 941
Lockheed 10-B (May 41 - Jun 42) 7633 7634 7648 7650
Northrop Delta Mk.I & II (May - Nov 41) 667 676 685
AN-P 690
Bristol Bolingbroke Mk.IV (Oct 41 - Jun 42)[3] 9033
AN-E 9034 to 9037 9057
Supermarine Stranraer (Oct 41 - Nov 42)
907 921 935 955
Cessna Crane (Feb - Jun 42) 8670 8672 8687 8694

Operational History Never employed operationally.
Summary Flying Hours: 11,805. Casualties: 3 aircraft; 9 aircrew killed. **Honours and Awards** 1 BEM

---

[1] The squadron's personnel and aircraft formed the nucleus of No. 3 (BR) Operational Training Unit.
[2] On loan to No. 120 (BR) Squadron from 10 May to 1 December 1941.
[3] Four Bolingbrokes (9033, 9034, 9035 and 9037) were on loan to No. 115 (F) Squadron from 27 November 1941 to 28 January 1942.

Formed as the Photographic Flight at Rockcliffe (Ottawa), Ontario on 14 January 1943 at the request of the British Air Ministry to carry out photographic research, the unit was equipped with Spitfire and Hurricane aircraft.[1] From 15 May 1944 onward, as an unofficial No. 13 (Photographic) Squadron, it flew mainly Mitchell aircraft on tri-camera high altitude aerial photography. The unit was formally designated No. 13 (P) Squadron on 15 November 1946[2] and renumbered No. 413 (P) Squadron on 1 April 1947.

Brief Chronology Formed as Photo Flight, Rockcliffe, Ont.

14 Jan 43. Known as No. 13 (P) Sqn 15 May 44.
Designated No. 13 (P) Sqn 15 Nov 46. Renumbered No.
413 (P) Sqn 1 Apr 47.
Commanders
S/L G.V. Miscampbell 15 Jan 43 - 15 May 43 *KIFA*.
S/L J.A. Wiseman, AFC 16 May 43 - 31 Mar 47.
Higher Formations and Squadron Location
Air Force Headquarters:
No. 7 (Photographic) Wing (20 May 44),
Rockcliffe, Ont. 15 Jan 43 - 31 Mar 47.
Representative Aircraft (Unit code AP)
North American Mitchell Mk.II (1944-47) 891 to 894
Consolidated Canso A (1944-47) 9815   11079
Avro Lancaster Mk.XP (1944-47)
FM292   KB884   KB916[3]   KB917
Noorduyn Norseman Mk.IV (1944-47) 368   372   2495
2496
de Havilland Mosquito B.Mk.25 (Apr-Jul 45, one only)
KA999

[1] Three of each. The Spitfires were R7143, X4492 and X4555.
[2] The organizational order stated in part: "This squadron became a component of 7 Photographic Wing on this Wing's formation (secret Organization Order 194 dated 15 May 1944). It has now been decided to retain 7 Photographic Wing as a peacetime organization with headquarters based RCAF Station Rockcliffe. As there is no existing individual organization covering 13 Photogrpahic Squadron, it is necessary to issue an organization order for this squadron.   13 Photographic Squadron is to be organized and retained as a component of 7 Photographic Wing with effect 15 November 1946. Function is to perform tri-camera and high altitude aerial photography."
[3] Crashed on 9 August 1946.

97   *Lockheed Hudson Mk.I 775 on take-off, in the markings of Photo Flight following service with No. 13 (OT) Squadron.*

98   *Grumman Goose 924 in the camouflage finish worn with No. 13 (OT) Squadron, photographed on 12 September 1943 at Rockcliffe, Ont. while with Photo Flight.*

95   *Fairchild 71B 647, seen here on beaching gear, served No. 13 (OT) Squadron during 1941.*

99   *Lockheed 10B 7648 in modified civilian dress and coded AN Y, with No. 13 (OT) Squadron on 29 December 1941 at Patricia Bay, B.C.*

96   *Noorduyn Norseman Mk.IV 2480 at Patricia Bay, B.C. on 13 December 1941 in the markings of No. 13 (OT) Squadron.*

100   *Bristol Bolingbroke Mk.IV 9034 following a wheels-up landing at Patricia Bay, B.C. on 21 May 1942. The markings of No. 13 (OT) Squadron are clearly displayed with the characteristic Canadian underlining bars.*

101 Northrop Delta Mk.II 685 in the service of No. 13 (P) Squadron.

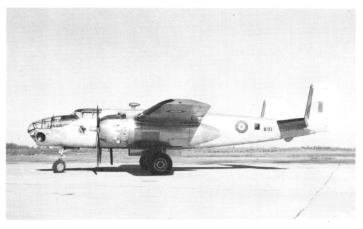

104 North American Mitchell B-25D with No. 13 (P) Squadron at Trenton, Ont. in 1946.

102 Lockheed 10B 7648 displaying the camouflage finish applied to aircraft on Canada's west coast following 7 December 1941. This photo, taken on 12 September 1943 with Photo Flight at Rockcliffe, Ont., is in sharp contrast to the appearance of 7648 in 1941.

105 Consolidated Canso A 11079 on the shore of a northern lake while serving with No. 13 (P) Squadron in 1945.

103 Supermarine Spitfire P.R.Mk.V X4492 based at Rockcliffe, Ont. with Photo Flight and later with No. 13 (P) Squadron. One of only three active Spitfires in Canada during wartime, it was used on photographic operations.

106 Noorduyn Norseman 372 and 2496 of No. 13 (P) Squadron at Corcoran's Camp, Windy Lake, Ont., in August 1945.

No. 14 Squadron

Formed as a Fighter unit at Rockcliffe (Ottawa), Ontario on 2 January 1942, the squadron flew Kittyhawk aircraft on West Coast air defence. From March to September 1943 it was part of the RCAF reinforcement to the United States Army Air Forces in Alaska, and completed two tours of offensive operations against Japanese forces on Kiska Island in the Aleutians. Selected in late 1943 as one of six home fighter units for overseas duty, it was renumbered No. 442 (Fighter) Squadron at Digby, Lincolnshire, England on 8 February 1944.

Brief Chronology Formed as No. 14 (F) Sqn, Rockcliffe, Ont. 2 Jan 42. Renumbered No. 442 (F) Sqn, Digby, Lincs, Eng. 8 Feb 44.

Commanders
S/L B.D. Russel, DFC 2 Jan 42 - 27 Nov 42.
S/L B.R. Walker, DFC 28 Nov 42 - 8 Feb 44.

Higher Formations and Squadron Locations
No. 3 Training Command, BCATP:
Rockcliffe, Ont. 2 Jan 42 - 23 Mar 42.
Western Air Command:
Sea Island, B.C. 27 Mar 42 - 16 Feb 43.
No. 4 Group (for adm),
US Alaskan Command (for ops),
RCAF "X" Wing,
Umnak Island, Aleutians 3 Mar 43 - 15 Sep 43.
Kiska operations: First tour (14 missions, 88 sorties),
Adak Island 31 Mar - 16 Apr 43.
Amchitka 17 Apr - 15 May 43.
Second tour (16 missions, 102 sorties),
Adak Island 3-8 Jul 43.
Amchitka 9 Jul - 29 Aug 43.
Western Air Command:
Boundary Bay, B.C. 24 Sep 43 - 23 Dec 43.
En route overseas 20 Jan 44 - 8 Feb 44.

Representative Aircraft (Unit Code BF, YA)[1]
North American Harvard Mk.I (Jan 42 - Feb 43) 2996 3238 YA-F 3239
Curtiss Kittyhawk Mk.I (Feb 42 - Dec 43) 1050 B 1053 K 1059 YA-K 1063 H 1064 D 1068 T 1072 M 1080 YA-O
Curtiss P40K-1 (Jul - Dec 43)[2] 42-45003 42-45004 42-45944 42-45945 42-45921 42-45951 42-45952 42-45954 42-45977

Operational History: First Mission 18 June 1942, 4 Kittyhawks from Sea Island — search for 2 unidentified aircraft reported 15 miles southeast of Victoria. No aircraft sighted. **First Offensive Mission** 18 April 1943, 4 Kittyhawks from Amchitka, each armed with a 500-pound bomb, dive-bombed and strafed the main Japanese camp on Kiska. Anti-aircraft fire was light; all aircraft returned safely. 26 April 1943, 12 Kittyhawks from Amchitka, on the first all-Canadian mission, dive-bombed a radar installation on Kiska; no direct hits observed. **Last Mission** 21 November 1943, 2 Kittyhawks from Boundary Bay — dawn patrol.

Summary Sorties: 190 in 30 missions against Kiska. Details of other sorties unavailable. Operational/Non-operational Flying Hours: 813/6803. Victories: nil. Casualties: Operational: nil. Non-operational: 7 aircraft; 7 pilots killed.

Honours and Awards 8 Air Medals (USA), 2 MiD's. **Battle Honours** See No. 442 (Fighter) Squadron.

---

[1] BF was used from January to May 1942, then changed to YA, and discontinued in October.
[2] During the squadron's two tours of operations in the Aleutians it flew P40K-1's purchased from the USAAF. One, 42,45997, was flown by S/L Walker on three sorties. The other serials were: 42-45160 42-45195 42-45297 42-45319 42-45731 42-45735 42-45781 42-45871 42-45902 42-45953 42-45995.

Re-formed unofficially as a Photographic unit at Rockcliffe, (Ottawa) Ontario in the summer of 1944, the unit flew Anson and Dakota aircraft on vertical aerial photographic operations. It was officially designated No. 14 (Photographic) Squadron on 15 November 1946 and renumbered No. 414 on 1 April 1947.

Brief Chronology Formed at Rockcliffe, Ont. (ad hoc) 12 Jun 44.[1] Designated No. 14 (P) Sqn 15 Nov 46. Renumbered No. 414 (P) Sqn 1 Apr 47.

Commanders
S/L G.E. Cherrington 12 Jun 44 - 4 Aug 46 *ret.*
S/L R.F. Milne, DFC 4 Aug 46 - 31 Mar 47.

Higher Formations and Squadron Location
Air Force Headquarters:
No. 7 (Photographic) Wing,
Rockcliffe, Ont. 12 Jun 44 - 31 Mar 47.

Representative Aircraft (Unit Code AQ)
Avro Anson Mk. VP (Jun 44 - Jan 47)
11900 12126 12416 12447
Douglas Dakota Mk.III & IV (Sep 46 - Mar 47) KG580 KG635 973 989

---

[1] Developed as an offshoot of No. 13 (Photographic) Squadron, it was listed as a component of No. 7 (Photographic) Wing when the wing was formed on 15 May 1944, but no individual squadron organization order was issued. The date 12 June 1944 is based on the appointment of S/L Cherrington to command the squadron.

107   Avro Anson Mk.V 11919 of No. 13 (P) Squadron at Rockcliffe, Ont. on 4 July 1944.

108   Curtiss Kittyhawk Mk.I, coded YA, of No. 14 (F) Squadron at Sea Island, B.C., on 1 July 1942.

109   Avro Anson Mk.VP 12450 in the service of the survey flight of No. 14 (P) Squadron in 1946.

No. 110 "City of Toronto" Squadron (Auxiliary)

Authorized as No. 10 (Army Co-operation) Squadron (Auxiliary) at Toronto, Ontario on 5 October 1932, the unit commenced flying training in October 1934 when it received four Moth aircraft. Affiliation with Toronto was recognized on 15 April 1935 when the title "City of Toronto" was officially incorporated into its designation, and on 15 November 1937 the unit was renumbered No. 110 Squadron. Mobilized on 10 September 1939, it was assigned to the Canadian Active Service Force for overseas duty with the 1st Division; moved to Rockcliffe (Ottawa), Ontario, where it trained on Lysander aircraft; it left for England in February 1940. The unit anticipated early action, but with the fall of France it was relegated to further training. The squadron was renumbered No. 400 Squadron at Odiham, Hampshire on 1 March 1941.

Brief Chronology Authorized as No. 10 (AC) Sqn NPAAF, Toronto, Ont. 5 Oct 32. Titled No. 10 "City of Toronto" (AC) Sqn NPAAF 15 Apr 35. Renumbered No. 110 "City of Toronto" (AC) Sqn NPAAF/AAAF 15 Nov 37. Mobilized 10 Sep 39. Renumbered No. 400 (AC) Sqn, Odiham, Hants., Eng. 1 Mar 41.

Title "City of Toronto"

Adoption City Council of Toronto, Ont.

Commanders
S/L G.S. O'Brian, AFC 5 Oct 32 - 4 Oct 35 *ret.*
S/L W.A. Curtis, DSC 5 Oct 35 - 31 Dec 38.
S/L A.H.K. Russel 1 Jan 39 - 28 Oct 39.
W/C W.D. Van Vliet 17 Dec 39 - 13 Sep 40.
W/C E.H. Evans 14 Sep 40 - 9 Nov 40 *repat.*
W/C R.M. McKay 10 Nov 40 - 28 Feb 41.

Higher Formations and Squadron Locations
Military District No. 2 (for adm),
RCAF Headquarters (for ops),
Air Training Command (1 Dec 38):
De Lesseps Aerodrome, Weston, Ont. 5 Oct 32 - 16 Dec 39.
Air Force Headquarters (for adm),
Canadian Active Service Force (for ops):
Rockcliffe, Ont. 17 Dec 39 - 13 Feb 40.
En route overseas 16 Feb 40 - 25 Feb 40.
RAF Army Co-operation Command:
No. 22 Group,
Old Sarum, Wilts. 26 Feb 40 - 9 Jun 40.
Odiham, Hants. 10 Jun 40 - 28 Feb 41.

Representative Aircraft
de Havilland DH-60 Moth (Oct 34 - Dec 38) 69   81   158   162
Fleet Fawn Mk.I (Jun 36 - Jan 40) 200   202   206   208
Avro 621 Tutor (Dec 37 - Dec 38) 186   188   189
Avro 626 (Dec 38 - Dec 39) 227   268   269
de Havilland DH-82 Tiger Moth (Dec 38 - Jan 40) 241   247   255   258
Westland Lysander Mk.II (Canada, Dec 39 - Feb 40, Unit Code AY) 428   429   432   433
Westland Lysander Mk.II (England, Mar-Aug 40, Unit Code SP) K6127   L4788   N1265   P1694
Westland Lysander Mk.III (Aug 40 - Feb 41) R9001   R9005 to R9008   R9113

110 De Havilland Gipsy Moth 158, used by No. 110 (AC) Squadron at Toronto, Ont. in the summer of 1938.

111 Fleet Fawn Mk.I 208 in the orderly flight-line of No. 110 (AC) Squadron, based at De Lesseps Aerodrome, Weston, Ont. in 1937.

112 Avro 621 Tutor 189, used by No. 110 (AC) Squadron at summer camp, Camp Borden, Ont., in 1938.

113 Avro 626 227 prior to departure on a typical exercise in the summer of 1939 with No. 110 (AC) Squadron.

114 De Havilland Tiger Moth 258 equipped for operations with No. 110 (AC) Squadron during the winter of 1938-39.

115 Westland Lysanders Mk.II 426 and 428, which were used extensively by No. 110 (AC) Squadron at Rockcliffe, Ont. and shown here on 8 January 1940.

116 Westland Lysander Mk.II P1694 of No. 110 (AC) Squadron at Odiham, Hants. in June 1940.

No. 111 Squadron (Auxiliary)

Authorized as No. 11 (Army Co-operation) Squadron (Auxiliary) at Vancouver, British Columbia on 5 October 1932, the unit commenced flying training in October 1934 when it received four Moth aircraft. It was renumbered and redesignated No. 111 (Coast Artillery Co-operation) Squadron on 15 November 1937. Mobilized on 10 September 1939, the unit moved to Patricia Bay (Vancouver), British Columbia in May 1940 where, with a strength of four Lysander aircraft, it was redesignated Fighter on 14 June but not converted. The squadron was disbanded on 1 February 1941.[1]

Brief Chronology Authorized as No. 11 (AC) Sqn NPAAF Vancouver, B.C. 5 Oct 32. Redesignated No. 111 (CAC) Sqn NPAAF/AAAF 15 Nov 37. Mobilized 10 Sep 39. Redesignated No. 111 (F) Sqn, Patricia Bay, B.C. 14 Jun 40. Disbanded 1 Feb 41.

Commanders
S/L A.D. Bell-Irving, MC 5 Oct 32 - 4 Apr 37.
W/C A.H. Wilson 5 Apr 37 - 30 Sep 40.
S/L W.J. McFarlane 1 Oct 40 - 14 Nov 40.
S/L G.W. Du Temple 17 Nov 40 - 1 Feb 41.

Higher Formations and Squadron Locations
Military District No. 11 (for adm),
RCAF Headquarters (for ops),
Western Air Command (1 Mar 38):
Vancouver City airport, B.C. 5 Oct 32- 30 May 35.
Sea Island, B.C. 31 May 35 - 14 May 40.
2 aircraft, Patricia Bay, B.C. 22 Oct 39 - 14 May 40.
Patricia Bay, B.C. 15 May 40 - 1 Feb 41.

Representative Aircraft (Unit Code 1939-41 TM)
de Havilland DH-60 Moth (Oct 34 - May 40) 152   154
155   156
Avro 621 Tutor (Oct 37 - Mar 40) 185   187
Avro 626 (Mar 38 - Jan 40) 225
de Havilland DH- 82A Tiger Moth (Aug 38 - Aug 40) 253
Armstrong Whitworth Atlas (Sep 39 - Mar 40) 405
Westland Lysander Mk.II (Dec 39 - Jan 41)
416   424 to 426   428
Blackburn Shark Mk.II (Jul 40 - Jan 41) 501   503
Fairey Battle Mk.I (Jul 40 - Jan 41) 1565
Fairchild 71 (Aug 40 - Jan 41) 647

Operational History: First Mission 29 June 1940, Lysander 428 from Patricia Bay with F/L W.J. McFarlane — search for a submarine reported in the area between Otter Point and the international boundary. **Last Mission** 31 January 1941, Lysander 428 from Patricia Bay with F/O H.F. Monnon — army co-operation exercise. **Summary** Sorties: 232. Operational/Non-operational Flying Hours: 203/3915. Casualties: nil.

---

[1] Provided nucleus for No. 3 (CAC) Detachment.

Re-formed as a Fighter unit at Rockcliffe (Ottawa), Ontario on 3 November 1941, the squadron flew Kittyhawk aircraft on West Coast air defence. From June 1942 to August 1943 it was part of the RCAF reinforcement to the United States Army Air Forces in Alaska, and completed two tours of offensive operations against Japanese forces on Kiska Island in the Aleutians. Selected in late 1943 as one of six home fighter units for overseas duty it was redesignated No. 440 (Fighter Bomber) Squadron at Ayr, Ayrshire, Scotland on 8 February 1944.

Brief Chronology Formed as No. 111 (F) Sqn, Rockcliffe, Ont. 3 Nov 41. Redesignated No. 440 (F/B) Sqn, Ayr, Scot. 8 Feb 44.

Nickname "Thunderbird"[1]

Commanders
S/L A.D. Nesbitt, DFC 3 Nov 41 - 12 Jun 42.
S/L J.W. Kerwin 13 Jun 42 - 16 Jul 42 *KIFA.*
S/L K.A. Boomer, DFC 20 Aug 42 - 27 Jun 43.
S/L D.L. Ramsay 28 Jun 43 - 6 Nov 43.
S/L G.J. Elliott 7 Nov 43 - 31 Jan 44.

Higher Formations and Squadron Locations
No. 3 Training Command, BCATP:
Rockcliffe, Ont. 3 Nov 41 - 10 Dec 41.
Western Air Command:
Sea Island, B.C. 14 Dec 41 - 17 Feb 42.
Patricia Bay, B.C. 18 Feb 42 - 3 Jun 42.
No. 4 Group (for adm),
US Alaskan Command (for ops),
RCAF "X" Wing,
Anchorage, Alaska 8 Jun 42 - 30 Oct 42.
12 aircraft, Umnak Island, Aleutians 13 Jul - 12 Oct 42.
Kodiak, Alaska 31 Oct 42 - 10 Aug 43.
6 aircraft, Chiniak Point, Alaska 5 Nov 42 - 24 Apr 43.
8 aircraft, Umnak Island, Aleutians 2 May - 12 Jun 43.
Western Air Command:
Patricia Bay, B.C. 15 Aug 43 - 15 Jan 44.
En route overseas 20 Jan 44 - 7 Feb 44.

Representative Aircraft (Unit Code 1941-42 TM, 1942 LZ)[2]
Curtiss Kittyhawk Mk.I (Nov 41 - Jun 43)AL124 S
AL138 LZ-S*   AL166 LZ-O*   AL201 LZ-H*   AL194 E
AL214 X   AL875 T   AL893 B   AK905 D   AK954 LZ-F*
AK996 LZ-S*
Curtiss P40K-1 Warhawk (Sep 42 - Jul 43)[3]
42-45003   42-45004   42-45944   42-45945   42-45921
42-45951   42-45977   42-46003
Curtiss Kittyhawk Mk.IV (Sep - Oct 43)[4]
861   862   867   870
Curtiss Kittyhawk Mk.I (Sep - Dec 43) 729   1051
1080   1098

Operational History: First Mission 1 July 1942, 6 Kittyhawks from Elmendorf Field, Anchorage, Alaska — to intercept unidentified aircraft, wrong vector given, intruder turned out to be a Bolingbroke. **First Offensive Mission** 25 September 1942, 4 Kittyhawks from Umnak Island (RCAF's first mission against Kiska) were part of the fighter cover of 12 P-39's and 20 P-40's for 9 B-24's attacking Japanese installations. After the B-24's completed their bombing run, the fighters strafed naval craft and gun emplacements, and S/L Boomer was credited with a Zero floatplane fighter destroyed.[5] **Last Mission** 12 November 1943, 2 Kittyhawks from Patricia Bay — intercepted an unidentified aircraft which proved to be friendly. **Summary** Sorties: 598. Operational/Non-operational Flying Hours: 1238/5858.

---

*Lost on 16 July 1940 while on a ferry flight from Cold Lake to Umnak Island. Shortly after passing Dutch Harbour, these aircraft encountered sudden fog and were advised to return to Cold Lake. All crashed on a hillside on Unalaska Island.

Victories: Aircraft: 1 destroyed. Casualties: Operational: nil. Non-operational: 6 aircraft; 5 pilots killed. **Honours and Awards** 1 DFC, 4 Air Medals (USA). **Battle Honours** see No. 440 (Fighter) Squadron.

---

[1] Adopted 17 March 1942, when the squadron was presented with a "thunderbird" totem pole by West Coast Indians.
[2] LZ was used in Alaska from June to October 1942.
[3] Purchased from USAAF in Alaska. These aircraft turned over to No. 14 (F) Squadron when No. 111 departed Alaska.
[4] Also purchased from the USAAF in Alaska were the following P40N Warhawks, re-registered by the RCAF as Kittyhawk Mk.IV's (original USAAF registration numbers shown in brackets): 861 (42-105839), 862 (42-105840), 867 (42-105867) and 870 (42-105878).
[5] This was the only victory over an enemy aircraft credited to a member of the Home War Establishment, and S/L Boomer was awarded an immediate DFC. He was later killed in action while serving as a flight commander with No. 418 (Intruder) Squadron.

119   *De Havilland Tiger Moth 253 receives a compass swing with No. 111 (CAC) Squadron at Vancouver, B.C. on 31 May 1939.*

117   *Avro Tutor 185 and three de Havilland Gipsy Moths lined up for inspection at Vancouver, B.C., in the service of No. 111 (CAC) Squadron in 1938.*

120   *Armstrong Whitworth Atlas Mk.I dual trainer, used by No. 111 (CAC) Squadron at Vancouver, B.C., as it appeared on 21 October 1939.*

118   *Avro 626 225 is washed down at Vancouver, B.C. by airmen of No. 111 (CAC) Squadron on 2 June 1939.*

121   *On 14 August 1940 the station at Patricia Bay, B.C. was prepared for an inspection visit by the Inspector General, Air Marshal G.M. Croil. The aircraft of the two resident squadrons were freshly groomed for the event and included Blackburn Shark 501 (still wearing the markings of No. 6 (BR) Squadron), Fairey Battle 1656 and Westland Lysander 425, all of No. 111 (CAC) Squadron. In the background are three Northrop Deltas of No. 120 (BR) Squadron and the Grumman Goose which had transported A/M Croil to the station.*

122 *Fairchild 71B 647 at Trenton, Ont. prior to delivery to No. 111 (CAC) Squadron in 1940.*

123 *Westland Lysander Mk.II 416 of No. 111 (CAC) Squadron at Patricia Bay, B.C. on 11 January 1941.*

125 *Curtiss Kittyhawk Mk.I AK940, coded LZ-E, of No. 111 (F) Squadron in Alaska during September 1942.*

124 *Curtiss Kittyhawk Mk.I AK863, coded TM-N, of No. 111 (F) Squadron suffers from excessive braking at Patricia Bay, B.C. on 5 March 1942.*

No. 112 Squadron (Auxiliary)

Authorized as No. 12 (Army Co-operation) Squadron (Auxiliary) at Winnipeg, Manitoba on 5 October 1932, the unit began flying training in September 1934 when it received four Moth aircraft. On 15 November 1937 it was renumbered No. 112 Squadron. Mobilized on 10 September 1939, and assigned to the Canadian Active Service Force for overseas duty with the 1st Division, the unit moved to Rockcliffe (Ottawa), Ontario where it trained on Lysander aircraft, and left for England in June 1940. The squadron had anticipated early action but, with the fall of France, was relegated to further training until December. In view of a shortage of fighter units for the defence of Britain, it was redesignated No. 2 (Fighter) Squadron at Digby, Lincolnshire on 9 December 1940, and on 1 March 1941 it was renumbered No. 402 (Fighter) Squadron.

Brief Chronology Authorized as No. 12 (AC) Sqn NPAAF, Winnipeg, Man. 5 Oct 32. Renumbered No. 112 (AC) Sqn NPAAF/AAAF 15 Nov 37. Mobilized 10 Sep 39. Redesignated No. 2 (F) Sqn, Digby, Lincs., Eng. 9 Dec 40. Renumbered No. 402 (F) Sqn 1 Mar 41.

Title or Nickname "City of Winnipeg"[1]

Adoptions City Council of Winnipeg, Man., Women's Air Force Auxiliary of Winnipeg.

Commanders
S/L J.A. Sully, AFC 1 Mar 33 - 30 Sep 37.
S/L H.P. Crabb 1 Oct 37 - 30 Sep 39.
S/L R.H. Little 1 Oct 39 - 14 Apr 40.
S/L W.F. Hanna 15 Apr 40 - 9 Dec 40.

Higher Formations and Squadron Locations
Military District No. 10 (for adm),
RCAF Headquarters (for ops),
Western Air Command (15 Oct 38):
Winnipeg, Man. 5 Oct 32 - 5 Feb 40.
Air Force Headquarters:
Canadian Active Service Force,
Rockcliffe, Ont. 7 Feb 40 - 7 Jun 40.
En route overseas 9 Jun 40 - 20 Jun 40.
RAF Army Co-operation Command:
No. 22 Group,
High Post, Wilts. 22 Jun 40 - 10 Nov 40.
Halton, Bucks. 11 Nov 40 - 9 Dec 40.

Representative Aircraft
de Havilland DH-60 Moth (Apr 34 - Feb 40) 119   164
Avro 621 Tutor (Aug 37 - Feb 40) 184   224
Avro 626 (Aug 37 - Feb 40) 226   266   267
Westland Lysander Mk.II (Canada, Mar - Jun 40, Unit Code XO) 416   420   436 to 440
Westland Lysander Mk.II (England, Jun - Dec 40, Unit Code AE) 436   to   440   P1729   P9178

---

[1] The Squadron's association with Winnipeg was recognized on 5 February 1940, when the mayor presented the unit with its colours and referred to it as the "City of Winnipeg Squadron". Air Force Headquarters, however, never incorporated this title in the unit's designation.

126   *De Havilland Gipsy Moth 164 of No. 112 (AC) Squadron.*

127   *Avro 621 Tutor 184 of No. 112 (AC) Squadron at Winnipeg, Man. in 1938.*

128   *Westland Lysander Mk.II 421 in the foreground of a line-up of No. 112 (AC) Squadron aircraft in April 1940.*

No. 113 Squadron (Auxiliary)

Authorized as No. 13 (Army Co-operation) Squadron (Auxiliary) at Calgary, Alberta on 1 January 1937, the unit was redesignated No. 113 (Fighter) Squadron on 15 November 1937. Called out on voluntary full-time duty in August 1939, the squadron's organization was incomplete and it was disbanded on 1 October.
Brief Chronology Authorized as No. 13 (AC) Sqn NPAAF, Calgary, Alta. 1 Jan 37. Redesignated No. 113 (F) Sqn NPAAF/AAAF 15 Nov 37. Disbanded 1 Oct 39.
Commanders
S/L S.F. Heard 1 Apr 38 - 1 Oct 39.
Higher Formations and Squadron Location
Military District No. 13 (for adm),
RCAF Headquarters (for ops),
Western Air Command (1 Mar 38):
Calgary, Alta. 1 Jan 37 - 1 Oct 39.

Re-formed as a Bomber Reconnaissance unit at Yarmouth, Nova Scotia on 15 February 1942, the squadron flew Hudson and Ventura aircraft on East Coast anti-submarine duty until disbanded at Torbay, Newfoundland on 23 August 1944.[1]
Brief Chronology Re-formed at Yarmouth, N.S. 15 Feb 42. Disbanded at Torbay, Nfld. 23 Aug 44.
Commanders
S/L A.G. Kenyon 15 Feb 42 - 25 Jun 42.
S/L N.E. Small, DFC, AFC 26 Jun 42 - 7 Jan 43 *KIFA*.
W/C A. Laut 27 Jan 43 - 27 Sep 43.
S/L I.M. Black 28 Sep 43 - 30 May 44.
W/C P.S. Delaney 31 May 44 - 23 Aug 44.
Higher Formations and Squadron Locations
Eastern Air Command:
Yarmouth, N.S. 15 Feb 42 - 10 May 43.
5 aircraft, Mont-Joli, Que. 6 Jul - 3 Aug 42, and 16 Sep - 23 Dec 42.
5 aircraft, Chatham, N.B. 9 Sep - 13 Dec 42.
Sydney, N.S. 11 May 43 - 20 Jun 44.
No. 1 Group,
Torbay, Nfld. 21 Jun 44 - 23 Aug 44.
Representative Aircraft (Unit Code 1942 BT, LM)[2]
Lockheed Hudson Mk.III (Feb 42 - Jan 43)
BW403 BW447 BT-T BW619 to BW627
Lockheed-Vega Ventura G.R. Mk.V (Apr 43 - Aug 44)
2142 C 2144 O 2145 H 2152 N 2157 W 2183 D
2235 E 2248 B 2267 K
Operational History: First Mission 25 March 1942, Hudson BW620 from Yarmouth with F/L W.J. Michalski and crew — special search. **Victory: U-Boat** 31 July 1942, Hudson BW625 from Yarmouth with S/L N.E. Small and crew — sank U-754 southeast of Cape Sable (4302N 6452W), Eastern Air Command's first kill.[3] **Last Mission** 16 August 1944, Ventura 2183 "O" from Torbay with F/O W.E. Campbell and crew — escort to US convoy 1.40 N "Jig" (2 motor vessels, 3 escorts). **Summary** Sorties: 2965. Operational/Non-operational Flying Hours: 14,764/5378. Victory: U-boat: 1 sunk; 13 attacks[4] on 22 sightings, dropped 45 250- and 1 600-pound depth charges. Casualties: Operational: 3 aircraft; 12 aircrew killed or missing. Non-operational: 11 killed, 3 died of natural causes. **Honours and Awards** 6 DFC's, 1 AFC, 16 MiD's **Battle Honours** see page xix

---

[1] Six of the squadron's aircraft and crews were transferred to No. 145 (BR) Squadron as the Torbay Detachment.
[2] BT used from February to May, LM until October.
[3] S/L Small was awarded the squadron's first DFC.
[4] Twelve of these attacks were recorded between June and November 1942, more than by all other squadrons of Eastern Air Command for the whole year.

No. 114 Squadron (Auxiliary)

Authorized as a Bomber unit at London, Ontario on 1 April 1938. When alerted for hostilities in August 1939, the squadron's organization was incomplete and it was disbanded on 20 October 1939.
Brief Chronology Authorized at London, Ont. 1 Apr 38. Disbanded 20 Oct 39.
Commanders
S/L J.M. Dobson 1 Oct 38 - 13 Sep 39 *ret.*
S/L G.S. O'Brian, AFC 14 Sep 39 - 20 Oct 39.
Higher Formations and Squadron Location
Military District No. 1 (for adm),
RCAF Headquarters (for ops),
Air Training Command (1 Apr 39):
London, Ont. 1 Apr 38 - 20 Oct 39.

No. 115 Squadron (Auxiliary)

Authorized as No. 15 (Fighter) Squadron (Auxiliary) at Montreal, Quebec on 1 September 1934, the unit commenced flying training in May 1936 when it received four Moth aircraft. It was renumbered No. 115 Squadron on 15 November 1937. Called out on voluntary full-time duty in September 1939, the squadron was disbanded at Montreal and its personnel absorbed by No. 1 (Fighter) Squadron at Dartmouth, Nova Scotia, on 26 May 1940.
Brief Chronology Authorized as No. 15 (F) Sqn NPAAF, Montreal, Que. 1 Sep 34. Renumbered No. 115 (F) Sqn NPAAF/AAAF 15 Nov 37. Disbanded 26 May 40.
Commanders
S/L F.S. McGill 1 Sep 34 - 8 Nov 38 *ret.*
S/L R.H. Foss 9 Nov 38 - 26 May 40.
Higher Formations and Squadron Location
Military District No. 4 (for adm),
RCAF/Air Force Headquarters (for ops):
Montreal, Que. 1 Sep 34 - 26 May 40.
Representative Aircraft
de Havilland DH-60 Moth (May 36 - Sep 38)
64   72   81[1]   110
Fleet Fawn Mk.I & II (Jul 38 - May 40)
198   262   263   264
North American Harvard Mk.I (Nov 39 - Jan 40)
1341   1342   1343
Fairey Battle Mk.I (Jan 40 - May 40)   1301   1303   1319

---

[1] Crashed at Camp Borden on 2 June 1938, at militia summer camp. P/O P.F. Birks was killed, the squadron's only casualty.

Badge A lynx's head affronte the mouth open
Motto Beware
Authority King George VI, January 1945
The Canadian lynx, an animal indigenous to Canada and noted for its fighting powers, is appropriate for a fighter squadron.
Re-formed as a Fighter unit at Rockcliffe (Ottawa), Ontario on 1 August 1941, the squadron flew Bolingbroke aircraft on West Coast air defence, and in April 1942 moved to Annette Island, Alaska as part of the RCAF reinforcement to the United States Army Air Forces. Redesignated Bomber Reconnaissance on 22 June 1942, the squadron returned to southern British Columbia where it was re-equipped with Ventura aircraft and employed on anti-submarine duty until disbanded at Tofino on 23 August 1944.
Brief Chronology Re-formed as No. 115 (F) Sqn, Rockcliffe, Ont. 1 Aug 41. Redesignated No. 115 (BR) Sqn, Annette Is., Alaska 22 Jun 42. Disbanded at Tofino, B.C. 23 Aug 44.
Commanders
S/L E.M. Reyno 22 Aug 41 - 9 Jun 42.
S/L R.A. Ashman 10 Jun 42 - 11 May 43.

S/L T.H. Christie 22 Jul 43 - 31 May 44.
S/L A.G. Hobbs 1 Jun 44 - 23 Aug 44.
Higher Formations and Squadron Locations
No. 3 Training Command, BCATP:
Rockcliffe, Ont. 22 Aug 41 - 10 Oct 41.
Western Air Command:
Patricia Bay, B.C. 15 Oct 41 - 25 Apr 42.
No. 4 Group (for adm),
US Alaskan Command (for ops),
RCAF "Y" Wing (14 Jun 42),
Annette Island, Alaska 27 Apr 42 - 18 Aug 43.
Western Air Command:
Patricia Bay, B.C. 21 Aug 43 - 16 Mar 44.
Tofino, B.C. 17 Mar 44 - 23 Aug 44.
Representative Aircraft (Unit Code 1939-42 BK, 1942 UV)
Bristol Bolingbroke Mk.I (Aug - Dec 41) 704   706
713   716
Bristol Bolingbroke Mk.IV (Nov 41 - Aug 43) 9030
9059 BK-J   9118 BK-V   9125 BK-W
Lockheed-Vega Ventura G.R. Mk.V (Aug 43 - Aug 44)
2218   to   2231 2222 2274
Operational History: First Mission 13 May 1942, Bolingbroke IV 9060 from Annette Island with F/L F.B. Curry and crew — search for a submarine reported to be 30 miles west of Annette. The "periscope" turned out to be a limb protruding from a floating log. **Last Mission** 17 August 1944, Ventura 2231 from Tofino with F/O A.W. Rogers and crew — anti-submarine patrol. **Summary** Sorties: 1580. Operational/Non-operational Flying Hours: 9680/3900. Victories: nil[1]. Casualties: Operational: nil. Non-operational: 2 aircraft; 1 killed, 6 missing. **Honours and Awards** 1 AFC, 4 MiD's. **Battle Honours** see page xix

---

[1] Wartime summaries credit the squadron with sinking a Japanese submarine, R.O. 32, on 7 July 1942, but Japanese records do not mention any submarines lost in the Aleutians on that date, and show R.O. 32 as still in service in August 1945.

130   North American Harvard Mk.I 1343, which later served with No. 115 (F) Squadron at Montreal, Que.

131   Bristol Bolingbrokes of No. 115 (BR) Squadron at Patricia Bay, B.C. on 19 August 1943.

129   Fleet Fawn 198 of No. 115 (F) Squadron, Montreal, Que.

132   Lockheed Vega Ventura G.R.Mk.V aircraft of No. 115 (BR) Squadron lined up at Tofino, B.C. on 10 May 1944.

No. 116 Squadron (Auxiliary)

Authorized as a Coast Artillery Co-operation unit at Halifax, Nova Scotia on 1 April 1938, the squadron was redesignated Fighter on 1 May 1939. When called out on voluntary full-time duty on 3 September, the squadron's organization was incomplete and it was disbanded on 2 November 1939.

Brief Chronology Authorized as No. 116 (CAC) Sqn NPAAF/AAAF, Halifax, N.S. 1 Apr 38. Redesignated No. 116 (F) Sqn AAAF 1 May 39. Disbanded 2 Nov 39.

Commanders
S/L G.E. Creighton 1 May 39 - 2 Nov 39.

Higher Formation and Squadron Location
Eastern Air Command:
Halifax, N.S. 1 Apr 38 - 2 Nov 39.

Re-formed as a Bomber Reconnaissance unit at Dartmouth, Nova Scotia on 28 June 1941,[1] the squadron flew Catalina and Canso A aircraft on East Coast and Gulf of St. Lawrence anti-submarine duty until disbanded at Sydney, Nova Scotia on 20 June 1945.

Brief Chronology Re-formed at Dartmouth, N.S. 28 Jun 41. Disbanded at Sydney, N.S. 20 Jun 45.

Commanders
W/C S.S. Blanchard 28 Jun 41 - 8 Jan 42.
W/C M.C. Doyle 9 Jan 42 - 7 Dec 42.
W/C W.G. Pate, AFC 15 Dec 42 - 3 Dec 43.
S/L W.G. Egan 4 Dec 43 - 12 Sep 44.
W/C A. Fleming 13 Sep 44 - 20 Jun 45.

Higher Formations and Squadron Locations
Eastern Air Command:
Dartmouth, N.S. 28 Jun 41 - 1 Aug 42.
No. 1 Group (10 Jul 41), 4 aircraft,
Botwood, Nfld. 7 Jul - 15 Nov 41,[2] and 31 May - 31 Jul 42.
No. 1 Group,
Botwood, Nfld. 1 Aug 42 - 11 Nov 42.
Eastern Air Command:
Shelburne, N.S. 13 Nov - 28 Dec 42.
4 aircraft, Dartmouth, N.S. 4 Nov 42 - 28 Dec 42.
Dartmouth, N.S. 29 Dec 42 - 13 Feb 43.
Shelburne, N.S. 14 Feb 43 - 10 Jun 43.
No. 1 Group,
Botwood, Nfld. 12 Jun 43 - 14 Nov 43.
Gander, Nfld. 15 Nov 43 - 7 Jun 44.
3 aircraft, Goose Bay, Lab. 15 Nov 43 - 7 Jun 44.
Aircraft at Botwood (Gander runways unserviceable)
4 May - 7 Jun 44.
Eastern Air Command:
Sydney, N.S. 8 Jun 44 - 20 Jun 45.

Representative Aircraft (Unit Code 1941-42 ZD, 1942 NO)
Consolidated Catalina Mk.I & IB (Jul 41 - Aug 43)
W8431 H   W8432 J   Z2134 A   Z2137 C   Z2138 D
Z2139 B   DP202 M   FP294 G   FP296 F   FP297 G
Consolidated Canso A (Sep 41 - Aug 43) 9701 to 9707
Consolidated Canso A (Jan 42 - Feb 42)[3]
9741   9742   9745   9746   9748 J   9749   9750   9781 M
Consolidated Canso A (Aug 43 - Jun 45) 9748 J   9757 C
9758 J   9772 Z   9777 R   9778 R   9798 K   9806 P
9808 J   9819 F   9823 A   9825 B   9828 C   9829 D
9830 P   9831 G   9832 H   9836 O   9839 G   9844 N
11034 A   11057 E   11061 M   11064 L

Operational History: First Mission, 22 July 1941, Catalina W8432 "J" from Dartmouth with W/C Blanchard and crew — convoy escort. Unable to locate convoy, it returned to base. **Last Mission** 2 June 1945, Canso A 9819 "F" from Sydney with F/O H.G. Cornwall and crew — escort of convoy HG9. **Summary** Sorties: 1555. Operational/Non-operational Flying Hours: 13,488/7210. Victories: U-boat: nil; 3 attacks on 8 sightings. Casualties: Operational: nil. Non-operational: 5 aircraft; 18 aircrew, 4 passengers killed. **Honours and Awards** 6 DFC's, 5 AFC's, 2 DFM's, 1 AFM, 3 BEM's, 16 MiD's. **Battle Honours** see page xix

---

[1] The squadron was formed by a division of No. 5 (BR) Squadron, and later provided the nucleus of crews and aircraft for No. 117 (BR) Squadron at North Sydney.
[2] The first aircraft to arrive at Botwood on 7 July 1941; followed by Z2136 on the 14th, Z2137 "C" on the 22nd, and W8431 "H" on the 23rd.
[3] These aircraft were transferred to No. 5 (BR) Squadron.

133   *Consolidated Catalina Mk.I Z2138 of No. 116 (BR) Squadron, with engine maintenance stands in place, at Dartmouth, N.S. on 18 September 1941.*

134   *Consolidated Canso A 9739 on patrol with No. 116 (BR) Squadron.*

No. 117 Squadron (Auxiliary)

Authorized as a Fighter unit at Saint John, New Brunswick on 1 April 1938, the squadron was redesignated Coast Artillery Co-operation on 1 May 1939. When ordered to mobilize in September, the squadron's organization was incomplete and it was disbanded on 28 October 1939.[1]

Brief Chronology Authorized as No. 117 (F) Sqn NPAAF/AAAF, Saint John, N.B. 1 Apr 38. Redesignated No. 117 (CAC) Sqn AAAF 1 May 39. Disbanded 28 Oct 39.

Commanders
S/L W.W. Rogers, MC 1 Oct 38 - 28 Oct 39.

Higher Formations and Squadron Location
Military District No. 7 (for adm),
RCAF Headquarters (for ops),
Eastern Air Command (17 Apr 39):
Saint John, N.B. 1 Apr 38 - 28 Oct 39.

---

[1] Squadron personnel provided the nucleus of "A" Flight of No. 118 (CAC) Squadron at Halifax, N.S..

Re-formed as a Bomber Reconnaissance unit at Sydney, Nova Scotia on 1 August 1941 with Stranraer flying boats, the squadron, less aircraft, was transferred to Western Air Command in October; there its members were distributed among other under-strength units. The squadron was temporarily disbanded at Jericho Beach (Vancouver), British Columbia on 20 November 1941. Reactivated at Sydney on 28 April 1942, the unit flew Catalina and Canso A aircraft on anti-submarine duty over the Gulf of St. Lawrence and the waters adjacent to Cape Breton Island until disbanded at Shelburne, Nova Scotia on 15 December 1943.

Brief Chronology Re-formed at Sydney, N.S. 1 Aug 41. Deactivated at Jericho Beach, B.C. 20 Nov 41. Reactivated at Sydney, N.S. 28 Apr 42. Disbanded at Shelburne, N.S. 15 Dec 43.

Commanders
S/L F.S. Carpenter 20 Aug 41 - 20 Nov 41.
Squadron inactive.
W/C S.R. McMillan 11 May 42 - 27 Feb 43.
W/C J.H. Roberts, AFC 28 Feb 43 - 26 Sep 43.
S/L J. Woolfenden 27 Sep 43 - 15 Dec 43.

Higher Formations and Squadron Locations
Eastern Air Command:
Kelly Beach, North Sydney, N.S. 1 Aug 41 - 27 Oct 41.
Western Air Command:
Jericho Beach, B.C. 31 Oct 41 - 20 Nov 41.
Squadron inactive.
Eastern Air Command:
Kelly Beach, North Sydney, N.S. 28 Apr 42 - 1 Dec 42.
2 aircraft, Gaspé, Que. 10 Jun - 23 Nov 42.
Dartmouth, N.S. 2 Dec 42 - 23 May 43.
No. 5 (Gulf) Group,
5 aircraft, Gaspé, Que. 18 May - 31 Oct 43.
Kelly Beach, North Sydney, N.S. 24 May 43 - 28 Nov 43.
Shelburne, N.S. 29 Nov 43 - 15 Dec 43.

Representative Aircraft (Unit Code 1941 EX, 1942 PQ)
Supermarine Stranraer (Sep - Oct 41) 907 914 929 938
Consolidated Canso A (May 42 - Aug 43) 9701 N 9702 K 9704 O 9705 L 9706 P 9707 Q 9709 R
Consolidated Catalina Mk.I (Mar - Aug 43) W8432 W Z2134 A Z2137 C
Consolidated Catalina Mk.IB & IVA (Mar - Aug 43)
FP209 S FP291 T FP292 U JX207 C JX209 D
JX211 F JX212 G JX213 B JX219 A JX572 Z
JX578 Z JX579 Y JX580 X
(The squadron had a Canso A for transport duties, 9806 "M" (named "Princess Alice"), from 25 April to 17 June 1943. It went aground on a sand bar at Gaspé and was replaced by 9818.)

Operational History: First Mission 13 May 1942, Canso A 9702 "K" from North Sydney with F/L J.H. Roberts and crew — area sweep from Cabot Strait to the western tip of Anticosti Island. Sighted 5 vessels, all friendly. **Last Mission** 27 November 1943, 2 Catalinas from North Sydney — escort of convoy TU4A (7 troopships, 1 tanker, 2 motor vessels, 7 destroyers). **Summary** Sorties: 1236. Operational/Non-operational Flying Hours: 12,391/867. Victories: U-boat: nil; 1 attack. Casualties: nil. **Honours and Awards** 1 DFC, 1 AFM, 2 MiD's. **Battle Honours** see page xix

135 Vickers Supermarine Stranraer 938 at Ottawa, Ont. on 23 August 1941 immediately prior to its delivery to No. 117 (BR) Squadron.

136 Consolidated Canso 9707 of No. 117 (BR) Squadron at Gaspé, Quebec May 1942.

137 Consolidated Catalina Mk.IA of No. 117 (BR) Squadron at North Sydney, N.S. in the summer of 1942.

138 Consolidated Catalina Mk.IVA JX212 aircraft "G" of No. 117 (BR) Squadron, based at Kelly Beach, N.S. in 1943.

No. 118 Squadron (Auxiliary)

Authorized as No. 18 (Bomber) Squadron (Auxiliary) at Montreal, Quebec on 1 September 1934, the unit commenced flying training in May 1936 when it received four Moth aircraft. It was renumbered No. 118 Squadron on 15 November 1937. Called out on voluntary full-time duty on 3 September 1939, the unit was redesignated Coast Artillery Co-operation on 28 October and ordered to Saint John, New Brunswick; there it was equipped with Atlas and Lysander aircraft. Redesignated Fighter on 8 August 1940, but not converted, the squadron was disbanded on 27 September 1940.

Brief Chronology Authorized as No. 18 (B) Sqn NPAAF, Montreal, Que. 1 Sep 34. Renumbered No. 118 (B) Sqn NPAAF/AAAF 15 Nov 37. Redesignated No. 118 (CAC) Sqn, Saint John, N.B. 28 Oct 39. Redesignated No. 118 (F) Sqn 8 Aug 40. Disbanded 27 Sep 40.

Commanders
S/L M.C. Dubuc 12 May 36 - 31 Aug 39.
S/L A. Raymond 1 Sep 39 - 29 Jul 40.
F/L G. Vadboncoeur 30 Jul 40 - 27 Sep 40.

Higher Formations and Squadron Locations
Military District No. 4 (for adm),
RCAF/Air Force Headquarters (for ops):
Montreal, Que. 1 Sep 34 - 31 Oct 39.
Eastern Air Command:
Saint John, N.B. 1 Nov 39 - 27 Sep 40.
"A" Flight[1]
Halifax, N.S. 28 Oct 39 - 30 Mar 40.
Dartmouth, N.S. 31 Mar - 27 Sep 40.

Representative Aircraft
de Havilland DH-60 (May 36 - Sep 39) 67 120 122 166
Armstrong Whitworth Atlas Mk.I (Nov 39 - Jun 40)
402 407 410 415
Blackburn Shark Mk.III (Nov 39 - Sep 40) 526
Westland Lysander Mk.II (Dec 39 - Sep 40)
422 423 430 to 432

Operational History: First Mission 11 November 1939, Atlas 404 from Saint John with F/L McKay and crew — reconnaissance of north shore of the Bay of Fundy to investigate a reported submarine; aircraft returned after 15 minutes because of engine trouble (faulty plugs) and adverse weather. **Last Mission** 27 April 1940, Lysander 423 from Saint John with F/O J.W. St. Pierre — 30-minute reconnaissance of the Bay of Fundy. **Summary** Sorties: 18 Operational/Non-operational Flying Hours: 21/1002. Victories: nil. Casualties: nil. **Honours and Awards** nil.

---

[1] Formed 1 December 1939 with personnel of the recently disbanded No. 117 (CAC) Squadron. When No. 118 Squadron was disbanded, "A" Flight became No. 2 (CAC) Detachment and "B" Flight No. 1 (CAC) Detachment.

Authorized on 8 August 1940,[1] the squadron was re-formed as a Fighter unit at Rockcliffe (Ottawa), Ontario, beginning with "A" Flight on 13 December 1940, and completed on 13 January 1941. Equipped with Grumman Goblin aircraft, it moved to Dartmouth, Nova Scotia in July, being the only fighter unit then available for East Coast defence. In November 1941 it was re-equipped with Kittyhawk aircraft, and in June 1942 it was transferred to Annette Island, Alaska as part of the RCAF reinforcement to the United States Army Air Forces. The pilots made the 4000-mile trip by air — the first fighter unit to fly from coast to coast. Selected as one of six home fighter squadrons for overseas duty in October 1943, the unit was redesignated No. 438 (Fighter Bomber) Squadron at Digby, Lincolnshire, England on 18 November 1943.

Brief Chronology Re-formed ("A" Flight only), Rockcliffe, Ont. 13 Dec 40. Completed 13 Jan 41. Redesignated No. 438 (FB) Sqn, Digby, Lincs., Eng. 18 Nov 43.
Title or Nickname "Cougar" (19 March 1943)
Commanders
F/L E.W. Beardmore 13 Dec 40 - 4 Apr 41.
W/C E.A. McNab, DFC 5 Apr 41 - 22 Jul 41.
S/L H. deM. Molson 23 Jul 41 - 14 Jun 42.
S/L A.M. Yuile 15 Jun 42 - 27 Feb 43.
S/L F.G. Grant 28 Feb 43 - 28 Jul 43.
S/L J.R. Beirnes 29 Jul 43 - 18 Nov 43.
Higher Formations and Squadron Locations
No. 3 Training Command, BCATP:
Rockcliffe, Ont. 13 Dec 40 - 15 Jul 41.
Eastern Air Command:
Dartmouth, N.S. 16 Jul 41 - 6 Jun 42.
Western Air Command:
No. 4 Group (for adm),
US Alaskan Command (for ops),
RCAF "Y" Wing,
Annette Island, Alaska 21 Jun 42 - 15 Aug 43.
Western Air Command:
Sea Island, B.C. 20 Aug 43 - 27 Oct 43.
En route overseas 1 Nov 43 - 18 Nov 43.
Representative Aircraft (Unit Code 1940-42, RE, 1942 VW)
Grumman Goblin (Dec 40 - Dec 41) 341 RE-N 344 RE-W
347 RE-V 335 RE-Y 338 RE-M
Curtiss Kittyhawk Mk.I (Nov 41 - Oct 43) AK803 RE-K
AK857 RE-H AL152 AL222
Operational History: First Mission 22 July 1941, 4 Goblins from Dartmouth — uneventful scramble. **Last Mission,** 29 September 1943, 2 Kittyhawks from Sea Island — dawn patrol. **Summary** Sorties: 251. Operational/Non-operational Flying Hours: 267/8332. Victories: nil; attacked U-boat with machine-gun fire.[2] Casualties: Operational: nil. Non-operational: 5 aircraft; 5 pilots killed or missing.
Honours and Awards nil. **Battle Honours** nil.

---

[1] The squadron was originally to have been formed by the conversion of No. 118 (CAC) Squadron NPAAF, but this was never carried out.
[2] The highlight of the squadron's East Coast operations occurred on 16 January 1942 when two Kittyhawks from Dartmouth machine-gunned a surfaced U-boat some ten miles east of Halifax. F/O W.P. Roberts in AK851 was able to fire six bursts and obtain a number of hits around the conning tower.

139 De Havilland Gipsy Moth 122 of No. 118 (B) Squadron, flown to summer camp at Camp Borden, Ont. from its home base in Montreal, Que.

140 Armstrong Whitworth Atlas Mk.I AC at Camp Borden on 19 June 1938, as it was to appear when issued to No. 118 (CAC) Squadron in the autumn of 1939.

141 While No. 118 (CAC) Squadron was based at Saint John, N.B. this line-up of mixed types was photographed on 20 July 1940. Westland Lysander 420, Blackburn Shark 526, and Fairchild 51A 624 join a transient Fairey Battle on the flight-line.

142 Grumman Goblins of No. 118 (F) Squadron based at Rockcliffe, Ont. practise formation flying in the spring of 1941.

143 Curtiss Kittyhawk Mk.I in the markings of No. 118 (F) Squadron at Dartmouth, N.S. on 4 April 1942.

No. 119 Squadron

Badge On an ogress a tiger's head affronte
Motto Noli me tangere (Touch me not)
Authority King George VI, October 1942
The tiger represents Hamilton, where the squadron was formed, and also symbolizes, by its springing attack, the dive bombing operations of a bomber reconnaissance unit.
Authorized as No. 19 (Bomber) Squadron (Auxiliary) at Hamilton, Ontario on 15 May 1935, the squadron commenced flying training in May 1937 when it received four Moth aircraft. It was renumbered No. 119 Squadron on 15 November 1937. Called out on voluntary full-time duty on 3 September 1939, and redesignated Bomber Reconnaissance on 31 October, the squadron flew Bolingbroke and Hudson aircraft on anti-submarine duty over the Gulf of St. Lawrence and the waters adjacent to Cape Breton Island until disbanded at Sydney, Nova Scotia on 15 March 1944.
Brief Chronology Authorized as No. 19 (B) Sqn NPAAF, Hamilton, Ont. 15 May 35. Renumbered No. 119 (B) Sqn NPAAF/AAAF 15 Nov 37. Mobilized 10 Sep 39. Redesignated No. 119 (BR) Sqn 31 Oct 39. Disbanded at Sydney, N.S. 15 Mar 44.
Title or Nickname "City of Hamilton", "Hamilton Tigers"
Commanders
S/L D.U. McGregor, MC 1 Jan 36 - 11 Jan 39.
W/C N.S. MacGregor, DFC 12 Jan 39 - 11 Jan 42.
S/L R.O. Shaw 12 Jan 42 - 22 Aug 42.
W/C D.H. Wigle 23 Aug 42 - 10 Aug 43.
S/L P.H. Douglas 11 Aug 43 - 15 Mar 44.
Higher Formations and Squadron Locations
Military District No. 2 (for adm),
RCAF Headquarters (for ops),
Air Training Command (1 Dec 38):
Hamilton, Ont. 15 May 35 - 4 Jan 40.
Western Air Command:
Jericho Beach, B.C.[1] 9 Jan 40 - 15 Jul 40.
Eastern Air Command:
Yarmouth, N.S. 21 Jul 40 - 10 Jan 42.
2 aircraft, Sydney, N.S. 7 Jun - 8 Aug 41.
2 aircraft, Dartmouth, N.S. 7 Jun - 14 Aug 41.
4 aircraft, Sydney, N.S. 30 Dec 41 - 9 Jan 42.
Sydney, N.S. 10 Jan 42 - 3 May 43.
3 aircraft, Mont-Joli, Que. 2 Aug - 1 Nov 42, and 24 Apr - 4 May 43.
2 aircraft, Chatham, N.B. 24 Apr - 2 Dec 43.
No. 5 (Gulf) Group,
Mont-Joli, Que. 4 May 43 - 1 Dec 43.
Eastern Air Command:
Sydney, N.S. 2 Dec 43 - 15 Mar 44.
Representative Aircraft (Unit Code 1939-42 DM, 1942 GR)
de Havilland DH-60 Moth (May 37 - Jun 38)
70 81 160 167
de Havilland DH-82A Tiger Moth (May 38 - Nov 39)
249 251 252 279
Fleet Fawn (Jun 38 - Nov 39) 190
Northrop Delta (Mar - May 40) 670
Bristol Bolingbroke Mk.I (Jul 40 - Aug 41)[2].

702 DM-A 703 to 707 711 to 716
Bristol Bolingbroke Mk.IVW (Aug - Nov 41) 9043 F
9061 D 9062 G 9070 N 9077 K 9112 O
Bristol Bolingbroke Mk.IV (Nov 41 - Jun 42)
9063 to 9068 9127
Lockheed Hudson Mk.III (Mar 42 - Mar 44) BW616 Y
BW622 M BW655 R BW657 L BW683 H BW688 S
BW623 J

Operational History: First Mission, 16 March 1941, 4 Bolingbroke I's from Yarmouth — escort to HMS *Ramillies* en route to Saint John, N.B. **Last Mission** 11 March 1944, six sorties — 1 ice patrol, 3 inner anti-submarine patrols, and 2 patrols from Sydney to Port aux Basques. **Summary** Sorties: 3417. Operational/Non-operational Flying Hours: 15,792/8143. Victories: U-boat: nil; 4 attacks on 11 sightings. Casualties: Operational: 6 aircraft; 5 aircrew killed. Non-operational: 8 aircraft; 6 aircrew killed.
Honours and Awards 2 DFC's, 1 AFC, 1 BEM, 3 MiD's.
Battle Honours see page xix

---

[1] Trained with No. 6(BR) Squadron.
[2] Bolingbroke Mk.I's were converted to a fighter configuration by the addition of a gun pack under the fuselage. When Mk.IVW's were received, the Mk.I's were transferred to No. 115 (F) Squadron on 11 August 1941.

146 *Fleet Fawn 190, the first of the type received by No. 119 (B) Squadron, at Camp Borden, Ont. in June 1938.*

147 *Northrop Delta Mk.II 670 at Camp Borden, Ont. This Delta was flown by No. 119 (B) Squadron from March to May 1940.*

144 *De Havilland Tiger Moth 251 of No. 119 (B) Squadron in June 1939.*

145 *No. 119 (B) Squadron aircraft at Camp Borden, Ont. in June 1939 include de Havilland Tiger Moths and a Fleet Fawn.*

148 *Bristol Bolingbroke Mk.I detachment of No. 119 (BR) Squadron based at Sydney, N.S., with three Northrop Deltas of No. 8 (BR) Squadron in the spring of 1941.*

No. 120 Squadron (Auxiliary)

Authorized as No. 20 (Bomber) Squadron (Auxiliary) at Regina, Saskatchewan on 1 June 1935, the unit commenced flying training in April 1937 when it received four Moth aircraft. It was renumbered No. 120 Squadron on 15 November 1937. Called out on voluntary full-time duty in September 1939, and redesignated Bomber Reconnaissance on 31 October, the squadron flew Delta, Hudson, Stranraer, Canso A and Catalina aircraft on West Coast anti-submarine duty until disbanded at Coal Harbour, British Columbia on 1 May 1944.

Brief Chronology Authorized as No. 20 (B) Sqn NPAAF, Regina, Sask. 1 Jun 35. Renumbered No. 120 (B) Sqn NPAAF/AAAF 15 Nov 37. Mobilized 10 Sep 39. Redesignated No. 120 (BR) Sqn 31 Oct 39. Disbanded at Coal Harbour, B.C. 1 May 44.

Commanders
S/L D.U. McGregor, MC 1 Oct 35 - 31 Dec 35.
W/C R.A. Delhaye, DFC 1 Jan 36 - 19 May 40.
S/L J.E. Jellison 20 May 40 - 20 Jan 41.
S/L G.W. Jacobi 21 Jan 41 - 29 Mar 41.
S/L F.J. Ewart 30 Mar 41 - 17 May 42.
S/L P.B. Cox 18 May 42 - 23 Sep 42.
S/L R.I. Thomas 24 Sep 42 - 8 Dec 42.
S/L R.J.E. Benton 9 Dec 42 - 16 Nov 43.
S/L J.T. Arnold 17 Nov 43 - 1 May 44.

Higher Formations and Squadron Locations
Military District No. 12 (for adm),
RCAF Headquarters (for ops),
Western Air Command (15 Oct 38):
Regina Sask. 1 Jun 35 - 6 Nov 39.
Sea Island, B.C.[1] 7 Nov 39 - 31 Jul 40.
Patricia Bay, B.C. 1 Aug 40 - 10 Dec 41.
Coal Harbour, B.C. 11 Dec 41 - 1 May 44.
Movement exercise, Ucluelet, B.C. 27 Jan - 8 Feb 43.

Representative Aircraft (Unit Code 1939-42 MX, 1942 RS)
de Havilland DH-60 Moth (Apr 37 - Aug 38) 65 109 153 161
de Havilland DH-82A Tiger Moth (Jun 38 - Nov 39) 275 276 277 278
Northrop Delta Mk.II (May 40 - Jul 41) 667 670 676 MX-C 685
Lockheed Hudson Mk.I (Mar - Jul 41) 764 to 766 775 MX-R 776
Supermarine Stranraer (Nov 41 - Oct 43) 909 913 924 950
Consolidated Canso A (Apr - Sep 43) 9752 9753 9792 9805
Consolidated Catalina Mk.IVA (Sep 43 - Apr 44) JX293 JX294 JX571 JX579

Operational History: First Mission 30 June 1940, Delta 675 from Sea Island with F/O M.P. Fraser and crew — patrol and search over Juan de Fuca Strait. (Although the squadron was to fly a total of 18 patrols in June and July 1940 (9 for the Department of Fisheries), it did not commence regular patrols until 11 December 1941, after Japan had entered the war. On that day, Stranraer 950 from Coal Harbour with F/L Addington and crew flew a reconnaissance patrol.) **Last Mission** 21 April 1944, Catalina JX571 from Coal Harbour with FS I.A.H. McFarlane and crew — patrol. **Summary** Sorties: 1058. Operational/Non-operational Flying Hours: 5617/6451. Victories: nil. Casualties: Operational: nil. Non-operational: 3 aircraft; 5 aircrew killed. **Honours and Awards** 1 AFC, 1 AFM, 1 BEM. **Battle Honours** see page xix

---

[1] Service training with No. 4 (BR) Squadron.

149 De Havilland Tiger Moth 276 in immaculate condition at Trenton, Ont. prior to delivery to No. 120 (B) Squadron in August 1938.

150 Northrop Delta Mk.II 676 on floats and beaching gear, and coded MX-C, of No. 120 (BR) Squadron.

151 Lockheed Hudson Mk.I 776 of No. 120 (BR) Squadron at Patricia Bay, B.C. on 15 October 1941.

No. 121 Squadron (Auxiliary)

Authorized as No. 21 (Bomber) Squadron (Auxiliary) at Quebec City on 1 January 1937, the unit was renumbered No. 121 Squadron on 15 November. When called out on voluntary full-time duty in September 1939, the squadron's organization was incomplete and it was disbanded on 30 September.
Brief Chronology Authorized as No. 21 (B) Sqn NPAAF, 1 Jan 37. Renumbered No. 121 (B) Sqn NPAAF/AAAF 15 Nov 37. Disbanded 30 Sep 39.
Commanders None appointed.
Higher Formations and Squadron Location
Military District No. 5 (for adm),
RCAF/Air Force Headquarters (for ops),
Quebec City 1 Jan 37 - 30 Sep 39.

152   Grumman Goose Mk.II 925 based at Dartmouth, N.S. with No. 121 (K) Squadron. Note code EN-B without the customary underlining. Probably photographed in the summer of 1942.

Re-formed as a Composite unit at Dartmouth, Nova Scotia on 10 January 1942 by amalgamating Eastern Air Command's Communications Flight[1] and Target Towing Flight, the unit added two more flights: Rescue and Salvage in July 1942, and Calibration in August. The squadron was disbanded on 30 September 1945.[2]
Brief Chronology Re-formed at Dartmouth, N.S. 10 Jan 42. Disbanded 30 Sept 45.
Commanders
F/L J.A.M.G. Gagnon 11 Jan 42 - 26 Mar 42.
S/L E.R. Gardner 27 Mar 42 - 16 Jul 43.
S/L A.L. Michalski 17 Jul 43 - 14 Aug 43.
S/L E. Henderson 15 Aug 43 - 30 Sep 45.
Higher Formation and Squadron Location
Eastern Air Command:
Dartmouth, N.S. 10 Jan 42 - 30 Sep 45.
4 aircraft, Bermuda 11 Mar - 8 Jun 45.
3 aircraft, Mountain View, Ont. 9 Jun - 29 Aug 45.
Representative Aircraft (Unit Code 1942 JY, EN)[3]
Boeing 247D (Jan - May 42) 7635   7636
Grumman Goose (Jan 42 - Jul 43) 797   925   926
Westland Lysander Mk.IITT (Jan 42 - Mar 44)
418   450   1559   V9519
Avro Anson Mk.I (Mar 42 - Nov 43) AX367
AX423   6252   6872
Noorduyn Norseman Mk.IV (Jul 42 - Jul 43)
2455   3524   3537
Bristol Bolingbroke Mk.IVTT (Aug 42 - May 44)
9010   9112   10086   10087
Lockheed Hudson Mk.III (Apr 44 - Sep 45)
BW617   BW626   BW647   BW706
Summary Non-operational Flying Hours: 14,903.
Casualties: 3 aircraft; 5 aircrew, 4 passengers killed.

153   Noorduyn Norseman MK.IV 3537 as used by No. 121 (K) Squadron in 1942-43.

154   Boeing 247D 7635 making a landing approach to Dartmouth, N.S., in the service of No. 121 (K) Squadron in May 1942.

---
[1] Detached on 15 July 1943 to form No. 167 (Communications) Squadron.
[2] The squadron's aircraft and personnel were used to form No. 4 (Composite) Flight.
[3] JY from January to May, EN to October.

No. 122 Squadron

Formed as a Composite unit at Patricia Bay (Vancouver), British Columbia on 10 January 1942 by amalgamating Western Air Command's Coast Artillery Co-operation Flight and Communications Flight,[1] the unit added an Air-Sea Rescue Flight in November 1944 when it received two modified Hudson aircraft with airborne lifeboats. The squadron was disbanded on 15 September 1945.[2]
Brief Chronology Formed at Patricia Bay, B.C. 10 Jan 42. Disbanded 15 Sep 45.
Title or Nickname "Flying Joe Boys"[3]
Commanders
S/L G.G. Diamond 10 Jan 42 - 27 Sep 42.
S/L G.D. Preston, AFC 28 Sep 42 - 2 Sep 43.
S/L E.L. Miners 3 Sep 43 - 12 Feb 44.
S/L R.H. Morris 13 Feb 44 - 15 Sep 44.
S/L J.W. Gledhill 16 Sep 44 - 9 Apr 45.
S/L W.G. Gardiner 10 Apr 45 - 15 Sep 45.
Higher Formation and Squadron Locations
Western Air Command:
Patricia Bay, B.C. 10 Jan 42 - 30 Apr 45.
Communications Flight,
Sea Island, B.C. 15 Mar - 14 Jul 43.
Calibration Flight,
Port Hardy, B.C. 14 May 43 - 30 Apr 45.
Port Hardy, B.C. 1 May 45 - 26 Aug 45.
Patricia Bay, B.C. 27 Aug 45 - 15 Sep 45.
Representative Aircraft (Unit Code 1942 AG)
Blackburn Shark Mk.II & III (Jan 42 - Sep 43) 502 506 523 549
Noorduyn Norseman (Jan 42 - Sep 43) 694 2470 2480 3539
Grumman Goose (Jan 42 - Sep 43) 798 917 924 942
Westland Lysander Mk.IITT (Jan 42 - Mar 44) 416 446 483 485
Lockheed Electra (Aug 42 - Sep 43) 7650
Bristol Bolingbroke Mk.IV TT (Aug 42 - Sep 45) 9006 9040 9088 9140
Avro Anson Mk.V (Mar 44 - Sep 45) 12020 12021 12384 12386
Lockheed-Vega Ventura G.R.Mk.V (Mar 44 - Jun 45) 2178 2252 2272
Lockheed Hudson Mk.III (Nov 44 - Sep 45) BW628 BW642 BW698 V9069
Summary Non-operational Flying Hours: 23,778. Casualties: 3 aircraft; 6 killed, 5 injured. **Honours and Awards** 1 AFC, 1 GM.

---

[1] On 16 September 1943 the Communications Flight was detached to form No. 166 (Communications) Squadron.
[2] The squadron's aircraft and personnel were used to form No. 3 (Composite) Flight.
[3] A "Joe boy" was someone who was expected to perform a wide variety of menial tasks.

155 *Noorduyn Norseman Mk.IV 2480 in camouflage finish and bearing the codes AG R of No. 122 (K) Squadron.*

156 *Westland Lysander Mk.II 445 of No. 122 (K) Squadron at Patricia Bay, B.C. on 5 June 1942.*

157 *Blackburn Shark Mk.II 504, coded AG-D, of No. 122 (K) Squadron at Patricia Bay, B.C. in the summer of 1942. Vickers Stranraer AN-B, of No. 13 (OTU) Squadron, is in the background.*

158 *Grumman Goose Mk.II 917, used by No. 122 (K) Squadron in 1942.*

No. 123 Squadron

Formed as the School of Army Co-operation at Rockcliffe (Ottawa), Ontario on 22 October 1941 with Lysander and Harvard aircraft, and redesignated No. 123 (Army Co-operation Training) Squadron on 15 January 1942, the unit provided training in close support and reconnaissance for Canadian troops.[1] Selected in the fall of 1943 as one of six home fighter units for overseas duty, it was redesignated No. 439 (Fighter Bomber) Squadron at Wellingore, Lincolnshire, England on 1 January 1944.
Brief Chronology Formed as SAC, Rockcliffe, Ont. 22 Oct 41. Redesignated No. 123 (ACT) Sqn 15 Jan 42. Redesignated No. 439 (FB) Sqn, Wellingore, Lincs., Eng. 1 Jan 44.
Commanders
S/L W.W.S. Ross 22 Oct 41 - 24 Jan 43.
S/L L.C. Rankin 25 Jan 43 - 31 Oct 43.
F/L W.H. Walker 1 Nov 43 - 1 Jan 44.
Higher Formations and Squadron Locations
No. 3 Training Command BCATP (for adm),
Air Force Headquarters (for ops):
Rockcliffe, Ont. 22 Oct 41 - 16 Feb 42.
Eastern Air Command:
Debert, N.S. 18 Feb 42 - 29 Nov 43.
3 aircraft, Sydney, N.S. 8 Oct 42 - 27 Jan 43.[2]
En route overseas 30 Nov 43 - 1 Jan 44.
Representative Aircraft (Unit Code 1941-42 VD)
Westland Lysander Mk.II (Oct 41 - Jun 43) 421   454   477   488
Grumman Goblin (Jan - Feb 42) 339   337   342
North American Harvard Mk.IIB (Nov 42 - Oct 43) FE522   FE523   FE791   FE792 F
Hawker Hurricane Mk.I & XII (Nov 42 - Oct 43) 5636 Q   5710 K   5723 N
Operational History: First Mission 10 October 1942, Lysander 448 from Sydney with P/O W.K. Scharff — harbour entrance patrol. **Last Mission** 19 January 1943, Lysander 488 from Sydney with P/O Dadson — harbour entrance patrol. **Summary** Sorties: 98 (all Lysander). Operational/Non-operational Flying Hours: 194/10,098. Casualties: nil. **Honours and Awards** 1 BEM. **Battle Honours** nil.

---

[1] Trained with the Canadian Army's 4th and 7th Divisions.
[2] Although employed in a training role, the squadron provided a detachment of three Lysander aircraft at Sydney for harbour entrance patrols.

No. 124 Squadron

Formed as the Air Force Headquarters Ferry Squadron at Rockcliffe (Ottawa), Ontario on 1 January 1942 for the inter-command ferrying of aircraft, the unit was numbered No. 124 (Ferry) Squadron on 14 February 1942. On 15 November, it was organized into an Eastern Division with headquarters at Rockcliffe and a Western Division with headquarters at Winnipeg, Manitoba; the latter became No. 170 (Ferry) Squadron on 1 March 1944. When the war ended, No. 170 Squadron was disbanded and became the Western Detachment of No. 124 Squadron, which itself was disbanded on 30 September 1946.
Brief Chronology Formed as AFHQ Ferry Sqn, Rockcliffe, Ont. 1 Jan 42. Redesignated No. 124 (Fy) Sqn 14 Feb 42. Disbanded 30 Sep 46.
Commanders
S/L H.O. Madden 1 Jan 42 - 29 Apr 43.
S/L E.O.W. Hall 30 Apr 43 - 19 Jan 44.
S/L F.V. Robinson, DFC 24 Feb 44 - 11 Jul 44.
S/L A.W. Hooper 12 Jul 44 - 25 Jan 45.
S/L E.A. Bland, AFC 26 Jan 45 - 4 Feb 46.
S/L W.E. Edser 5 Feb 46 - 30 Sep 46.
Higher Formations and Squadron Locations
Air Force Headquarters,
No. 9 (Transport) Group (5 Feb 45):
Rockcliffe, Ont. 1 Jan 42 - 29 Feb 44.
Eastern Division, H.Q. Rockcliffe: Detachments at Malton, North Bay and Kapuskasing, Ont; Montreal and Megantic, Que; Moncton, N.B.
Western Division, H.Q. Winnipeg: Detachments at Armstrong Station, Ont; Lethbridge, Alta; Regina, Sask; Cranbrook and Penticton, B.C.
St Hubert, Que. 1 Mar 44 - 24 Mar 46.
Detachments at Malton, Lethbridge, Kapuskasing and Winnipeg.
Rockcliffe, Ont. 25 Mar 46 - 30 Sep 46.
Detachments at Malton (to 15 Apr), Lethbridge (to 30 Apr), Kapuskasing (to 22 Jul).
Summary Non-operational Flying Hours: 126,402. Casualties: 21 aircraft; 48 aircrew and 4 passengers killed. (The squadron's most serious loss occurred on 15 September 1946 when Dakota 962 crashed at Estevan, Saskatchewan, killing 19 pilots who had ferried Cornell aircraft to Estevan and were returning to Rockcliffe.)

159   Hawker Hurricane Mk.I 1352 in a line-up of No. 125 Squadron's aircraft at Torbay, Nfld. on 9 June 1942.

No. 125 Squadron

Formed as a Fighter unit at Sydney, Nova Scotia on 20 April 1942, the squadron flew Hurricane aircraft on East Coast air defence. Selected in late 1943 as one of six home fighter units for overseas duty, it was renumbered No. 441 (Fighter) Squadron at Digby, Lincolnshire, England on 8 February 1944.
Brief Chronology Formed at Sydney, N.S. 20 Apr 42. Renumbered No. 441 (F) Sqn, Digby, Lincs., Eng. 8 Feb 44.
Commanders
F/L C.W. Trevena 13 May 42 - 2 Jun 42.
S/L R.W. Norris 3 Jun 42 - 8 Feb 44.
Higher Formations and Squadron Locations
Eastern Air Command:
Sydney, N.S. 20 Apr 42 - 7 Jun 42.
No. 1 Group,
Torbay, Nfld. 9 Jun 42 - 24 Jun 43.
Eastern Air Command:
Sydney, N.S. 25 Jun 43 - 22 Dec 43.
En route overseas 20 Jan 44 - 7 Feb 44
Representative Aircraft (Unit Code 1942 BA)
Hawker Hurricane Mk.I (Apr 42 - May 43) 1352   1359   1360   1375
Hawker Hurricane Mk.XII (Nov 42 - Dec 42) 5488   5495   5501   5690
Operational History: First Mission 23 June 1943, 2 Hurricane XII's from Torbay — dusk patrol. **Last Mission** 2 November 1943, 2 Hurricane XII's from Sydney — dusk patrol. **Summary** Sorties: 519. Operational/Non-operational Flying Hours: 570/5357. Victories: nil. Casualties: Operational: nil. Non- operational: 5 aircraft; 1 pilot killed, 1 airman died. **Honours and Awards** nil. **Battle Honours** nil.

No. 126 Squadron

Formed as a Fighter unit at Dartmouth, Nova Scotia on 27 April 1942 with Hurricane aircraft, the squadron was employed on East Coast air defence until disbanded on 31 May 1945.
Brief Chronology Formed at Dartmouth, N.S. 27 Apr 42. Disbanded 31 May 45.
Title or Nickname "Flying Lancers" (September 1942)
Commanders
F/L A.M. Yuile 27 Apr 42 - 8 Jun 42.
S/L H. deM. Molson 9 Jun 42 - 6 Sep 42.
F/L P.A. Gilbertson 7 Sep 42 - 16 Nov 42.
S/L R.C. Weston 17 Nov 42 - 25 Jan 44.
S/L N.R. Johnstone 9 Feb 44 - 20 Jul 44.
S/L F.W. Kelly, DFC 21 Jul 44 - 31 May 45.
Higher Formations and Squadron Locations
Eastern Air Command:
Dartmouth, N.S. 20 Apr 42 - 23 Jul 43.
2 aircraft, Bagotville, Que. 14 Jun - 31 Jul 42.
No. 1 Group,
Gander, Nfld. 24 Jul 43 - 1 Jun 44.
Eastern Air Command:
Dartmouth, N.S. 4 Jun 44 - 31 May 45.
Representative Aircraft (Unit Code 1942 BV)
Hawker Hurricane Mk.XIIA (Apr - Dec 42) BW835 F   BW844 O   BW852 J   BW853 L   BW854 X   BW855 E   BW867 Z   BW882 H
Hawker Hurricane Mk.XII (Dec 42 - May 45) 5430 L   5476 B   5489 D   5489 E   5496 X   5640 G   5653 F   5664 N 5665 M   5668 H   5672 Z   5699 P   5700 T   5709 V   5712 R   5717 S
Operational History: First Mission 11 July 1942, 2 Hurricanes from Dartmouth — dawn patrol. **Last Mission** 13 May 1945, 2 Hurricanes from Dartmouth — dawn patrol. **Summary** Sorties: 1919. Operational/Non-operational Flying Hours: 2596/12,238. Victories: nil. Casualties: Operational: 2 aircraft; 2 pilots killed. Non-operational: 7 aircraft; 4 pilots killed, 1 airman died. **Honours and Awards** nil. **Battle Honours** nil.

160   Hawker Hurricane Mk.XIIA and the No. 125 (F) Squadron pilots at Torbay, N.S. on 2 October 1942.

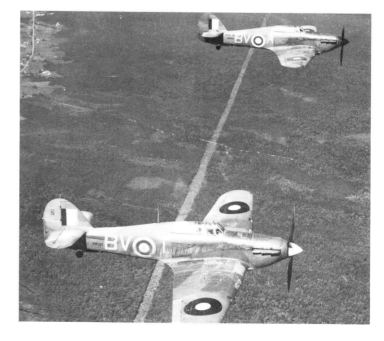

161   Hawker Hurricane Mk.XIIA BW850 and BW859 of No. 126 (F) Squadron patrol from their base at Dartmouth, N.S.

No. 127 Squadron

Formed as a Fighter unit at Dartmouth, Nova Scotia on 1 July 1942 with Hurricane aircraft, the squadron was employed on East Coast air defence. Selected in late 1943 as one of six home fighter units for overseas duty, it was renumbered No. 443 (Fighter) Squadron at Digby, Lincolnshire, England on 8 February 1944.
Brief Chronology Formed at Dartmouth, N.S. 1 Jul 42. Renumbered No. 443 (F) Sqn, Digby, Lincs., Eng. 8 Feb 44.
Commanders
F/L W.P. Roberts 1 Jul 42 - 26 Nov 42.
S/L P.A. Gilbertson 27 Nov 42 - 22 Dec 43.
S/L H.W. McLeod, DFC and Bar 23 Dec 43 - 8 Feb 44.
Higher Formations and Squadron Locations
Eastern Air Command:
Dartmouth, N.S. 1 Jul 42 - 17 Aug 42.
No. 1 Group,
Gander, Nfld. 20 Aug 42 - 23 Jul 43.
Eastern Air Command:
Dartmouth, N.S. 24 Jul 43 - 23 Dec 43.
2 aircraft, Pennfield Ridge, N.B. 25 Jul - 17 Dec 43.
En route overseas 20 Jan 44 - 7 Feb 44.
Representative Aircraft (Unit Code 1942 TF)
Hawker Hurricane Mk.XIIA (Jul 42 - Mar 43)
BW839   BW844   BW853   BW863 V
Hawker Hurricane Mk.XII (Oct 42 - Dec 43) 5459   5497 Y   5655   5729
Operational History: First Mission 24 August 1942, 2 Hurricanes from Gander — uneventful scramble **Last Mission** 14 December 1943, 2 Hurricanes from Dartmouth — dawn patrol. **Summary** Sorties: 788. Operational/Non-operational Flying Hours: 661/5939. Victories: nil. Casualties: Operational: 1 aircraft; 1 pilot killed. Non-operational: 1 aircraft. **Honours and Awards** nil. **Battle Honours** nil.

No. 128 Squadron

Formed as a Fighter unit at Sydney, Nova Scotia on 7 June 1942 with Hurricane aircraft, the squadron was employed on East Coast air defence until disbanded at Torbay, Newfoundland on 15 March 1944.
Brief Chronology Formed at Sydney, N.S. 7 Jun 42. Disbanded at Torbay, Nfld. 15 Mar 44.
Title or Nickname "Dragon" (September 1942)
Commanders
F/L C.C. Moran 7 Jun 42 - 26 Aug 42.
S/L E.C. Briese 27 Aug 42 - 18 Apr 43.
S/L N.R. Johnstone 19 Apr 43 - 23 May 43.
S/L A.E.L. Cannon 24 May 43 - 15 Mar 44.
Higher Formations and Squadron Locations
Eastern Air Command:
Sydney, N.S. 7 Jun 42 - 15 Jun 43.
No. 1 Group,
Torbay, Nfld. 24 Jun 43 - 15 Mar 44.
Representative Aircraft (Unit Code 1942 RA)
Hawker Hurricane Mk.I (Jun 42 - Jan 43) 1351   1355   1366   1379
Hawker Hurricane Mk.XII (Dec 42 - Mar 44) 5429   5488   5700   5717
Operational History: First Mission 24 September 1942, 2 Hurricanes from Sydney — dusk patrol. **Last Mission** 13 March 1944, 2 Hurricanes from Torbay — dawn patrol.
Summary Sorties: 760. Operational/Non-operational Flying Hours: 927/6647. Victories: nil. Casualties: Operational: nil. Non-operational: 7 aircraft; 2 pilots missing.

162   Hawker Hurricane Mk.XII 5658 of No. 127 (F) Squadron.

No. 129 Squadron

Formed as a Fighter unit at Dartmouth, Nova Scotia on 28 August 1942 with Hurricane aircraft, the squadron was employed on East Coast air defence until disbanded at Gander, Newfoundland on 30 September 1944.
Brief Chronology Formed at Dartmouth, N.S. 28 Aug 42. Disbanded at Gander, Nfld. 30 Sep 44.
Title or Nickname "Micmac" (September 1942)
Commanders
S/L C.C. Moran 28 Aug 42 - 3 Nov 42.
S/L M. Lipton 4 Nov 42 - 10 Feb 43.
S/L W.F. Napier 14 Feb 43 - 12 Jan 44.
S/L P.A. Gilbertson 13 Jan 44 - 30 Sep 44.
Higher Formations and Squadron Locations
Eastern Air Command:
Dartmouth, N.S. 28 Aug 42 - 7 Apr 43.
No. 1 Group,
Goose Bay, Lab. 8 Apr 43 - 15 Oct 43.
Eastern Air Command:
Bagotville, Que. 26 Oct 43 - 28 Dec 43.
Dartmouth, N.S. 30 Dec 43 - 1 Jun 44.
No. 1 Group,
Gander, Nfld. 3 Jun 44 - 30 Sep 44.
10 aircraft, Quebec City 2 -23 Sep 44.[1]
Representative Aircraft (Unit Code 1942 HA)
Hawker Hurricane Mk.XIIA (Aug 42 - Jan 43)
BW839 BW842 BW844 BW877
Hawker Hurricane Mk.XII (Jan 43 - Sep 44) 5450 W
5452 A 5455 T 5456 B 5457 H 5462 D 5470 L
5478 V 5500 X 5687 P 5697 R
Operational History: First Mission 17 February 1943, 2 Hurricanes from Dartmouth — dawn patrol. **Last Mission** 18 September 1944, 2 Hurricanes from Quebec City — patrol. **Summary** Sorties: 1166. Operational/Non-operational Flying Hours: 1429/8609. Victories: nil. Casualties: Operational: nil. Non-operational: 5 aircraft; 1 pilot killed.

---

[1] The Quebec City Detachment covered the Quebec Conference of war leaders, including Prime Minister Churchill and President Roosevelt.

No. 130 Squadron

Formed as a Fighter unit at Mont-Joli, Quebec on 1 May 1942, the squadron flew Kittyhawk and Hurricane aircraft on East Coast air defence until disbanded at Goose Bay, Labrador on 15 March 1944.
Brief Chronology Formed at Mont-Joli, Que. 1 May 42. Disbanded at Goose Bay, Lab. 15 Mar 44.
Title or Nickname "Panther" (September 1942)
Commanders
S/L J.A.J. Chevrier 1 May 42 - 6 Jul 42 *KIFA!*
F/L A.E.L. Cannon 7 Jul 42 - 25 Nov 42.
S/L E.L. Neal, DFC 26 Nov 42 - 14 Dec. 42.
S/L A.E.L. Cannon 19 Dec 42 - 19 May 43.
S/L N.R. Johnstone 20 May 43 - 1 Nov 43.
S/L H.J.L. Merritt 2 Nov 43 - 15 Mar 44.
Higher Formations and Squadron Locations
Eastern Air Command:
Mont-Joli, Que. 1 May 42 - 13 Jul 42.
Bagotville,[2] Que. 14 Jul 42 - 24 Oct 43.
No. 1 Group,
Goose Bay, Lab. 26 Oct 43 - 15 Mar 44.
Representative Aircraft (Unit Code 1942 AE)
Curtiss Kittyhawk Mk.I (May - Oct 42) AK865
AK915 AK930 AL136
Hawker Hurricane Mk.XII (Sep 42 - Mar 44) 5452 A
5455 T 5456 B 5458 E 5460 X 5461 Y 5462 D
5464 E 5466 Z 5467 G 5470 L 5478 V
Operational History: First Mission 7 June 1943, 2 Hurricanes from Bagotville — dusk patrol (Note: one previous mission was flown on 6 July 1942. See note 1) **Last Mission** 11 March 1944, 6 Hurricanes from Goose Bay — dawn patrol. **Summary** Sorties: 723. Operational/Non-operational Flying Hours: 619/8198. Victories: nil. Casualties: Operational: 1 aircraft; 1 pilot killed. Non-operational: 5 aircraft; 5 pilots killed, 1 missing, 1 injured.

---

[1] On 6 July 1942, four Kittyhawks from Mont-Joli were on a special anti-submarine search after a freighter was torpedoed and sunk ten miles off Ste Anne-des-Monts. Two lifeboats with survivors were spotted but no U-boats. S/L Chevrier, on the return to Mont-Joli, ran out of fuel 1½ miles off Ste Anne and was killed when AK915 ditched in the St. Lawrence.
[2] Bagotville was then known as Saguenay.

163 Curtiss Kittyhawk Mk.IA of No. 130 (F) Squadron at Mont-Joli, Que. in the period from 1 May to 14 July 1942.

No. 132 Squadron

Formed as a Fighter unit at Rockcliffe (Ottawa), Ontario on 14 April 1942 with Kittyhawk aircraft, the squadron was employed on West Coast air defence until disbanded at Sea Island (Vancouver), British Columbia on 30 September 1944.
Brief Chronology Formed at Rockcliffe, Ont. 14 Apr 42. Disbanded at Sea Island, B.C. 30 Sep 44.
Commanders
F/L A.E.L. Cannon 21 Apr 42 - 26 Apr 42.
S/L K.A. Boomer 27 Apr 42 - 6 Aug 42.
S/L G.J. Elliot 7 Aug 42 - 6 Nov 43.
S/L J.A. Thompson 8 Nov 43 - 30 Sep 44.
Higher Formations and Squadron Locations
No. 3 Training Command, BCATP:
Rockcliffe, Ont. 14 Apr 42 - 3 Jun 42.
Western Air Command:
Sea Island, B.C. 4 Jun 42 - 17 Jul 42.
Patricia Bay, B.C. 18 Jul 42 - 15 Oct 42.
Tofino, B.C. 16 Oct 42 - 30 Jun 43.
Boundary Bay, B.C. 1 Jul 43 - 9 Mar 44.
Tofino, B.C. 10 Mar 44 - 7 Aug 44.
Patricia Bay, B.C. 8 Aug 44 - 21 Aug 44.
Sea Island, B.C. 22 Aug 44 - 30 Sep 44.
Representative Aircraft (Unit Code 1942 ZR)
Curtiss Kittyhawk Mk.I, IA & III (Apr 42 - Sep 44)
1062 P   AK851 H   ET849 P   ET858 D
Operational History: First Mission 4 September 1942, 4 Kittyhawks from Patricia Bay — patrol, sighted two friendly corvettes. **Last Mission** 21 September 1944, 2 Kittyhawks from Sea Island — dusk patrol. **Summary** Sorties: 1426. Operational/Non-operational Flying Hours: 1267/13,058. Victories: nil. Casualties: Operational: nil. Non-operational: 13 aircraft; 8 killed, 1 injured. **Honours and Awards** nil. **Battle Honours** nil.

No. 133 Squadron

Formed as a Fighter unit at Lethbridge, Alberta on 3 June 1942, the squadron flew Hurricane, Kittyhawk and Mosquito aircraft on West Coast air defence until disbanded at Patricia Bay (Vancouver), British Columbia on 10 September 1945.
Brief Chronology Formed at Lethbridge, Alta. 3 Jun 42. Disbanded at Patricia Bay, B.C. 10 Sep 45.
Title or Nickname "Falcon" (September 1942)
Commanders
S/L W.T. Brooks 3 Jun 42 - 20 Mar 43.
S/L B.E. Christmas 21 Mar 43 - 27 Apr 43.
S/L J.B. Doak 28 Apr 43 - 21 Oct 43.
S/L W.C. Connell 2 Nov 43 - 24 Apr 44.
S/L I.C. Ormston, DFC 12 Jun 44 - 1 Jan 45.
S/L J.E. Sheppard, DFC 31 Jan 45 - 22 Apr 45.
S/L H.S. Lisson, AFC 23 Apr 45 - 10 Sep 45.
Higher Formations and Squadron Locations
No. 4 Training Command, BCATP:
Lethbridge, Alta. 3 Jun 42 - 4 Oct 42.
Western Air Command:
Boundary Bay, B.C. 5 Oct 42 - 30 Jun 43.
Tofino, B.C. 1 Jul 43 - 9 Mar 44.
Sea Island, B.C. 10 Mar 44 - 20 Aug 44.
Patricia Bay, B.C. 21 Aug 44 - 10 Sep 45.
4 aircraft, Tofino, B.C. 11 Oct 44 - 30 Jun 45.[1]
Representative Aircraft (Unit Code 1942 FN)
Hawker Hurricane Mk.XII (Jul 42 - Mar 44) 5377 S
5378 W   5381 T   5384 P   5387 D   5389 M   5392 K
5394 J   5398 L[2]   5401 E
Curtiss Kittyhawk Mk.I (Mar 44 - Jul 45) 729 S
864 T   876 A   1028 H   1034 H   1035 L   1038 D
1041 C   1043 W   1045 J   1047 D   1052 G   1057 B
1058 X   1064 P   1067 M   1068 T   1074 M   1078 X
1084 Y   1086 T
de Havilland Mosquito F.B.Mk.26 (Apr - Sep 45)[3]
KA103 N   KA111 P   KA112 L   KA113 I   KA118 H
KA123 M   KA124 B   KA125 O   KA126 C   KA127 T
KA129 G   KA131 D   KA133 A   KA143 R
Operational History: First Mission 25 February 1943, 2 Hurricanes from Boundary — dawn patrol. **Victories, Japanese Fire Balloons** 21 February 1945, P/O E.E. Maxwell in Kittyhawk 866 "R" from Patricia Bay shot down one at an altitude at 25,000 feet near Patricia Bay. 10 March 1945, P/O J.G. Patten in Kittyhawk 858 "F" from Patricia Bay shot down one at an altitude of 13,500 feet near Patricia Bay. **Last Mission** 9 August 1945, Mosquito KA132 "E" from Patricia Bay with F/O J.A.T. Behan and P/O F.P. McKernan — unsuccessful attempt to intercept a fire balloon. **Summary** Sorties: 3036. Operational/Non-operational Flying Hours: 4240/14,471. Victories: 2 fire balloons. Casualties: Operational: 4 aircraft; 3 pilots killed, 1 missing. Non-operational: 12 aircraft; 5 killed, 2 missing. **Honours and Awards** nil. **Battle Honours** nil.

---

[1] Alternating every four weeks with a detachment from No. 135 (F) Squadron.
[2] A presentation aircraft named "March of Dimes", donated to the squadron on 16 July 1942.
[3] The conversion to Mosquito aircraft was in order to attain the speed and altitude required to intercept Japanese fire balloons, but no victories were scored by Mosquito crews.

164 Curtiss Kittyhawk Mk.I AK752, later renumbered 1028 and flown by No. 133 (F) Squadron in the same finish as illustrated here.

165 De Havilland Mosquito F.B.Mk.26 KA133 as aircraft "A" of No. 133 (F) Squadron.

No. 135 Squadron

Formed as a Fighter unit at Mossbank, Saskatchewan on 15 June 1942, the squadron flew Hurricane and Kittyhawk aircraft on West Coast air defence until disbanded at Patricia Bay (Vancouver), British Columbia on 10 September 1945.
Brief Chronology Formed at Mossbank, Sask. 15 Jun 42. Disbanded at Patricia Bay, B.C. 10 Sep 45.
Title or Nickname "Bulldog" (July 1942)
Commanders
S/L E.M. Reyno 15 Jun 42 - 18 Jul 42.
S/L W.C. Connell 19 Jul 42 - 30 Oct 43.
S/L D.J. Smith 9 Nov 43 - 10 Apr 45.
F/L A.R. MacKenzie, DFC 11 Apr 45 - 4 Jun 45.
S/L J.E. Sheppard, DFC 5 Jun 45 - 10 Sep 45.
Higher Formations and Squadron Locations
No. 4 Training Command, BCATP:
Mossbank, Sask. 15 Jun 42 - 2 Oct 42.
Western Air Command:
Patricia Bay, B.C. 4 Oct 42 - 14 Aug 43.
No. 4 Group (for adm),
US Alaskan Command (for ops),
RCAF "Y" Wing,
Annette Island, Alaska 16 Aug 43 - 16 Nov 43.
No. 4 Group,
Terrace, B.C. 17 Nov 43 - 9 Mar 44.
3 aircraft Training Flight, Smithers, B.C. 20 Jan - 9 Mar 44.
Western Air Command:
Patricia Bay, B.C. 12 Mar 44 - 10 Sep 45.
Movement exercise, Tofino, B.C. 18-24 Aug 44.
4 aircraft, Tofino, B.C. 8 Oct 44 - 5 Aug 45.[1]
Representative Aircraft (Unit Code 1942 XP)
Hawker Hurricane Mk.XII (Jul 42 - May 44)
5377   5422   5503   5579 E
Curtiss Kittyhawk Mk.IV (May 44 - Sep 45) 847 B
848 P   850 D   852 M   857 V   858 J   859 K   860 H
862 C   864 F   865 U   866 M   872 S   875 A   879 X
880 N
Operational History: First Mission 9 November 1942, 2 Hurricanes from Patricia Bay — uneventful scramble. **Last Mission** 11 August 1945, 2 Kittyhawks from Patricia Bay — dusk patrol. **Summary** Sorties: 3542. Operational/Non-operational Flying Hours: 3420/15,285. Victories: nil. Casualties: Operational: 1 aircraft; 1 pilot killed. Non-operational: 9 aircraft; 7 pilots killed. **Honours and Awards** nil. **Battle Honours** nil.

---

[1] Alternating every four weeks with a detachment from No. 133 (F) Squadron.

166 Curtiss Kittyhawk Mk.IV 854 aircraft "J" of No. 135 (F) Squadron at Patricia Bay, B.C. in a line-up of its fighters on 10 March 1945.

167 Hawker Hurricane Mk.XII fighters of No. 135 (F) Squadron at Patricia Bay, B.C. on 14 August 1943.

No. 145 Squadron

Badge A lion rampant, holding in the forepaws a balance
Motto Furor non sine frenis (Fury with balance)
Authority: King George VI, September 1944.
The lion represents the fury referred to in the motto, the balance speaks for itself.
Formed as a Bomber Reconnaissance unit at Torbay, Newfoundland on 30 May 1942, the squadron flew Hudson and Ventura aircraft on East Coast anti-submarine duty until disbanded at Dartmouth, Nova Scotia on 30 Jun 1945.
Brief Chronology Formed at Torbay, Nfld. 30 May 42. Disbanded at Yarmouth, N.S. 30 Jun 45.
Commanders
S/L R.H. Batty 30 May 42 - 23 Oct 42.
S/L E.M. Williams, AFC 24 Oct 42 - 18 Apr 43.
S/L R.L. Lee 19 Apr 43 - 2 Oct 43 KIFA.
W/C J.F. Green 13 Oct 43 - 1 Sep 44.
W/C P.S. Delaney 20 Sep 44 - 28 Feb 45.
W/C J.D. Pattison, DFC 1 Mar 45 - 30 Jun 45.
Higher Formations and Squadron Locations
Eastern Air Command:
No. 1 Group, Torbay, Nfld. 30 May 42 - 26 Oct 43.
Eastern Air Command:
Dartmouth, N.S. 30 Oct. 43 - 30 Jun 45.
6 aircraft, Torbay, Nfld. 17 Aug 44 - 10 May 45.[1]
6 aircraft, Yarmouth, N.S. 11 May - 8 Jun 45.
Representative Aircraft (Unit Code 1942 EA)
Lockheed Hudson Mk.I & II (May 42 - Jun 43)
BW430 BW441 BW618 766
Lockheed-Vega Ventura G.R. Mk.V (May 43 - Jun 45)
2146 M 2152 K 2158 R 2159 P 2161 O 2162 U
2163 B 2164 H 2165 X 2166 T 2167 W 2168 E
2169 A 2170 G 2171 Y 2172 I 2184 C 2207 Q
2208 L 2212 J 2214 F 2240 I 2241 Z
Operational History: First Mission 2 June 1942, Hudson 771 from Torbay with FS James and crew — patrol.
Victory, U-Boat 30 October 1942, Hudson 784 from Torbay with F/O E.L. Robinson and crew — anti-submarine sweep; sank U-658 320 miles east of St. John's, Newfoundland (5032N 4632W). This was the squadron's fourth attack on a U-boat, and Eastern Air Command's second kill. **Last Mission** 9 June 1945, Ventura 2239 from Dartmouth with F/L G.P. Aitken and crew — patrol. **Summary** Sorties: 3085. Operational/Non-operational Flying Hours: 16,851/8443. Victories: U-boat: 1 sunk; 7 attacks on 9 sightings. Casualties: Operational: 4 aircraft; 12 personnel, of whom 4 killed, 8 missing. Non-operational: 8 aircraft; 27 personnel, of whom 8 were killed, 17 (9 passengers) missing, 2 injured. **Honours and Awards** 7 DFC's, 12 MiD's
Battle Honours see page xix

---

[1] This detachment was formerly part of No. 113 (BR) Squadron which had been disbanded on 17 August 1944.

168 Lockheed Hudson Mk.1 762 on North Atlantic patrol with No. 145 (BR) Squadron on 1 October 1942.

169 Lockheed-Vega Ventura G.R. Mk.V and the personnel of No. 145 (BR) Squadron at Torbay, Nfld. on 23 August 1943.

170 Bristol Bolingbroke Mk.IV 9066, coded SZ-K, in a line-up with other No. 147 (BR) Squadron aircraft at Sea Island, B.C. on 11 July 1942.

No. 147 Squadron

Formed as a Bomber Reconnaissance unit at Sea Island (Vancouver), British Columbia on 1 July 1942, the squadron flew Bolingbroke aircraft on West Coast anti-submarine duty until disbanded at Tofino, British Columbia on 15 March 1944.

Brief Chronology Formed at Sea Island, B.C. 1 Jul 42. Disbanded at Tofino, B.C. 15 Mar 44.

Commanders
W/C A.L. Anderson 10 Jul 42 - 26 Mar 43.
W/C G.S. Austin 27 May 43 - 14 Dec 43.
S/L J.M. Stroud 15 Dec 43 - 15 Mar 44.

Higher Formation and Squadron Locations
Western Air Command:
Sea Island, B.C. 1 Jul 42 - 6 Mar 43.
4 aircraft, Tofino, B.C. 9 Jan - 6 Mar 43.
Tofino, B.C. 7 Mar 43 - 15 Mar 44.

Representative Aircraft (Unit Code 1942 SZ)
Bristol Bolingbroke Mk.I & IV (Jul 42 - Mar 44)
702   713   9055   9123

Operational History: First Mission 7 November 1942, Bolingbroke 9123 from Tofino with F/L G.R.M. Hunt and crew — patrol. **Last Mission** 29 February 1944, Bolingbroke 9089 from Tofino with F/L A. Gee and crew — patrol. **Summary** Sorties: 560. Operational/Non-operational Flying Hours: 1610/5585. Victories: nil. Casualties: Operational: nil. Non-operational: 4 aircraft; 13 aircrew, of whom 8 were killed, 3 missing, 2 injured. **Honours and Awards** 1 DFC. **Battle Honours** see page xix

No. 149 Squadron

Formed as a Torpedo Bomber unit at Patricia Bay (Vancouver), British Columbia on 26 October 1942, the squadron was the only home unit to be equipped with the Bristol Beaufort to meet the Japanese naval threat from the Aleutians. When the Japanese withdrew in the summer of 1943, the squadron was redesignated Bomber Reconnaissance, re- equipped with Ventura aircraft, and employed on West Coast anti-submarine duty until disbanded at Terrace, British Columbia on 15 March 1944.

Brief Chronology Formed as No. 149 (TB) Sqn, Patricia Bay, B.C. 26 Oct 42. Redesignated No. 149 (BR) Sqn 1 Jul 43. Disbanded at Terrace, B.C. 15 Mar 44.

Title or Nickname "Sea Wolf" (February 1943)

Commanders
S/L J.T. Wilson 9 Nov 42 - 17 Jan 43.
W/C R.R. Dennis 18 Jan 43 - 15 Mar 44.

Higher Formations and Squadron Locations
Western Air Command:
Patricia Bay, B.C. 26 Oct 42 - 16 Aug 43.
No. 4 Group (for adm),
US Alaskan Command (for ops),
RCAF "Y" Wing,
Annette Island, Alaska 18 Aug 43 - 12 Nov 43.
No. 4 Group,
Terrace, B.C. 18 Nov 43 - 15 Mar 44.
3 aircraft, Smithers, B.C. 20 Jan - 1 Mar 44.

Representative Aircraft (Unit Code 1942 ZM)
Bristol Beaufort Mk.I (Nov 42 - Aug 43) L9967 K
L9968 L   N1007 A   N1021 B   N1027 C   N1029 D
N1030 N   N1045 F   N1078 G   N1107 H   W6848 M
Lockheed-Vega Ventura G.R.Mk.V (Jul 43 - Mar 44)
2191 to 2193   2195 to 2204

Operational History: First Mission 1 September 1943, Ventura 2198 from Annette Island with F/O Higham and crew — patrol. **Last Mission** 19 February 1944, Ventura 2204 from Terrace with F/L Watts and crew — patrol. **Summary** Sorties: 188 (all Ventura). Operational/Non-operational Flying Hours: 654/5118. Victories: nil. Casualties: Operational: nil. Non-operational: 1 aircraft; 4 aircrew missing.

Honours and Awards nil. **Battle Honours** see page xix

171   Bristol Beaufort Mk.I N1030 aircraft "N" of No. 149 Squadron at Patricia Bay, B.C. on 18 June 1943.

172   Lockheed-Vega Ventura G.R.Mk.V at Terrace, B.C. in the service of No. 149 (BR) Squadron.

No. 160 Squadron

Formed as a Bomber Reconnaissance unit at Sea Island (Vancouver), British Columbia on 3 May 1943, the squadron moved to Yarmouth, Nova Scotia in July and flew Canso A aircraft on East Coast anti-submarine duty until disbanded at Torbay, Newfoundland on 15 June 1945.
Brief Chronology Formed at Sea Island, B.C. 1 May 43. Disbanded at Torbay, Nfld. 15 Jun 45.
Commanders
W/C F.S. Carpenter, AFC 4 May 43 - 27 May 43.
S/L A. Vanhee 28 May 43 - 24 Mar 44.
S/L J.C. Mulvihill, AFC 25 Mar 44 - 10 May 44.
W/C F.C. Colborne, DFC 11 May 44 - 14 Feb 45.
W/C H.C. Vinnicombe 15 Feb 45 - 15 Jun 45.
Higher Formations and Squadron Locations
Western Air Command:
Sea Island, B.C. 3 May 43 - 28 Jun 43.
Eastern Air Command:
Yarmouth, N.S. 4 Jul 43 - 2 Aug 44.
No. 1 Group,
Torbay, Nfld. 3 Aug 44 - 15 Jun 45.
Representative Aircraft
Consolidated Canso A (May - Jun 43)
9752   9785   9786   9789
Consolidated Canso A (Jul 43 - Jun 45) 9743 B
9768 G   9780 D   9793 F   9794 G   9797 V   9798 J
9799 P   9812 K   9813 L   9814 M   9817 N   9818 O
9822 Q   9840 X   11023 A   11047 C   11048 R   11078 S
11080 U
Operational History: First Mission 9 August 1943, Canso 9817 "N" from Yarmouth with F/O J.R. Bell and crew — escort to the ferry *Princess Helene* through the Bay of Fundy. (While operating from Yarmouth, the squadron's prime duty was escorting *Princess Helene* between Digby, N.S. and Saint John, N.B.) **Last mission** 17 May 1945, Canso 9813 "L" from Torbay with F/O G.O.L. Hill and crew — escort of convoy ON301. **Summary** Sorties: 1547. Operational/Non-operational Flying Hours: 13,251/6580. Victories: U-boat: nil; 1 attack on 5 sightings. Casualties: nil. **Honours and Awards** 8 DFC's 7 MiD's **Battle Honours** see page xix

No. 161 Squadron

Formed as a Bomber Reconnaissance unit at Dartmouth, Nova Scotia on 28 April 1943, the squadron flew Digby and Canso aircraft on East Coast anti-submarine duty over the Gulf of St. Lawrence until disbanded at Yarmouth, Nova Scotia on 31 May 1945.
Brief Chronology Formed at Dartmouth, N.S. 28 Apr 43. Disbanded at Yarmouth, N.S. 31 May 45.
Commanders
W/C E.B. Hale 28 Apr 43 - 19 Dec 44.
S/L J.W. Clarke 20 Dec 44 - 28 Feb 45.
S/L H.F. Monnon 1 May 45 - 31 May 45.
Higher Formations and Squadron Locations
Eastern Air Command:
Dartmouth, N.S. 28 Apr 43 - 8 May 44.
No. 5 (Gulf) Group,
Gaspé, Que. 9 May 44 - 3 Nov 44.
3 aircraft, Summerside, P.E.I. 30 Apr - 30 Nov 44.
2 aircraft, Yarmouth, N.S. 26 May - 9 Jun 44.
3 aircraft, Sept-Iles, Que. 4 Jul - 1 Nov 44.
Eastern Air Command:
Yarmouth, N.S. 5 Nov 44 - 31 May 45.
4 aircraft, Sydney, N.S. 1 Nov 44 - 25 Apr 45.
Representative Aircraft
Douglas Digby (May 43 - Jan 44) 741   751 Y   754 B
756 M
Consolidated Canso A (Oct 43 - May 45)
9750   9840   11004   11040
Operational History: First Mission 24 May 1943, 1 Digby from Dartmouth — harbour entrance patrol at Halifax. **Last Mission** 22 May 1945, Canso 9782 from Yarmouth with F/O H.B. Sutherland and crew — patrol. **Summary** Sorties: 1847. Operational/Non-operational Flying Hours: 13,639/6588. Victories: U-boat: nil; 4 attacks on 6 sightings. Casualties: nil. **Honours and Awards** 2 DFC's, 3 DFM's, 6 MiD's. **Battle Honours** see page xix.

174   Consolidated Canso A 9750 in original factory finish prior to service with No. 161 (BR) Squadron.

173   Consolidated Canso A 9818, 9813 and 9814 of No. 160 (BR) Squadron on the day of their departure from Yarmouth, N.S. to their new base at Torbay, Nfld. on 1 August 1944.

175   Douglas Digby, serial unknown, visits Trenton, Ont. on 20 August 1943 from No. 161 (BR) Squadron.

No. 162 Squadron

Badge Above barry wavy charged with five billets an osprey volant to dexter holding in its claws a sixth billet
Motto Sectabimur usque per ima (We will hunt them even through the lowest deeps)
Authority Queen Elizabeth II, February 1960

The osprey is a great fisherman and is used in this instance to allude to the war-time role of the squadron. The billets indicate depth charges and could also refer to the six submarines sent to the bottom by the squadron. The barry wavy illustrates the ocean over which the squadron operated.

Formed as a Bomber Reconnaissance unit at Yarmouth, Nova Scotia on 19 May 1942 with Canso A aircraft, the squadron spent an uneventful eighteen months on East Coast anti-submarine duty. In January 1944 it was lent to RAF Coastal Command and stationed in Iceland to cover the mid-ocean portion of the North Atlantic shipping route. During June and July, the squadron operated from Wick, Scotland and scored a series of brilliant successes by sinking four German submarines, and sharing a fifth, that were attempting to break through the North Transit Area (Shetland Islands) to attack the Allied D-Day invasion fleet. In one of these engagements F/L D.E. Hornell won the Victoria Cross. The squadron was disbanded at Sydney, Nova Scotia on 7 August 1945.

Brief Chronology Formed at Yarmouth, N.S. 19 May 42. Disbanded at Sydney, N.S. 7 Aug 45.

Commanders
S/L N.E. Small 19 May 42 - 17 Jun 42.
S/L D.S. Turner 18 Jun 42 - 2 Sep 43.
W/C C.G.W. Chapman, DSO 3 Sep 43 - 4 Sep 44.
W/C W.F. Poag 5 Sep 44 - 29 Nov 44.
W/C J.K. Sully 30 Nov 44 - 7 Aug 45.

Higher Formations and Squadron Locations
Eastern Air Command:
Yarmouth, N.S. 19 May 42 - 30 Sep 43.
2 aircraft, Mont-Joli, Que. 10 Oct - 31 Nov 42.
No. 1 Group,
2 aircraft, Gander, Nfld. 27 Dec 42 - 28 Feb 43.
6 aircraft, Stephenville, Nfld. 24 Sep - 6 Oct 43.
Dartmouth, N.S. 3 Oct 43 - 4 Jan 44.
No. 1 Group,
9 aircraft, Goose Bay, Lab. 6-23 Oct 43.
Mont-Joli, Que 24 Oct - 26 Nov 43.
RAF Coastal Command:
Air Headquarters Iceland,
Reykjavik, Iceland 7 Jan 44 - 13 Jun 45.
No. 18 Group,
Wick, Caith., Scot. 24 May - 7 Aug 44.
Eastern Air Command:
Sydney, N.S. 14 Jun 45 - 7 Aug 45

Representative Aircraft (Unit Codes 1942 GK, DZ)[1]
Consolidated Canso A (May 42 - Aug 45) 9764 D 9766 K 9768 M 9769 N 9812 E 11023 E 11033 R 11039 F 11056 Z 11067 B 11075 L 11077 J 11089 C 11091 G 11093 D 11097 M

Operational History: First Mission 25 May 1942, Canso 9748 from Yarmouth with S/L Small and crew — anti-submarine patrol; returned early because of weather. **Victories, U-Boats** 17 April 1944, Canso 9767 "S" from Iceland with F/O T.C. Cooke and crew sank U-342 at 6023N 2920W. 3 June 1944, Canso 9816 "T" from Wick with F/L R.E. MacBride and crew sank U-477 at 6359N 0137E in the face of intense return fire. 11 June 1944, Canso 9842 "B" from Wick with F/O L. Sherman and crew sank U-980 at 6307N 0026E. 13 June 1944, Canso 9816 "T" from Wick with W/C C.G.W. Chapman and crew sank U-715 at 6245N 0259W. Forced to ditch as a result of return fire, the crew spent nine hours in the water; one member drowned but the other seven were rescued. W/C Chapman awarded the squadron's first DSO for this action. 24 June 1944, Canso 9754 "P" from Wick with F/L D.E. Hornell and crew sank U-1225 at 6300N 0050W. Badly damaged by the U-boat's defensive fire, the Canso was ditched and the crew spent 21 hours in the water with one dinghy. Two members died while awaiting rescue. Hornell died shortly after rescue and was posthumously awarded the Victoria Cross for inspiring leadership, valour, and devotion to duty.[2] 30 June 1944, Canso 9841 "A" from Wick with F/L R.E. MacBride and crew damaged U-478 at 6327N 0050W; submarine later sunk by a Liberator of No. 86 Squadron RAF. 4 August 1944, Canso 9759 "W" from Wick with F/O W.O. Marshall and crew damaged U-300. **Last Mission** 31 May 1945, Canso 11074 "A" from Iceland with F/O R.J. Mills and crew — patrol. **Summary** Sorties: 2100. Operational/Non-operational Flying Hours: 22,856/7541. Victories: U-boat: 5 sunk, 1 shared sinking, 1 damaged. Casualties: Operational: 6 aircraft; 34 aircrew, of whom 17 were killed, 17 missing. Non-operational: 3 aircraft; 8 killed. **Honours and Awards** 1 VC, 2 DSO's, 2 MBE's, 16 DFC's, 3 AFC's, 4 DFM's, 1 BEM, 21 MiD's. **Battle Honours** see page xix

---

[1] Allotted, but no record of them being applied to the aircraft.
[2] Hornell's VC was the first awarded to a member of the RCAF, although after the war P/O A.C. Mynarski of No. 419 (B) Squadron was posthumously awarded the VC for an earlier action.

176 Consolidated Canso A of No. 162 (BR) Squadron landing at Skerja Fiord, Reykjavik, Iceland on 25 July 1944.

No. 163 Squadron

Formed as an Army Co-operation unit at Sea Island (Vancouver), British Columbia on 1 March 1943, the squadron flew Bolingbroke aircraft on West Coast photographic work and Harvard aircraft in close air support training for Canadian troops at Wainwright, Alberta. Converted to Hurricane aircraft in June 1943, and redesignated Fighter on 14 October, the unit re-equipped with Kittyhawk aircraft and was employed on West Coast air defence until disbanded at Patricia Bay (Vancouver), British Columbia on 15 March 1944.

Brief Chronology Formed as No. 163 (AC) Sqn, Sea Island, B.C. 1 Mar 43. Redesignated No. 163 (F) Sqn 14 Oct 43. Disbanded at Patricia Bay, B.C. 15 Mar 44.

Commanders
W/C R.M. McKay 1 Mar 43 - 4 May 43.
S/L J.T. Wilson 5 May 43 - 11 Nov 43.
S/L D.L. Ramsay 12 Nov 43 - 7 Mar 44.
F/L D.W.P. Connolly 8 Mar 44 - 15 Mar 44.

Higher Formation and Squadron Locations
Western Air Command:
Sea Island, B.C. 1 Mar 43 - 9 Jan 44.
5 aircraft, Wainwright, Alta. 27 Jul - 15 Oct 43.
Patricia Bay, B.C. 10 Jan 44 - 15 Mar 44.

Representative Aircraft
Bristol Bolingbroke Mk.IV (Mar - Jun 43)
9071   9091   9093   9099
North American Harvard Mk.II (Mar 43 - Feb 44)
2572 Q   2637 P   2726 S   2786 R   2787 L   2798 M
2803 F   5565 O
Hawker Hurricane Mk.XII (Jun - Nov 43) 5584   to
5588   5590
Curtiss Kittyhawk Mk.I & III (Oct 43 - Mar 44)[1]
731   838   1029   1044

Summary Non-operational Flying Hours: 3808. Casualties: 2 aircraft; 2 pilots killed.

---

[1] Received from 118 (F) Squadron when it went overseas.

177   Curtiss Kittyhawk Mk.III 838, originally coded PK before the lower cowling from PD had been fitted to this No. 163 (F) Squadron aircraft.

178   Hawker Hurricane Mk.XII 5584, flown by No. 163 (F) Squadron in June 1943.

No. 164 Squadron

Formed as a Transport unit at Moncton, New Brunswick on 23 January 1943, the squadron flew Lodestar and Dakota aircraft on East Coast transport duty. It was the RCAF's premier transport squadron and the cornerstone of the peacetime Air Transport Command, in that it provided trained crews as the nucleus of other air transport units formed both in Canada and overseas.[1] Retained on the peacetime establishment, it was reorganized on 1 August 1946 into two transport units, the squadron at Dartmouth becoming No. 426 Squadron and the detachment at Edmonton No. 435 Squadron.

Brief Chronology Formed at Moncton, N.B. 23 Jan 43. Reorganized as No. 426 (T) Sqn, Dartmouth, N.S. and No. 435 (T) Sqn, Edmonton, Alta. 1 Aug 46.

Commanders
W/C R.B. Middleton, AFC 23 Jan 43 - 30 Jun 43.
W/C C.W. Hoyt 31 Jul 43 - 1 Feb 45.
W/C R.W. Goodwin 7 Feb 45 - 28 Sep 45.
W/C W.U. Michalski AFC 29 Sep 45 - 29 Apr 46.
W/C C.A. Willis, DFC 30 Apr 46 - 1 Aug 46.

Higher Formations and Squadron Locations
Eastern Air Command (for adm),
Air Force Headquarters (for ops),
No. 9 (Transport) Group (5 Feb 45):
Moncton, N.B. 23 Jan 43 - 30 Sep 45.
Detachments at Edmonton, Alta. 8 Apr - 15 May 43.
Rivers, Man. 22 Apr - 15 May 43.[2]
Dartmouth, N.S. 1 Oct 45 - 1 Aug 46.
Detachment at Edmonton with sub-detachment at Winnipeg 1 Nov 45 - 1 Aug 46.[3]

Representative Aircraft
Lockheed Lodestar (Jan 43 - Sep 45) 551 to 555 564 to 566
Douglas Dakota (May 43 - Jul 46) 650 to 658 660 to 662

Operational History: First Mission 25 January 1943, Digby 750 from Moncton with F/L A. Cirko and crew airlifted 2634 pounds of freight to Goose Bay, Labrador.[4] **Last Mission** 31 July 1946, Edmonton detachment flew 4 passengers, 4901 pounds cargo; destination not recorded. **Summary** Operational/Non-operational Flying Hours: 42,004/5402. Airlifted 70,500 passengers, 15,600 tons of freight, 1,840 tons of mail. Casualties: 1 aircraft; 3 aircrew killed. **Honours and Awards** 3 AFC's, 7 MiD's.

179 *Lockheed Lodestars and Douglas Dakotas of No. 164 (T) Squadron lined up at Moncton, N.B. in 1943.*

180 *Douglas Dakota Mk.I 652 operated by No. 164 (T) Squadron standing by to load stretcher cases at Halifax, N.S. in 1943.*

---

[1] The squadron provided the nucleus of crews for No. 165 (T) Squadron at Sea Island, B.C. in May 1943; No. 168 (HT) Squadron at Rockcliffe, Ont., in October 1943, and Nos. 435 and 436 (T) Squadrons in India in July-August 1944.
[2] The first Lodestar in Edmonton was 553 (F/L W.R. Lavery). The first in Rivers, 552 (F/L D. Daniels), carried out the first paradrop over Canadian soil on 4 May 1943, a "stick" of ten paratroopers. These two detachments later formed No. 165 (T) Squadron.
[3] Both formerly of No. 165 (T) Squadron.
[4] To assist the squadron during its formation, three Digby aircraft (740, 741 and 750) and crews were detached from No. 10 (BR) Squadron from 24 January to 22 March 1943.

No. 165 Squadron

Formed as a Transport unit at Sea Island (Vancouver), British Columbia on 13 July 1943, the squadron flew Lodestar and Dakota aircraft on West Coast transport duty until disbanded on 1 November 1945.
Brief Chronology Formed at Sea Island, B.C. 13 Jul 43. Disbanded 1 Nov 45.
Commanders
W/C H.O. Madden, AFC 13 Jul 43 - 4 Jul 45.
W/C E.B. Hale, DFC 5 Jul 45 - 1 Nov 45.
Higher Formations and Squadron Location
Western Air Command (for adm),
Air Force Headquarters (for ops),
No. 9 (Transport) Group (5 Feb 45):
Sea Island, B.C. 13 Jul 43 - 1 Nov 45.
Detachments at Edmonton, Alta. 13 Jul 43 - 31 Oct 45[1]
Rivers, Man. 13 Jul 43 - 30 Jun 45.[2]
Winnipeg, Man. 13 Jul 43 - 1 Nov 45.[3]
Representative Aircraft
Lockheed Lodestar (Jul 43 - Oct 45) 553   558   to 560   568
Douglas Dakota (Jul 43 - Oct 45) 654   to   656   664   968
Summary Flying Hours: 25,759 (operational only; non-operational not recorded). Casualties: 1 aircraft; 6 killed, 8 injured. **Honours and Awards** 2 AFC's.

---

[1] Responsible for transport of freight and personnel along the North West Staging Route.
[2] Engaged in the training of the 1st Canadian Parachute Battalion, the first parachute jumps were made on 4 May 1943, and the first class graduated on 14 September; training ceased on 30 June 1945.
[3] Supported the Rivers detachment and provided transport for No. 3 Training Command, BCATP.

181   *Lockheed Lodestars 555, 565 and 566 in wartime camouflage, serving No. 165 (T) Squadron in western Canada.*

No. 166 Squadron

Formed as a Communications unit at Sea Island (Vancouver), British Columbia on 16 September 1943,[1] the squadron flew various types of aircraft to transport staff officers of Western Air Command until disbanded on 31 October 1945.[2]
Brief Chronology Formed at Sea Island, B.C. 16 Sep 43. Disbanded 31 Oct 45.
Commanders
S/L G.D. Preston, AFC 16 Sep 43 - 30 Jan 44.
S/L J.E. Cosco, AFC 31 Jan 44 - 30 Sep 45.
S/L D.F. Ritzel 1 Oct 45 - 31 Oct 45.
Higher Formation and Squadron Location
Western Air Command:
Sea Island, B.C. 16 Sep 43 - 31 Oct 45.
Representative Aircraft
Supermarine Stranraer (Sep 43 - Feb 44)
919   938   953   954
Cessna Crane (Sep 43 - Mar 44) 8670
Lockheed Electra (Oct 43 - Mar 44) 7650
Grumman Goose (Sep 43 - Oct 45) 388   392   798   940
Noorduyn Norseman (Sep 43 - Oct 45)
361   696   2488   3536
North American Harvard Mk.IIB (Oct 43 - Mar 44)
FE312   FE313
Beechcraft Expeditor (May 44 - Oct 45) 1384   1386
Avro Anson Mk.V (Jun 44 - Oct 45) 12014   12015
Summary Flying Hours: 13,583. Casualties: 4 aircraft; 5 personnel killed, 1 injured. **Honours and Awards** 1 DFC, 2 AFC's, 1 BEM.

---

[1] Nucleus provided by the Communications Flight of No. 122 (Composite) Squadron.
[2] The squadron's aircraft and personnel were used to form Western Air Command Communications Flight.

182   *Cessna Crane Mk.I 8670, which served No. 166 (T) Squadron in September 1943.*

No. 167 Squadron

Formed as a Communications unit at Dartmouth, Nova Scotia on 15 August 1943,[1] the squadron flew various types of aircraft to transport staff officers of Eastern Air Command until disbanded on 1 October 1945.[2]

Brief Chronology Formed at Dartmouth, N.S. 15 Aug 43. Disbanded 1 Oct 45.

Commanders
S/L W.J. Michalski, AFC 15 Aug 43 - 12 Jul 43.
S/L P.J. Phelan 13 Jul 44 - 27 May 45.
S/L F. Butler 28 May 45 - 4 Jun 45.
S/L J.R. Barclay 5 Jun 45 - 1 Oct 45.

Higher Formation and Squadron Location
Eastern Air Command:
Dartmouth, N.S. 15 Aug 43 - 1 Oct 45.

Representative Aircraft
Avro Anson Mk.I (Aug 43 - Sep 45) AX367   6090   6252   6293
Grumman Goose (Aug 43 - Sep 45) 382   387   796   924
Noorduyn Norseman (Aug 43 - Sep 45) 370   788   2455   2469
Douglas Digby (Aug 43 - Sep 45) 740   747   748   755
Lockheed Hudson Mk.III (Apr 44 - Sep 45) BW625[3]   BW632   BW644   BW702
Beechcraft Expeditor (Mar 44 - Jun 44) 1382   1383

Summary Flying Hours: 8057. Casualties: 2 aircraft.

---

[1] Nucleus provided by the Communications Flight of No. 121 (Composite) Squadron.
[2] The squadron's aircraft and personnel were used to form Eastern Air Command Communications Flight.
[3] Assigned to the Air Officer Commanding (AOC) Eastern Air Command.

183   Noorduyn Norseman Mk.VI 370 of No. 167 (Comm) Squadron visiting Debert, N.S. in the spring of 1945.

184   Douglas Digby 745 aircraft "R" of No. 167 (Comm) Squadron at Summerside, P.E.I. on 9 February 1945.

185   Lockheed Hudson Mk.III aircraft "T" of No. 167 (Comm) Squadron at Torbay, Nfld. in 1944.

No. 168 Squadron

Formed as a Heavy Transport unit at Rockcliffe (Ottawa), Ontario on 18 October 1943, the squadron flew Fortress and Liberator aircraft in delivering mail to Canadian servicemen in the United Kingdom and on the Continent until disbanded on 21 April 1946.

Brief Chronology Formed at Rockcliffe, Ont. 18 Oct 43. Disbanded 21 Apr 46.

Commanders
W/C R.B. Middleton, AFC 18 Oct 43 - 11 Mar 44.
W/C L.G.D. Fraser, DFC 13 Mar 44 - 11 Sep 45 *ret*.
W/C R.W. Goodwin, AFC 12 Sep 45 - 28 Feb 46 *ret*.
W/C J.R. Frizzle 1 Mar 46 - 21 Apr 46.

Higher Formations and Squadron Location
Air Force Headquarters,
No. 9 (Transport) Group (5 Feb 45):
Rockcliffe, Ont. 18 Oct 43 - 21 Apr 46,
Detachments at Prestwick, Scot. (later Istres, Fr.); Blakehill Farm, Wilts. (later Biggin Hill, Kent); Gibraltar (later Rabat-Sale, Morocco.)

Representative Aircraft
Lockheed Lodestar (Oct 43 - Jan 44) 555 559 560 564
Boeing Fortress Mk.IIA (Dec 43 - Apr 46) 9202 to 9207
Douglas Dakota Mk.I, III & IV (Feb 44 - Apr 46,
UK detachments) 653 971 973 988
Consolidated Liberator G.R.Mk.VIT (Aug 44 - Apr 46)
570 574 to 579 11101

Operational History: First Mission 15 December 1943, Fortress 9204 from Rockcliffe to Prestwick, via Dorval and Gander, with 189 bags (5,502 pounds) of mail. The captain was the squadron commander, W/C Middleton. At Prestwick, the mail was transferred to Fortress 9203, whose route was Morocco - Algeria - Tunisia - Sicily - Libya - Egypt. The aircraft reached Cairo on 3 January 1944 and began the return flight on the 6th, via Malta - Sicily - Italy - North Africa, and reached Prestwick on the 11th. The squadron's first complete round trip was flown by Fortress 9202. It left Rockcliffe on 22 December 1943 with F/L W.R. Lavery as captain; after deliveries at Prestwick, it returned on 11 January 1944 with F/L W.H. McIntosh as captain. In all, 1,400,000 pieces of service mail had been carried.

Summary Trans-Atlantic Mail Flights: 636, of which 332 were by Liberator, 240 by Fortress and 64 by Dakota aircraft. Flying Hours: 26,417. Airlifted: Canada to UK: 2,245,269 pounds of mail, including 9,125,000 letters. UK to Continent: 8,977,600 pounds of mail, 2,762,771 pounds of freight, 42,057 passengers. Casualties: 5 aircraft; 18 personnel killed or missing, 7 injured. **Honours and Awards** 1 DFC, 8 AFC's, 1 AFM, 1 BEM, 1 Croix de Guerre (French).

186 *Lockheed Lodestar 555, known popularly as "Muggs," at Rockcliffe, Ont. with No. 168 (HT) Squadron on 23 November.*

187 *Boeing Flying Fortress Mk.IIA 9204 taking off in December 1943 from its base at Rockcliffe, Ont. on the transatlantic mail run operated by No. 168 (HT) Squadron.*

188 Douglas Dakota Mk.III 973 of No. 168 (HT) Squadron with two other squadron aircraft in Britain, as part of the UK Detachment in the summer of 1944.

189 Consolidated Liberator G.R.Mk.VIT 570 of No. 168 (HT) Squadron, based at Rockcliffe, Ont. in 1945.

No. 170 Squadron

Formed as a Ferry unit at Winnipeg, Manitoba on 1 March 1944,[1] the squadron ferried training and operational aircraft in western Canada until disbanded on 1 October 1945.[2]
Brief Chronology Formed at Winnipeg, Man. 1 Mar 44. Disbanded 1 Oct 45.
Commanders
W/C D.W. Russell 1 Mar 44 - 20 Jul 45.
S/L J.H. Sanderson, DFC 21 Jul 45 - 1 Oct 45.
Higher Formations and Squadron Location
No. 2 Training Command, BCATP (for adm),
Air Force Headquarters (for ops),
No. 9 (Transport) Group (5 Feb 45):
Winnipeg, Man. 1 Mar 44 - 1 Oct 45.
Summary Flying Hours: 26,529.

[1] From the Western Division of No. 124 (Ferry) Squadron.
[2] The squadron's personnel were used to form the Western Detachment of No. 124 (Ferry) Squadron.

The RCAF Overseas

Personnel in England
When war began, the RCAF was represented at the British Air Ministry in London by a liaison officer and fifteen officers attending training courses.[1] In addition, three officers were serving in RAF squadrons under an exchange system, and two of them were soon to record "firsts."

Squadron Leader W.I. Clements, while serving with No. 53 Squadron RAF, flew a Blenheim aircraft on a deep penetration of the Hamm-Hanover sector on the night of 19/20 September 1939, to become the first member of the RCAF to fly over enemy territory. Squadron Leader F.M. Gobeil, commanding No. 242 (Canadian) Squadron RAF, scored the first victories by a member of the RCAF in May 1940.[2]

Proposal to Form Overseas Units
As early as the fall of 1939, senior RCAF officers were pressing for the formation of overseas units. On 23 November Air Vice-Marshal G.M. Croil, Chief of the Air Staff, in a memorandum to the Minister of National Defence, argued that it was essential that the RCAF should take part in overseas war activities and not be entirely restricted to home defence and the British Commonwealth Air Training Plan. He pointed out that plans were already drafted for an army co-operation wing to work with the 1st Canadian Division, destined for France, and now recommended that additional formations and units be formed as the BCATP developed. He specifically proposed the formation of an RCAF overseas command, to operate under an RAF headquarters in the field, and both a bomber and a fighter group, each group to consist of three wings, and each wing of two squadrons.

In response to these and other proposals, Article 15 of the BCATP agreement, signed on 17 December 1939, stated:

> "...pupils of Canada, Australia and New Zealand shall, after training is completed, be identified with their respective Dominions, either by the method of organizing Dominion units and formations or in some other way."

A supplementary agreement on 7 January 1941, between Canada and Britain, stipulated that 25 RCAF squadrons[3] would be formed in the United Kingdom in the next 18 months — not counting the three already overseas. At the same time, Canadian officials approached the British Air Ministry as to the possibility of forming Canadian fighter and bomber groups. The reply was that it would be too difficult to form a fighter group because of the geographical nature of RAF Fighter Command's organization and the frequent moves of squadrons between groups. To remain all-Canadian, such a group would require between 40 and 50 squadrons. As for a bomber group, however, the RAF Bomber Command organization was more stable and suitable to Canadian needs, and No. 6 (RCAF) Bomber Group was eventually formed.

First Squadrons Overseas
Preoccupation with the BCATP and home defence made it necessary for the RCAF to retain the greater part of its strength in Canada. In the early months only three squadrons could be spared for overseas duty.[4]

The first RCAF unit in England was No. 110 (Army Co-operation) Squadron which, arriving in February 1940, began training with the intention of going to France with the 1st Division. Four months later No. 112 (Army Co-operation) Squadron arrived in England for the same purpose. The fall of France and the cessation of land operations in western Europe, however, relegated both units to a long period of waiting and training. Also in England by this time was No. 1 (Fighter) Squadron which, after a short but intensive training period, was declared operational in time to take part in the closing stages of the Battle of Britain.

The "400-Block" Squadrons
Because of the large number of Dominion squadrons which were to be formed in the United Kingdom under RAF control, and to avoid confusion with low-numbered RAF squadrons, the British Air Ministry assigned the numbers 400-445 to Canadian squadrons in the UK, or yet to come.[5] Accordingly, on 1 March 1941, No. 110 Squadron became No. 400; No. 1 became No. 401; and No. 2 (formerly No. 112) became No. 402.

The first of the RCAF's 25 (later increased to 35) "Article 15" squadrons to be formed overseas was No. 403 (Fighter) Squadron, on 1 March 1941. Seventeen more were formed in 1941, ten in 1942, four in 1943 and three in 1944. Additionally, six more were transferred from Canada, complete with air and ground crews, so that by the end of the war the number of squadrons in the 400-block had grown to 44. In addition, No. 162 (Bomber Reconnaissance) Squadron was detached from Eastern Air Command and operated under RAF Coastal Command during the last 17 months of the war, while three Air Observation Post squadrons (Nos. 664, 665 and 666) were formed with RCAF and Royal Canadian Artillery personnel. The total number of RCAF squadrons that served overseas was 48, of which 15 were bomber, 11 day-fighter, three fighter-bomber, three fighter-reconniassance, three night-fighter, one intruder, six coastal, three transport and three AOP.[6]

RCAF Units in RAF Commands

Army Co-operation Command
When Canadian army requirements for France were drawn up, one of the units was to have been an army co-operation wing (No. 101) of three squadrons (Nos. 2, 110 and 112) with Lysander aircraft. To maintain even two squadrons at full strength, however, it was necessary to disband No. 2 Squadron and redistribute its personnel and equipment, and to dispense with the proposed wing headquarters. Consequently, only Nos. 110 and 112 were sent to England under this plan.

No. 110 Squadron arrived in February 1940 and was assigned to RAF Army Co-operation Command for training; No. 112 Squadron followed in June. As we have seen, however, the fall of France deprived both squadrons of an immediate operational role. In December, in view of the pressing need for more fighter units, No. 112 became No. 2 (Fighter) Squadron and joined Fighter Command for conversion to Hurricane aircraft and tactical training.

No. 400 Squadron, formerly No. 110, was re-equipped with Tomahawk aircraft in April and continued to train for army co-operation work. In August it was joined by No. 414 Squadron, also equipped with Tomahawks. In September 1942 No. 39 (Army Co-operation) Wing (RCAF) was formed and attached to the First Canadian Army, to control the two squadrons in their role of tactical photographic reconnaissance for the ground forces. A third squadron was added to the wing with the formation of No. 430 Squadron in January 1943. By this time the three squadrons had been re-equipped with Mustang aircraft and, in addition to their role of photographic reconnaissance, carried out two-plane low-level sorties strafing targets of opportunity — most notably German rail traffic — in northern France.[7] On 1 June 1943 RAF Army Co-operation Command was disbanded and the four RCAF units were transferred to the newly-created Second Tactical Air Force.

Fighter Command

A total of twelve RCAF squadrons served with Fighter Command during the war — eight day fighter, three night fighter and one intruder.

Day Fighter Squadrons

It was in the critical period of June 1940, before the Battle of Britain, that No. 1 (Fighter) Squadron arrived in England under the command of Squadron Leader E.A. McNab. After a short period of intensive training the squadron began battle operations on 19 August. The first few days resulted only in fruitless scrambles. Then, on 26 August, the unit finally encountered a formation of Dornier 215 bombers, accounting for three destroyed and three damaged. Eight weeks later, when the squadron flew to Scotland for a well-earned rest, its score stood at 31 enemy aircraft destroyed and 43 probably destroyed or damaged. Three pilots were lost in action — the RCAF's first combat casualties — and the three most successful pilots, Squadron Leader McNab and Flight Lieutenants G.R. McGregor and B.D. Russel, were awarded the Distinguished Flying Cross, the first members of the RCAF to be decorated for gallantry in action.

During the Battle of Britain, Fighter Command had been involved only in defensive action. When the German bomber offensive changed from daylight to night attacks in October 1940, it lost little time in converting its day fighter force to the offensive by reorganizing them into two-squadron wings led by experienced wing commanders. The RCAF squadrons operated as part of RAF wings until the spring of 1941, when, as battle-proven RCAF leaders became available, all-Canadian RCAF fighter wings emerged having geographical names.[8]

The first was the Canadian Digby Wing, formed on 14 April 1941.[9] Led by Wing Commander G.R. McGregor, it was initially composed of Nos. 401 and 402 Squadrons, with Hurricane aircraft. On 15 April the wing carried out the RCAF's first offensive mission over enemy-occupied territory when twelve Hurricanes of No. 402 Squadron, supported by two RAF Spitfire squadrons (Nos. 65 and 266) of the Wittering Wing, flew an uneventful fighter sweep over the Boulogne sector of France.

The second, and most outstanding, RCAF wing was the Canadian Kenley Wing formed on 25 November 1942 when Squadron Leader J.C. Fee, commanding No. 412 Squadron, was promoted to succeed Wing Commander C.B. Kingcombe (RAF).[10] Composed of four squadrons equipped with Spitfire VB's and IX's, the wing took part in a variety of missions over the Pas de Calais sector where many of its pilots distinguished themselves.

Night Fighter and Intruder Squadrons

When the war began, night fighting was still in the early stages of development, but, thanks to radar, it became a very efficient science. The RCAF contributed three squadrons, No. 406, 409 and 410, to night fighting. These squadrons became operational late in the summer of 1941 on Beaufighter aircraft equipped with intercept radar, and were involved in the closing states of the German night bombing of England.[11] After the Allied invasion of Europe in June 1944, and now equipped with Mosquito aircraft, they patrolled over Second Tactical Air Force airfields and the enemy's rear areas to intercept German night raiders.[12]

As an intruder unit, No. 418 Squadron, equipped first with Boston and later Mosquito aircraft, engaged in counter-offensive operations by patrolling over enemy airfields to attack returning German bombers, or German night fighters attempting to intercept Allied bombers. In the absence of such targets, they harrassed the enemy's airfields with bombing and strafing attacks.

When German V-1 flying bombs began to cross the English Channel in June 1944, Nos. 409 and 418 Squadrons were detailed to patrol the night skies as part of the first line of defence. In the short time that they were so engaged, No. 409 Squadron was credited with ten "buzz-bombs" destroyed, and No. 418 Squadron with 77 shot down over the Channel and five more over the English coast. One crew alone, Squadron Leader R. Bannock and Flying Officer R.R. Bruce of No. 418 Squadron, was credited with 19 V-1's.

Bomber Command

In discussion between Canadian and British officials during the summer of 1941 on the question of forming Canadian groups and stations in the United Kingdom, it was agreed in principle to form a Canadian bomber group.[13] This agreement was formalized in the early summer of 1942 with the understanding that the group would be assembled as soon as enough RCAF bomber squadrons were available.[14] At this time there were five, and plans called for six more by the end of the year.

In August 1942, RCAF bomber squadrons began to concentrate at stations of No. 4 Group RAF in Yorkshire in preparation for being taken over by the still unformed Canadian group.[15] It was to be the most northerly group in Bomber command, with its headquarters located at Allerton Hall, Allerton Park, in Yorkshire's North Riding. The advance party arrived on 25 October.

No. 6 (RCAF) Group

No. 6 (RCAF) Group assumed operational status at 0001 hours on 1 January 1943.[16] It then consisted of six RCAF bomber squadrons located at four stations: Croft (No. 427 Squadron), Dalton (No. 428), Dishforth (Nos. 425 and 426), and Middleton St. George (Nos 419 and 420). Leeming, with No. 408 Squadron, joined the group on 2 January, and Topcliffe, with No. 424, on the 3rd. Skipton-on-Swale was under construction as the group's seventh station. By the end of the war the group was to grow to 14 squadrons at eight stations.

Throughout the whole of the war, the bomber offensive was highly centralized and closely controlled by Bomber Command Headquarters. The group headquarters, in addition to its administrative function and concern with forming additional squadrons, was responsible for ensuring that its squadrons were fully briefed in accordance with Bomber Command's instructions as to the number of aircraft to be despatched, their bomb loads, departure times, route and altitude to and from the target. The stations provided the squadrons with housing and messing facilities, and with aerodrome security. The squadrons themselves were independent units responsible for their own administration and aircraft maintenance.

On 25 March 1943, the group began to reorganize under Bomber Command's newly-devised Bomber Operational Base system. The new base organization consisted of a parent station (usually a pre-war permanent station, from which the base took its geographical name) and either one or two satellite or sub-stations (wartime-built), each station housing one to two bomber squadrons. Station headquarters assumed full administrative responsibility for the squadrons, plus organizing a central maintenance section to service aircraft. Squadron establishments were reduced to consist only of the aircrew and aircraft, with a small ground staff to handle operational matters and the daily inspection of aircraft.

Topcliffe Operational Base, formed on 15 March 1943, consisted of RCAF Stations Topcliffe, Dishforth and Dalton, and on 30 April it was redesignated No. 6 Group Training Base. During April and May, three additional stations were added to No. 6 Group and, on 1 June, the Linton-on-Ouse Operational Base was formed consisting of RCAF Stations Linton-on-Ouse, East Moor and Tholthorpe

On 16 September 1943 Bomber Command issued a directive stating that bases were henceforth to be known not by geographical name but by number. The number was to be a two-figure combination, the first figure identifying the parent group and the second the base itself. The Training (or Conversion) Base was to be number one in each group. Thus Topcliffe Training Base became No. 61 (Training) Base and Linton-on-Ouse Operational Base became No. 62 (Operational) Base. On 1 May 1944 the group had added enough new stations and squadrons to complete its organization, and Nos. 63 and 64 (Operational) Bases were formed.

No. 331 Wing in North Africa

In May 1943 three RCAF bomber squadrons — Nos. 420, 424 and 425, all equipped with Wellington B.Mk.X's — were detached from No. 6 Group and sent on loan to North Africa. There, as No. 331 (Medium Bomber) Wing (RCAF), they took part in a heavy bombardment in preparation for, and in support of, the Allied landings in Sicily and Italy. For over three months the wing sent out aircraft almost nightly to bomb airfields, harbours, freight yards and rail junctions. They became known for their ability, even under cover of darkness, to locate obscure battlefield targets. As a consequence, they were employed at crucial times during the assault phase to hamper enemy road movement in areas adjacent to the beachhead. After handing over its aircraft to the RAF Wellington wings of No. 205 Group in late October, the wing returned to England where its headquarters was disbanded, and the squadrons redeployed within No. 6 (RCAF) Group.

Coastal Command

As implied in its motto "find the enemy; strike the enemy; protect our ships," Coastal Command's major task was to wage war against the enemy's submarines and surface ships, including merchant vessels, in close co-operation with the Royal Navy. To this end Canada contributed large numbers of air and ground personnel and, at one time or another, seven squadrons. Three of the squadrons (Nos. 404, 407 and 415) were equipped with landplanes and four with flying boats (Nos. 413, 422, 423 and 162, the latter on detachment from the RCAF's Eastern Air Command.)

South East Asia Command

Three RCAF squadrons served in South East Asia; two as transport, and one as coastal reconnaissance.

Formed in Britain in late 1941, No. 413 Squadron at first flew coastal reconnaissance sorties over the North Sea with Catalina flying boats. Early in 1942 it was hastily transferred to South East Asia where the Japanese flood had broken loose. The squadron arrived in Ceylon just in time, for on one of the first sorties Squadron Leader L.J. Birchall and his crew discovered and reported the approach of an invasion fleet. Although the Catalina was shot down by Japanese carrier-borne fighter aircraft and the crew became prisoners of war, the warning alerted the island's defences, and the invasion fleet was repulsed. This was the first check to Japanese expansion up to that time.

The two transport squadrons, Nos. 435 and 436, were formed in India and flew Dakota aircraft in support of the British Fourteenth Army in India and Burma. They dropped supplies by parachute, usually on small clearings. In addition to the hazards of the jungle, the sudden storms and the tropical diseases, the crews had to run a gauntlet of intense ground fire from Japanese positions located close to the perimeter of the dropping zones or landing strips. These two squadrons were the last RCAF units to be actively engaged in operations, as their work continued after the end of hostilities in Europe, in May 1945, to when the war in the Pacific ended in August. They were then transferred to the United Kingdom where they joined No. 437 Squadron in ferrying supplies, mail and personnel to Canadian occupation units in Germany.

Second Tactical Air Force

Before the war, RAF strategy had been based on the use of massive bombing attacks against the enemy homeland and little consideration had been given to close support of army operations. The success of the *Luftwaffe* in supporting ground operations in its *Blitzkrieg* (lightning) campaigns in Europe, however, led to a reappraisal of this strategy.

The RAF's first effective close support operations were those of the Desert Air Force, in support of the Eighth Army.[17] Its reconnaissance squadrons obtained valuable information concerning enemy movements, while its day-fighter and fighter-bomber squadrons maintained air superiority over the battlefield and provided the necessary assistance to the ground forces in overcoming enemy strongpoints and breaking up concentrations of infantry and armour. For the invasion of Europe, it was planned to form a similar tactical air force to support British and Canadian troops.[18]

As envisaged, Second Tactical Air Force, the majority of its squadrons to be drawn from Fighter Command, would support the British 21st Army Group (British Second Army and First Canadian Army) and would consist of No. 2

(Bomber) Group, with light and medium bomber units for tactical bomber support; Nos. 83 and 84 (Composite) Groups, with day-fighter, fighter-bomber and fighter-reconnaissance units for tactical close support; and No. 85 (Base) Group, with day and night fighters for the defence of Second Tactical Air Force airfields in Europe.

It was hoped that the First Canadian Army would be supported by an all-Canadian composite group. Of the 30 squadrons required for such a group, however, only seven day fighter and three fighter reconnaissance squadrons were readily available, and the RCAF would have to find an additional 20 squadrons at short notice. The forming of new squadrons was out of the question, as 29 of Canada's quota of 35 "Article15" squadrons had been formed and Bomber Command had priority on the remaining six. Consideration was then given to converting the seven miscellaneous RCAF squadrons then in the United Kingdom to a new role as day fighter or fighter bomber.[19] While this conversion would have had advantages for Canada, there were overriding objections from the RAF that these units were more urgently required in their present roles, especially in Coastal Command, which would be responsible for the safety of the Allied armada from attack by German surface ships and submarines.

Although the all-Canadian composite group never came into existence, discussions concerning it resulted in an increase to the RCAF day fighter force overseas by the transfer of six squadrons, complete with air and ground crews, from Canada to the United Kingdom;[20] and in June 1943 it was agreed to combine these and the seven established squadrons in No. 83 (Composite) Group, and assign that group to the First Canadian Army. No. 84 Group was to have supported the British Second Army in the assault on the beaches, but on 26 January 1944 it was decided that for such a demanding role Second Army should have the support of the more experienced No. 83 Group; and No. 84 Group, which contained no Canadian squadrons, supported the First Canadian Army. The dream of Canadian ground and air forces working together was never realized, much to the regret of both.

Transport Command
In the late summer of 1944, No. 437 Squadron was formed in Transport Command, equipped with Dakota aircraft, and began operations almost immediately when it towed gliders for the airborne landings at Arnhem in September. In the weeks and months that followed, the squadron's aircraft dropped supplies and ferried troops, equipment and gasoline to Continental bases, returning with casualties. In March 1945 the squadron again towed gliders for the Rhine crossing at Wesel. After the German surrender, it moved to the Continent and extended its operations as far afield as Oslo, Vienna, Naples and Athens. Its aircraft brought home released prisoners of war and displaced civilians, carried food supplies for the relief of starving peoples of the once Nazi-occupied lands, and flew supplies and mail to Canadian servicemen dispersed over the Continent.

British Air Forces of Occupation (Germany)
To the British Air Forces of Occupation (Germany) the RCAF contributed No. 126 (Fighter) Wing, composed of Nos. 411, 412, 416 and 443 Squadrons, with Spitfire Mk.XVI's, and No. 664 (Air Observation Post) Squadron with Austers for army communication flights.

Four bomber squadrons (Nos. 424, 427, 429 and 433) were retained in the United Kingdom as part of RAF Bomber Command's Strike Force and were also employed for a time on troop transport flights between Britain and Italy.

Another bomber squadron, No. 426, was transferred to Transport Command, re-equipped with the transport version of the Liberator aircraft, and employed on troop runs between England and India for several months.

Also within Transport Command was formed No. 120 (RCAF) Wing, consisting of Nos. 435, 436 and 437 Squadrons. All were equipped with Dakota aircraft and employed on transport and communication flights within Europe until June 1946.

"Tiger Force" (Pacific)
From the first days of the alliance between the United States and Britain in the war against Germany and Japan, it had been agreed that the overall strategy was to defeat Germany first and then turn, with full weight, on Japan. The defeat of Germany was referred to as Phase One and the period thereafter until the defeat of Japan as Phase Two. By the late summer of 1944 an Allied victory in Europe was assured and planning was begun on the Commonwealth contribution to Phase Two.

On 20 October 1944 a very long range bomber force was proposed. Named "Tiger Force," it was to consist of three bomber groups: one RAF, one RCAF, and the third a composite group of British, Australian, New Zealand and South African squadrons. In each group were to be 22 squadrons: twelve bomber, six fighter, three transport and one air-sea rescue. On 21 March 1945 this was scaled down to two groups — the composite group to be built around No. 5 (RAF) Group, and the Canadian group around No. 6 (RCAF) Group — and the bomber strength of each group reduced to ten squadrons. On 5 April the Canadian group was further reduced to eight bomber and three transport squadrons.[21]

When Germany surrendered on 8 May 1945, plans for the formation of "Tiger Force" were stepped up. The bomber squadrons selected for the RCAF group were converted to Canadian-built Lancaster B.Mk.X's and returned to Canada for training and reorganization. But the atomic bombing of Hiroshima on 6 August 1945 and Nagasaki on the 9th resulted in Japan's acceptance of the Allied terms of surrender on 15 August (signed on 2 September) and the RCAF "Tiger Force" units were disbanded early in September without having commenced their training.

Summary
A summary of the work performed by the squadrons at home and overseas is but one part of the story of the RCAF's contribution in the Second World War. The other part of the story concerns the 249,662 men and women who wore the uniform of the RCAF. Of this total, 93,844 personnel served overseas, the majority with British rather than with Canadian units. A report, dated 8 August 1944, stated that of the 27,104 RCAF aircrew then serving overseas, nearly 60 per cent were in RAF squadrons — 17,111 as compared to 9993. Throughout the war, the Canadian contribution to the RAF was significant. At least one in four fighter pilots in the Battle of Malta came from

Canada, as did one-fifth of Coastal Command's aircrew. At the end of the war, almost a quarter of Bomber Command's aircrew were from the RCAF — approximately 1250 pilots, 1300 navigators, 1000 air bombers, 1600 air gunners and 750 wireless operators — and these figures excluded those serving in No. 6 (RCAF) Group.

Recognition of the services performed by the RCAF is to be found in the list of honours and awards conferred upon its personnel. More than 8000 officers, airmen and airwomen received decorations from the British and Allied governments including two Victoria Crosses (Flight Lieutenant D.E. Hornell and Pilot Officer A.C. Mynarski, both posthumously) and four George Crosses.

The RCAF's Roll of Honour contains the names of 17,100 personnel who gave their lives in the service of Canada. Of these fatalities, 14,544 occurred overseas — among them 12,266 on operations and 1906 in training accidents. The major overseas casualties were within Bomber Command; a total of 9980, 8240 of which were on operations.

---

[1] There were more Canadian officers serving as RAF aircrew than the total number of officers in the RCAF, including the Auxiliary. They were to be found in 12 fighter, at least 18 bomber, and five coastal squadrons.

[2] No. 242 (Canadian) Squadron, formed on 30 October 1939, became operational with Hurricanes on 23 March 1940. It served with the RAF's Advance Air Striking Force in the Low Countries and France, and was one of the last British air units to return to England on 18 June 1940 after the fall of France. S/L Gobeil was credited with a Bf. 109 probably destroyed on 23 May 1940 near Berck and a Bf. 110 destroyed on the 25th near Menin, Belgium.

[3] While Canada would provide the air and ground crews to man these squadrons, the RAF would bear the cost of equipment and pay and allowances, at Canadian rates, thus offsetting part of the enormous cost to Canada of the BCATP. Early in 1943 Canada voluntarily assumed full financial responsibility for the maintenance and administration of its overseas formations, the RAF continuing to provide suitable aircraft and to exercise operational control.

One of the greatest problems with these squadrons was getting the RAF to "Canadianize" them. By the end of 1941, 18 had been formed, and of their 1037 aircrew only 499 were RCAF, although 8959 Canadian graduates of the BCATP had been sent to the United Kingdom. The aircrew of the single-engine fighter squadrons were almost entirely Canadian; the difficulty was with the multi-engine units. The problem appeared to be that sufficient facilities were not available to establish operational training units on a national basis. It was not until Canada assumed full financial responsibility for its overseas formations, early in 1943, that Canadianization progressed satisfactorily.

[4] The three squadrons were Nos. 110 and 112 (Army Co-operation), formerly of the Auxiliary, and No. 1 (Fighter) from the Permanent Force.

[5] Australian squadrons were allotted the 450-467 block, New Zealand 485-490.

[6] In 1941 the emphasis was on day fighter squadrons for the defence of Britain, and later for offensive air operations over northwestern Europe; in 1942-43 it shifted to bomber squadrons for the strategic bombing of Germany's war-making machinery; and by late 1943 the emphasis had re-shifted to fighter and fighter bomber squadrons for the invasion of Europe.

[7] "Train busting", or "loco busting." Low-level attacks by pairs of aircraft on road and rail traffic, and other ground targets of opportunity, were known as "rhubarbs."

[8] A fighter wing took its geographical name from its parent station. Redhill, for example, was a satellite aerodrome of Kenley and a part of the Kenley Wing. The title "Wing Commander Flying" was created for the air leader who was responsible to the sector commander for air operations, and carried no administrative responsibility for the wing or its parent station.

[9] On this same date, RAF Station Digby and it satellite aerodrome at Wellingore were transferred to the RCAF and became RCAF Station Digby. It was here that all subsequent RCAF day-fighter squadrons were formed and received their operational training in squadron and wing tactics.

[10] Although RAF Station Kenley and its satellite aerodrome at Redhill had been manned by RCAF day-fighter squadrons since 2 November, it was not considered an all-Canadian wing until the appointment of a Canadian WCF. Similarly, on 22 January 1943, Redhill became an RCAF station with G/C W.R. MacBrien as station commander.

[11] The RCAF's first night fighter victory was scored on 12 September 1941 by F/O R.C. Fumerton and Sgt. L.P.S. Bing of No. 406 Squadron.

[12] RAF Fighter Command's first daylight Mosquito intruder mission over Germany was flown by P/O M.A. Cybulski and P/O H.H. Ladbrook (RAF, navigator) of No. 410 Squadron on 27 March 1943.

[13] Mainly at a meeting held on 8 July 1941 at the British Air Ministry on the employment of RCAF personnel and formations with the RAF.

[14] At a meeting in Ottawa during the period of 22 May to 5 June 1942 on the extension of the original BCATP agreement. The provision is contained in the New Agreement concerning British Commonwealth Air Training Plan, 5 June 1942, Appx IV, paras 15-17.

[15] Formed under authority RAF Bomber Command Organization Order No. 151, 17 October 1942.

[16] RAF Bomber Command Order of Detail No. 30.

[17] One RCAF day-fighter squadron, No. 417, served with the Desert Air Force on operations from the Nile Valley to the plains of northern Italy.

[18] In March 1943 Exercise "Spartan" was held in England to rehearse the planned role of the First Canadian Army of breaking out from a beachhead established by an assaulting force, the British Second Army. An experimental composite air group was formed for this exercise to test close support operations, and became the model for Second Tactical Air Force.

[19] The seven squadrons were Nos. 404 (Coastal Fighter), 407 (General Reconnaissance), 415 (Torpedo Bomber), 418 (Intruder), and 406, 409 and 410 (Night Fighter).

[20] These six units, their 400-block renumberings shown in brackets, were (from Eastern Air Command) Nos. 123 (439), 125 (441) and 127 (443) Squadrons, and (from Western Air Command) Nos. 14 (442), 111 (440) and 118 (438) Squadrons.

[21] The fighter escort force was to be provided by British, Australian and New Zealand units already in the Pacific.

UNIT HISTORIES
"400 BLOCK" OVERSEAS SQUADRONS

No. 400 Squadron

Badge In front of two tomahawks in saltire an eagle's head erased
Motto Percussuri vigiles (On the watch to strike).
Authority King George VI, September 1942

The eagle's head indicates the squadron's role as a reconnaissance unit on army co-operation work; the tomahawks indicate the type of aircraft with which it was once equipped.

Formed in Canada as No. 110 "City of Toronto" (Army Co-operation) Squadron (Auxiliary) on 5 October 1932, the squadron arrived in England with Lysander aircraft in February 1940, was renumbered No. 400 (Army Co-operation) Squadron at Odiham, Hampshire on 1 March 1941, and redesignated Fighter Reconnaissance on 28 June 1943. The unit flew Mustang and Spitfire aircraft on photographic reconnaissance work, collecting photographic intelligence for Allied invasion planners, and before-and-after photographs of Allied air attacks against German "Noball" (V-1 flying bomb) launching sites; following the Allied invasion of Europe in June 1944, it provided tactical photographic reconnaissance for the British Second Army in North-West Europe. The squadron was disbanded at Luneburg, Germany on 7 August 1945.

Brief Chronology Formed as No. 110 "City of Toronto" (AC) Sqn (Aux), Toronto, Ont. 5 Oct 32. Renumbered No. 400 (AC) Sqn, Odiham, Hants., Eng. 1 Mar 41. Redesignated No. 400 (FR) Sqn, Dunsfold, Surrey 28 Jun 43. Disbanded at Luneburg, Ger. 7 Aug 45.

Title "City of Toronto"
Adoption City Council of Toronto, Ont.
Commanders
W/C R.M. McKay 1 Mar 41 - 11 Aug 41.
W/C H.W. Kerby 12 Aug 41 - 9 May 42 *repat*.
S/L R.C.A. Waddell, DSO, DFC, 10 May 42 - 14 Jul 43.
S/L W.B. Woods, DFC 15 Jul 43 - 11 Sep 43 *repat*.
W/C R.A. Ellis, DFC 12 Sep 43 - 13 Nov 44 *repat*.
S/L M.G. Brown, DFC and Bar 14 Nov 44 - 1 Jul 45 *repat*.
S/L J.A. Morton, DFC 2 Jul 45 - 7 Aug 45.

Higher Formations and Squadron Locations
Army Co-operation Command:
No. 35 (Army Co-operation) Wing (RAF),
No. 39 (Army Co-operation) Wing (RCAF) (12 Sep 42),
Odiham, Hants. 1 Mar 41 - 3 Dec 42.
Fighter Command: No. 10 Group,
"A" Flight, Middle Wallop, Hants. (training and operations in fighter tactics) 27 Oct - 28 Dec 42.[1]
Dunsfold, Surrey 4 Dec 42 - 27 Dec 42.
Coastal Command: No. 19 Group,
6 aircraft (fighter escort over the Bay of Biscay)
Portreath, Cornwall 3-23 Dec 42.
Trebelzue, Cornwall 24 Dec 42 - 14 Jan 43.
Fighter Command:
No. 10 Group,
Middle Wallop, Hants. 28 Dec 42 - 13 Jan 43.
Army Co-operation Command:
No. 39 (Army Co-operation) Wing (RCAF),
Dunsfold, Surrey 14 Jan 43 - 27 Jul 43.

Second Tactical Air Force:
No. 83 (Composite) Group,
No. 39 (RCAF)Sector (disbanded 1 Jul 44),
No. 128 (RCAF) Wing (disbanded 1 Jul 44),
Woodchurch, Kent 28 Jul 43 - 14 Oct 43.
Redhill, Surrey 15 Oct 43 - 17 Feb 44.
Kenley, Surrey (Redhill runways under repair) 2-30 Dec 43.
Odiham Hants. 18 Feb 44 - 30 Jun 44.
No. 39 (RCAF) Wing,
B(Base) 8 Sommervieu, Fr. 1 Jul 44 - 14 Aug 44.
"B" Flight remained at Odiham until 10 Aug 44.
B.21 Ste-Honorine-de-Ducy, Fr. 15 Aug 44 - 31 Aug 44.
B.34 Avrilly, Fr. 1 Sep 44 - 19 Sep 44.
B.66 Blakenburg, Bel. 20 Sep 44 - 2 Oct 44.
B.78 Eindhoven, Neth. 3 Oct 44 - 6 Mar 45.
B.90 Petit-Brogel, Bel. 7 Mar 45 - 9 Apr 45.
B.108 Rheine, Ger. 10 Apr 45 - 15 Apr 45.
B.116 Wunstorf, Ger. 16 Apr 45 - 27 Apr 45.
B.154 Soltau, Ger. 28 Apr 45 - 6 May 45.
B.156 Luneburg, Ger. 7 May 45 - 17 Jul 45.
B.160 Copenhagen, Den. 18 Jul 45 - 1 Aug 45.
B.156 Luneburg, Ger. 2 Aug 45 - 7 Aug 45.
Representative Aircraft (Unit Code SP)
Westland Lysander Mk.III (Mar - Apr 41)
R9001   R9009   R9119   R9125
Curtiss Tomahawk Mk.I, IIA & IIB (Apr 41 - Jul 42)
AH789 L   AH806 W   AH824 F   AH831 N   AH841 K
AH862 J   AH895 B   AK324 S   AK481 S   AK484 Y
AK528 B
North American Mustang Mk.I (Jun 42 - Feb 44)[2]
AG488 B   AG521 P   AG528 B   AG583 G   AG587 L
AG591 A   AG615 Q   AG641 V   AG658 T   AG659 U
AG661 X   AL971 S   AM126 D   AM129 M   AM184 N
AM187 J   AM237 E   AM256 Y   AP191 O
de Havilland Mosquito P.R. Mk.XVI (Dec 43 - May 44, 6 aircraft)[3]   MM275   MM284   MM306   MM353
Supermarine Spitfire P.R.Mk.XI (Dec 43 - Aug 45)
PL799   PM124   PM133   PM158
Operational History: First Mission 6 November 1941, 2 Tomahawks from Odiham — reconnaissance Le Tréport to Courtrai; not completed, insufficient cloud cover. **First Victory** Aircraft: 7 November 1942, F/O F.E. Hanton in Mustang AG660 SP-W from Middle Wallop — returning from a rhubarb near Caen, was attacked five miles from the French coast and credited with a Bf.109 probably destroyed. Ground: 27 November 1942, F/O D.M. Grant in Mustang AP173 SP-M from Middle Wallop — rhubarb over the Cherbourg Peninsula, credited with destroying 3 locomotives. Aircraft: 13 April 1943, F/O D.M. Grant in Mustang AP259 SP-R from Dunsfold — night ranger to Paris, joined the circuit of an enemy night flying school 15 miles south of Paris and credited with a Do.217 destroyed. **Last Mission** 8 May 1945, 2 Spitfires from Luneburg — sea patrols. **Summary** Sorties: 3,000. Operational/Non-operational Flying Hours: 4833/13,907. Victories: Aircraft: 9 destroyed, 2 probably destroyed, 9 damaged. Ground: 15 locomotives destroyed, 1 probably destroyed, 75 damaged; 34 trains, 23 miscellaneous targets damaged. Casualties: Operational: 12 pilots killed or missing, 2 wounded. Non-operational: 17 killed, 2 died. **Top Scores** F/L F.E.W. Hanton, DFC 2-0-0 aircraft, 35 locomotives. F/L D.M. Grant, DFC 2-0-0 aircraft, 30 locomotives. F/O A.T. Carlson, DFC 2-0-0 aircraft, 9 locomotives. **Honours and Awards** 1 Bar to DFC, 10 DFC's, 1 BEM, 3 MiD's. **Battle Honours** Fortress Europe 1941 -1944: *Dieppe*. France and Germany 1944-1945: *Normandy 1944, Arnhem, Rhine*. Biscay 1942-1943.

---

[1]The flight was employed on "rhubarb" operations (attacks on ground targets of opportunity, in pairs) and acquired a reputation for destroying railway locomotives.
[2]While flying under Fighter Command, Mustang aircraft carried a full set of code letters, later reduced to the single aircraft letter under Second Tactical Air Force.
[3]Forty-three Mosquito high-level photographic sorties were flown between 26 March and 2 May 1944.

190   Westland Lysander Mk.III   R9003 in the foreground of a line of No. 400 (AC) Squadron aircraft while stationed at Odiham, Hants., England.

191   Curtiss Tomahawk Mk.I   AH806 and AH785, aircraft "W" and "S" respectively, of No. 400 (AC) Squadron on patrol from Odiham, Hants.

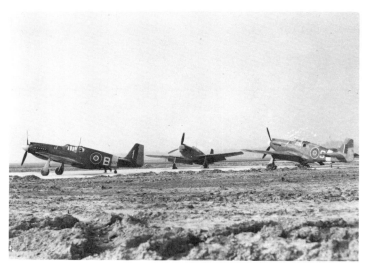

192 North American Mustang Mk.I AG528 aircraft "B" of No. 400 (FR) Squadron. Note the oblique camera mounted aft of the cockpit canopy.

193 De Havilland Mosquito P.R.Mk.XVI MM307 of No. 400 (FR) Squadron.

194 Supermarine Spitfire Mk.XI PL975 of No. 400 (FR) Squadron on take-off with landing gear tucking up. The modified photo reconnaissance markings and waxed finish disclose the time as the spring of 1945.

No. 401 Squadron

Badge A Rocky Mountain sheep's head caboshed
Motto Mors celerrima hostibus (Very swift death for the enemy)
Authority King Goerge VI, September 1944
The mountain sheep is known for its great stamina and fighting power and is indigenous to many part of Canada.
Formed in Canada as No. 1 (Fighter) Squadron on 17 May 1937, the squadron arrived in England in June 1940 and was renumbered No. 401 (Fighter) Squadron at Digby, Lincolnshire on 1 March 1941. It flew Hurricane and Spitfire aircraft on offensive and defensive air operations; and in support of Allied ground forces in North-West Europe. The squadron's final score of 195 enemy aircraft destroyed made it the top-scorer among RCAF fighter squadrons, and it was also the leading fighter unit in the Second Tactical Air Force with 112 air and 15 ground victories. It also held the record for the number of sorties flown. On 5 October 1944 the unit scored the first RAF/RCAF victory over a German Me.262 jet fighter.[1] The squadron was disbanded at Fassberg, Germany on 10 July 1945.
Brief Chronology Formed as No. 1 (F) Sqn, Trenton, Ont. 17 May 37. Absorbed No. 115 (F) Sqn (Aux), Halifax, N.S. 28 May 40. Renumbered No. 401 (F) Sqn, Digby, Lincs., Eng. 1 Mar 41. Disbanded at Fassberg, Ger. 10 Jul 45.
Title or Nickname "Ram"
Adoption No. 2 Service Flying Training School, Uplands, Ont.
Ancestry No. 1 Squadron and No. 115 (Fighter) Squadron (Auxiliary)
Commanders
S/L P.B. Pitcher 1 Mar 41 - 11 Mar 41.
S/L A.D. Nesbitt 12 Mar 41 - 28 Aug 41.
S/L N.R. Johnstone 29 Aug 41 - 24 Jan 42.
S/L A.G. Douglas (RAF), DFC 25 Jan 42 - 2 Jun 42.
S/L K.L.B. Hodson, DFC 3 Jun 42 - 21 Jan 43.
S/L E.L. Neal, DFC 22 Jan 43 - 17 Dec 43 *repat.*
S/L L.M. Cameron, DFC 18 Dec 43 - 2 Jul 44 *POW.*
S/L I.F. Kennedy, DFC and Bar 3 Jul 44 - 26 Jul 44 *MIA.*[2]
S/L H.C. Trainor, DFC 26 Jul 44 - 19 Sep 44 *POW.*
S/L R.I.A. Smith, DFC and Bar 28 Sep 44 - 30 Nov 44 *2OTE.*
S/L H.J. Everard, DFC 3 Dec 44 - 25 Dec 44 *POW.*
S/L W.T. Klersy, DFC and Bar 4 Jan 45 - 22 May 45 *KIFA.*
S/L E.A. Ker, DFC 23 May 45 - 10 Jul 45.
Higher Formations and Squadron Locations
Fighter Command:
No. 12 Group,
Canadian Digby Wing,
Digby, Lincs. 1 Mar 41 - 19 Oct 41.
No. 11 Group,
Biggin Hill, Kent 20 Oct 41 - 18 Mar 42.
Gravesend, Kent 19 Mar 42 - 2 Jul 42.
Eastchurch, Kent 3 Jul 42 - 2 Aug 42.
No. 15 Armament Practice Camp, Martlesham Heath, Suffolk 28 Jul - 2 Aug 42.
Biggin Hill, Kent 3 Aug 42 - 13 Aug 42.
Lympne, Kent 14 Aug 42 - 20 Aug 42.

Biggin Hill, Kent 21 Aug 42 - 23 Sep 42.
Canadian Kenley Wing,
Kenley, Surrey 24 Sep 42 - 22 Jan 43.
No. 13 Group,
Catterick, Yorks. 23 Jan 43 - 28 May 43.
4 aircraft, Thornaby, Yorks. 25 Jan - 27 May 43.
No. 11 Group,
Canadian Kenley Wing,
Redhill, Surrey 29 May 43 - 5 Jul 43.
Second Tactical Air Force:
No. 83 (Composite) Group,
No. 17 (RCAF) Sector (disbanded 13 Jul 44),
No. 126 (RCAF) Wing,
Redhill, Surrey 6 Jul 43 - 6 Aug 43.
No. 15 Armament Practice Camp, Martlesham Heath, Suffolk 20-30 Jul 43.
Staplehurst, Kent 7 Aug 43 - 12 Oct 43.
Biggin Hill, Kent 13 Oct 43 - 7 Apr 44.
No. 11 Armament Practice Camp, Fairwood Common, S. Wales 8 Apr 44 - 17 Apr 44.
Tangmere, Sussex 18 Apr 44 - 17 Jun 44.
B. (Base) 4 Beny-sur-Mer, Fr. 18 Jun 44 - 7 Aug 44.
B.18 Cristot, Fr. 8 Aug 44 - 31 Aug 44.
B.28 Evreux, Fr. 1 Sep 44.
B.24 St André, Fr. 2 Sep 44.
B.44 Poix, Fr. 3 Sep 44 - 6 Sep 44.
B.56 Evère, Bel. 7 Sep 44 - 20 Sep 44.
B.68 Le Culot, Bel. 21 Sep 44 - 2 Oct 44.
B.84 Rips, Neth. 3 Oct 44 - 13 Oct 44.
B.80 Volkel, Neth. 14 Oct 44 - 5 Dec 44.
No. 17 Armament Practice Camp, Warmwell, Dorset, Eng. 24 Oct - 4 Nov 44.
B.88 Heesch, Neth. 6 Dec 44 - 11 Apr 45.
B.108 Rheine, Ger. 12 Apr 45 - 14 Apr 45.
B.116 Wunstorf, Ger. 15 Apr 45 - 12 May 45.
B.152 Fassberg, Ger. 13 May 45 - 1 Jul 45.
No. 127 (RCAF) Wing,
B.152 Fassberg, Ger. 2 Jul 45 - 10 Jul 45.
Representative Aircraft (Unit Code YO)
Hawker Hurricane Mk.I (Mar - May 41) P2647 B
P3041 S   P3069 A   P3080 C   P3534 S   P3647 U
P3670 E   P3873 H   P3883 Q   P3963 K   V6603 V
V6605 N   V6609 X   V6670 Z   V6671 K   V6697 L
V7287 Y   V7288 W
Hawker Hurricane Mk.II (May - Sep 41) Z3018 C
Z3020 X   Z3222 Q   Z3577 J   Z3585 K   Z3649 L
Z3655 B   Z3658 N   Z3659 P   Z3969 Y   Z5048 T
Z5050 D   AE960 W   AE966 A   AE974 C   AG667 L
Supermarine Spitfire Mk.IIA (Sep - Oct 41) P8783 A
W3603 P   W3178 Q
(Only the above three aircraft were flown on operations)
Supermarine Spitfire Mk.VB (Oct 41 - Aug 42) AA973 V
AB926 E   AD228 V   AD232 D   AD234 X   AD248 K
AD253 C   AD355 J   AD418 Y   AD421 H   AD451 P
AD506 J   AR320 W   BL685 P   BL753 H   BL756 G
BL782 Z   BL829 C   BL832 Q   BL849 S   BL797 X
BM372 F   BM481 T   BM525 K
Supermarine Spitfire Mk.IX (Jul - Dec 42) BR626 T
BS107 C   BS177 H   BS307 T
Supermarine Spitfire Mk.VB (Dec 42 - Oct 43)
Supermarine Spitfire Mk.IXB (Oct 43 - Apr 45) EN569 B
MH456 Z   MJ131 T   MJ180 C   MJ300 P   MJ565 S
MJ794 G   MJ854 A   MK195 F   MK196 E   MK300 S
MK311 D   MK590 H   MK780 R   MK845 F   MK864 V
MK888 K   ML141 E   ML142 C   PL344 H
Supermarine Spitfire Mk.XIVE (May - Jun 45; not on operations)
Supermarine Spitfire Mk.XVI (Jun - Jul 45)
Operational History: First Mission 1 March 1941, 2 Hurricanes from Digby — dusk patrol.[3] **First Offensive Mission** 18 June 1941, 12 Hurricane II's from Digby — with escort of Spitfires from the West Malling Wing (No. 257 (RAF) and No. 310 (Polish) Squadron), wing patrol over the Channel. **First Victory** 8 August 1941, 2 Hurricane II's from Digby — coast patrol off Skegness, F/O E.L. Neal in Z3577 YO-J credited with a Ju.88 damaged. Hit by return fire, Neal safely crash landed in a wheat field near Horncastle. 21 October 1941, 12 Spitfire VB's from Biggin Hill — wing sweep over northern France. F/O C.A.B. Wallace in AB991 credited with a Bf.109 destroyed and a second probably destroyed south of Hardelot. **Triple Victories** 1 January 1945, 10 Spitfire IXB's were on the runway at Heesch waiting to take off when the airfield was attacked by 40-plus Bf.109's and Fw.190's; the squadron was immediately scrambled. F/O G.D. Cameron in MJ 448 YO-A credited with 3 Bf.109's destroyed over Heesch. F/L J. MacKay in NH240 YO-Z credited with 1 Bf.109 and 2 Fw.190's destroyed over the Reichswald. 14 January 1945, 10 Spitfire IXB's from Heesch — fighter sweep; attacked Fw.190's taking off from Twente and credited with 5 destroyed for the loss of 1 Spitfire. F/L J. MacKay in MJ980 YO-M credited with 3 Fw.190's destroyed over Rheine-Munster, his second triple victory. 1 March 1945, 12 Spitfire IXB's from Heesch — armed reconnaissance; attacked by 40-plus Bf.109's and Fw.190's and credited with 4 destroyed, 1 probably destroyed, for the loss of 2 Spitfires and 1 pilot. S/L W.T. Klersy in MH847 YO-H credited with 2 Bf.109's and 1 Fw.190 destroyed over Dorsten.
Last Mission 5 May 1945, 8 Spitfire IXB's from Wunstorf — two 4-plane patrols in the Hamburg area. **Summary** Sorties: 12,087. Operational/Non-operational Flying Hours: 17,211/13,747. Victories: Aircraft: 195 destroyed, 35 probably destroyed, 106 damaged. Ground: dropped 278 tons of bombs. Casualties: Operational: 61 pilots, of whom 6 were killed, 28 presumed dead, 18 POW, 9 evaded capture. Non-operational: 10 Killed. **Squadron Aces** S/L W.T. Klersy, DFC and Bar[4] 16 ½-0-3. F/L J. MacKay, DFC and Bar 11 1/5-0-5 1/2. F/L D.R. Morrison, DFC, DFM 5 5/6-4-5. F/O G.D. Cameron, DFC 5 1/5-0-2. F/O J.P.W. Francis, DFC 5-1-0. F/O R.R. Bouskill 5-0-3. F/L G.W. Johnson, DFC and Bar 5-0-2. S/L L.M. Cameron, DFC 5-0-1. F/O D.B. Dack 5-0-0. **Honours and Awards** 4 Bars to DFC, 15 DFC's, 1 DFM. **Battle Honours** Battle of Britain 1940. Defence of Britain 1940-1944. English Channel and North Sea 1942. Fortress Europe 1941-1944: *Dieppe*. France and Germany 1944-1945: *Normandy 1944, Arnhem, Rhine*.

---

[1]The five pilots, all flying Spitfire IXB's, who scored the shared victory over the Arnhem-Nijmegen area were S/L R.I.A. Smith (MK577 YO-F), F/L J.H. Everard (MK852 YO-S), F/L R.M. Davenport (ML260 YO-D), F/O J. MacKay (MJ726 YO-Z) and F/O G.D. Cameron (MJ202 YO-A).
[2]Evaded capture and returned to England.
[3]This was the squadron's first entry in its combat diary since 7 October 1940 when it was known as No. 1 (F) Squadron.
[4]The second ranking scorer in Second TAF for the period 6 June 1944 - 5 May 1945, with 13½ destroyed; the top scorer was S/L Laubman of No. 412 (F) Squadron.

195 Hawker Hurricane Mk.I V6657 aircraft "D" of No. 401 (F) Squadron at Digby, Lincs. in the spring of 1941.

196 Hawker Hurricane Mk.II Z3658 with servicing panels removed, is run-up by fitters of No. 401 (F) Squadron.

197 Supermarine Spitfire Mk.VB W3834, a presentation aircraft inscribed "Corps of Imperial Frontiersmen," with No. 401 (F) Squadron.

198 Supermarine Spitfire Mk.IXB aircraft "A" of No. 401 (F) Squadron on the Continent in 1944.

No. 402 Squadron

Badge none

Formed as No. 112 (Army Co-operation) Squadron (Auxiliary) at Winnipeg, Manitoba on 5 October 1932, the squadron arrived in England in June 1940, was redesignated No. 2 (Fighter) Squadron at Digby, Lincolnshire on 9 December 1940, and No. 402 (Fighter) Squadron on 1 March 1941. The unit flew Hurricane aircraft, and from October 1941 to March 1942 was the only RCAF unit with the fighter-bomber version of this aircraft. Converted to Spitfire aircraft in March 1942, it was employed on offensive and defensive air operations, and in the support of Allied ground forces in North-West Europe. The squadron was disbanded at Fassberg, Germany on 10 July 1945.

Brief Chronology Formed as No. 112 (AC) Sqn (Aux), Winnipeg, Man. 5 Oct 32. Redesignated No. 2 (F) Sqn, Digby, Lincs., Eng. 9 Dec 40. Renumbered No. 402 (F) Sqn 1 Mar 41. Disbanded at Fassberg, Ger. 10 Jul 45.

Titles or Nicknames "City of Winnipeg", "Winnipeg Bears"

Adoption City Council of Winnipeg, Man., and Women's Air Force Auxiliary, Winnipeg.

Commanders
S/L W.F. Hanna 9 Dec 40 - 6 Jan 41.
S/L G.R. McGregor, DFC 7 Jan 41 - 13 Apr 41.
S/L V.B. Corbett 14 Apr 41 - 13 Dec 41.
S/L R.E.E. Morrow, DFC 14 Dec 41 - 16 Aug 42.
S/L N.H. Bretz, DFC 17 Aug 42 - 26 Sep 42.
S/L D.G. Malloy, DFC 27 Sep 42 - 14 May 43.
S/L L.V. Chadburn, DFC 15 May 43 - 12 Jun 43.
S/L P.L.I. Archer, DFC 13 Jun 43 - 17 Jun 43 *KIA*.[1]
S/L G.W. Northcott, DSO, DFC and Bar 18 Jun 43 - 25 Jul 43.
S/L W.G. Dodd, DFC 26 Jul 43 - 28 Oct 44.
S/L J.B. Lawrence 29 Oct 44 - 21 Feb 45.
S/L L.A. Moore, DFC 22 Feb 45 - 25 Mar 45 *KIA*.
S/L D.C. Laubman, DFC and Bar 6 Apr 45 - 14 Apr 45 *MIA*.[2]
S/L D.C. Gordon, DFC and Bar 15 Apr 45 - 10 Jul 45.

Higher Formations and Squadron Locations
Fighter Command renamed
Air Defence Great Britain (15 Nov 43):
No. 12 Group,
Digby, Lincs. 9 Dec 40 - 22 Jun 41.
No. 11 Group,
Martlesham Heath, Suffolk 23 Jun 41 - 9 Jul 41.
No. 13 Group,
Ayr, Scot. 10 Jul 41 - 18 Aug 41.
No. 11 Group,
Southend, Essex 19 Aug 41 - 5 Nov 41.
No. 10 Group,
Warmwell, Dorset. 6 Nov 41 - 1 Mar 42.
Colerne, Wilts. 2 Mar 42 - 16 Mar 42.
Fairwood Common, S. Wales 17 Mar 42 - 13 May 42.
No. 11 Group,
Canadian Kenley Wing,
Kenley, Surrey, 14 May 42 - 30 May 42.
Redhill, Surrey 31 May 42 - 12 Aug 42.
Firing practice, Ipswich, Suffolk, 29 Jun - 6 Jul 42.
No. 17 Armament Practice Camp, Martlesham Heath, Suffolk 3-9 Aug 42.
Kenley, Surrey 13 Aug 42 - 20 Mar 43,
No. 12 Group,
Canadian Digby Wing,
Digby, Lincs. 21 Mar 43 - 6 Aug 43.
No. 11 Group,
Merston, Sussex 7 Aug 43 - 18 Sep 43

No. 12 Group
Canadian Digby Wing,
Digby, Lincs. 19 Sep 43 - 11 Feb 44.
No. 3 Armament Practice Camp, Ayr, Scot. 19 Dec 43 - 2 Jan 44.
Wellingore, Lincs. 12 Feb 44 - 29 Apr 44.
No. 15 Armament Practice Camp, Peterhead, Scot. 12-21 Apr 44.
Second Tactical Air Force:
No. 85 (Base) Group,
No. 142 (RAF) Wing,
Horne, Yorks. 30 Apr 44 - 18 Jun 44.
Westhampnett, Sussex 19 Jun 44 - 26 Jun 44.
Merston, Sussex 27 Jun 44 - 7 Aug 44.
Air Defence Great Britain:
No. 11 Group,
Hawkinge, Kent 8 Aug 44 - 29 Sep 44.
Second Tactical Air Force:
No. 83 (Composite) Group,
No. 125 (RAF) Wing,
B.(Base) 70 Antwerp, Bel. 30 Sep 44 - 1 Oct 44.
B.82 Grave, Neth. 1 Oct 44 - 31 Oct 44.
B.64 Diest, Bel. 1 Nov 44 -26 Dec 44.
No. 126 (RCAF) Wing,
B.88 Heesch, Neth 27 Dec 44 - 11 Apr 45.
No. 17 Armament Practice Camp, Warmwell, Dorset., Eng. 14 Jan - 2 Feb 45.
B.108 Rheine, Ger. 12 Apr 45 - 14 Apr 45.
B.116 Wunstorf, Ger, 15 Apr 45 - 11 May 45.
B.152 Fassberg, Ger. 12 May 45 - 1 Jul 45.
No. 127 (RCAF) Wing,
B.152 Fassberg, Ger. 2 Jul 45 - 10 Jul 45.
Representative Aircraft (Unit Code AE)
Hawker Hurricane Mk.I (Mar - May 41) P3021 X
Hawker Hurricane Mk.IIA (May - Jul 41)
Hawker Hurricane Mk.IIB (Jun - Oct 41)
Hawker Hurricane Mk.IIB(B) (Oct 41 - Mar 42)
BE417 K   BE419 P   BE424 R   BE477 S   BE478 O   BE485 W
Supermarine Spitfire Mk.VB (Mar - May 42) AD113 Y
BM134 F   BM272 D   BM276 S   BM480 J   EP115 F
Supermarine Spitfire Mk.IX (May 42 - Mar 43)
BS127 V   BS306 A   BS353 G   BS438 U   BS430 N   BS434 J
Supermarine Spitfire Mk.VC (Apr 43 - Jun 44)
Supermarine Spitfire Mk.IXB (Jul - Aug 44)
Supermarine Spitfire Mk.XIVE (Aug 44 - Jun 45)
Supermarine Spitfire Mk.XVI (Jun 45)
Operational History: First Mission 1 March 1941, 2 Hurricane I's from Digby — aerodrome patrol. **First Offensive Mission** 15 April 1941, 12 Hurricane II's from Digby — led by W/C G.R. McGregor of the Digby Wing and escorted by Spitfires from the Wittering Wing (Nos. 65 and 266 Squadrons), fighter sweep over the Boulogne Sector of France. This was the RCAF's first offensive mission over enemy-held territory. **First Victory** 26 June 1941, 2 Hurricane IIB's from Martlesham Heath — convoy patrol. Sgt. G.D. Robertson in Z2409 credited with a Ju.88 damaged. 18 September 1941, 12 Hurricane IIB's from Southend — as part of the Canadian Digby Wing, escorted 11 Blenheims bombing Rouen, France, engaged enemy aircraft and the squadron was credited with 3 destroyed, 2 damaged, without loss. Sgt. G. McClusky in Z3341 credited with a Bf.109 destroyed.[3] **Last Mission** 4 May 1945, 12 Spitfire XVIE's from Wunstorf — fighter sweep of southern Denmark, claimed 2 vehicles destroyed and 110 damaged.
Summary Sorties: 10,504. Operational/Non-operational Flying Hours: 17,643/12,027. Victories: Aircraft: 49 destroyed, 10 probably destroyed and 37 damaged, plus 5 V-1 (flying bombs) destroyed.[4] Casualties: Operational: 47 pilots, of whom 36 were killed or missing, 3 POW, 1 evaded capture, 7 wounded. Non-operational: 12 personnel killed, 1 injured. **Squadron Aces** S/L G.W. Northcott, DSO, DFC and Bar 7-7-1. P/O J.D. Mitchner 5 1/2-5/6-1. **Honours and Awards** 1 Bar to DFC, 4 DFC's. **Battle Honours** Defence of Britain 1941-1944. English Channel and North Sea 1941-1944. Fortress Europe 1941-1944: *Dieppe*. France and Germany 1944-1945: *Normandy 1944, Arnhem, Rhine.*

---

[1] Was to have taken command of No. 421 Squadron, and this was to have been his last sortie as commander of No. 402 Squadron.
[2] Evaded capture and returned to England.
[3] Killed in a training accident on 18 October 1941 while flying the squadron Magister.
[4] The V-1 (flying bomb) victories were scored in August 1944.

199   Hawker Hurricane Mk.IIB BE485, coded AE-W, crossing the Channel with two 250-pound bombs beneath the wings, on an intruder sortie into occupied France in the service of No. 402 (F) Squadron.

200   Supermarine Spitfire Mk.IX BS306, coded AE-A, of No. 402 (F) Squadron at Kenley, Surrey in May 1942.

201   Supermarine Spitfire Mk.XIVE RN119 of No. 402 (F) Squadron. This aircraft was flown by F/O C.B. McConnell on 19 April 1945, when he was credited with destroying a Ju.88 while based at Wunstorf, Germany.

No. 403 Squadron

Badge A wolf's head erased
Motto Stalk and strike
Authority King George VI, October 1943
The wolf is a fierce and powerful antagonist, indigenous to most parts of Canada.
Formed at Baginton, Warwickshire, England on 1 March 1941 as the first of 35 RCAF squadrons to be formed overseas — third Fighter squadron in service — the unit flew Spitfire aircraft on offensive and defensive air operations, and in support of Allied ground forces in North-West Europe. The squadron was disbanded at Fassberg, Germany on 10 July 1945.
Brief Chronology Formed at Baginton, War., Eng. 1 Mar 41. Disbanded at Fassberg, Ger. 10 Jul 45.
Title or Nickname "Wolf"
Adoption City of Calgary, Alta.
Commanders
S/L B.G. Morris (RAF) 6 Mar 41 - 21 Aug 42 *POW*.
S/L R.A. Lee-Knight (RAF), DFC 23 Aug 41 - 27 Sep 41 *KIA*.
S/L C.E. Gray (RAF), DFC and Bar 28 Sep 41 - 29 Sep 41.
S/L A.G. Douglas (RAF), DFC 30 Sep 41 - 11 Jan 42.
S/L C.N.S. Campbell (RAF), DFC 12 Jan 42 - 27 Apr 42 *POW*.
S/L A.C. Deere (RAF), DFC and Bar 30 Apr 42 - 12 Aug 42.
S/L L.S. Ford, DFC and Bar 13 Aug 42 - 21 Apr 43.
S/L C.M. Magwood, DFC 22 Apr 43 - 12 Jun 43.
S/L H.C. Godefroy, DFC and Bar 13 Jun 43 - 11 Aug 43.
S/L W.A.G. Conrad, DFC 12 Aug 43 - 17 Aug 43 *MIA*.[1]
S/L F.E. Grant 27 Aug 43 - 4 Sep 43 *KIA*.
S/L N.R. Fowlow, DFC 5 Sep 43 - 5 Oct 43.
S/L R.A. Buckham, DFC and Bar 6 Oct 43 - 14 Jun 44.
S/L E.P. Wood, DFC 16 Jun 44 - 15 Nov 44.
S/L J.E. Collier 26 Nov 44 - 15 Feb 45.[2]
S/L H.P.M. Zary, DFC 16 Feb 45 - 16 May 45.
S/L A.E. Fleming 17 May 45 - 10 Jul 45.
Higher Formations and Squadron Locations
Fighter Command:
No. 9, Group,
Baginton, War. 1 Mar 41 - 29 May 41.
Ternhill, Salop. 30 May 41 - 3 Aug 41.
No. 11 Group,
Hornchurch, Essex 4 Aug 41 - 14 Aug 41.
Debden, Essex 25 Aug 41 - 2 Oct 41.
Martlesham Heath, Suffolk 3 Oct 41 - 21 Dec 41.
North Weald, Essex 22 Dec 41 - 1 May 42.
Southend, Essex 2 May 41 - 2 Jun 42.
Martlesham Heath, Suffolk 3 Jun 42 - 18 Jun 42.
No. 13 Group,
Catterick, Yorks. 19 Jun 42 - 22 Jan 43.
4 aircraft, West Hartlepool, Durham 19 Jun 42 - 22 Jan 43.
No. 11 Group,
Canadian Kenley Wing,
Kenley, Surrey 23 Jan 43 - 4 Jul 43.
Second Tactical Air Force:
No. 83 (Composite) Group,
No. 17 (RCAF) Sector (disbanded 13 Jul 44).
No. 127 (RCAF) Wing,
Kenley, Surrey 5 Jul 43 - 6 Aug 43.
Lashenden, Kent 7 Aug 43 - 19 Aug 43.
Headcorn, Kent 20 Aug 43 - 13 Oct 43.
Kenley, Surrey 14 Oct 43 - 17 Apr 44.
No. 16 Armament Practice Camp, Hutton Cranswick, Yorks. 24-29 Feb 44.
Tangmere, Sussex 18 Apr 44 - 15 Jun 44.
B.(Base) 2 Bazenville, Fr. 16 Jun 44 - 15 Aug 44.
B.26 Illiers l'Eveque, Fr. 26 Aug 44 - 21 Sep 44.
B.68 Le Culot, Bel. 22 Sep 44 - 30 Sep 44.
No. 11 Armament Practice Camp, Fairwood Common, S. Wales 22 Sep - 3 Oct 44.
B.82 Grave, Neth. 1 Oct 44 - 20 Oct 44.
B.58 Melsbroek, Bel. 21 Oct 44 - 2 Nov 44.
B.56 Evère, Bel. 3 Nov 44 - 1 Mar 45.
Pilots to United Kingdom to re-equip with Spitfire XVI's 2-4 Dec 44.
No. 17 Armament Practice Camp, Warmwell, Dorset., Eng. 4-14 Jan 45.
B.90 Petit-Brogel, Bel. 2 Mar 45 - 30 Mar 45.
B.78 Eindhoven, Neth. 31 Mar 45 - 10 Apr 45.
B.100 Goch, Ger. 11 Apr 45 - 27 Apr 45.
B.152 Fassberg, Ger. 28 Apr 45 - 10 Jul 45.
Representative Aircraft (Unit Code KH)
Curtiss Tomahawk Mk.I & IIA (Mar - Jun 41) AK878 H
Supermarine Spitfire Mk.I (May - Jul 41) N3066 N
P7129 A   R6611 T   R6984 F   R7058 O   R7065 X
R7066 E   R7068 G   R7140 D   X4026 J   X4319 M
X4329 C   X4674 R   X4766 H   X4856 L
Supermarine Spitfire Mk.IIA (Jul - Sep 41) P7280 N
P7352 B   P7355 W   P7368 J   P7422 O   P7438 G
P7505 C   P7529 C   P7552 Q   P7622 M   P7743 Q
P7744 E   P7746 K   P7756 U   P7776 S   P7818 T
P7905 P   P7911 B   P7915 W   P7979 Z   P8017 Z
P8090 H   P8171 A   P8233 L
Supermarine Spitfire Mk.VB & VC (4-21 Aug 41)
P7220 F   P7235 C   P7310 B   P7342 A   P7343 D
P8740 E   P8744 P   P8792 Y   R7260 T   R7266 J   R7273 L
R7279 S   W3114 R   W3436 X   W3438 G   W3446 V
W3453 M   W3502 K   W3573 K   W3630 Z
Supermarine Spitfire Mk.VB (Sep 41 - Jan 43)
P7438 G   R6890 H   W3170 V   W3318 N   W3426 D
W3421 F   W3564 K   W3650 X   W3822 Q   W3823 S
W3938 J   AA834 B   AA927 H   AB190 Z   AB364 U
AB799 J   AB865 T   AB981 Z   AD114 W   AD191 W
AD199 H   AD206 R   AD207 O   AD208 L
Supermarine Spitfire Mk.IX (Jan 43 - Feb 44)
BS509 H   MA575 P
Supermarine Spitfire Mk.IXB (Feb - Dec 44)
MJ352 Q   MJ355 H   MJ480 R   MK859 T
Supermarine Spitfire Mk.XVI (Dec 44 - Jul 45)
Operational History: First Mission 11 May 1941, 2 Tomahawks from Baginton — base patrol at 25,000 feet. **First Offensive Mission** 5 August 1941, 11 Spitfire VB's from Hornchurch — low squadron of high cover wing for Blenheims bombing St Omer, France. **First Victory** 9 August 1941, 11 Spitfire VB's from Hornchurch — target support wing for Blenheims bombing Gosnay. P/O K.H. Anthony in K3573 KH-K credited with a Bf.109F probably destroyed. 19 August 1941, 12 Spitfire VB's from Hornchurch — close-cover target support (independent wing) for 6 Blenheims bombing Gosnay, engaged 15 Bf.109's. Squadron credited with 4 destroyed, 1 probable and 2

damaged for the loss of 2 Spitfires. (One pilot listed as missing in action; the other, P/O Anthony, rescued from the Channel.) S/L B.G. Morris (RAF), in K3438 KH-G, credited with 1 destroyed, 1 damaged. **Triple Victory** 2 July 1944, 12 Spitfire IXB's from Bazenville — front line patrol east of Caen, France; encountered 20-plus Bf.109's and credited with 9 destroyed, 3 damaged, without loss. F/L J.D. Lindsay leading a flight assigned to high cover; while climbing through cloud, engaged a second group of 15-plus Bf.109's and credited with 3 destroyed. **Last Mission** 8 May 1945, 6 Spitfire XVI's from Fassberg — escorted Dakotas to Copenhagen. Two of the Spitfires had mechanical trouble and landed at Kastrup. **Summary** Sorties: 13,004 (29 on Tomahawks). Operational/Non-operational Flying Hours: 17,728/13,253. Victories: Aircraft: 123 (plus 7 shared) destroyed, 10 probably destroyed, 72 (plus 1 shared) damaged. Ground: dropped 70 tons of bombs, credited with 17 rail cuts; destroyed or damaged 50 locomotives, 150 freight cars, and almost 130 vehicles (including 30 armoured). Casualties: Operational: 85 aircraft; 76 pilots, of whom 4 were killed, 40 presumed dead, 21 POW (1 escaped, 2 died), 11 evaded capture. Non-operational: 19 personnel killed, 1 seriously injured. **Squadron Aces** F/L H.D. MacDonald, DFC and Bar 7½-1-2. F/L J.D. Lindsay, DFC 6½-0-5. S/L L.S. Ford, DFC and Bar 6-0-2. S/L H.C. Godefroy, DFC and Bar 5-0-1 **Honours and Awards** 4 Bars to DFC, 16 DFC's, 1 Military Medal,[3] 3 MiD's. **Battle Honours** Defence of Britain 1941-1944. English Channel and North Sea 1942. Fortress Europe 1941-1944: *Dieppe*. France and Germany 1944-1945: *Normandy 1944, Rhine*.

---

[1] Returned safely to Britain on 10 October 1943.
[2] Credited with the RCAF's first solo victory over a Me.262 jet on 25 December 1944 flying a Spitfire Mk.XVI SM338.
[3] To Sgt. (later P/O) C.E. McDonald, for escaping from POW camp. Sgt. McDonald returned to France with the aid of the Polish underground, and eventually to England.

203 Curtiss Tomahawk Mk.IIA AH896 aircraft "H" and AH882 aircraft "R" of No. 403 (F) Squadron in April 1941.

204 Supermarine Spitfire Mk.IX of No. 403 (F) Squadron starting up on a snow-covered field on the Continent while a No. 349 (F) Squadron Spitfire is armed behind it.

202 Curtiss Tomahawk Mk.I in the service of No. 403 (F) Squadron in the spring of 1941.

205 Supermarine Spitfire Mk.XVI, coded KH-C, of No. 403 (F) Squadron on take-off from a wire mesh runway on the Continent in 1945.

No. 404 Squadron

Badge A buffalo's head
Motto Ready to fight
Authority King George VI, March 1943.
The buffalo is a fierce and powerful fighter.
Formed as a Coastal Fighter unit at Thorney Island, Hampshire, England on 15 April 1941 as the RCAF's second — first coastal — squadron formed overseas, the unit flew Blenheim, Beaufighter and Mosquito aircraft against enemy shipping along the Norwegian and Dutch coasts, and provided long-range fighter escort for Coastal Command's anti-submarine aircraft operating over the Bay of Biscay. On the afternoon of D-Day (6 June 1944) three German destroyers were sighted heading for the Allied invasion fleet in the English Channel and the squadron, then equipped with rocket-firing Beaufighters, was part of a strike force that attacked the destroyers with rocket and cannon for the next two days until all three were sunk. The squadron was disbanded at Banff, Scotland on 25 May 1945.
Brief Chronology Formed at Thorney Island, Hants., Eng. 15 Apr 41. Disbanded at Banff, Banffs., Scot. 25 May 45.
Title or Nickname "Buffalo"
Adoption Business group in Toronto, Ont.
Commanders
W/C H.P. Woodruff (Can/RAF) 2 May 41 - 17 Jun 42.
W/C J.A. Nixon (RAF) 18 Jun 42 - 13 Jul 42 *KIA*.
W/C E.H. McHardy (NZ/RAF), DFC and Bar 14 Jul 42 - 16 Oct 42.
W/C G.G. Truscott 17 Oct 42 - 7 Sep 43.
W/C C.A. Willis 8 Sep 43 - 30 Mar 44 *POW*.
W/C A.K. Gatwood (RAF), DSO, DFC and Bar 1 Apr 44 - 23 Aug 44.
W/C E.W. Pierce 24 Aug 44 - 25 May 45.
Higher Formations and Squadron Locations
Coastal Command:
No. 16 Group,
Thorney Island, Hants. 15 Apr 41 - 19 June 41.
No. 18 Group,
Castletown, Caith., Scot. 20 Jun 41 - 26 Jul 41.
Skitten, Caith., Scot. 27 Jul 41 - 7 Oct 41.
Dyce, Aber., Scot. 8 Oct 41 - 2 Dec 41.
"B" Flight, Sumburgh, Shetland Islands 9 Oct - 2 Dec 41.
Sumburgh, Shetland Islands 3 Dec 41 - 25 Mar 42.
Dyce, Aber., Scot. 26 Mar 42 - 5 Aug 42.
Sumburgh, Shetland Islands 6 Aug 42 - 23 Sep 42.
Dyce, Aber., Scot. 24 Sep 42 - 21 Jan 43.
No. 19 Group,
Chivenor, Devon. 22 Jan 43 - 1 Apr 43.
No. 18 Group,
Tain, Ross., Scot. 2 Apr 43 - 19 Apr 43.
Wick, Caith., Scot. 20 Apr 43 - 7 May 44.
No. 19 Group,
Davidstow Moor, Cornwall 8 May 44 - 30 Jun 44.
No. 16 Group,
No. 154 (General Reconnaissance) Wing,
Strubby, Lincs., 1 Jul 44 - 2 Sep 44.
No. 18 Group,
Banff, Scot. 3 Sep 44 - 21 Oct 44.
Dallachy, Moray., Scot. 22 Oct 44 - 2 Apr 45.
Banff, Scot. 3 Apr 45 - 25 May 45.
Representative Aircraft (Unit Code EE)[1]
Bristol Blenheim Mk.IV (Apr 41 - Jan 43) L9337 Q
L9454 G   N3542 S   N3600 K   N3611 H   P4845 D
P4847 A   T1808 N   T1869 O   T1949 D   T1950 C
V5430 V   V5433 F   V5729 R   Z5747 P   Z5753 T
Z5963 J   Z5972 X   Z6175 H   Z6181 B   Z6190 E
Z6245 L   Z6341 Z   Z6343 W
Bristol Beaufighter Mk.IIF (Sep 42 - Apr 43)
T3155 N   T3196 U   T3430 K   T3431 O   T3436 J
T3440 E   V8131 T   V8144 S   V8152 W   V8157 P
V8168 B   V8189 F   V8191 C   V8202 L   V8203 H
V8205 R   V8212 Q   V8214 G   NT916 S
Bristol Beaufighter Mk.XIC (Mar - Dec 43) JL944 L
JL947 D   JM105 B   JM106 M   JM111 T   JM112 A
JM113 P   JM114 C   JM115 O   JM117 F   JM119 E
JM121 G   JM122 V   JM123 H   JM124 S   JM130 J
JM132 U   JM136 Z   JM160 Q   JM166 K   JM173 N
JM174 R   JM179 M
Bristol Beaufighter T.F. Mk.XC (Sep 43 - Mar 45)
LX940 Y   LZ173 W   LZ174 H   LZ176 X   LZ177 B
LZ189 Q   LZ190 V   LZ196 N   LZ297 A   LZ443 T
LZ444 E   NE199 D   NE318 G   NE339 U   NE354 F
NE426 K   NE729 M   NE766 L   NE916 S   NE922 V
NV177 Z   NV183 P   NV292 O   NV423 R
de Havilland Mosquito P.R. Mk.VIC (Mar - May 45)
RF777 L   RF838 A   RF842 C   RF844 D   RF848 G
RF849 J   RF850 M   RF851 H   RF852 E   RF853 N
RF856 P   RF857 Q   RF879 S   RF880 Y   RF882 Z
RF895 R
Operational History: First Mission 22 September 1941, 4 Blenheims from Castletown — convoy escort. **First Victory** Aircraft: 18 December 1941, Blenheim IV Z5753 EE-T from Sumburgh with W/C Woodruff (pilot), FS I.R. Sims (observer) and P/O Mathews (air gunner) — scrambled and credited with 1 Ju.88 destroyed 50 miles east of Sumburgh. Shipping: 19 April 1943, Beaufighter XIC from Tain with FS K.S. Miller (pilot) and Sgt J. Young RAF, (navigator) — Norwegian coastal shipping reconnaissance, attacked and credited with damaging a 1500 to 2000 ton vessel. **Last Mission** 4 May 1945, 7 Mosquitos (5 with rocket projectiles and 2 as fighter cover) as part of the Banff Wing of 41 Mosquitos, plus 18 Mustangs and 3 air-sea rescue Warwicks — anti-shipping sweep of the Kattegat near Kiel; wing credited with 1 vessel probably destroyed, 7 damaged, without loss. **Summary** Sorties: 3,144. Operational/Non-operational Flying Hours: 12,460/10,268. Victories: Aircraft: 8 destroyed, 6 probably destroyed, 10 damaged. Shipping: dropped 32 tons of bombs and 319 18-inch torpedos, and fired 1602 25-pound rockets; credited with 4 vessels (9878 tons) sunk, 4 (4065 tons) damaged, and a share in 37 (61,586 tons) sunk and in 14 (51,296 tons) damaged. U-boats: 3 damaged (shared). Casualties: Operational: 35 aircraft; 79 aircrew, of whom 77 were killed or missing, 2 POW. Non-operational: 16 aircraft; 42 personnel killed or injured. **Honours and Awards** 2 DSO's, 3 Bars

to DFC, 45 DFC's, 3 DFM's, 1 GM, 19 MiD's. **Battle Honours** Atlantic 1941-1945. English Channel and North Sea 1941-1945. Baltic 1944-1945. *Normandy 1944.* Biscay 1943-1944.

---

[1] While at Davidstow in May-June 1944, as part of a strike wing, squadron aircraft carried the code figure "2" in place of "EE".

206 Bristol Blenheim Mk.IV Z6343 aircraft "W" of No. 404 (CF) Squadron is readied for a sortie in the wet dawn of the Shetland Islands.

207 Bristol Beaufighter T.F.Mk.X LZ451, aircraft "M" in the foreground, with the No. 404 (CF) Squadron bison emblem on the nose and bearing the Wing Commander's pennant. At left is NE425 aircraft "E".

No. 405 Squadron

Badge An eagle's head erased faced to the sinister and holding in the beak a sprig of maple
Motto Ducimus (We lead)
Authority King George VI, September 1946
The motto refers to the fact that this was the first RCAF bomber squadron overseas, and the only RCAF pathfinder squadron. The eagle's head, facing to the sinister to suggest leadership, is derived from the pathfinder badge.

Formed at Driffield, Yorkshire, England on 23 April 1941 as the RCAF's third — first Bomber — squadron formed overseas, the unit flew Wellington and Halifax aircraft on night bombing operations. On the night of 12/13 June, ten weeks after its formation, it carried out the RCAF's first bombing mission. It also flew the RCAF's first four-engined heavy bomber mission when it took part in the first 1000-plane raid against Cologne, on 30/31 May 1942.

From October 1942 to March 1943 the squadron was on loan to Coastal Command and flew anti-submarine patrols over the Bay of Biscay, protecting Allied convoys sailing for the North African invasion. On its return to Bomber Command, it had a brief stay with No. 6 (RCAF) Group before being transferred to No. 8 (Pathfinder) Group, as the only RCAF squadron to serve with that famous force.

After hostilities in Europe, the squadron was selected as part of "Tiger Force" for duty in the Pacific, converted to Canadian-built Lancaster aircraft, and returned to Canada for reorganization and training. The sudden end of hostilities in the Far East resulted in the squadron being disbanded at Greenwood, Nova Scotia on 5 September 1945.

Brief Chronology Formed at Driffield, Yorks., Eng. 23 Apr 41. Disbanded at Greenwood, N.S. 5 Sep 45.
Title or Nickname "Vancouver"
Adoption City of Vancouver, B.C. (February 1943) and Vancouver Women's Canadian Club.
Commanders
W/C P.A Gilchrist (Can/RAF), DFC 20 May 41 - 24 Jul 41 POW.[1]
W/C R.M. Fenwick-Wilson (Can/RAF), AFC 13 Aug 41 - 16 Feb 42.
W/C J.E. Fauquier, DFC 17 Feb 42 - 6 Aug 42.
W/C L.G.D. Fraser, DFC 7 Aug 42 - 19 Nov 42.
W/C A.C.P. Clayton (Can/RAF), OBE, DFC and Bar 20 Nov 42 - 19 Apr 43.
G/C J.E. Fauquier, DSO and 2 Bars, DFC 20 Apr 43 - 21 Jun 44.
G/C R.J. Lane, DSO, DFC and Bar 22 Jun 44 - 22 Aug 44.
W/C C.W. Palmer, DFC 23 Aug 44 - 26 Sep 44 KIA.
W/C H.A. Morrison, DSO, DFC 27 Sep 44 - 31 Oct 44.
W/C W.F.M. Newson, DSO, DFC and Bar 1 Nov 44 - 5 Sep 45.
Higher Formations and Squadron Locations
Bomber Command:
No. 4 Group,
Driffield, Yorks. 23 Apr 41 - 19 Jun 41.
Pocklington, Yorks. 20 Jun 41 - 6 Aug 42.
Topcliffe, Yorks. 7 Aug 42 - 24 Oct 42.

Coastal Command:
No. 18 Group,
Beaulieu, Hants. 25 Oct 42 - 28 Feb 43.
Bomber Command:
No. 6 (RCAF) Group,
Topcliffe, Yorks. 1 Mar 43 - 5 Mar 43.
Leeming, Yorks. 6 Mar 43 - 18 Apr 43.
No. 8 (Pathfinder) Group,
Gransden Lodge, Beds. 19 Apr 43 - 25 May 45.
No. 6 (RCAF) Group,
No. 62 (RCAF) Base,
Linton-on-Ouse, Yorks. 26 May 45 - 15 Jun 45.
En route to Canada 16 Jun 45 - 20 Jun 45.
RAF "Tiger Force" (for ops),
RCAF Eastern Air Command (for training):
No. 6 (RCAF) Group,
No. 664 (RCAF) Wing,
Greenwood, N.S. 21 Jun 45 - 5 Sep 45.
Representative Aircraft (Unit Code LQ)
Vickers Wellington Mk.II (May 41 - Apr 42)
W5368 K   W5421 T   W5476 H   W5495 W   W5515 R/Y
W5564 Q   Z8358 B   Z8412 P   Z8414 U
Z8419 V   Z8421 T   Z8428 N
Handley Page Halifax B.Mk.II (Apr 42 - Sep 43)
W1094 B   W1095 N   W1096 O   W1097 P   W1110 C
W1111 D   W1112 E   W1113 G   W7703 Q   W7704 M
W7707 K   W7708 H   W7709 J   W7710 R   W7714 K
DT576 U
Avro Lancaster B. Mk.I & III (Aug 43 - May 45)
ME304 J   ME370 K   ME379 B   ND709 G   ND980 W
ND982 E   PA970 Y   PA972 D   PA981 K   PA982 U
PB183 E   PB229 H   PB267 X   PB288 D   PB413 K
PB451 G   PB848 X   PB555 O   PB585 P   PB614 V
PB627 J   PB628 W   PB681 M   SW225 A
Avro Lancaster B.Mk.X (May - Jun 45, not on operations)[2]
FM110 R   FM115 Z   FM120 J   FM122 L   KB945 T
KB949 U   KB950 O   KB952 X   KB995 V   KB956 N
KB957 W   KB959 Y   KB961 A   KB964 B   KB965 D
KB967 H   KB968 P   KB973 F   KB967 K   KB977 E
KB991 G   KB997 C   KB999 M[3]
Operational History: First Mission, Bombing 12/13 Jun 1941, 4 Wellingtons from Driffield despatched to bomb marshalling yards at Schwerte, near Dortmund, Germany; 3 bombed the primary target (one 1000- and four 500-pound bombs and two 750-pound canisters of incendiaries), 1 aborted. All aircraft returned safely. **First Mission, Pathfinder** 26 April 1943, 11 Halifaxes from Gransden Lodge despatched to mark Duisburg, Germany; 10 marked the primary target (one 2000-pound high concussion and seven 100-pound general purpose markers); 1 failed to return. **Last Mission, Bombing** 25 April 1945, 13 Lancaster I's and III's from Gransden Lodge; 9 bombed Hitler's refuge at Berchtesgaden and 4 bombed gun positions on the Island of Wangerooge. All returned safely. **Last Mission, Pathfinder** 7 May 1945, 8 Lancaster I's and III's from Gransden Lodge marked the aiming point for a night drop of supplies to the people of Rotterdam. **Summary** Sorties: 4427 (including 349 with Coastal Command, 41 on the airlift of 947 POW's). Operational/non-operational Flying Hours: 24,843/12,089. Bombs dropped: 12,856 tons. Victories: Aircraft: 6 destroyed, 4 damaged. U-boat: U-263 damaged. Casualties: Operational: 167 aircraft; 937 aircrew, of whom 309 were killed, 28 presumed dead, 415 missing, 133 POW (6 died, 3 escaped), 12 wounded, 14 interned in Sweden, 26 evaded capture. Non-operational: 8 aircraft; 43 personnel killed, 2 injured. **Honours and Awards** 9 DSO's, 24 Bars to DFC, 161 DFC's, 38 DFM's, 2 CGM's, 2 BEM's, 11 MiD's. **Battle Honours** Fortress Europe 1941-1944. France and Germany 1944-1945: *Biscay Ports 1941-1945, Ruhr 1941-1945, Berlin 1941; 1943-1944, German Ports 1941-1945, Normandy 1944, Walcheren, Rhine.* Biscay 1942-1943.

---

[1] Later escaped and returned to England.
[2] The squadron was the first unit to receive a Canadian-built Lancaster B.Mk.X, KB700 christened in Canada "Ruhr Express," Coded LQ-Q, it served with the squadron from 30 October 1943 to 20 December 1943 and flew two sorties before being transferred to No. 419 Squadron.
[3] The last production line aircraft, christened in Canada "Malton Mike."

208   *Vickers Wellington Mk.II W5553 receiving its complement of bombs and fuel in readiness for the night's activities with No. 405 (B) Squadron.*

209   *Handley Page Halifax Mk.II W7710, coded LQ-R, slides into loose formation over the English countryside with No. 405 (B) Squadron.*

210 Avro Lancaster B.Mk.III ND616, aircraft "A" of No. 405 (B) Squadron.

211 Avro Lancaster Mk.X KB700, coded LQ-Q, was the first Lancaster manufactured in Canada and served with No. 405 (B) Squadron as the "Ruhr Express."

No. 406 Squadron

Badge A lynx salient affronte
Motto We kill by night
Authority King George VI, December 1942
The lynx, an animal possessing keen sight at night, typifies the unit's operational role as a night fighter squadron.
Formed at Acklington, Northumberland, England on 10 May 1941 as the RCAF's fifth — first Night Fighter — squadron formed overseas, the unit flew Blenheim, Beaufighter and Mosquito aircraft in the night air defence of Britain. On 27 November 1944 it was redesignated Intruder and converted to offensive operations over Europe. Listed as the top-scoring RAF/RCAF intruder unit at the war's end,[1] the squadron was disbanded at Predannack, Cornwall on 1 September 1945.
Brief Chronology Formed as No. 406 (NF) Sqn, Acklington, Northumb., Eng. 10 May 41. Redesignated No. 406 (I) Sqn, Manston, Kent 27 Nov 44. Disbanded at Predannack, Cornwall 1 Sep 45.
Title or Nickname "Lynx"
Adoption Women's Air Force Auxiliary, Saskatoon, Sask. (August 1944)
Commanders
W/C D.G. Norris (RAF), DFC 28 May 41 - 6 Aug 42.
W/C R.A. Wills (RAF) 7 Aug 42 - 31 Jan 43.
W/C I.R. Stephenson (RAF) 1 Feb 43 - 24 Aug 43.
W/C R.C. Fumerton, DFC and Bar 25 Aug 43 - 26 Jul 44.
W/C D.J. Williams, DSO, DFC 27 Jul 44 - 3 Nov 44.
W/C R. Bannock, DFC and Bar 23 Nov 44 - 14 May 45.
W/C R.G. Gray, DFC 15 May 45 - 1 Sep 45.
Higher Formations and Squadron Locations
Fighter Command, renamed
Air Defence Great Britain (15 Nov 43 - 15 Oct 44):
No. 13 Group,
Acklington, Northumb. 10 May 41 - 31 Jan 42.
Ayr, Scot. 1 Feb 42 - 15 Jun 42.
4 aircraft, Scorton, Yorks. 2 Feb - 15 Jun 42.
Scorton, Yorks. 16 Jun 42 - 3 Sep 42.
No. 10 Group,
Predannack, Cornwall 4 Sep 42 - 7 Dec 42.
Middle Wallop, Hants., 8 Dec 42 - 30 Mar 43.
No. 9 Group,
Valley, Anglesey, Wales 31 Mar 43 - 14 Nov 43.
No. 10 Group,
Exeter, Devon. 15 Nov 43 - 13 Apr 44.
Winkleigh, Devon. 14 Apr 44 - 16 Sep 44.
Colerne, Wilts. 17 Sep 44 - 26 Nov 44.
No. 11 Group,
Manston, Kent 27 Nov 44 - 12 Jun 45.
No. 10 Group,
Predannack, Cornwall 14 Jun 45 - 1 Sep 45.
Representative Aircraft (Unit Code HU)
Bristol Blenheim Mk.I & IV (May - Jun 41)
Bristol Beaufighter Mk.IIF (Jun 41 - Aug 42)
Bristol Beaufighter Mk.VIF (Jun 41 - Aug 44) KW103 T
de Havilland Mosquito Mk.XII (Apr - Jul 44)
de Havilland Mosquito F.B.Mk.XXX (Jul 44 - Sep 45)
MM699 P MM727 B MM734 K MM739 D MM741 T
MM744 G MM745 A MM751 S NT283 V NT312 M
NT325 N NT423 O NT433 Y NT477 H NT453 E

NT478 R   NT495 C   NT498 P   NT539 G   NT544 Z
Also had the following F.B.Mk.VI
RS525 F   RS531 S

Operational History: First Mission, Night Fighter 17 June 1941, 2 Blenheims from Acklington — uneventful scramble. **First Victory** 1/2 September 1941, Beaufighter II R2336 from Acklington with F/O R.C. Fumerton and Sgt. L.P.S. Bing (observer) — night readiness, scrambled, and credited with a Ju.88 destroyed over Bedlington. This was the RCAF's first night fighter victory. (Although the squadron itself was not operational at this time, several of the crews were declared operational and were standing night readiness.) **First Mission, Intruder** 5 December 1944, 5 Mosquito XXX's from Manston — 4 to Leeuwarden and Steenwijk, 1 to Biblis — unable to complete mission owing to weather. **Last Mission** 9 May 1945, 6 Mosquito XXX's from Manston covered the liberation of the Channel Islands. **Summary** Sorties: 1835. Operational/Non-operational Flying Hours: 4552/19,274. Victories: Aircraft: 64 destroyed, 7 probably destroyed, 47 damaged.[2] Ground: destroyed 88 locomotives, at least 8 freight cars 32 vehicles and 7 small vessels. Casualties: Operational: 11 aircraft; 20 aircrew killed or missing, 2 POW. Non-operational: 17 personnel killed. **Squadron Aces** F/O C.J. Kirkpatrick (navigator), DFC 7-0-1. F/L W.A. Boak (navigator) 6-1-1. W/C D.J. Williams (pilot) DSO, DFC 5-0-0. **Honours and Awards** 3 DSO's, 1 second bar to DFC, 1 bar to DFC, 14 DFC's, 2 DFM's, 4 MiD's. **Battle Honours** Defence of Britain 1941-1945. English Channel and North Sea 1944. Fortress Europe 1943-1944. France and Germany 1944-1945: *Biscay Ports 1944, Normandy 1944, Rhine.* Biscay 1944.

---

[1]For the period 27 Nov 44 to the end of hostilities in Europe. The overall RCAF top-scoring unit was No. 418 Squadron
[2]Breakdown of victories: Night Fighter: 31-5-11; Intruder: Air 23-1-13.

212 *Bristol Beaufighter Mk.II aircraft "M" and "K" of No. 406 (NF) Squadron being serviced on a rainy afternoon in June 1941.*

213 *De Havilland Mosquito F.B.Mk.XXX of No. 406 (I) Squadron at Manston, Kent in February 1945.*

No. 407 Squadron

Badge A winged trident piercing the shank of an anchor
Motto To hold on high
Authority King George VI, March 1943
The badge represents the blows struck against enemy shipping by the Demon squadron.
Formed at Thorney Island, Hampshire, England on 8 May 1941 as the RCAF's fourth — second coastal — squadron formed overseas, its wartime history falls into two distinct parts. As a Coastal Strike unit it flew Blenheim and Hudson aircraft for seventeen months on attacks against enemy shipping between Heligoland and the Bay of Biscay. Redesignated General Reconnaissance on 29 January 1943, it then flew Wellington aircraft equipped with Leigh lights on anti-submarine duty. The squadron was disbanded at Chivenor, Devonshire on 4 June 1945.
Brief Chronology Formed as No. 407 (CS) Sqn, Thorney Island, Hants., Eng. 8 May 41. Redesignated No. 407 (GR) Sqn, Docking, Norfolk 29 Jan 43. Disbanded at Chivenor, Devon. 4 Jun 45.
Title or Nickname "Demon"
Adoption No. 9 Bombing and Gunnery School, Mont-Joli, Que.
Commanders
W/C H.M. Styles (RAF), DSO 15 May 41 - 6 Jan 42.
W/C A.C. Brown (Can/RAF), DSO, DFC 7 Jan 42 - 28 Sep 42.
W/C C.F. King (RAF) 29 Sep 42 - 3 Nov 42 *KIA*.
W/C J.C. Archer (RAF) 7 Nov 42 - 27 Sep 43 *KIA*.
W/C R.A. Ashman 2 Nov 43 - 31 Oct 44.
W/C K.C. Wilson 1 Nov 44 - 4 Jun 45.
Higher Formations and Squadron Locations
Coastal Command:
No. 16 Group,
Thorney Island, Hants. 8 May 41 - 8 Jul 41.
North Coates, Lincs. 9 Jul 41 - 17 Feb 42.
Thorney Island, Hants. 18 Feb 42 - 30 Mar 42.
Bircham Newton, Norfolk 31 Mar 42 - 30 Sep 42.
No. 19 Group,
St. Eval, Cornwall 1 Oct 42 - 9 Nov 42.
No. 16 Group,
Docking, Norfolk 10 Nov 42 - 15 Feb 43.
No. 18 Group,
Skitten, Caith., Scot. 16 Feb 43 - 31 Mar 43.
No. 19 Group,
Chivenor, Devon. 1 Apr 43 - 2 Nov 43.
St. Eval, Cornwall 3 Nov 43 - 1 Dec 43.
Chivenor, Devon. 2 Dec 43 - 26 Jan 44.
No. 15 Group,
Limavady, Derry, N. Ire. 29 Jan 44 - 26 Apr 44.
No. 19 Group,
Chivenor, Devon 28 Apr 44 - 22 Aug 44.
No. 18 Group,
Wick, Caith., Scot. 24 Aug 44 - 9 Nov 44.
No. 19 Group,
Chivenor, Devon 11 Nov 44 - 4 Jun 45.

Representative Aircraft (Unit Code 1941-43 RR, 1943-45 C1)
Bristol Blenheim Mk.IV (May - Jul 41, not on operations)
Lockheed Hudson Mk.III & V (Jun 41 - Apr 43)[1]
V9095 C	V9102 X	V9107 W	AM525 A	AM551 C
AM586 T	AM597 F	AM598 P	AM602 M	AM614 N
AM619 Q	AM626 K	AM627 J	AM649 R	AM650 D
AM679 B	AM701 V	AM731 H	AM732 S	AM811 H
AM812 S	AM838 L	AM841 E	AM906 O	

Vickers Wellington Mk.XI (Feb - Apr 43) MP534 E
Vickers Wellington Mk.XII(L/L) (Mar 43 - Feb 44)
MP503 O	MP541 G	MP542 C	MP587 A	MP593 P
MP596 B	MP618 S	MP622 V	MP632 E	MP634 F
MP652 S	MP688 D	MP754 H		

Vickers Wellington Mk.XIV(L/L) (Jun 43 - Jun 45)
HF124 Q	HF131 P	HF132 A	HF142 H	HF144 G
HF149 C	HF171 J	HF182 T	HF187 S	HF207 O
HF228 M	HF186 L	HF306 E	HF412 P	NB811 S
NB821 B	NB830 Z	NB838 V	NB839 R	NB856 X
NC512 J	NC513 G	NC775 F	NC848 P	

Operational History: First Mission 7 September 1941, 3 Hudson V's from North Coates — unsuccessful search for a Whitley bomber. **First Offensive Mission** 7/8 September 1941, Hudson V AM556 RR-E from North Coates with W/C Styles and crew attacked a 1500-ton motor vessel six miles north of Borkum; no observed results. **Last Mission: Coastal Strike** 29 January 1943, 2 Hudson V's from Docking — convoy strike; unable to bomb because of mist. **First Mission, General Reconnaissance** 7 March 1943, Wellington XI MP534 C1-E from Skitten with F/O M.P. Jordan and crew — anti-submarine patrol. **Victories, U-Boat** 6/7 September 1943, Wellington XII(L/L) HF115 C1-W from Chivenor with P/O E.M. O'Donnell and crew sank U-669 in the Bay of Biscay (4536N 1013W). 10/11 February 1944, Wellington XII(L/L) MP578 C1-D from Limavady with F/O P.W. Heron and crew convoy cover, sank U-283 (6045N 1250W). 3/4 May 1944, Wellington XIV(L/L) HF134 C1-M from Chivenor with F/O L.J. Bateman and crew sank U-846 in the Bay of Biscay (4604N 0920W). 29/30 December 1944, Wellington XIV(L/L) NB855 C1-L from Chivenor with S/L C.I.W. Taylor and crew sank U-772 in the central Channel (5005N 0231W). **Last Mission** 1 June 1945, Wellington XIV(L/L) HF302 C1-J from Chivenor with F/L L.W. Manuel and crew — convoy escort. **Summary** Sorties: 2900 (including 913 anti-shipping, 1987 anti-submarine). Operational/Non-operational Flying Hours: 11,926/3759. Victories: Aircraft: 3 destroyed, 1 damaged. Shipping: dropped 173 tons of bombs, credited with 10 ships (24,020 tons) sunk, 4 ships (8970 tons) damaged, plus 2 shared sinkings (10,396 tons). U-boat: 4 sunk, 3 damaged; dropped 331 250-pound depth charges. Casualties: Operational: 42 aircraft; 197 aircrew, of whom 24 were killed, 151 presumed dead, 8 POW, 6 evaded capture, 8 wounded. Non-operational: 64 personnel, of whom 38 killed, 20 presumed dead, 6 injured. **Honours and Awards** 3 DSO's, 1 bar to DFC, 18 DFC's, 1 AFC, 6 DFM's, 35 MiD's. **Battle Honours** Atlantic 1943-1945. English Channel and North Sea 1941-1945. Fortress Europe 1942: *German Ports 1942, Normandy 1944. Biscay 1942-1945.*

---

[1] Hudson Mk.III's were withdrawn from service in May 1942.

214 Lockheed Hudson Mk.III N3106 aircraft "D" of No. 407 (GR) Squadron in early Coastal Command finish.

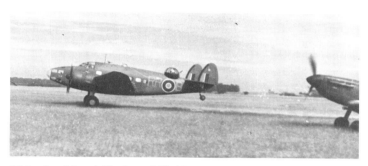

215 Lockheed Hudson Mk.V AM679, coded RR-B, taking off from a grass strip to patrol the eastern approaches with No. 407 (GR) Squadron.

216 Vickers Wellington Mk.XII HF113/G aircraft "P" of No. 407 (GR) Squadron. The "G" following the serial number indicates that the aircraft contained special secret equipment requiring that a guard be present whenever the aircraft was on the ground.

217 Vickers Wellington Mk.XIV NB858, in white Coastal Command finish, with No. 407 (GR) Squadron.

No. 408 Squadron

Badge A Canada goose volant
Motto For freedom
Authority King George VI, October 1942
The Canada goose is at home in Canada, England and Scotland; its speed and power of flight are indicative of the operational functions of the Squadron.
Formed at Lindholme, Yorkshire, England on 24 June 1941 as the RCAF's eighth — second Bomber — squadron formed overseas, the unit flew Hampden, Halifax and Lancaster aircraft on strategic and tactical bombing operations. After hostilities in Europe, it was selected as part of "Tiger Force" for duty in the Pacific, converted to Canadian-built Lancaster aircraft, and returned to Canada for reorganization and training. The sudden end of hostilities in the Far East resulted in the squadron being disbanded at Greenwood, Nova Scotia on 5 September 1945.
Brief Chronology Formed at Lindholme, Yorks., Eng. 24 Jun 41. Disbanded at Greenwood, N.S. 5 Sep 45.
Title or Nickname "Goose"
Adoption Kingsville, Ont.
Commanders
W/C N.W. Timmerman (Can/RAF), DSO, DFC 24 Jun 41 - 25 Mar 42.
W/C A.C.P. Clayton (Can/RAF), DFC and Bar 26 Mar 42 - 14 Apr 42.
W/C J.D. Twigg 18 May 42 - 28 Aug 42 *KIA*.
W/C W.D.S. Ferris, DFC 1 Sep 42 - 27 Oct 43.
W/C A.C. Mair, DFC 28 Oct 43 - 26 Nov 43 *KIA*.
W/C D.S. Jacobs, DFC 27 Nov 43 - 22 May 44 *KIA*.
W/C A.R. McLernon, DFC 24 May 44 - 13 Oct 44.
W/C J.F. Easton, DFC 14 Oct 44 - 25 Nov 44.
W/C F.R. Sharp, DFC 26 Nov 44 - 5 Sep 45.
Higher Formations and Squadron Locations
Bomber Command:
No. 5 Group,
Lindholme, Yorks. 24 Jun 41 - 19 Jul 41.
Syerston, Notts. 20 Jul 41 - 8 Dec 41.
Balderton, Notts. 9 Dec 41 - 12 Sep 42.
North Luffenham, Notts. (Balderton runways under repair) 25 Jan - 17 Mar 42.
No. 4 Group,
Leeming, Yorks. 14 Sep 42 - 31 Dec 42.
No. 6 (RCAF) Group,
Leeming, Yorks. 1 Jan 43 - 26 Aug 43.
No. 62 (RCAF) Base,
Linton-on-Ouse, Yorks. 27 Aug 43 - 13 Jun 45.
En route to Canada 14 Jun 45 - 17 Jun 45.
RAF "Tiger Force" (for ops),
RCAF Eastern Air Command (for training):
No. 6 (RCAF) Group,
No. 664 (RCAF) Wing,
Greenwood, N.S. 18 Jun 45 - 5 Sep 45.
Representative Aircraft (Unit Code EQ)
Handley Page Hampden Mk.I (Jul 41 - Sep 42)[1]
P1165 B   P5334 Q   P5392 W   AD754 Y   AD829 E
AD980 Y   AE133 X   AE148 B   AE197 N   AE219 R
AE244 P   AE245 C   AE432 S   AE433 D   AT113 A
AT139 A   AT180 B   AT189 G   AT191 A   AT220 G
AT224 A   AT227 L   AT228 T   AT437 U
Handley Page Halifax Mk.V (Sep - Dec 42, not on operations)
Handley Page Halifax Mk.II (Dec 42 - Oct 43)
BB311 L   BB343 X   DT676 B   DT679 J   DT749 C
HR658 V   HR662 H   JB363 A   JB893 U   JB898 Q
JB909 G   JB913 F   JB925 R   JB969 D   JB971 W
JD107 Y   JD164 K   JD216 P   JD276 T   JD278 O
JD323 S   JD332 E   JD333 Q   JD374 M
Avro Lancaster B.Mk.II (Oct 43 - Sep 44) LL621 Y
LL633 L   LL634 F   LL636 G   LL637 P   LL642 B
LL643 Q   LL675 M   LL699 C   LL700 J   LL717 W   LL718 E   LL720 R   LL722 N   LL723 H   LL725 Z   DS639 A
DS657 L   DS675 K   DS692 S   DS838 I   DS841 X
DS845 T   DS849 X
Handley Page Halifax B.Mk.III & A.Mk.VII (Sep 44 - May 45) NP711 O   NP712 R   NP713 X   NP714 V
NP718 Z   NP719 W   NP745 H   NP746 E   NP747 N
NP749 Y   NP750 F   NP751 L   NP754 P   NP756 T
NP757 B   NP769 J   NP772 Q   NP773 M   NP777 S
NP778 U   NP780 C   NP804 K   NP809 G   NP810 H
NR126 Z   NR199 F   NR209 A   MZ907 D
Avro Lancaster B.Mk.X (May - Sep 45, not on operations)
FM130 M   KB877 S   KB904 Q   KB905 V   KB907 U
KB913 X   KB919 J   KB925 E   KB929 O   KB939 W
KB947 Z   KB948 V   KB951 A   KB960 F   KB963 H
KB972 C   KB979 L   KB993 U   KB994 K   KB995 B
KB996 P   KB998 G
Operational History: First Mission 11/12 August 1941, 4 Hampdens from Syerston despatched to bomb docks at Rotterdam; 2 (AE197 EQ-N and AE267) bombed the primary target, 2 (AE196 and AE244 EQ-P) were unable to locate the target. All aircraft returned safely. **Last Mission** 25 April 1945, 17 Halifaxes from Linton-on-Ouse despatched to bomb gun positions on the Island of Wangerooge; 16 bombed the primary target, 1 failed to return. **Summary** Sorties: 4610 (1217 on Hampdens, 690 on Halifaxes, 2703 on Lancaster II's). Operational/Non-operational Flying Hours: 25,568/11,054. Victories: Aircraft: 11 destroyed, 6 damaged. Bombs dropped: 11,430 tons. Casualties: Operational: 146 aircraft; 897 aircrew, of whom 877 were killed, missing or POW, 3 evaded capture, 17 wounded. Non-operational: 12 aircraft; 32 personnel killed, 7 injured. **Honours and Awards** 1 MBE, 6 bars to DFC, 161 DFC's, 32 DFM's, 10 MiD's. **Battle Honours** English Channel and North Sea 1941-1943. Baltic 1941-1943. Fortress Europe 1941-1944. France and Germany 1944-1945. *Biscay Ports 1941-1944. Ruhr 1941-1945, Berlin 1943-1944, German Ports 1941-1945, Normandy 1944, Rhine.* Biscay 1942-1943.

---

[1]During May-June 1942 the squadron had one Manchester aircraft, L7401 EQ-N, for evaluation as a possible replacement for the Hampden. A Manchester Conversion Flight was set up under S/L L.B.B. Brice, who flew this aircraft on its two sorties with the squadron: the 1,000-plane raid on Cologne, 30/31 May, and a bombing of Essen on 1/2 June. The Manchester had to turn back from Cologne owing to hydraulic failure; in the raid on Essen it dropped 126 four-pound incendiaries.

The last Bomber Command sorties by Hampden aircraft were flown by No. 408 Squadron on the night of 14/15 September 1942 against Wilhelmshaven.

218 Handley Page Hampden Mk.I AE196 is made ready for operations with No. 408 (B) Squadron.

219 Handley Page Halifax A.Mk.VII NP790 is bombed-up in its rustic dispersal while serving with No. 408 (B) Squadron at Linton-on-Ouse, Yorks.

220 Avro Lancaster Mk.II DS858 aircraft "D" of No. 408 (B) Squadron is serviced prior to night operations. The inscription under the cockpit reads "My hope is constant in thee".

No. 409 Squadron

Badge In front of a cloak a crossbow
Motto Media nox meridies noster (Midnight is our noon)
Authority King George VI, March 1944
The crossbow in front of a dark cloak indicates the unit's functions as a night fighter squadron.
Formed at Digby, Lincolnshire, England on 7 June 1941 as the RCAF's seventh — second Night Fighter — squadron formed overseas, the unit flew Beaufighter and Mosquito aircraft in the night air defence of Britain and Allied forces in North-West Europe. It was the first night fighter squadron to cross over to Normandy following the Allied landings and operate from the Continent, and the first to be stationed in Belgium and Germany. For the period June 1944 to May 1945, it was the top-scoring RAF/RCAF night fighter unit, with 58½ aircraft and 12 V-1 flying bombs destroyed. The squadron was disbanded at Twente, in The Netherlands, on 1 July 1945.
Brief Chronology Formed at Digby, Lincs., Eng. 17 Jun 41. Disbanded at Twente, Neth. 1 Jul 45.
Title or Nickname "Nighthawk"
Adoption City of Victoria, B.C.
Commanders
W/C N.B. Petersen 17 Jun 41 - 2 Sep 41 *KIFA*.
W/C P.Y. Davoud, DFC 5 Sep 41 - 3 Feb 43 *OTE*.
W/C J.W. Reid 4 Feb 43 - 31 Jul 44 *OTE*.
W/C M.W. Beveridge, DFC 1 Aug 44 - 20 Sep 44 *KIFA*.
W/C J.D. Somerville, DSO, DFC 10 Oct 44 - 11 Mar 45 *OTE*.
W/C R.F. Hatton 12 Mar 45 - 1 Jul 45.
Higher Formations and Squadron Locations
Fighter Command renamed
Air Defence Great Britain (15 Nov 43):
No. 12 Group,
Digby, Lincs. 17 Jun 41 - 24 Jul 41.
Coleby Grange, Lincs. 25 Jul 41 - 27 Feb 43.
Acklington, Northumb. 28 Feb 43 - 29 Feb 44.
Second Tactical Air Force:
No. 85 (Base) Group,
No. 147 (RAF) Wing,
Acklington, Northumb. 1 Mar 44 - 13 May 44.
No. 148 (RAF) Wing,
West Malling, Kent 14 May 44 - 18 Jun 44.
Hunsdon, Herts. 19 Jun 44 - 23 Aug 44.
B.(Base) 17 Carpiquet, Fr. 24 Aug 44 - 10 Sep 44.
B.24 St André, Fr. 11 Sep 44 - 26 Sep 44.
B.48 Glisy, Fr. 27 Sep 44 - 3 Oct 44.
B.68 Le Culot, Bel. 4 Oct 44 - 11 Oct 44.
B.51 Vendeville, Fr. 12 Oct 44 - 18 Apr 45.
B.108 Rheine, Ger. 19 Apr 45 - 11 May 45.
B.77 Gilze-Rijen, Neth. 12 May 45 - 2 Jun 45.
B.106 Twente, Neth. 3 Jun 45 - 1 Jul 45.
Representative Aircraft (Unit Code KP)
Boulton Paul Defiant Mk.I (Jul - Sep 41)
Bristol Beaufighter Mk.IIF (Aug 41 - Jun 42)
Bristol Beaufighter Mk.VIF (Jun 42 - Apr 44)
V8717 P   X8104 T   X8111 V   X8231 M   X8918 N
BT301 Z   MM843 C   MM845 Y   MM853 T   MM854 S

MM865 D   MM866 W   MM883 R   MM914 B
MM915 F   MM919 U   MM946 A   ND268 H
de Havilland Mosquito N.F. Mk.XIII (Mar 44 - Jun 45)
HK366 U   HK368 F   HK381 C   HK421 V   HK425 D
HK430 P   HK473 K   HK506 H   HK512 A   MM437 W
MM454 F   MM456 M   MM458 N   MM459 Z
MM466 G   MM502 A   MM508 K   MM509 B
MM513 J   MM517 S   MM522 S   MM567 E   MM588 T
MM590 H

Operational History: First Mission 3 August 1941, Defiant from Coleby Grange — aerodrome patrol. **First Victory** 1 November 1941, Beaufighter II from Coleby Grange with W/C Davoud and Sgt T. Carpenter (RAF, navigator) — night readiness, scrambled and credited with a Do.217 destroyed 70 miles east of Digby. **Triple Victory** 23/24 April 1945, Mosquito XIII HK429 KP-D from Rheine with F/O E. Hermanson and F/L D. Hamm (navigator) — patrol; credited with 2 Ju.87's and 1 Fw.190 destroyed. **Last Mission** 2 May 1945, 6 Mosquito XIII's from Rheine — front line patrol; 2 returned early with radar problems, 2 called back early owing to lack of enemy activity, and 2 completed uneventful patrols. **Summary** Sorties: 2313. Operational/Non-operational Flying Hours: 7514/24,424. Victories: Aircraft: 65½ destroyed, 7 probably destroyed, 24 damaged. V-1: 12 destroyed. Casualties: Operational: 17 aircraft; 52 aircrew (including 6 RAF) killed or missing. **Squadron Ace** F/L R.I.E. Britten, DFC 5-0-1. **Honours and Awards** 2 MBE, 1 DSO, 1 bar to DFC, 13 DFC's, 2 AFC's, 2 BEM's, 7 MiD's, 1 DFC (USA). **Battle Honours** Defence of Britain 1941-1944. Fortress Europe 1942-1944. France and Germany 1944-1945: *Normandy 1944, Rhine.*

221 De Havilland Mosquito N.F.Mk.XIII MM466 of No. 409 (NF) Squadron is serviced on a frosty winter morning in France, 1945.

No. 410 Squadron

Badge In front of a decrescent, a cougar's face
Motto Noctivaga (Wandering by night)
Authority King George VI, May 1945
The cougar is a Canadian animal noted for its speed and power in striking down its prey. The waning moon indicates the squadron's night operations.
Formed at Ayr, Scotland on 30 June 1941 as the RCAF's ninth — third and last Night Fighter — squadron formed overseas, the unit flew Beaufighter and Mosquito[1] aircraft in the night air defence of Britain and of Allied forces in North-West Europe. The squadron was disbanded at Gilze-Rijen, in the Netherlands, on 9 June 1945.
Brief Chronology Formed at Ayr, Scot. 30 Jun 41. Disbanded at Gilze-Rijen, Neth. 9 Jun 45.
Title or Nickname "Cougar"
Adoption City of Saint John, N.B.
Commanders
S/L P.Y. Davoud 30 Jun 41 - 4 Sep 41.
W/C M. Lipton 5 Sep 41 - 30 Jul 42.
W/C F.W. Hillock 19 Aug 42 - 19 May 43 *OTE.*
W/C G.H. Elms 20 May 43 - 18 Feb 44 *OTE.*
W/C G.A. Hiltz 19 Feb 44 - 1 Apr 45 *OTE.*
W/C E.P. Heybroek 2 Apr 45 - 9 Jun 45.
Higher Formations and Squadron Locations
Fighter Command renamed
Air Defence Great Britain (15 Nov 43):
No.13 Group,
Ayr, Scot. 30 Jun 41 - 5 Aug 41.
Drem, E. Loth., Scot. 6 Aug 41 - 14 Jun 42.
"A" Flight at Acklington, Northumb. 5 Aug - 1 Sep 41.
Ouston, Durham 2 Sep 41 - 1 Apr 42.
4 aircraft, Dyce, Aber., Scot. 27 Dec 41 - 8 Jan 42.
Ayr, Scot. 15 Jun 42 - 31 Aug 42.
5 aircraft, Drem, E. Loth., Scot. 15 Jun - 31 Aug 42.
Scorton, Yorks. 1 Sep 42 - 19 Oct 42.
Acklington, Northumb. 20 Oct 42 - 20 Feb 43.
No. 12 Group,
Coleby Grange, Lincs. 21 Feb 43 - 19 Oct 43.
4 aircraft, Predannack, Cornwall 12 June - 7 Jul 43.
No. 11 Group,
West Malling, Kent 20 Oct 43 - 7 Nov 43.
Hunsdon, Herts. 8 Nov 43 - 29 Dec 43.
Castle Camps, Cambs. 30 Dec 43 - 29 Feb 44.
Second Tactical Air Force:
No. 85 (Base) Group,
No. 149 (RAF) Wing,
Castle Camps, Cambs. 1 Mar 44 - 28 Apr 44.
Hunsdon, Herts. 29 Apr 44 - 17 Jun 44.
Zeals, Wilts. 18 Jun 44 - 27 Jul 44.
Colerne, Wilts. 28 Jul 44 - 8 Sep 44.
Hunsdon, Herts. 9 Sep 44 - 21 Sep 44.
B.(Base) 48 Glisy, Fr. 22 Sep 44 - 2 Nov 44.
No. 148 (RAF) Wing (from 4 Nov 44),
B.51 Vendeville, Fr. 3 Nov 44 - 6 Jan 45.
B.48 Glisy, Fr. 7 Jan 45 - 5 Apr 45.
B.77 Gilze-Rijen, Neth. 6 Apr 45 - 9 Jun 45.

Representative Aircraft (Unit Code RA)
Boulton Paul Defiant Mk.IF (Jul 41 - May 42) V1123 R
Bristol Beaufighter Mk.IIF (Apr 42 - Jan 43)
de Havilland Mosquito N.F.Mk.II (Nov 42 - Dec 43)
HJ919 B   DZ757 Q
de Havilland Mosquito F.B.Mk.VI (Jul - Sep 43)
de Havilland Mosquito N.F.Mk.XIII (Dec 43 - Aug 44)
HK366 Q   HK430 W   HK432 F   HK455 Q   HK456 H
HK458 S   HK459 A   HK462 E   HK463 R   HK465 P
HK466 J   HK467 T   HK470 X   HK476 O   HK500 I
HK521 L   HK523 G   MM456 D   MM457 Z   MM462 K
MM477 U   MM499 C   MM501 V   MM570 B
de Havilland Mosquito N.F.Mk.XXX (Aug 44 - Jan 45)
Operational History: First Mission 23 July 1941, Defiant V1183 from Ayr with P/O Lucas — night readiness, uneventful scramble. **First Victories** 6/7 September 1942, Beaufighter II T3428 from Drem with P/O R.R. Ferguson and P/O D. Creed (navigator) — while on a ground control intercept exercise, was vectored to, and credited with, a Ju.88 damaged over Whitby. 22 January 1943, Mosquito II DZ929 from Acklington with FS B.M. Haight and Sgt. T. Kipling (RAF, observer) — night readiness, scrambled along with two other Mosquitos, and credited with a Do.217 destroyed near Hartlepool. **Last Mission** 3 May 1945, 3 Mosquito XXX's from Gilze-Rijen — patrols, 1 in the Hamburg area and 2 over the Scheldt. One reported two radar contacts but no visual contact; the other two patrols were uneventful. **Summary** Sorties: 2972. Operational/Non-operational Flying Hours: 7157/20,993. Victories: Aircraft: 75¾ destroyed, 2 probably destroyed, 9 damaged. Casualties: Operational: 17 aircraft; 32 aircrew, of whom 10 were killed, 20 presumed dead, 2 POW. Non-operational: 14 aircraft; 30 personnel, of whom 29 killed, 1 injured. **Squadron Aces** F/L R.D. Schultz, DFC and Bar 8-0-0. F/O D.G. Tonque (RAF, navigator), DFC and Bar 8-0-0. Lt. A.A. Harrington (USAAF), DSO, DFC 7-0-0. F/L C.E. Edinger, DFC 6-0-1. F/O J.S. Christie (RAF, navigator), DFC 6-0-1. F/O C.L. Vaessen (navigator), DFC 6-0-1. F/L G.P.A. Bodard (navigator), DFC 6-0-0. S/L J.D. Somerville, DSO, DFC 5-0-1. F/O G.D. Robinson (navigator), DFC 5-0-1. F/L V.A. Williams (navigator), DFC 5-0-0. **Honours and Awards** 1 DSO, 1 MBE, 2 bars to DFC, 19 DFC's, 1 BEM, 17 MiD's. **Battle Honours** Defence of Britain 1941-1944. Fortress Europe 1943. France and Germany 1944 - 1945: *Normandy 1944, Rhine.* Biscay 1943.

---

[1]Fighter Command's first Mosquito daylight penetration into Germany was flown on 27 March 1944 by P/O M.A. Cybulski and his navigator P/O H.H. Ladbrook (RAF). Flying an intruder mission along the canal and rail line between Meppen and Papenburg, they damaged a tug and two barges, a locomotive and six freight cars, and two army buses.

222 Boulton Paul Defiant Mk.I F V1123 of No. 410 (NF) Squadron while the unit was based at Drem, East Lothian, Scotland.

223 De Havilland Mosquito N.F.Mk.II DZ757, coded RA-Q, of No. 410 (NF) Squadron, severely damaged after flying through the wreckage of an enemy night fighter on 25 September 1943.

No. 411 Squadron

Badge A bear rampant
Motto *Inimicus inimico* (Hostile to an enemy)
Authority King George VI, October 1942
The bear is a fierce fighter and is found in Canada
Formed at Digby, Lincolnshire, England on 16 June 1941 as the RCAF's sixth — fourth Fighter — squadron formed overseas, the unit flew Spitfire aircraft on offensive and defensive air operations, and in support of ground forces in North-West Europe. The squadron's leading scorer was F/L R.J. Audet, who set an RAF/RCAF record on 29 December 1944 by destroying five enemy aircraft in two minutes of actual combat. The unit also had the distinction of scoring the RCAF's last victory of the war when F/L's D.F. Campbell and T.L. O'Brien shared in the destruction of a He.111 northwest of Flensburg, Germany on 4 May 1945. After hostilities in Europe, the squadron served as one of four RCAF day fighter units in the British Air Forces of Occupation (Germany) until disbanded at Utersen, Germany on 21 March 1946.
Brief Chronology Formed at Digby, Lincs., Eng. 16 Jun 41. Disbanded at Utersen, Ger. 21 Mar 46.
Title or Nickname "Grizzly Bear"
Adoption Parkdale Lions Club, Toronto, Ont.
Commanders
S/L P.B. Pitcher 16 Jun 41 - 16 Dec 41.
S/L P.S. Turner (Can/RAF), DFC and Bar 17 Dec 41 - 11 Feb 42.
S/L R.B. Newton (RAF), DFC 12 Feb 42 - 27 Sep 42.
S/L N.H. Bretz, DFC 28 Sep 42 - 21 Mar 43.
S/L D.G.E. Ball 22 Mar 43 - 12 Apr 43 *KIA.*
S/L B.D. Russel, DFC 16 Apr 43 - 7 Jul 43.
S/L G.C. Semple 8 Jul 43 - 25 Sep 43 *OTE.*
S/L I.C. Ormston, DFC 26 Sep 43 - 20 Dec 43 *inj.*[1]
S/L J.D. McFarlane 21 Dec 43 - 9 Apr 44.
S/L N.R. Fowlow, DFC 10 Apr 44 - 19 May 44 *KIA.*
S/L G.D. Robertson 20 May 44 - 4 Aug 44 *OTE.*
S/L R.K. Hayward, DFC 5 Aug 44 - 9 Oct 44.
S/L E.G. Lapp, DFC 10 Oct 44 - 10 Dec 44.
S/L J.N. Newell 19 Dec 44 - 25 Jun 45.
S/L B.E. Innes, DFC 19 Sep 45 - 21 Mar 45.
Higher Formations and Squadron Locations
Fighter Command:
No. 12 Group,
Canadian Digby Wing,
Digby, Lincs. 16 Jun 41 - 18 Nov 41.
No. 11 Group,
Hornchurch, Essex 19 Nov 41 - 5 Mar 42.
Southend, Essex 7 Mar 42 - 29 Mar 42.
No. 12 Group,
Canadian Digby Wing,
Digby, Lincs. 30 Mar 42 - 21 Mar 41.
No. 11 Group,
Canadian Kenley Wing,
Kenley, Surrey 22 Mar 43 - 7 Apr 43.
Redhill, Surrey 8 Apr 43 - 4 Jul 43.
Second Tactical Air Force:
No. 83 (Composite) Group,
No. 17 (RCAF) Sector (disbanded 13 Jul 44),
No. 126 (RCAF) Wing,
Redhill, Surrey 5 Jul 43 - 6 Aug 43.
Staplehurst, Kent 7 Aug 43 - 12 Oct 43.
Biggin Hill, Kent 13 Oct 43 - 15 Apr 44.
No. 15 Armament Practice Camp, Peterhead, Scot. 24-29 Feb 44.
Tangmere, Sussex 16 Apr 44 - 18 Jun 44.
No. 11 Armament Practice Camp, Fairwood Common, Wales 17-22 Apr 44.
B.(Base) 4 Beny-sur-Mer, Fr. 19 Jun 44 - 7 Aug 44.
B.18 Cristot, Fr. 8 Aug 44 - 31 Aug 44.
B.24 St André, Fr. 1 Sep 44.
B.26 Illiers l'Evêque, Fr. 2 Sep 44.
B.44 Poix, Fr. 3 Sep 44 - 6 Sep 44.
B.56 Evère, Bel. 7 Sep 44 - 20 Sep 44.
B.68 Le Culot, Bel. 21 Sep 44 - 3 Oct 44.
B.84 Rips, Neth. 4 Oct 44 - 19 Oct 44.
No. 17 Armament Practice Camp, Warmwell, Dorset., Eng. 15-23 Oct 44.
B.80 Volkel, Neth. 20 Oct 44 - 5 Dec 44.
B.88 Heesch, Neth. 6 Dec 44 - 11 Apr 45.
B.108 Rheine, Ger. 12 Apr 45 - 14 Apr 45.
B.116 Wunstorf, Ger. 15 Apr 45 - 13 May 45.
B.152 Fassberg, Ger. 14 May 45 - 4 Jul 45.
No. 17 Armament Practice Camp, Warmwell, Dorset, Eng. 21 May - 7 Jun 45
B.174 Utersen, Ger. 5 Jul 45 - 15 Jul 45.
British Air Forces of Occupation (Germany)
No. 83 (Composite) Group
No.126 (RCAF) Wing
B.174 Utersen, Ger. 16 Jul 45 - 21 Mar 46.
Representative Aircraft (Unit Code DB)
Supermarine Spitfire Mk.IA (Jun - Jul 41, not on operations)
Supermarine Spitfire Mk.IIA (Jul - Oct 41) P7595 G
P7697 F   P7694 W   P7703 Z   P7880 Q   P7914 D
P7915 B   P7923 R   P7926 N   P7966 L/P
P7985 P   P8076 M/T   P8136 S   P8172 A
P8263 C   P9657 V   P8663 X
Supermarine Spitfire Mk.VB (Oct 41 - Oct 43)
P8715 O   R6897 Z   W3215 C   W3853 C   X4257 J
AA833 L   AA836 S   AA839 E   AA840 D   AB181 A
AB268 B   AB284 L   AB365 R   AD117 F   AD195 M
AD259 N   AD261 V   AD264 A   AD292 R   AD299 Q
AD356 R   AD509 W   AD557 G   AD847 T
Supermarine Spitfire Mk.IXB (Sep 43 - Sep 44)
Supermarine Spitfire Mk.IXE (Sep 44 - May 45)
MK434 R   EN574 S   JL373 L
Supermarine Spitfire L.F.Mk.IX (Apr - May 45)
Supermarine Spitfire Mk.XVI (May 45, not on operations)
Supermarine Spitfire Mk.XIV (Jun 45 - Mar 46, not on operations) RM873 R   RM928 N   RM864 P
Operational History: First Mission 21 August 1941, 2 Spitfire IIA's from Digby — uneventful scramble.[2] **First Offensive Mission** 20 September 1941, 12 Spitfire IIA's from Digby — part of the Canadian Digby Wing with Nos. 401 and 266 (RAF) Squadrons, rear cover to a bombing force. (Seven previous wing assignments had been cancelled because of weather conditions). **First Victory** 27 September 1941, 12 Spitfire IIA's from Digby — top cover to a Canadian Digby Wing fighter sweep by Nos. 412 and 266 (RAF)

Squadrons. While patrolling from Merks-de-Mer to Montreuil, squadron attacked a mixed gaggle of Fw.190's and Bf.109's. One Spitfire lost; pilot bailed out and was rescued. P/O R.W. McNair in P8263 DB-Y credited with one Bf.109 damaged. 13 October 1941, 10 Spitfire IIA's from Digby — part of a Canadian Digby Wing fighter sweep with Nos. 412 and 266 (RAF) Squadrons over France. P/O R.W. McNair in P7679 DB-F credited with a Bf.109 destroyed and a second probably destroyed 5 to 10 miles off Boulogne. McNair himself was shot down but bailed out over the Channel and was rescued, and returned to Digby. **Triple Victories** 29 December 1944, 11 Spitfire IXE's from Heesch — fighter sweep in the Rheine area. F/L R.J. Audet in RR201 DB-A credited with 3 Fw.190's and 2 Bf.109's destroyed.[3] **Last Mission** 4 May 1945, 10 Spitfire LF/IXB's from Wunstorf — armed reconnaissance in the Elmshorn-Flensburg-Sylt-Friedrichskoog area. **Summary** Sorties: 10,747. Operational/Non-operational Flying Hours: 15,502/16,335. Victories: Aircraft: 84 destroyed, 3 probably destroyed, 44 damaged. Ground: dropped 291 tons of bombs, destroyed/damaged 367/353 motor vehicles, 26/65 locomotives. Casualties: Operational: 48 aircraft; 48 pilots, of whom 4 were killed, 4 wounded, 19 presumed dead, 4 POW; 17 missing, proved safe. Non-operational: 4 aircraft; 5 personnel killed, 2 wounded. **Squadron Aces** F/L R.J. Audet, DFC and Bar 10½-0-1. F/L H.C. Trainor, DFC and Bar 6½-1-0. F/L J.J. Boyle, DFC 5½-0-1. S/L E.G. Lapp, DFC 5-0-1. F/O M.G. Graham, DFC 5-0-1. **Honours and Awards** 1 DSO, 2 MBE's, 2 bars to DFC, 19 DFC's, 1 MiD. **Battle Honours** Defence of Britain 1941-1944. English Channel and North Sea 1942-1943. Fortress Europe 1941-1944: *Dieppe*. France and Germany 1944-1945: *Normandy 1944, Arnhem, Rhine*.

---

[1] Crashed taking off on an operational flight.
[2] One of the pilots was P/O D.J.M. Blakeslee (P8657 DB-X), an American in the RCAF who later commanded No. 133 (Eagle) Squadron. Transferred with this squadron to the USAAF, he rose to the rank of lieutenant-colonel and commanded the famed 4th Fighter Group.
[3] F/L Audet was awarded an immediate DFC for this action. In January 1945, he scored an additional 4½ Fw.190's and 2 Me.262 (jets) destroyed and 1 Me.262 damaged, and was awarded a bar to his DFC. On 3 March 1945 he was killed while on a strafing mission.

224   Supermarine Spitfire Mk.IIA P7923 aircraft "R" of No. 411 (F) Squadron based at Digby, Lincs. in the summer of 1941.

225   Supermarine Spitfire Mk.IXB aircraft "R" of No. 411 (F) Squadron receives hastily applied invasion markings on 5 June 1944 at Tangmere, Sussex.

226   Supermarine Spitfire L.F.Mk.IX PV347 aircraft "E" of No. 411 (F) Squadron in dispersal on the Continent in the spring of 1945.

227 *Supermarine Spitfire Mk.IXE of No. 411 (F) Squadron in the foreground, with a No. 442 (F) Squadron Spitfire in the background.*

228 *Supermarine Spitfire Mk.IXE RR201 aircraft "A" of No. 411 (F) Squadron, illustrated here by artist Peter Mossman on 29 December 1944 when F/L R.J. Audet was credited with five confirmed victories in a single day.*

No. 412 Squadron

Badge A falcon volant
Motto Promptus ad vindictam (Swift to avenge)
Authority King George VI, September 1944
The falcon is indigenous to all parts of Canada. Known for its skill and aggressiveness in dealing with its enemies, it has been used for hunting from an early date in history.
Formed at Digby, Lincolnshire, England on 30 June 1941 as the RCAF's tenth — seventh Fighter — squadron formed overseas, the unit flew Spitfire aircraft on offensive and defensive air operations; and in support of ground forces in North-West Europe. After hostilities in Europe, the squadron served as one of four RCAF day fighter units assigned to the British Air Forces of Occupation (Germany) until disbanded at Utersen, Germany on 21 March 1946.
Brief Chronology Formed at Digby, Lincs., Eng. 30 Jun 41. Disbanded at Utersen, Ger. 21 Mar 46.
Title or Nickname "Falcon"
Adoption Parkdale Lions Club, Toronto, Ont. (13 May 1943)
Commanders
S/L C.W. Trevena 30 Jun 41 - 11 Nov 41.
S/L J.D. Morrison 12 Nov 41 - 24 Mar 42.
S/L R.C. Weston 1 Apr 42 - 27 Jul 42.
S/L J.C. Fee, DFC 28 Jul 42 - 26 Nov 42.
S/L F.W. Kelly, DFC 27 Nov 42 - 24 Jun 43.
S/L G.C. Keefer, DFC and Bar 25 Jun 43 - 11 Apr 44.
S/L J.E. Sheppard, DFC 12 Apr 44 - 1 Aug 44 MIA.[1]
S/L D.H. Dover, DFC and Bar 2 Aug 44 - 28 Jan 45 OTE.
S/L M.D. Boyd 29 Jan 45 - 29 May 45 2 OTE.
S/L D.J. Dewan 30 May 45 - 21 Mar 46.
Higher Formations and Squadron Locations
Fighter Command:
No. 12 Group,
Canadian Digby Wing,
Digby, Lincs. 30 Jun 41 - 19 Oct 41.
Wellingore, Lincs. 20 Oct 41 - 30 Apr 42.
No. 11 Group,
Martlesham Heath, Suffolk 1 May 42 - 3 Jun 42.
North Weald, Sussex 4 Jun 42 - 18 Jun 42.
Merston, Sussex 19 Jun 42 - 23 Aug 42.
Tangmere, Sussex 24 Aug 42 - 22 Sep 42.
Canadian Kenley Wing,
Redhill, Surrey 23 Sep 42 - 1 Nov 42.
Kenley, Surrey 2 Nov 42 - 28 Jan 43.
No. 10 Group,
Angle, S. Wales 29 Jan 43 - 7 Feb 43.
Fairwood Common, S. Wales 8 Feb 43 - 12 Apr 43.
Perranporth, Cornwall 13 Apr 43 - 20 Jun 43.
Friston, Sussex 21 Jun 43 - 13 Jul 43.
Second Tactical Air Force:
No. 83 (Composite) Group,
No. 17 (RCAF) Sector (disbanded 13 Jul 44),
No. 126 (RCAF) Wing,
Redhill, Surrey 14 Jul 43 - 7 Aug 43.
Staplehurst, Kent 8 Aug 43 - 13 Oct 43.
Biggin Hill, Kent 14 Oct 43 - 14 Apr 44.

Tangmere, Sussex 15 Apr 44 - 18 Jun 44.
B.(Base) 4 Beny- sur-Mer, Fr. 19 Jun 44 - 8 Aug 44.
B.18 Cristot, Fr. 9 Aug 44 - 28 Aug 44.
B.28 Evreux, Fr. 29 Aug 44 - 1 Sep 44.
B.26 Illiers l'Evêque, Fr. 2 Sep 44.
B.44 Poix, Fr. 3 Sep 44 - 5 Sep 44.
B.44 Evère, Bel. 6 Sep 44 - 20 Sep 44.
B.68 Le Culot, Bel. 21 Sep 44 - 3 Oct 44.
B.84 Rips, Neth. 4 Oct 44 - 14 Oct 44.
B.80 Volkel, Neth. 15 Oct 44 - 5 Dec 44.
B.88 Heesch, Neth. 6 Dec 44 - 12 Apr 45.
B.108 Rheine, Ger. 13 Apr 45 - 15 Apr 45.
B.116 Wunstorf, Ger. 16 Apr 45 - 12 May 45.
B.152 Fassberg, Ger. 13 May 45 - 5 Jul 45.
British Air Forces of Occupation (Germany):
No. 83 (Composite) Group,
No. 126 (RCAF) Wing,
B.174 Utersen, Ger, 6 Jul 45 - 21 Mar 46.
Bombing and gunnery training, Sylt, Isle of Sylt, Ger. 8-21 Dec 45.
Representative Aircraft (Unit Code VZ)
Supermarine Spitfire Mk.IIA (Jul - Oct 41) P8086 B
P8145 K   P8250 F   P8369 G   P8391 L
Supermarine Spitfire Mk.VB (Oct 41 - Nov 43)
Supermarine Spitfire Mk.IXB (Nov 43 - Sep 44)
Supermarine Spitfire Mk.IXE (Sep 44 - May 45)
Supermarine Spitfire Mk.XVI (May - Jun 45, not on operations)
Supermarine Spitfire Mk.XIVE (Jun 45 - May 46)
Operational History: First Mission 31 August 1941, 2 Spitfire IIA's from Digby — uneventful scramble. **First Offensive Mission** 21 September 1941, 12 Spitfire IIA's from Digby — Canadian Digby Wing fighter sweep with Nos. 411 and 266 (RAF) Squadrons over the north coast of France. Formation came under anti-aircraft fire and sighted enemy aircraft; no action, no casualties. **First Victory** 13 October 1941, 12 Spitfire IIA's from Digby — Canadian Digby Wing patrol with No. 266 (RAF) Squadron from Boulogne to south of Hardelot, engaged an enemy force. Sgt. E.N. Macdonell in P7856 VZ-E credited with 1 Bf.109 destroyed ten miles off Boulogne. **Triple Victory** 24 July 1944, 4 Spitfire IXB's from Beny-sur-Mer — weather recce, engaged a mixed force of 40-plus aircraft east of Lisieux and credited with 7 destroyed, no losses. F/L W.J. Banks credited with 2 Fw.190's and 1 Bf.109 destroyed. **Last Mission** 4 May 1945, 8 Spitfire IX's from Wunstorf — armed recce of Elmshorn-Flensburg-Friedrichskoog area, claimed 1 locomotive and 5 trains damaged, 1 automobile destroyed.
Summary Sorties: 12,761. Operational/Non- operational Flying Hours: 16,955/14,359. Victories: Aircraft: 106 destroyed, 11 probably destroyed, 46 damaged. Ground: dropped 203 tons of bombs, credited with 86 rail cuts and destroyed/damaged 282/721 motor vehicles, 42/135 railway cars, 22/84 locomotives, 1/10 tanks. Casualties: Operational: 63 aircraft; 62 pilots, of whom 21 were killed, 14 presumed dead, 11 POW (1 died), 1 missing, 14 safe, 1 evaded capture. Non-operational: 9 personnel, of whom 6 were killed, 2 injured, 1 died of injuries. **Squadron Aces** F/L D.C. Laubman, DFC and Bar 14-0-2.[2] F/L W.J. Banks, DFC and Bar 7-3-2. F/O D.R.C. Jamieson, DFC and Bar 6-0-0. F/O P.M. Charron 5-0-1. F/L R.I.A. Smith, DFC 5-0-0. **Honours and Awards** 7 bars to DFC, 16 DFC's, 4 MiD's. **Battle Honours** Defence of Britain 1941-1944.

English Channel and North Sea 1942-1943. Fortress Europe 1941-1944: *Dieppe*. France and Germany 1944-1945. *Normandy 1944, Arnhem, Rhine.*

---

[1] Evaded capture and returned to England.
[2] On 26 and 27 September 1944, in four sorties, S/L Laubman was credited with 4 Fw.190's and 3 Bf.109's destroyed and 1 Bf.109 damaged. All victories were scored flying Spitfire IXB MJ393 VZ-Z.

229 Supermarine Spitfire Mk.II of No. 412 (F) Squadron, coded VZ-L, at Digby, Lincs. in the summer of 1941.

230 Supermarine Spitfire Mk.VB, in the markings of No. 412 (F) Squadron, makes a low pass over the field.

231 Supermarine Spitfire Mk.IX MJ452 aircraft "J" of No. 412 (F) Squadron armed and ready for a sortie from its base on the Continent.

232 Supermarine Spitfire Mk.IXE aircraft "F" of No. 412 (F) Squadron running up prior to leading a flight on patrol from Rheine, Germany.

No. 413 Squadron

Badge In front of a maple leaf an elephant's head affronte
Motto Ad vigilamus undis (We watch the waves)
Authority King George VI, October 1943

The elephant head represents the squadron's operations from Ceylon, while the motto suggests its functions.
Formed at Stranraer, Scotland on 1 July 1941 as the RCAF's eleventh — third coastal and first flying boat General Reconnaissance — squadron formed overseas, the unit flew Catalina aircraft on north Atlantic reconnaissance and anti-submarine patrols. In March 1942, it was hurriedly moved to the Far East. As aircraft and crews arrived at Koggala, Ceylon, they were pressed into service making reconnaissance flights over the Indian Ocean to watch for the approach of a Japanese naval force. On 4 April, S/L L.J. Birchall, whose aircraft and crew had arrived in Ceylon only two days earlier, sighted and reported a large Japanese fleet 350 miles south of Ceylon. The Catalina (Mk.I AJ155 "A") was shot down by Japanese carrier aircraft, and S/L Birchall and the survivors of his crew were taken prisoner, but their warning had alerted Ceylon's defences and the Japanese attack was repulsed.[1] The squadron remained in Ceylon until early 1945,[2] was then returned to the United Kingdom for conversion to a Bomber unit, but was instead disbanded at Bournemouth, Hampshire on 23 February 1945.

Brief Chronology Formed at Stranraer, Scot. 1 Jul 41. Disbanded at Bournemouth, Hants., Eng. 23 Feb 45.
Title or Nickname "Tusker"
Commanders
W/C V.H.A. McBratney (RAF) 1 Jul 41 - 18 Aug 41.
W/C R.G. Briese 19 Aug 41 - 22 Oct 41 *KIA*.
W/C J.D. Twigg 15 Nov 41 - 19 Mar 42.
W/C J.L. Plant 20 Mar 42 - 21 Oct 42.
W/C J.C. Scott, DSO 22 Oct 42 - 20 Jun 43.
W/C L.H. Randall, DFC 21 Jun 43 - 14 Sep 44.
W/C S.R. McMillan 15 Sep 44 - 18 Feb 45.
Higher Formations and Squadron Locations
Coastal Command:
No. 15 Group,
Stranraer, Scot. 1 Jul 41 - 30 Sep 41.
No. 18 Group,
Sullom Voe, Shetland 1 Oct 41 - 3 Mar 42.
En route to Ceylon 4 Mar 42 - 29 May 42.
(First 4 aircraft arrived 2-4 Apr 42)
South East Asia Command:
No. 222 Group,
Koggala, Ceylon 2 Apr 42 - 19 Jan 45.
Detachments at Addu Atoll, Seychelles Is., Indian Ocean; Mombasa, Kenya; Bahrein, Persian Gulf; Langebaan, South Africa; Aden.
En route to England (minus aircraft) 21 Jan 45 - 20 Feb 45.
Bomber Command:
No. 6 (RCAF) Group (on paper only),
Bournemouth, Hants. 21 Feb 45 - 23 Feb 45.
(Officially transferred to Bomber Command 18 Feb 45, while still at sea, though "virtually non-existent" as a unit.)

Representative Aircraft (Unit code, UK only, QL)
Consolidated Catalina Mk.I, IB & IV (Jul 41 - Dec 44)
W8412 P   W8421 D   W8427 G   W8434 F   Z2135 H
Z2149 V   AH549 F   AH550 L   AH567 S   AJ161 W
FP182 G   FP282 Y   FP306 Z   FP323 A   JX276 Z
JX280 A   JX292 B   JX299 V   JX311 C   JX321 D
JX330 H   JX333 F   JX336 C   JX357 V
On 1 Aug 42, the following aircraft were relettered: W8412 P to B, W8421 D[3], AH550 L to Z, AH567 S to C
Operational History: First Mission 6 October 1941, Catalina AH561 from Sullom Voe with F/L Thomas and crew — convoy escort; the convoy did not sail, and the aircraft was recalled. **Last Mission** 3/4 December 1944, 4 Catalinas from Koggala — hunt for a submarine reported at position 1237N 8840E; nothing was seen. **Summary** Sorties: 871. Operational/Non-operational Flying Hours: 11,794/2783. Victories: U-boat: 1 possibly damaged; dropped 52 250-pound depth charges. Casualties: Operational: 3 aircraft; 27 aircrew were killed, missing or POW. Non-operational: 1 aircraft; 10 personnel, of whom 5 were killed, 5 died of other causes. **Honours and Awards** 1 DSO, 4 DFC's, 2 AFC's, 1 MiD. **Battle Honours** Atlantic 1941-1943. Ceylon 1942. Eastern Waters 1942-1944.

---

[1] S/L Birchall was awarded the DFC for his part in saving Ceylon from invasion, and later the OBE for his conduct while a prisoner of war. He also came to be known as the "Saviour of Ceylon."
[2] It was also involved in reopening air service between Ceylon and Australia in May 1943, using a specially modified and stripped-down BOAC Catalina, FP244 G-AGFM. The first of three flights was made on 3 May by W/C Scott, who took 34 flying hours to reach Perth, and returned to Ceylon on 12 May. The second flight, under F/L R.W. Fursman, departed from Ceylon on 31 May and returned on 9 June; the third, under F/L P.A.S. Rumbold, departed 16 June and returned 26 June.
[3] This aircraft had previously belonged to No. 2 (GR) Squadron RAF and carried that unit's code letters "AX".

233 Consolidated Catalina Mk.I W8406, coded AX-D, of No. 413 (GR) Squadron based in Ceylon.

234 Consolidated Catalina Mk.IB W8434, coded QL-F, of No. 413 (GR) Squadron based in the Shetland Islands.

No. 414 Squadron

Badge none

Formed as an Army Co-operation unit at Croydon, Surrey, England on 13 August 1941 as the RCAF's 12th — second army co-operation — squadron formed overseas, and redesignated Fighter Reconnaissance on 28 June 1943, the unit flew Mustang aircraft on air intelligence work. It obtained photographic reconnaissance for Allied invasion planners, and before-and-after photographs of air attacks against German "Noball" (V-1 flying bomb) launching sites. On 6 June 1944, its task was to spot for the naval bombardment of coastal defences from Le Havre to Cherbourg, France; then began a daily schedule of tactical reconnaissance flights over the battle area, reporting on enemy road and rail traffic. In August the unit, re-equipped with Spitfire aircraft, moved to the Continent to provide tactical photographic reconnaissance for ground forces. The squadron was disbanded at Luneburg, Germany on 7 August 1945.

Brief Chronology Formed as No. 414 (AC) Sqn, Croydon, Surrey, Eng. 13 Aug 41. Redesignated No. 414 (FR) Sqn, Dunsfold, Surrey 28 Jun 43. Disbanded at Luneburg, Ger. 7 Aug 45.

Title or Nickname "Sarnia Imperials"

Adoption City of Sarnia, Ont.

Commanders
W/C D.M. Smith 15 Aug 41 - 17 Jul 42.
W/C R.F. Begg 18 Jul 42 - 27 Jun 43.
S/L J.M. Godfrey 28 Jun 43 - 23 Jul 43.
S/L H.P. Peters, DFC 24 Jul 43 - 4 Nov 43 *KIA*.
S/L C.H. Stover, DFC 5 Nov 43 - 30 Jun 44.
S/L R.T. Hutchinson, DFC 1 Jul 44 - 1 Oct 44 *OTE*.
S/L G. Wonnacott, DFC and Bar 2 Oct 44 - 9 Mar 45 *OTE*.
S/L F.S. Gilbertson, DFC 10 Mar 45 - 9 Apr 45 *OTE*.
S/L J.B. Prendergast, DFC 18 Apr 45 - 7 Aug 45.

Higher Formations and Squadron Locations
Army Co-operation Command:
No. 35 (RAF) Wing,
Croydon, Surrey 13 Aug 41 - 4 Dec 42.
No. 39 (RCAF) Wing,
Dunsfold, Surrey 5 Dec 42 - 31 Jan 43.
Middle Wallop, Hants. 1 Feb 43 - 19 Feb 43.
Dunsfold, Surrey 20 Feb 43 - 8 Apr 43.
Middle Wallop, Hants. 9 Apr 43 - 25 May 43.
Fighter Command:
No. 11 Group,
Harrowbeer, Devon. 26 May 43 - 3 Jun 43.
Portreath, Cornwall 4 Jun 43 - 19 Jun 43.
Dunsfold, Surrey 20 Jun 43 - 4 Jul 43.
Second Tactical Air Force:
No. 83 (Composite) Group,
No. 39 (RCAF) Sector (disbanded 1 Jul 44),
No. 129 (RCAF) Wing,
Gatwick, Surrey 5 Jul 43 - 30 Jul 43.
Weston Zoyland, Som. 31 Jul 43 - 9 Aug 43.
Gatwick, Surrey 10 Aug 43 - 12 Aug 43.
Ashford, Kent 13 Aug 43 - 4 Oct 43.
No. 128 (RCAF) Wing,
Woodchurch, Kent 5 Oct 43 - 14 Oct 43.
Redhill, Surrey 15 Oct 43 - 2 Nov 43.
No. 129 (RCAF) Wing,
Gatwick, Surrey 3 Nov 43 - 4 Feb 44.
No. 15 Armament Practice Camp,
Peterhead, Scot. 5 Feb 44 - 19 Feb 44.
Odiham, Hants. 20 Feb 44 - 28 Feb 44.
Dundonald, Ayr., Scot. 29 Feb 44 - 10 Mar 44.
Gatwick, Surrey 11 Mar 44 - 31 Mar 44.
No. 128 (RCAF) Wing,
Odiham, Hants. 1 Apr 44 - 1 Jul 44.
No. 39 (RCAF) Wing,
Odiham, Hants. 2 Jul 44 - 14 Aug 44.
B.(Base) 21 Ste Honorine-de-Ducy, Fr. 15 Aug 44 - 28 Aug 44.
B.26 Illiers l'Evêque, Fr. 29 Aug 44 - 2 Sep 44.
B.44 Poix, Fr. 3 Sep 44 - 6 Sep 44.
B.56 Evère, Bel. 7 Sep 44 - 19 Sep 44.
B.66 Blakenburg, Bel. 20 Sep 44 - 2 Oct 44.
B.78 Eindhoven, Neth. 3 Oct 44 - 6 May 45.
B.90 Petit- Brogel, Bel. 7 Mar 45 - 10 Apr 45.
B.108 Rheine, Ger. 11 Apr 45 - 16 Apr 45.
B.116 Wunstorf, Ger. 17 Apr 45 - 27 Apr 45.
B.154 Soltau, Ger. 28 Apr 45 - 6 May 45.
B.118 Celle, Ger. 7 May 45.
B.156 Luneburg, Ger. 8 May 45 - 7 Aug 45.
No. 17 Armament Practice Camp, Warmwell, Dorset, Eng. 23 Jun - 6 Jul 45.

Representative Aircraft (Unit Code RU)
Westland Lysander Mk.III (Aug 41 - Jun 42, not on operations)
Curtiss Tomahawk Mk.I & II (Aug 41 - Sep 42, not on operations)
North American Mustang Mk.I (Jun 42 - Aug 44)
AG375 F AG376 R AG416 S AG444 V/Y AG459 C
AG543 E AG601 A AG612 B AG655 X AL980 C
AL984 H AM160 T/Y AM248 J AM251 O AP197 Z
AP204 X AP211 V
Supermarine Spitfire L.F.Mk.IX (Aug 44 - Apr 45)
MJ351 S MJ518 O MJ553 G MJ619 D MJ746 V
MJ780 B MJ896 A MJ910 F MJ966 J MK127 K
MK132 H MK153 P MK183 T MK202 Q MK249 C
MK290 U MK359 X MK374 R MK617 W MK924 L
Supermarine Spitfire F.R.Mk.XIV (Apr - Aug 45)

Operational History: First Mission 30 June 1942, 3 Mustangs from Croydon — defensive patrol over south coast. **First Victory** 19 August 1942, Dieppe Raid — Mustangs, operating from Croydon, flew nine 2-plane missions on tactical reconnaissance. Two aircraft were lost but both pilots safe. F/O H.H. Hills in AG470 RU-M, on his second mission over Dieppe as "weaver" to the "Tac/R" aircraft, attacked by 3 Fw.190's and credited with destroying 1. **Last Mission** 5 May 1945, 2 Spitfire XIV's from Soltau — tactical and weather reconnaissance in the Bremerhaven - Cuxhaven - Elbe estuary. **Summary** Sorties 6087. Operational/Non-operational Flying Hours: 7738/11,495. Victories: Aircraft: 29 destroyed, 1 probably destroyed, 11 damaged. Ground: destroyed or damaged 76 locomotives, 13 assorted vessels, 30 electrical pylons, plus a score of miscellaneous targets. Casualties: Operational: 23 pilots, of whom 19 were killed or missing, 4 POW. Non-operational: 9 personnel, of whom 6 were killed, 3 died. **Squadron Ace** F/L D.I. Hall, DFC 7-0-2. **Honours and Awards** 2 bars to DFC, 16 DFC's, 3 MiD's. **Battle Honours** Defence of Britain 1942-1943. Fortress Europe 1942-1944. *Dieppe*. France and Germany 1944-1945: *Normandy 1944, Arnhem, Rhine. Biscay 1943.*

235 *Curtiss Tomahawk Mk.I, aircraft "S," "R" and "M" of No. 414 (AC) Squadron fly formation above their base at Croydon, Surrey in the spring of 1942.*

237 *Supermarine Spitfire Mk.IX MJ966 aircraft "J" of No. 414 (FR) Squadron bearing the remains of the invasion stripes on the lower surface.*

236 *North American Mustang Mk.I AM251 bearing the RCAF operational roundel in its proper location ahead of the wing leading edge. Aircraft "O" served with No. 414 (FR) Squadron on cross-Channel sorties.*

238 *Supermarine Spitfire Mk.XIV NM896 aircraft "B" of No. 414 (FR) Squadron with camera ports and bubble canopy clearly illustrated.*

No. 415 Squadron

Badge A swordfish
Motto Ad metam (To the mark)
Authority King George VI, October 1942
The swordfish and motto indicate the squadron's operational duties in attacking enemy shipping.
Formed at Thorney Island, Hampshire, England on 20 August 1941 as the RCAF's 13th — fourth coastal and only Torpedo Bomber — squadron formed overseas, the unit flew Hampden, Wellington and Albacore aircraft on anti-shipping operations over the English Channel and along the Dutch coastline. Following the invasion of Europe in June 1944, and the resulting lack of enemy shipping activity in the English Channel, the squadron was redesignated Bomber at Bircham Newton, Norfolk on 12 July. Drawing from trained personnel within No. 6 (RCAF) Group, it was reorganized at East Moor, Yorkshire two weeks later as the group's 14th bomber unit, and flew Halifax aircraft. The squadron was disbanded on 15 May 1945.
Brief Chronology Formed as No. 415 (TB) Sqn, Thorney Island, Hants., Eng. 20 Aug 41. Redesignated No. 415 (B) Sqn, Bircham Newton, Norfolk 12 Jul 44. Disbanded, East Moor, Yorks. 15 May 45.
Title or Nickname "Swordfish"
Adoption San Antonio Mines, Bissett, Man. (April 1945).
Commanders
W/C E.L. Wurtele 26 Aug 41 - 30 Jul 42.
W/C R.R. Dennis 31 Jul 42 - 10 Nov 42 *repat.*
W/C W.W. Bean 11 Nov 42 - 15 Mar 43.
W/C G.H.D. Evans 16 Mar 43 - 1 Aug 43.
W/C C.G. Ruttan, DSO 2 Aug 43 - 11 Jul 44.
W/C J.G. McNeill, DFC 12 Jul 44 - 21 Aug 44 *KIFA.*
W/C J.H.L. Lecomte, DFC 22 Aug 44 - 30 Oct 44.
W/C F.W. Ball 31 Oct 44 - 15 May 45.
Higher Formations and Squadron Locations
Coastal Command:
No. 16 Group,
Thorney Island, Hants. 20 Aug 41 - 10 Apr 42.
No. 19 Group,
St. Eval, Cornwall 11 Apr 42 - 15 May 42.
No. 16 Group,
Thorney Island, Hants. 16 May 42 - 4 Jun 42.
North Coates, Lincs. 5 Jun 42 - 30 Jul 42.
No. 18 Group,
Wick, Caith., Scot. 31 Jul 42 - 31 Aug 42.
Tain, Ross., Scot. 1 Sep 42 - 9 Sep 42.
Leuchars, Fife., Scot. 10 Sep 42 - 10 Nov 42.
No. 19 Group, 6 aircraft, St. Eval, Cornwall 18-31 Oct 42.
No. 16 Group, 6 aircraft, Thorney Island, Hants. 18 Oct - 10 Nov 42.
No. 16 Group,
Thorney Island, Hants 11 Nov 42 - 14 Nov 43.
Bircham Newton, Norfolk 15 Nov 43 - 11 Jul 44.
Albacores at Manston, Kent and Thorney Island, Hants.; Wellingtons at Docking, Norfolk and North Coates, Lincs., 8 Albacores at Winkleigh, Devon. 8 May - 12 Jul 44.
RAF Bomber Command:
No. 6 (RCAF) Group,
Bircham Newton, Norfolk 12 Jul 44 - 25 Jul 44.
No. 62 (RCAF) Base,
East Moor, Yorks. 26 Jul 44 - 15 May 45.
Representative Aircraft (Unit Codes GX, NH, GU)
Bristol Beaufort Mk.I (Sep 41 - Feb 42, not on operations)
L9802 R L9819 Q L9893 B N1102 S N1082 A
AW219 C
Handley Page Hampden Mk.1 (Jan 42 - Sep 43) X3140 T
AD762 J AD767 Y AE360 H AE368 W AN124 K
AT114 D AT152 S AT193 R AT230 L AT232 A
AT233 B AT234 C AT235 P AT238 O AT240 D
AT242 F AT243 N AT244 M AT245 U AT247 W
AT248 J AT250 K
Vickers Wellington Mk.XIII(L/L) (Sep 43 - Jul 44)
Fairey Albacore Mk.I (Oct 43 - Jul 44, coded NH)[1]
9272 NH-J BF600 NH-P1
Handley Page Halifax B.Mk.III & VII (Jul 44 - May 45, coded GU) LW552 S LW595 Q MZ456 P MZ483 M
MZ590 C MZ603 E MZ861 Z MZ946 O MZ847 K
NA124 I NA181 D NA185 A NA517 R NA600 U
NA610 B NP199 N NP938 Y NR122 X NR156 K
NR206 F NR253 L PN236 J PN239 V PN240 W
(The squadron had only six B.Mk.VII's from March to May 1945)
Operational History: First Mission 27 April 1942, 2 Hampdens from St. Eval — anti-shipping patrol. AT232 GX-A sighted two unidentified aircraft and two Spanish trawlers; AT244 GX-M reported no sightings. **First Victory** 29 May 1942, 3 Hampdens from Thorney Island — anti-shipping patrol; sighted an enemy convoy of six motor vessels and five flak (anti-aircraft) ships at 5337N 0545E, attacked the largest ship (6000 tons); claimed as damaged. Flak was heavy and 1 Hampden failed to return. The aircraft and captains were: AT229 GX-V, FS Garfin; AT236 GX-R, P/O Sargent (KIA); AT239 GX-T, P/O Lawrence. **Last Mission, Torpedo Bomber** 11 July 1944, 21 Albacore sorties flown from Manston over the Channel — 12 smoke-laying, 6 armed and 3 strike patrols; several vessels attacked without results. **First Mission, Bomber** 28/29 July 1944, 16 Halifaxes from East Moor despatched to bomb Hamburg; 1 crashed on takeoff (crew safe), 14 bombed primary target, 1 failed to return. **Last Mission** 25 April 1945, 18 Halifaxes from East Moor bombed gun positions on the Island of Wangerooge. **Summary, Torpedo Bomber** Sorties: 1645 (including 681 Albacore). Operational/Non-operational Flying Hours: 6243/8510. Victories: Shipping: dropped 196 tons of bombs; credited with 5 ships (8836 tons) sunk, 2 (10,646 tons) damaged, plus 2 shared sinkings (8293 tons). Casualties: Operational: 33 aircraft; 116 aircrew, of whom 14 were killed, 97 missing, 4 POW, 1 wounded. Non-operational: 13 aircraft; 43 personnel, of whom 33 were killed, 4 missing, 5 injured, 1 died. **Bomber** Sorties: 1608 Operational/Non-operational Flying Hours: 9035/2669. Bombs dropped: 5041.7 tons of high explosive, 797.8 incendiary. Victories: Aircraft: 1 destroyed, 1 probably destroyed. Casualties: Operational: 22 aircraft; 151 aircrew killed, missing or POW. Non-operational: 1 aircraft; 12 personnel killed. **Honours and Awards** 1 DSO, 1 bar to

DFC, 86 DFC's, 1 GM, 9 DFM's. **Battle Honours** Atlantic 1942. English Channel and North Sea 1942-1944. France and Germany 1944-1945: *Biscay Ports 1944, Ruhr 1944-1945, German Ports 1944-1945, Normandy 1944, Rhine. Biscay 1942-1943*

---

¹The Albacore Flight became the nucleus of No. 119 Squadron RAF, formed at Manston, Kent.

239 Handley Page Hampden Mk.I X3115 of No. 415 (TB) Squadron in the winter of 1942 at Thorney Island, Hants.

240 Fairey Albacore Mk.I, with the squadron code NH of No. 415 (TB) Squadron, at Bircham Newton, Norfolk, 1943.

No. 416 Squadron

Badge In front of the maple leaf a lynx leaping down
Motto Ad saltum paratus (Ready for the leap)
Authority King George VI, September 1944

The lynx is a fierce and dangerous opponent, indigenous to Canada.
Formed at Peterhead, Aberdeen, Scotland on 22 November 1941 as the RCAF's 15th — sixth Fighter — squadron formed overseas, the unit flew Spitfire aircraft on offensive and defensive air operations; and in support of ground forces in North-West Europe. After hostilities in Europe, the squadron served as one of four RCAF day fighter units with the British Air Forces of Occupation (Germany), until disbanded at Utersen, Germany on 21 March 1946.
Brief Chronology Formed at Peterhead, Aber., Scot. 22 Nov 41. Disbanded at Utersen, Ger. 21 Mar 46.
Title or Nickname "City of Oshawa"
Adoption City of Oshawa, Ont. (October 1942).
Commanders
S/L P.C. Webb 22 Nov 41 - 8 Mar 42.
S/L L.V. Chadburn, DFC 9 Mar 42 - 7 Jan 43.
S/L F.H. Boulton, DFC 8 Jan 43 - 13 May 43 *POW*.
S/L R.W. McNair, DFC 17 May 43 - 18 Jun 43.
S/L F.E. Grant 19 Jun 43 - 27 Aug 43.
S/L R.H. Walker 29 Aug 43 - 19 Oct 43.
S/L F.E. Green, DFC 20 Oct 43 - 29 Jun 44.
S/L J.F. McElroy, DFC 30 Jun 44 - 31 Oct 44 *OTE*.
S/L J.D. Mitchner, DFC and Bar 1 Nov 44 - 15 Jan 46 *repat*.
F/L S.H. Straub 16 Jan 46 - 21 Mar 46.
Higher Formations and Squadron Locations
Fighter Command:
No. 14 Group,
Peterhead, Aber., Scot 22 Nov 41 - 13 Mar 42.
Dyce, Aber., Scot. 14 Mar 42 - 2 Apr 42.
"B" Flight Montrose, Scot. 14-27 Mar 42.
Peterhead 28 Mar - 2 Apr 42.
Peterhead, Aber., Scot. 3 Apr 42 - 16 Jul 42.
"A" Flight remained at Dyce until 25 Jun.
No. 11 Group,
Martlesham Heath, Suffolk 17 Jul 42 - 22 Sep 42.
Canadian Kenley Wing,
Redhill, Surrey 23 Sep 42 - 31 Jan 43.
Kenley, Surrey 1 Feb 43 - 28 May 43.
No. 12 Group,
Canadian Digby Wing,
Wellingore, Lincs. 29 May 43 - 6 Jun 43.
Digby, Lincs. 7 Jun 43 - 8 Aug 43.
No. 11 Group,
Merston, Sussex 9 Aug 43 - 18 Sep 43.
No. 12 Group;
Canadian Digby Wing,
Wellingore, Lincs. 19 Sep 43 - 1 Oct 43.
Digby, Lincs. 2 Oct 43 - 11 Feb 44.
Second Tactical Air Force:
No. 83 (Composite) Group,
No. 17 (RCAF) Sector (disbanded 13 Jul 44),

No. 127 (RCAF) Wing,
Kenley, Surrey 12 Feb 44 - 16 Apr 44.
Tangmere, Sussex 17 Apr 44 - 15 Jun 44.
B.(Base) 2 Bazenville, Fr. 16 Jun 44 - 27 Aug 44.
B.26 Illiers l'Evêque, Fr. 28 Aug 44 - 21 Sep 44.
B.68 Le Culot, Bel. 22 Sep 44 - 29 Sep 44.
B.82 Grave, Neth. 30 Sep 44 - 21 Oct 44.
B.58 Melsbroek, Bel. 22 Oct 44 - 3 Nov 44.
B.56 Evère, Bel. 4 Nov 44 - 2 Mar 45.
B.90 Petit-Brogel, Bel. 3 Mar 45 - 30 Mar 45.
B.78 Eindhoven, Neth. 31 Mar 45 - 11 Apr 45.
B.100 Goch, Ger. 12 Apr 45 - 13 Apr 45.
B.114 Diepholz, Ger. 14 Apr 45 - 25 Apr 45.
B.154 Soltau, Ger. 26 Apr 45 - 1 Jul 45
British Air Forces of Occupation (Germany):
No. 83 (Composite) Group,
No. 126 (RCAF) Wing,
B.152 Fassberg, Ger. 2 Jul 45 - 4 Jul 45.
B.174 Utersen, Ger. 5 Jul 45 - 21 Mar 46.
Representative Aircraft (Unit Code DN)
Supermarine Spitfire Mk.IIA & IIB (Nov 41 - Mar 42)
Supermarine Spitfire Mk.VB (Mar 42 - Mar 43)
Supermarine Spitfire Mk.IX (Mar - May 43) BS430 N[1]
Supermarine Spitfire Mk.VB & VC (May 43 - Feb 44) EN950 P
Supermarine Spitfire Mk.IXB (Jan - Dec 44)
Supermarine Spitfire Mk.XVI (Dec 44 - Sep 45) SM191 T SM200 N SM397 V SM466 Y SM470 R SM503 D TB272 W TB377 G TB392 B TB616 P TB756 H TB905 K TD129 J TD187 S TD251 F
Supermarine Spitfire Mk.XIVE (Sep 45 - Mar 46))
Operational History: First Mission 1 February 1942, 1 Spitfire IIA from Peterhead — uneventful scramble. **First Victory** 27 May 1942, FS J. Moul in Spitfire VB BL887 from Peterhead — scrambled, credited with a He.111 damaged northeast of Peterhead. 18 July 1942, F/L P.L.I. Archer in Spitfire VB EP113 from Martlesham Heath — convoy patrol, credited with a Do.217 destroyed 15 miles east of Orford Ness. This was Archer's fifth victory. **First Offensive Mission** 19 August 1942, 12 Spitfire VB's from Merston — four missions in support of the Dieppe Raid. The first and fourth missions were uneventful; during the second, the squadron made several attacks on enemy aircraft without results; and on the third, engaged 15 Fw.190's, claimed 3 destroyed, 1 probably destroyed, 7 damaged without loss. **Last Mission** 7 May 1945, 4 Spitfire XVI's from Eindhoven — escort of a Mosquito carrying VIP (very important person) from Copenhagen to Kastrup. **Summary** Sorties: 10,263. Operational/Non-operational Flying Hours: 15,109/17,971. Victories: Aircraft: 75 destroyed, 3 probably destroyed, 37 damaged. Ground: dropped 65 tons of bombs; destroyed 286 motor vehicles, 28 horsedrawn transport, 13 locomotives and 14 miscellaneous targets; damaged 600-plus targets. Casualties: Operational: 42 aircraft; 35 pilots, of whom 19 killed or missing, 13 POW (1 escaped), 1 evaded capture, 2 proved safe. Non-operational: 5 aircraft; 7 personnel killed. **Squadron Aces** F/L D.E. Noonan, DFC 5½-0-0. S/L F.H. Boulton, DFC 5-1-3.

Honours and Awards 1 bar to DFC, 11 DFC's, 1 DFM, 1 DFC(USA), 1 Flying Cross (Neth.). **Battle Honours** Defence of Britain 1942-1944. English Channel and North Sea 1943. Fortress Europe 1942-1944: *Dieppe*. France and Germany 1944-1945: *Normandy 1944, Arnhem, Rhine*.

---

[1]Presentation aircraft named "Canadian Pacific" and flown by S/L Boulton on 18 sorties. He was credited with 1 Bf.109 and 3 Fw.190's destroyed while flying this aircraft.

241 *Supermarine Spitfire Mk.IXB aircraft "G" of No. 416 (F) Squadron framed over the nose of aircraft "B" from the same squadron.*

242 *Supermarine Spitfire Mk.XVI SM191 aircraft "T" of No. 416 (F) Squadron in the summer of 1945 at Reinsehlen, Germany.*

No. 417 Squadron

Badge In front of a palm tree eradicated a sword and fasces in saltire
Motto Supporting liberty and justice
Authority King George VI, October 1945

The squadron operated with the 8th Army in North Africa, in the Sicilian landings, and in Italy in close support of the army. The palm tree suggests the desert, the sword air support to the army, and the fasces the fighting in Italy.

Formed at Charmy Down, Somersetshire, England on 27 November 1941 as the RCAF's 16th — seventh Fighter — squadron formed overseas, the unit was ordered to the Middle East in the spring of 1942.[1] Equipped with Hurricane and, later, Spitfire aircraft, it spent five months in the defence of the Suez Canal and the Nile Delta. In April 1943 it became the only Canadian squadron in the Desert Air Force[2] and was to provide air defence and close support to the British Eighth Army through the closing stages of the Tunisian campaign, and throughout the Sicilian and Italian campaigns. The squadron was disbanded at Treviso, Italy on 30 June 1945.

Brief Chronology Formed at Charmy Down, Som., Eng. 27 Nov 41. Disbanded at Treviso, Italy 30 Jun 45.
Title or Nickname "City of Windsor"
Adoption City of Windsor, Ont. (May 1943)
Commanders
S/L C.E. Malfroy (RAF) 27 Nov 41 - 27 Mar 42.
S/L P.B. Pitcher 28 Mar 42 - 17 Nov 42 *repat*.
S/L F.B. Foster 18 Nov 42 - 23 Jun 43 *repat*.
S/L P.S. Turner (Can/RAF), DFC and Bar 24 Jun 43 - 19 Nov 43.
S/L A.U. Houle, DFC and Bar 20 Nov 43 - 17 Feb 44 *repat*.
S/L K.L. Magee 18 Feb 44 - 12 Mar 44.
S/L W.B. Hay 13 Mar 44 - 23 Jun 44 OTE.
S/L O.C. Kallio, DFC 24 Jun 44 - 11 Nov 44 OTE.
S/L D. Goldberg, DFC 12 Nov 44 - 30 Jun 45.

Higher Formations and Squadron Locations
Fighter Command:
No. 10 Group,
Charmy Down, Som. 27 Nov 41 - 26 Jan 42.
Colerne, Wilts. 27 Jan 42 - 23 Feb 42.
No. 14 Group,
Tain, Ross., Scot. 25 Feb 42 - 12 Apr 42.
En route to Egypt 13 Apr 42 - 3 Jun 42.
Middle East Command:
Air Headquarters Egypt,
Kasfareet, Egypt 4 Jun 42 - 17 Jul 42.[1]
Deversoir, Egypt 18 Jul 42 - 4 Sep 42.
Shandur, Egypt 5 Sep 42 - 9 Oct 42.
Idku, Egypt 10 Oct 42 - 24 Jan 43.
4 aircraft each at Cairo, Egypt 7 Oct - 22 Nov 42. Al Kufra Oasis, Libyan Desert 3-25 Nov 42. Cyprus 7 Dec 42 - 7 Jan 43. Benghazi, Libya 7 Dec 42 - 7 Jan 43.
L.G. 175 Alexandria, Egypt 25 Jan 43 - 7 Feb 43.
En route to Tripoli, Libya 8 Feb 43 - 18 Feb 43[2].
Castel Benito, Tripoli, Libya 19 Feb 43 - 27 Feb 43.
Mellaha, Tripoli, Libya 28 Feb 43 - 11 Apr 43.
8 aircraft, Ben Gardane, Tunisia 19-29 Mar 3-10 Apr 43.
Ben Gardane, Tunisia 12 Apr 43 - 14 Apr 43.
Aircraft at La Fauconnerie.
Desert Air Force.[3]
No. 211 Group,
No. 244 Wing,
Goubrine South, Tunisia 15 Apr 43 - 5 May 43.
Hergla, Tunisia 6 May 43 - 14 May 43.
Ben Gardane, Tunisia 15 May 43 - 14 Jun 43.
Luqa, Malta (Sicily landings) 15 Jun 43 - 15 Jul 43.
Cassibile, Syracuse, Sicily 16 Jul 43 - 25 Jul 43.
Lentini West, Sicily 26 Jul 43 - 18 Sep 43.
Grottaglie, Italy 19 Sep 43 - 24 Sep 43.
Gioia delle Colle, It. 25 Sep 43 - 4 Oct 43.
Foggia, It. 5 Oct 43 - 17 Oct 43.
Triolo, It. 18 Oct 43 - 25 Nov 43.
Canne, It. 26 Nov 43 - 16 Jan 44.
Marcianise Landing Ground, Anzio, It. 17 Jan 44 - 23 Apr 44.
Venafro, Cassino, It. 24 Apr 44 - 12 Jun 44.
Littorio, Rome, It. 13 Jun 44 - 16 Jun 44.
Fabrica, It. 17 Jun 44 - 3 Jul 44.
Perugia, It. 4 Jul 44 - 25 Aug 44.
Loreto, It. 26 Aug 44 - 4 Sep 44.
Fano, It. 5 Sep 44 - 4 Dec 44.
Bellaria, It. 5 Dec 44 - 3 May 45.
Treviso, It. 4 May 45 - 30 Jun 45.

Representative Aircraft (Unit Code AN)
England
Supermarine Spitfire Mk.IIA & IIB (Nov 41 - Feb 42)
P8570 R
Supermarine Spitfire Mk.VB (Feb - Mar 42) AB797 Z
AD319 U
Middle East
Hawker Hurricane Mk.IIB (Sep - Oct 42)
Hawker Hurricane Mk.IIC (Sep 42 - Jan 43) BD779 B
BP590 R   HL835 A   HL843 C
Supermarine Spitfire Mk.VB & VC (Oct 42 - Sep 43)
BR483 V   EP315 F   EP893 R   ER134 Y   ER364 T
ER634 H   ER944 C   ES124 J
Supermarine Spitfire Mk.VIII (Aug 43 - Apr 45) JF336 O
JF403 E   JF469 M   JF473 X   JF557 D   JF565 L   JF579 J
JF586 C   JF672 M   JF781 J   JF881 H   HF932 W   JF952 Y
JF956 B   JF964 T   JG184 A   JG185 U   JG240 R   JG294 F
JG317 V   JG337 X   JG475 Z   MT546 Z   MT770 A
Supermarine Spitfire Mk.IXB (Apr - Jun 45) MJ818 U

Operational History: First Mission, England 17 February 1942, 3 Spitfire VB's from Charmy Down — uneventful scramble. **First Mission, Middle East** 13 September 1942, 2 Hurricane IIB's from Shandur — patrol over the town of Suez at 25,000 feet. **First Victory** 26 September 1942, FS J.H.G. Leguerrier in Hurricane IIC HL891 AN-N from Shandur — patrol over the town of Suez, credited with a Ju.88 destroyed southeast of the town. **First Offensive Mission** 24 March 1943, 8 Spitfire VB's from Mellaha, Tripoli — top cover for 18 Baltimores bombing targets on the Mareth Line. **Last Mission** 5 May 1945, 2 Spitfire IX's from Treviso — patrol of Villach - St. Michael - Linz - Dobbiaco, in the Austro-Italian Alps. **Summary** Sorties: 12,116 (including 11 in England). Operational/Non-operational Flying Hours: 15,088/4974. Victories: Aircraft: 29 destroyed, 8 probably destroyed, 22 damaged. Ground:

dropped 1080 tons of bombs; destroyed/damaged 11/111 locomotives, 64/261 freight cars, 14/27 passenger cars; 79/137 motor vehicles; 7/11 tanks; 6/18 barges, plus miscellaneous targets. Casualties: Operational: 32 aircraft; 28 pilots, of whom 7 were killed, 11 presumed dead, 8 POW, 2 evaded capture. Non-operational: 4 personnel, of whom 1 killed, 3 died. **Squadron Ace** S/L A.U. Houle, DFC and Bar 7-0-4. **Honours and Awards** 1 DSO, 1 bar to DFC, 9 DFC's, 5 MiD's. **Battle Honours** Defence of Britain 1942. Egypt and Libya 1942 - 1943. North Africa 1943. Sicily 1943. Italy 1943-1945: *Salerno, Anzio and Nettuno, Gustav Line, Gothic Line.*

---

[1] No aircraft were available to the squadron from its arrival in Egypt until midsummer of 1942. Between June and August, under British command, pilots were employed on ferry duty and groundcrew on maintenance.
[2] The squadron remained under Air Force Headquarters Egypt, but was attached to Air Headquarters Western Desert for operational control.
[3] Until July 1943, still known officially as Air Headquarters Western Desert.

243 *Supermarine Spitfire Mk.VIII JF526 waiting to start up on a field in Italy with No. 417 (F) Squadron.*

No. 418 Squadron

Badge none
Formed at Debden, Essex, England on 15 November 1941 as the RCAF's 14th — only Intruder — squadron formed overseas, the unit flew Boston and Mosquito aircraft on day- and night-intruder operations deep into enemy territory. Its claim of 178 enemy aircraft and 79½ V-1 flying bombs destroyed made it the top-scoring unit of the RCAF. The leading individual score was S/L R. Bannock, with 11 aircraft and 18½ V-1's. On 21 November 1944 it was transferred to close support work[1] with the Second Tactical Air Force in the Low Countries. The squadron was disbanded at Volkel, in The Netherlands, on 7 September 1945.
Brief Chronology Formed as No. 418 (I) Sqn, Debden, Essex, Eng. 15 Nov 41. Disbanded at Volkel, Neth. 7 Sep 45.
Title or Nickname "City of Edmonton"
Adoption City of Edmonton, Alta (March 1944)
Commanders
W/C G.H. Gatheral (RAF) 22 Nov 41 - 14 May 42.
W/C A.E. Saunders (RAF) 15 May 42 - 11 Dec 42.
W/C J.H. Little (RAF), DFC 12 Dec 42 - 13 Jun 43 *KIA*.
W/C P.Y. Davoud, DFC 15 Jun 43 - 7 Jan 44.
W/C D.C.S. MacDonald, DFC 8 Jan 44 - 24 Feb 44.
W/C R.J. Bennell, DFC 25 Feb 44 - 9 Mar 44 *KIA*.
W/C A. Barker 30 Mar 44 - 9 Oct 44.
W/C R. Bannock, DFC and Bar 10 Oct 44 - 22 Nov 44.
W/C J.C. Wickett 23 Nov 44 - 22 Feb 45 *POW*.
W/C D.B. Annan 23 Feb 45 - 23 May 45 *OTE*.
W/C H.D. Cleveland, DFC 24 May 45 - 7 Sep 45.
Higher Formations and Squadron Location
Fighter Command renamed
Air Defence Great Britain (15 Nov 43):
No. 11 Group,
Debden, Essex 15 Nov 41 - 14 Apr 42.
Bradwell Bay, Essex 15 Apr 42 - 14 Mar 43.
Ford, Sussex 15 Mar 43 - 7 Apr 44.
Holmsley South, Hants. 8 Apr 44 - 13 Jul 44.
Hurn, Hants. 14 Jul 44 - 28 Jul 44.
Middle Wallop, Hants. 29 Jul 44 - 26 Aug 44.
Hunsdon, Herts. 28 Aug 44 - 20 Nov 44.
Second Tactical Air Force:
No. 2 (Bomber) Group,
No. 136 (RAF) Wing,
Hartford Bridge, Hants. 21 Nov 44 - 14 Mar 45.
B.(Base) 71 Coxyde, Bel. 15 Mar 45 - 25 Apr 45.
B.80 Volkel, Neth. 26 Apr 45 - 7 Sep 45.
Representative Aircraft (Unit Code TH)
Douglas Boston Mk.III (Nov 41 - Jul 43) W8263 P
W8268 O   W8317 V   W8321 G   W8356 D   Z2192 Z
Z2226 K
de Havilland Mosquito Mk.II (Mar 43 - Nov 44)
de Havilland Mosquito F.B.Mk.VI (Nov 44 - Sep 45)[2]
HR148 B   HR184 Z   HR324 N   HR358 K   HX953 X
NS823 W   NS857 L   NS930 V   NT115 J   NT153 Y
PZ219 E   PZ235 M   PZ414 P   PZ454 Y   RS454 F
RS560 G   RS561 F   RS569 V   RS594 L   SZ962 U
SZ964 X   SZ965 T   SZ967 V   TA374 C

Operational History: First Mission, Bombing 27 March 1942, 8 Bostons from Ford despatched to bomb oil refineries and tanks at Ertvelde, near Ghent, Belgium; 7 bombed the primary target, 1 had a "bomb hang-up." **First Mission, Intruder** 28 March 1942, 6 Bostons from Ford — night patrols of enemy airfields in France (Lille, Vendeville, Rennes, Le Touquet and Abbeville) and The Netherlands (Gilze-Rijen) in co-operation with No. 23 Squadron RAF. **First Victory** 26 April 1942, Sgts. G.W.C. Harding (RAF), R.P. Shannon (observer) and H.J.H. Irving (air gunner) in a Boston from Holmsley South — night intruder over Evreux airfield in France, credited with 1 unidentifiable enemy aircraft damaged. 7 May 1942, P/O A. Lucas (RAF), Sgts W.S. Randolph (observer) and H. Haskell (air gunner) in a Boston from Holmsley South — night intruder over Gilze-Rijen airfield in The Netherlands, shared with a Hurricane aircraft of No. 3 Squadron RAF 1 unidentifiable enemy aircraft destroyed. **Last Mission** 3 May 1945, Mosquito VI from Volkel — reconnaissance of the battle area. **Summary** Sorties: 3492 (including 402 on anti-flying bomb patrols). Operational/Non-operational Flying Hours: 11,248/12,255. Victories: Aircraft: 178 destroyed (73 on ground), 9 probably destroyed, 103 damaged. V-1: 76 destroyed over water, 7 over England.[3] Ground: dropped 56 tons of bombs; credited with 17 locomotives destroyed and 59 damaged, 52 freight and passenger cars destroyed or derailed, 200 motor vehicles destroyed. Casualties: Operational: 59 aircraft; 143 aircrew, of whom 94 killed or presumed dead, 27 missing, 14 POW, 8 evaded capture. Non-operational: 13 aircraft; 31 personnel killed, 2 injured. **Squadron Aces** W/C R. Bannock, DFC and Bar 25½-7-0-1-18½.* S/L R. Gray, DFC 12-10-0-12-2. F/O S.P. Reid, DFC 11½-8-0-7-3½. S/L H.D. Cleveland, DFC 10-10-0-1-0. F/L C.M. Jasper, DFC 9-6-1-1-3. F/L C.J. Evans 9-1½-1-0-7½. F/L S.H.R. Cotterill, DFC 8-4-0-1-4. F/L D.E. Forsyth 8-4-0-0-4. S/L J.B. Kerr 6-5-0-3-1. F/L H.E. Miller 5-2-0-0-3. F/L P.S. Leggat 5-0-0-0-5. **Honours and Awards** 3 DSO's, 1 second bar to DFC, 9 bars to DFC, 42 DFC's, 5 DFM's, 1 DFC(USA), 1 Air Medal (USA). **Battle Honours** Defence of Britain 1944. Fortress Europe 1942-1944. *Dieppe*. France and Germany 1944-1945. *Normandy 1944*, *Rhine*.

*The five categories indicate, for this squadron: Total; confirmed; probable; damaged; and V-1's destroyed.

---

[1] In the intruder role, it was replaced by No. 406 Squadron.
[2] Aircraft were named and decorated after characters of the comic strip "Li'l Abner."
[3] A V-1 destroyed over England was counted as only half a victory.

244 Douglas Boston Mk.III W8268 aircraft "O" for Ottawa, Ont. departing on an intruder sortie while serving with No. 418 (I) Squadron in 1941.

245 De Havilland Mosquito Mk.VI HR147 aircraft "Z" of No. 418 (I) Squadron. This was W/C R. Bannock's aircraft and bore the insignia "Hairless Joe" on the nose. Below the circle, a tally of 8 aircraft and 19 flying bombs is displayed.

No. 419 Squadron

Badge A moose attacking
Motto Moosa aswayita (Beware of the moose)
Authority King George VI, June 1944
The moose, representing the squadron's nickname, is a fierce fighter indigenous to Canada. The motto is in Cree.
Formed at Mildenhall, Suffolk, England on 15 December 1941 as the RCAF's 17th — third Bomber — squadron formed overseas, the unit flew Wellington, Halifax and Canadian-built Lancaster aircraft on strategic and tactical bombing operations. Among the many decorations awarded to its members was the Victoria Cross — posthumously, to P/O A.C. Mynarski.[1] After hostilities in Europe, the squadron was selected as part of "Tiger Force" for duty in the Pacific, and returned to Canada for reorganization and training. The sudden end of the war in the Far East resulted in the squadron being disbanded at Yarmouth, Nova Scotia on 5 September 1945.
Brief Chronology Formed at Mildenhall, Suffolk, Eng. 15 Dec 41. Disbanded at Yarmouth, N.S. 5 Sep 45.
Title or Nickname "Moose"[2]
Adoption City of Kamloops, B.C.
Commanders
W/C J. Fulton (Can/RAF), DSO, DFC, AFC 21 Dec 41 - 28 Jul 42 *KIA*.
W/C A.P. Walsh (Can/RAF), DFC, AFC 5 Aug 42 - 2 Sep 42 *KIA*.
W/C M.M. Fleming (Can/RAF), DSO, DFC 8 Sep 42 - 8 Oct 43.
W/C G.A. McMurdy 11 Oct 43 - 22 Oct 43 *KIA*.
W/C W.P. Pleasance, DFC and Bar 25 Oct 43 - 21 Aug 44.
W/C D.C. Hagerman, DFC and Bar 22 Aug 44 - 25 Jan 45.
W/C M.E. Ferguson 26 Jan 45 - 6 Aug 45.
W/C H.R.F. Dyer, DFC 7 Aug 45 - 5 Sep 45.
Higher Formations and Squadron Locations
Bomber Command:
No. 3 Group,
Mildenhall, Suffolk 21 Dec 41 - 12 Aug 42.
No. 4 Group,
Leeming, Yorks. 13 Aug 42 - 17 Aug 42.
Topcliffe, Yorks. 18 Aug 42 - 30 Sep 42.
Croft, Yorks. 1 Oct 42 - 9 Nov 42.
Middleton St. George, Durham. 10 Nov 42 - 31 Dec 42.
No. 6 (RCAF) Group,
No. 64 (RCAF) Base,
Middleton St. George, Durham. 1 Jan 43 - 31 May 45.
En route to Canada 1 Jun 45 - 4 Jun 45.
RAF "Tiger Force" (for ops)
RCAF Eastern Air Command (for training):
No. 6 (RCAF) Group,
No. 661 (RCAF) Wing,
Yarmouth, N.S. 5 Jun 45 - 5 Sep 45.

Representative Aircraft (Unit Code VR)
Vickers Wellington Mk.IC (Jan - Nov 42) Z1053 F
Z1067 C   Z1077 P   Z1083 O   Z1091 A   Z1095 Q
Z1146 E   Z8967 E   Z8981 H   Z9757 N   Z9894 G
Z9920 D   DV509 S
Vickers Wellington Mk.III (Feb - Nov 42) X3277 F
X3357 T   X3416 J   X3467 N   X3470 E   X3483 S
X3484 O   X3486 U   X3488 H   X3563 T   X3592 E
X3659 B   X3699 V   X3711 R   X3712 D   X3715 G
X3717 C   X3796 C   Z1572 Q   Z1623 V   Z1626 G
BJ602 J   BJ604 A   BJ668 X
Handley Page Halifax B.Mk.II (Nov 42 - Apr 44) JB900 E
JB929 J   JD114 O   JD158 D   JD163 N   JD325 X
JD410 V   JD420 T   JD456 B   JD457 F   JD458 C
JD459 Q   JP112 R   JP121 S   JP125 L   JP200 G
JP201 P   JP202 T   JP203 M   JP204 E   LW239 R
LW243 Y   LW325 H   LW327 A
Avro Lancaster B.Mk.X (Mar 44 - Sep 45) KB700 Z
KB704 Y   KB707 W   KB717 E   KB718 J   KB720 P
KB712 B/E KB722 A   KB731 S   KB733 G   KB738 D
KB748 O   KB761 H   KB767 U   KB722 R   KB776 F
KB800 C   KB814 N   KB845 L   KB854 T   KB866 M
KB869 Q   KB878 I   KB884 K
Operational History: First Mission 11 January 1942, 2 Wellington IC's from Mildenhall bombed Brest, France; W/C Foulton and crew in X9748 VR-B, P/O Cottier and crew in Z1145 VR-A. **Last Mission** 25 April 1945, 15 Lancaster X's from Middleton St. George bombed gun positions on the Island of Wangerooge. **Summary** Sorties: 4002. Operational/Non-operational Flying Hours: 25,386/8613. Bombs dropped: 13,417 tons. Victories: Aircraft: 129 destroyed, 6 damged. Casualties: Operational: 129 aircraft; 1404 aircrew, of whom 310 were killed, 867 missing, 187 POW, 40 evaded capture. Non-operational: 2 aircraft; 15 personnel, of whom 10 killed, 3 injured, 2 died.
Honours and Awards 1 VC, 4 DSO's, 1 MC, 3 bars to DFC, 150 DFC's, 1 CGM, 35 DFM's. **Battle Honours** English Channel and North Sea 1942-1944. Baltic 1942-1944. Fortress Europe 1942-1944. France and Germany 1944-1945: *Biscay Ports 1942-1944, Ruhr 1942-1945, Berlin 1943-1944, German Ports 1942-1945, Normandy 1944, Rhine.* Biscay 1942; 1944.

---

[1] On 12 June 1944, during a raid on Cambrai, France, Lancaster B.Mk.X KB726 VR-A, of which Mynarski was the mid-upper gunner, was set aflame by a German night fighter. The captain ordered the crew to abandon the aircraft, but the rear gunner was trapped and Mynarski made every effort to free him, but was unsuccessful. The rear gunner finally persuaded him to bail out and, although still trapped in his turret, survived the crash and became a prisoner of war. Mynarski died in a German hospital of severe burns. After the war, when details of his efforts to free the rear gunner became known, he was posthumously awarded the VC.
[2] Derived from the nickname of the squadron's first commander, W/C John "Moose" Fulton, of Kamloops, B.C.

246 *Vickers Wellington Mk.III of No. 419 (B) Squadron being serviced for an evening mission over Germany in November 1942.*

247 *Handley Page Halifax Mk.II of No. 419 (B) Squadron with a group of the squadron's ground and air crew.*

248 *Avro Lancaster Mk.X KB866 and KB875 aircraft "M" and "Z" respectively of No. 419 (B) Squadron in mint condition at Middleton St. George, Durham.*

No. 420 Squadron

Badge A snowy owl, wings elevated and adorsed
Motto Pugnamus finitum (We fight to the finish)
Authority King George VI, March 1943
The snowy owl is indigenous to Canada and hunts by night.
Formed at Waddington, Lincolnshire, England on 19 December 1941 as the RCAF's 18th — fourth Bomber — squadron formed overseas, the unit flew Hampden, Wellington and Halifax aircraft on strategic and tactical bombing operations. From June to October 1943, it flew tropicalized Wellington aircraft from North Africa in support of the invasions of Sicily and Italy. After hostilities in Europe, it was selected as part of "Tiger Force" for duty in the Pacific, converted to Canadian-built Lancaster aircraft, and returned to Canada for reorganization and training. The sudden end of the war in the Far East resulted in the squadron being disbanded at Debert, Nova Scotia on 5 September 1945.
Brief Chronology Formed at Waddington, Lincs., Eng. 19 Dec 41. Disbanded at Debert, N.S. 5 Sep 45.
Title or Nickname "Snowy Owl"
Adoption Women's Air Force Auxiliary, London, Ont. (November 1943)
Commanders
W/C J.D.D. Collier (RAF), DFC 19 Dec 41 - 29 May 42.
W/C D.A.R. Bradshaw 30 Mar 42 - 11 Apr 43.
W/C D. McIntosh (Can/RAF), DFC 12 Apr 43 - 7 Apr 44.
W/C A.G. McKenna, DFC 8 Apr 44 - 23 Oct 44 OTE.
W/C G.J.J. Edwards, DFC 24 Oct 44 - 23 Nov 44 OTE.
W/C W.G. Phelan, DFC 24 Nov 44 - 27 Jan 45 OTE.
W/C F.S. McCarthy 28 Jan 45 - 23 Apr 45 OTE.
W/C R.J. Gray, MBE 24 Apr 45 - 5 Sep 45.
Higher Formations and Squadron Locations
Bomber Command:
No. 5 Group,
Waddington, Lincs. 19 Dec 41 - 6 Aug 42.
No. 4 Group,
Skipton-on-Swale, Yorks. 7 Aug 42 - 15 Oct 42.
Middleton St. George, Durham. 16 Oct 42 - 31 Dec 42.
No. 6 (RCAF) Group,
Middleton St. George, Durham. 1 Jan 43 - 15 May 43.
En route to North Africa[1] 16 May 43 - 18 Jun 43.
Mediterranean Air Command:
No. 205 Group,
No. 331 (RCAF) Wing,
Kairouan/Zina, Tunisia 19 Jun 43 - 29 Sep 43.
Hani East Landing Ground, Tunisia 29 Sep 43 - 16 Oct 43.
En route to England (minus aircraft) 17 Oct 43 - 5 Nov 43.
Bomber Command:
No. 6 (RCAF) Group,
No. 61 (RCAF) Base,
Dalton, Yorks. 6 Nov 43 - 11 Dec 43.
No. 62 (RCAF) Base,
Tholthorpe, Yorks. 12 Dec 43 - 11 Jun 45.
En route to Canada 12 Jun 45 - 16 Jun 45.
RAF "Tiger Force" (for ops),

RCAF Eastern Air Command (for training):
No. 6 (RCAF) Group,
No. 663 (RCAF) Wing,
Debert, N.S. 16 Jun 45 - 5 Sep 45.
Representative Aircraft (Unit Code PT)
Handley Page Hampden Mk.I (Dec 41 - Aug 42)
P1187 X   P1314 S   P2094 Q   P4400 J   P5332 T
X3057 C   AD915 F   AE202 K   AE248 A   AE258 W
AE260 O   AE267 V   AE314 Q   AE355 A   AE384 M
AE389 D   AE390 Z   AE853 T   AT130 S   AT132 U
AT134 K   AT135 R   AT136 N   AT228 P
Vickers Wellington B.Mk.III (Aug 42 - Apr 43)
X3800 Z   X3808 B   X3926 A   X3963 D   Z1679 B
BJ717 Q   BJ966 R   BK235 T   BK295 H   BK331 W
BK365 L   DF636 S   DF637 F   HK330 K
Vickers Wellington B.Mk.X (England, Feb - May 43)
HE157 U   HE294 P   HE375 H   HE417 L   HE422 Q
HE457 S   HE458 W   HE481 O   HE550 C   HE555 A
HE569 P   HE630 V   HE632 R   HE682 T   HE693 W
HE732 J   HE771 F   HE863 G   HE873 D   HE965 R
HE975 U   MS479 F   MS482 Y
Vickers Wellington B.Mk.X (North Africa, May - Oct 43)
HE640 E   HE975 U
Handley Page Halifax B.Mk.III (Dec 43 - May 45)
LW122 Z   LW392 S   LW419 F   LW421 K   LW476 J
LW645 T   LW692 V   MZ473 G   MZ502 U   MZ503 L
MZ504 H   MZ505 X   MZ569 R   MZ594 W   NA169 O
NA509 B   NP951 Y   NR135 E   NR138 T   NR139 A
NR717 P   NR207 C   NR208 D   NR258 I
Avro Lancaster B.Mk.X (Apr - Sep 45, not on operations)
KB871 E   KB885 Y   KB886 H   KB896 O   KB898 W
KB901 Q   KB902 C   KB908 P   KB909 R   KB910 V
KB914 A   KB923 N   KB927 I   KB928 K   KB993 J
KB937 G   KB938 D   KB941 U   KB942 M   KB946 Z
Operational History: First Mission 21/22 January 1942, 5 Hampdens from Waddington despatched to bomb targets at Emden; 2 bombed the primary target, 2 bombed the alternative, 1 failed to return. A sixth laid mines in the "Nectarines" (Frisian Islands) area. **Last Mission** 22 April 1945, 17 Halifaxes from Tholthorpe despatched to Bremen but on arriving over the target the master bomber ordered them not to bomb; explanation not recorded. **Summary** Sorties: 4186. Operational/Non-operational Flying Hours: 22,820/9137. Bombs dropped (1944-45): 9771 tons. Victories: Aircraft: 6 destroyed, 2 damaged. Casualties: Operational: 65 aircraft; 324 aircrew, of whom 84 were killed or presumed dead, 228 missing, 6 POW, 4 injured, 2 proved safe. Non-operational: 3 aircraft; 12 personnel were killed, 6 injured, 1 died. **Honours and Awards** 1 bar to DFC, 38 DFC's, 9 DFM's. **Battle Honours** English Channel and North Sea 1942-1944. Baltic 1942. Fortress Europe 1942-1944. France and Germany 1944-1945: *Biscay Ports 1942-1944, Ruhr 1942-1945, Berlin 1944, German Ports 1942-1945, Normandy 1944, Rhine.* Biscay 1942-1943. Sicily 1943. Italy 1943: *Salerno.*

249 Handley Page Halifax B.Mk.III MZ620 of No. 420 (B) Squadron about to take off from Tholthorpe, Yorks. on 6 January 1945.

250 Avro Lancaster Mk.X KB896 aircraft "O" of No. 420 (B) Squadron at Tholthorpe, Yorks.

---

¹Ground crew departed by ship on 16 May; aircraft took off 5 June. Two aircraft were shot down over the Bay of Biscay by German fighters and their crews listed as missing. See also No. 331 (Medium Bomber) Wing.

No. 421 Squadron

Badge In front of two tomahawks in saltire, a Red Indian warrior's head-dress
Motto Bellicum cecinere (They have sounded the war trumpet)
Authority King George VI, September 1944
The Canadian Red Indian is well known for his courage and fighting qualities and with the tomahawks, his traditional weapon, makes an appropriate device for a fighter squadron.
Formed at Digby, Lincolnshire, England on 9 April 1942 as the RCAF's 20th — eighth and last Fighter — squadron formed overseas, the unit flew Spitfire aircraft on offensive and defensive air operations; and in close support of ground forces in North-West Europe. The squadron was disbanded at Utersen, Germany on 10 July 1945.
Brief Chronology Formed at Digby, Lincs., Eng. 9 Apr 42. Disbanded at Utersen, Ger. 10 Jul 45.
Title or Nickname "Red Indian"[1]
Adoption McColl-Frontenac Oil Company of Canada
Commanders
S/L F.W. Kelly 9 Apr 42 - 13 Jul 42.
S/L F.C. Willis (Can/RAF) 14 Jul 42 - 8 Nov 42 *KIA.*
S/L F.E. Green, DFC 17 Nov 42 - 12 Apr 43 *OTE.*
S/L J.D. Hall, DFC 13 Apr 43 - 16 Jun 43.
S/L R.W. McNair, DFC and 2 Bars 19 Jun 43 - 16 Oct 43.
S/L C.M. Magwood, DFC 17 Oct 43 - 12 Dec 43.
S/L J.F. Lambert 13 Dec 43 - 20 Dec 43 *KIA.*
S/L W.A.G. Conrad, DFC 1 Jan 44 - 27 Jul 44 *2OTE.*
S/L W.A. Prest 28 Jul 44 - 22 Nov 44 *OTE.*
S/L J.D. Browne, DFC 23 Nov 44 - 10 Jul 45.
Higher Formations and Squadron Locations
Fighter Command:
No. 12 Group,
Canadian Digby Wing,
Digby, Lincs. 9 Apr 42 - 2 May 42.
No. 10 Group,
Fairwood Common, S. Wales 3 May 42 - 14 Oct 42.
No. 17 Armament Practice Camp, Warmwell, Dorset 14-28 Jun 42.
14 aircraft, Exeter and Bolt Head, Devon. 30 Jun - 8 Jul 42.
14 aircraft, Kenley, Surrey 8-10 Oct 42.
Angle, S. Wales 15 Oct 42 - 28 Jan 43.
14 aircraft, Zeals, Wilts. 21-24 Oct 42.
Exercise, Charmy Down, Som. 30 Nov - 3 Dec 42.
No. 11 Group,
Canadian Kenley Wing,
Kenley, Surrey 29 Jan 43 - 22 Mar 43.
Exercise, Croughton, Beds. 1-12 Mar 43.
8 aircraft, night fighter duty, Manston, Kent 19-25 Mar 43.
Redhill, Surrey 23 Mar 43 - 16 May 43.
No. 15 Armament Practice Camp, Martlesham Heath, Suffolk 10-21 Apr 43.
Kenley, Surrey 17 May 43 - 4 Jul 43.
Second Tactical Air Force:
No. 83 (Composite) Group,
No. 17 (RCAF) Sector (disbanded 13 Jul 44),
No. 127 (RCAF) Wing,
Kenley Surrey 5 Jul 43 - 5 Aug 43.
Lashenden, Kent 6 Aug 43 - 19 Aug 43.
Headcorn, Kent 20 Aug 43 - 13 Oct 43.
Kenley, Surrey 14 Oct 43 - 17 Apr 44.
No. 16 Armament Practice Camp, Hutton Cranswick, Yorks. 2-9 Mar 44.
Tangmere, Sussex 18 Apr 44 - 15 Jun 44.
B.(Base) 2 Bazenville, Fr. 16 Jun 44 - 28 Aug 44
B.26 Illiers l'Evêque, Fr. 29 Aug 44 - 21 Sep 44.
B.68 Le Culot, Bel 22 Sep 44 - 30 Sep 44.
B.82 Grave, Neth. 1 Oct 44 - 22 Oct 44.
B.58 Melsbroek, Bel. 23 Oct 44 - 2 Nov 44.
B.56 Evère, Bel. 3 Nov 44 - 1 Mar 45.
No. 17 Armament Practice Camp, Warmwell, Dorset, Eng. 6-20 Dec 44.
B.90 Petit-Brogel, Bel. 2 Mar 45 - 30 Mar 45.
B.78 Eindhoven, Neth. 31 Mar 45 - 10 Apr 45.
B.100 Goch, Ger. 11 Apr. 45 - 13 Apr 45.
B.114 Diepholz, Ger. 13 Apr 45 - 27 Apr 45.
B.154 Soltau, Ger. 28 Apr 45 - 1 Jul 45.
B.152 Fassberg, Ger. 2 Jul 45 - 7 Jul 45.
B.174 Utersen, Ger. 8 Jul 45 - 10 Jul 45.
Representative Aircraft (Unit Code AU)
Supermarine Spitfire Mk.VA (May 42, not on operations)
Supermarine Spitfire Mk.VB (May 42 - May 43)
AR430 Z BL450 M BL658 Y BL772 S
Supermarine Spitfire Mk.IX (May 43 - Feb 44)
Supermarine Spitfire Mk.IXB (Feb - Dec 44)
Supermarine Spitfire Mk.XVI (Dec 44 - Jul 45)
Operational History: First Mission 16 May 1942, 2 Spitfire VB's from Fairwood Common — convoy patrol. **First Victory** 6 July 1942, F/L G.D. Robertson in Spitfire VB BL772 AU-S from Fairwood Common — scrambled, credited with a Bf.109 damaged over the Channel. (On a second scramble later that day, he was credited with a Ju.88 damaged over the Channel.) 27 July 1942 Sgt. C.D. Myers and Sgt. J.A. Omand in Spitfire VB's from Fairwood Common — scrambled, shared in destruction of a Ju.88 south of Pembroke, Wales. **First Offensive Mission** 31 July 1942, 8 Spitfire VB's from Fairwood Common, operating from Harrowbeer for the day — rear cover for Bostons returning from a raid over St. Malo, France. **Triple Victory** 25 July 1944, 12 Spitfire IXB's from Bazenville — armed reconnaissance in the Pont l'Evêque-Seine area, engaged 40-plus Bf.109's, credited with 5 destroyed, 1 probably destroyed and 3 damaged for the loss of 1 Spitfire. F/L H.P.M. Zary in MK920 credited with 3 Bf.109's destroyed southeast of Rouen, France. **Last Mission** 4 May 1945, 4 Spitfire XVI's from Soltau — front line patrol along Hamburg-Luneburg, engaged Fw.190's flying south. No claims, no losses.
Summary Sorties: 10,915. Operational/Non-operational Flying Hours: 14,609/8216. Victories: Aircraft: 79 destroyed, 2 probably destroyed, 27 damaged. Ground: dropped 45 tons of bombs, no reliable record of damage. Casualties: Operational: 34 aircraft; 33 pilots, of whom 4 were killed, 25 missing, 4 injured. Non-operational: 2 aircraft; 2 pilots killed. **Squadron Aces** S/L R.W. McNair, DFC and 2 Bars 8 destroyed. F/L P.G. Johnson 5 destroyed. **Honours and Awards** 1 second bar to DFC, 2 bars to DFC, 5 DFC's. **Battle Honours** Defence of Britain 1942-1943. Fortress Europe 1942-1944. France and Germany 1944-1945. *Normandy 1944, Arnhem. Rhine.*

---

[1] The squadron was unique in that its aircraft carried the familiar Indian Head emblem of the McColl-Frontenac Oil Company.

251 Supermarine Spitfire Mk.IXB of No. 421 (F) Squadron undergoing maintenance checks while squadron personnel unload a harmonium under the watchful eye of the padre.

252 Supermarine Spitfire Mk.XVI AU-Y of No. 421 (F) Squadron on the Continent in the winter of 1944-45.

No. 422 Squadron

Badge A cubit arm holding in the hand a tomahawk
Motto This arm shall do it
Authority King George VI, October 1943
The painted arm indicates that the Red Indian brave is at war. The motto, from Shakespeare, refers to the squadron's striking power.
Formed as a General Reconnaissance unit at Lough Erne, Fermanagh, Northern Ireland on 2 April 1942 as the RCAF's 19th — fifth coastal — squadron formed overseas, the unit flew Catalina and Sunderland flying boats on convoy escort and anti-submarine patrol over the North Atlantic shipping route and the Bay of Biscay. When hostilities ended in Europe, there was a need for more long-range transport units to support the proposed "Tiger Force" in the Pacific, and the squadron was redesignated Transport on 5 June 1945 and began conversion training to Liberator aircraft. The sudden end of the war in the Far East found the squadron still in the early stages of conversion training, and it was disbanded on 3 September 1945.
Brief Chronology Formed as No. 422 (GR) Sqn, Lough Erne, Ferm., N. Ire. 2 Apr 42. Redesignated No. 422 (T) Sqn, Pembroke Dock, Wales 5 Jun 45. Disbanded at Bassingbourn, Cambs., Eng. 3 Sep 45.
Title or Nickname "Flying Yachtsmen"
Commanders
W/C L.W. Skey (Can/RAF), DFC 1 Jul 42 - 26 Oct 43.
W/C J.R. Frizzle 27 Oct 43 - 30 Oct 44.
W/C J.R. Sumner 31 Oct 44 - 3 Sep 45.
Higher Formations and Squadron Locations
Coastal Command:
No. 15 Group,
Lough Erne, Ferm., N. Ire. 2 Apr 42 - 29 Oct 42.
Kesh, Ferm. 30 Oct 42 - 3 Nov 42.
Oban, Argyll, Scot. 5 Nov 42 - 7 May 43.
Bowmore, Argyll 8 May 43 - 1 Nov 43.
St. Angelo, Ferm., N. Ire. 3 Nov 43 - 12 Apr 44.
Castle Archdale, Ferm. 13 Apr 44 - 3 Nov 44.
No. 19 Group,
Pembroke Dock, Wales 4 Nov 44 - 4 Jun 45.
Transport Command:
No. 47 Group,
No. 301 Wing,
Pembroke Dock, Wales 5 Jun 45 - 24 Jul 45.
Bassingbourn, Cambs. 25 Jul 45 - 3 Sep 45.
Representative Aircraft (Unit Code DG, 2)[1]
Saro Lerwick Mk.I (Jul - Nov 42, not on operations)
L7250 U   L7256 V   L7258 R   L7259 Q   L7260 P
L7264 N   L7266 Y   L7267 B
Consolidated Catalina Mk.IB, III & VB (Jul - Nov 42)
FP103 A   FP105 B   FP106 C   FP529 B   FR533 C
Short Sunderland Mk.III (Nov 42 - Jun 45) W6026 A
W6027 N   W6028 C   W6029 D   W6030 F   W6031 G
W6032 H   W6033 J   DD831 K   DD838 X   DD845 B
DD855 Y   DV990 R   EK567 Q   JM679 E   ML777 M
ML778 S   ML884 Z   NJ172 O   NJ173 W   NJ174 U
NJ175 T   NJ176 P   NJ189 V
Consolidated Liberator C.Mk.VI & VIII (Aug 45)

Operational History: First Mission 1 March 1943,[2] Sunderland W6026 2-C from Oban with W/C Skey and crew — anti-submarine patrol to assist a convoy joining up at position 5940N 1225W. **Victory, U-Boat** 10 March 1944, Sunderland EK591 2-U from St. Angelo with WO2 W.F. Morton and crew sank U-625 at position 5253N 2019W. Morton was on his first operational sortie as captain and was being screened by F/L S.W. Butler (RAF). **Last Mission** 1/2 June 1945, 5 Sunderlands from Pembroke Dock — escort to a repatriation convoy. **Summary** Sorties: 1116. Operational/Non-operational Flying Hours: 13,346/5842. Victories: U-boat: 1 sunk, 1 damaged; dropped 99 250-pound depth charges. Casualties: Operational: 9 aircraft; 70 aircrew, of whom 11 were killed, 31 presumed dead, 6 injured, 22 rescued. Non-operational: 2 aircraft; 33 personnel, of whom 5 were killed, 5 presumed dead, 23 injured. **Honours and Awards** 1 OBE, 1 MBE, 6 DFC's, 1 BEM, 1 Air Medal (USA), 22 MiD's. **Battle Honours** Atlantic 1942-1945. English Channel and North Sea 1944-1945: *Normandy 1944.* Biscay 1944-1945. Arctic 1942.

---

[1]During 1943, aircraft carried single digit "2" as unit code. See Appendix 4.

[2]This mission was actually the squadron's 28th sortie, but the first true general reconnaissance sortie. The first fifteen, between 30 August and 24 September 1942, involved Catalinas FP103 DG-A (S/L R.E. Hunter), FP105 DG-B (F/O J.W. Bellis) and FP106 DG-C (WO L.W.C. Linpert) on escort to Convoys PQ18 and PQ14, transporting personnel and equipment from Sullom Voe to Murmansk and Archangel, and sorties between the two Russian bases. On sorties 16 to 26, 29 October to 15 November 1942, Catalinas were ferried from Gander, Newfoundland to Lough Erne. Sortie 27 was flown on 25 February 1943 and involved Sunderland W6032 2-H, with F/O P.T. Sargent and crew, on a training anti-submarine patrol to Convoy ONS169 (38 motor vessels and 2 escorts bound for Canada) at 5642N 1245W.

254 Short Sunderland Mk.III EK591 aircraft "U" of No. 422 (GR) Squadron captured in that critical moment that spells success or failure when landing on calm waters. This aircraft scored the only confirmed submarine sinking for the squadron when piloted by WO2 W.F. Morton.

253 Saro Lerwick L7260 aircraft "P" of No. 422 (GR) Squadron at Lough Erne, Northern Ireland on 3 August 1942.

No. 423 Squadron

Badge A bald-headed eagle volant
Motto Quaerimus et petimus (We search and strike)
Authority King George VI, March 1943.
The bald-headed eagle is a powerful bird of prey from the Canadian side of the Atlantic. The motto refers to the squadron's role as a coastal unit.
Formed as a General Reconnaissance unit at Oban, Argyll, Scotland on 18 May 1942 as the RCAF's 21st — sixth and last coastal — squadron formed overseas, the unit flew Sunderland flying boats on convoy escort and anti-submarine patrols over the Atlantic shipping routes. When hostilities ended in Europe, there was a need for more long-range transport units to support the proposed "Tiger Force" in the Pacific, and the squadron was redesignated Transport on 5 June 1945 and began conversion training to Liberator aircraft. The sudden end of the war in the Far East found the squadron still in the early stages of conversion training and it was disbanded on 4 September 1945.
Brief Chronology Formed as No. 423 (GR) Sqn, Oban, Argyll, Scot. 18 May 42. Redesignated No. 423 (T) Sqn, Castle Archdale, Ferm., N. Ire. 5 Jun 45. Disbanded at Bassingbourn, Cambs., Eng. 4 Sep 45.
Commanders
W/C F.J. Rump (RAF) 18 May 42 - 9 Jul 43.
W/C L.G.G.J. Archambault 10 Jul 43 - 16 Jul 44.
W/C P.J. Grant 17 Jul 44 - 28 Feb 45.
W/C S.R. McMillan 1 Mar 45 - 4 Sep 45.
Higher Formations and Squadron Locations
Coastal Command:
No. 15 Group,
Oban, Argyll, Scot. 18 May 42 - 1 Nov 42.
Castle Archdale, Ferm., N. Ire. 2 Nov 42 - 4 Jun 45.
Transport Command:
No. 47 Group,
No. 301 Wing,
Castle Archdale, Ferm., N. Ire. 5 Jun 45 - 7 Aug 45.
Bassingbourn, Cambs. 8 Aug 45 - 4 Sep 45.
Representative Aircraft (Unit Code AB, 3)[1]
Short Sunderland Mk.III (Jul 42 - May 45) W6000 A
W6001 B W6007 G W6008 H W6009 J W6061 K
DD849 M DD853 K DD867 G DP181 D DP198 J
DW111 S EJ157 K EK583 J ML783 H ML784 L
ML883 F NJ182 N NJ183 G NJ184 C NJ185 E
NJ186·A NJ187 B
Consolidated Liberator C.Mk.VI & VII (Aug 45)
Operational History: First Mission 23 August 1942, Sunderland W6053 AB-E from Oban with F/L J. Musgrave and crew — 13-hour submarine search. **Victories**, U-boat: 12 May 1943, Sunderland W6006 3-G from Castle Archdale with F/L J. Musgrave and crew — escort to Convoy HX.237, shared with 2 destroyers (HMCS *Drumheller* and HMS *Lagan*) in sinking U-456 in the Western Approaches at 4837N 2239W. 4 August 1943, Sunderland DD859 3-G from Castle Archdale with F/O A.A. Bishop and crew — sank U-489 in the Western Approaches at 6111N 1438W. Shot down by return fire, five of the eleven-man crew were lost; the other six, all wounded, were rescued by a destroyer along with 23 survivors of the submarine. 8 October 1943, Sunderland DD863 3-J from Castle Archdale with F/O A.H. Russell and crew — escort to Convoy SC.143, sank U-610 in the Western Approaches at 5545N 2433W. 24 April 1944, Sunderland DD862 AB-A from Castle Archdale with F/L F.G. Fellows and crew — sank U-311 at 5036N 1836W. 11 September 1944, Sunderland ML825 AB-D from Castle Archdale with F/O J.N. Farren and crew — joined two of HMC Ships, *Dunver* and *Hespeler*, in sinking U-484 at 5651N 0804W. **Last Mission** 31 May 1945, Sunderland ML777 AB-F from Castle Archdale with F/L Magor and crew — patrol to the southwest of Ireland.
Summary Sorties 1392. Operational/Non-operational Flying Hours: 16,277/5122. Victories: U-boat: 3 sunk, 2 shared sinkings, 1 damaged; 25 actual submarine, periscope or schnorkel sightings, 10 of schnorkel smoke, 6 disturbances, swirls or suspicious oil slicks; delivered 201 250-pound depth charges in 26 attacks, an average of 1 attack for every 684 operational hours flown. Casualties: Operational: 6 aircraft; 43 aircrew, of whom 40 were killed or presumed dead, 2 wounded, 1 injured. Non-operational: 14 personnel, of whom 9 were killed, 5 injured. **Honours and Awards** 4 DFC's, 1 DFM. **Battle Honours** Atlantic 1942-1945. English Channel and North Sea 1944—1945: *Normandy 1944. Biscay 1944.*

---

[1] During 1943, aircraft carried single digit "3" as Unit code. See Appendix 4.

255 Short Sunderland Mk.III ML825 aircraft "D" of No. 423 (GR) Squadron scored the last of the squadron's five submarine sinkings in the hands of F/O J.N. Farren and crew on 11 September 1944.

No. 424 Squadron

Badge A heraldic tiger's head erased
Motto Castigandos castigamus (We chastise those who deserve to be chastised)
Authority King George VI, June 1945
The squadron was adopted by the City of Hamilton and adopted the tiger's head in reference to the Hamilton rugby team.
Formed at Topcliffe, Yorkshire, England on 15 October 1942 as the RCAF's 23rd — sixth Bomber — squadron formed overseas, the unit flew Wellington, Halifax and Lancaster aircraft on strategic and tactical bombing operations. From June to October 1943, it flew tropicalized Wellington aircraft from North Africa in support of the invasions of Sicily and Italy. After hostilities in Europe, the unit remained in England as part of Bomber Command's strike force, under which it lifted British and Canadian troops from Italy to the United Kingdom. The squadron was disbanded at Skipton-on-Swale, Yorkshire on 15 October 1945.
Brief Chronology Formed at Topcliffe, Yorks., Eng. 15 Oct 42. Disbanded at Skipton-on-Swale, Yorks. 15 Oct 45.
Title or Nickname "Tiger"
Adoption City of Hamilton, Ont.
Commanders
W/C H.M. Carscallen 20 Oct 42 - 16 Apr 43.
W/C G.A. Roy, DFC 17 Apr 43 - 2 Oct 43.
W/C J.P. McCarthy, DFC 3 Oct 43 - 29 Nov 43.
W/C A.N. Martin 18 Dec 43 - 21 Jan 44 KIA.
W/C J.D. Blane 27 Jan 44 - 28 Jul 44 KIA.
W/C G.A. Roy, DFC 15 Aug 44 - 9 Oct 44 POW.
W/C C.C.W. Marshall, DFC 19 Oct 44 - 26 Mar 45.
W/C R.W. Norris 27 Mar 45 - 30 Sep 45.
W/C R.D.P. Blagrave 1 Oct 45 - 15 Oct 45.
Higher Formations and Squadron Locations
Bomber Command:
No. 4 Group,
Topcliffe, Yorks. 15 Oct 42 - 31 Dec 42.
No. 6 (RCAF) Group,
Topcliffe, Yorks. 1 Jan 43 - 7 Apr 43.
Leeming, Yorks. 8 Apr 43 - 2 May 43.
Dalton, Yorks. 3 May 43 - 15 May 43.
En route to North Africa[1] 16 May 43 - 22 Jun 43.
Mediterranean Air Command:
No. 205 Group,
No. 331 (RCAF) Wing,
Kairouan/Pavillier, Tunisia 23 Jun 43 - 29 Sep 43.
Hani East Landing Ground, Tunisia 30 Sep 43 - 18 Oct 43.
En route to England (minus aircraft) 26 Oct 43 - 5 Nov 43.
Bomber Command:
No. 6 (RCAF) Group,
No. 63 (RCAF) Base,
Skipton-on-Swale, Yorks. 6 Nov 43 - 29 Aug 45.
No. 1 Group,
Skipton-on-Swale, Yorks. 30 Aug 45 - 15 Oct 45.
Representative Aircraft (Unit Code QB)
Vickers Wellington B.Mk.III (Oct 42 - Apr 43) X3401 E
X3409 J   X3426 D   X3789 N   X3790 H   Z1674 G
Z1691 R   Z1692 L   BJ658 Q   BJ714 F   BK144 V
BK348 J   BK398 W   BK435 U   BK436 P   BK560 C
DF613 X   DF618 B   DF621 O   DF671 K
Vickers Wellington B.Mk.X (England, Feb - May 43)
HE159 P   HE222 E   HE367 F   HE369 P   HE438 A
HE554 M   HE589 N   HE594 C   HE656 L   HE684 R
HE687 T   HE689 X   HE691 K   HE703 B   HE705 H
HE737 V   HE738 O   HE864 W   HZ272 S   HZ273 G
LN402 U   LN439 D   MS481 Q
Vickers Wellington B.Mk.X (North Africa, May - Nov 43)
HE273 B   HE296 V   HE482 W   HE513 A   HE515 T
HE535 R   HE536 E   HE794 D   HE795 F   HE929 V
HE962 N   HE963 U   HE967 O   HE968 Q   HE971 G
HE974 K   HF477 C   HZ371 E   HZ515 B   LN423 S
LN433 L   LN441 X   MS476 H   MS477 J
Handley Page Halifax B.Mk.III (Dec 43 - Jan 45)
HX313 B   HX318 O   HX319 P   LV879 A   LV959 R
LV991 U   LW113 F   LW131 J   LW170 I   LW347 X
LW416 L   LW433 W   LW438 T   LW444 H   MZ418 U
MZ458 G   MZ802 G MZ813 B   MZ822 F   MZ901 N
NP930 W   NP945 D   NR227 A   NR228 Q
Avro Lancaster B.Mk.I & III (Jan - Oct 45) ME456 K
ME458 T   NG277 G   NG279 O   NG280 U   NG281 X
NG347 P   NG400 R   NG441 L   NG446 J   NG451 E
NG456 D   NG457 C   NG458 H   NG459 K   NG484 L
NN777 F   PA324 K   PA326 W   PB989 A   RA504 M
RA507 S   RF128[2] V   RF150 W
Operational History: First Mission 15/16 January 1943, 5 Wellington I's from Topcliffe bombed Lorient, France. **Last Mission** 25 April 1945, 10 Lancaster I/III's from Skipton-on-Swale despatched to bomb gun positions on the Island of Wangerooge; 9 bombed the primary target, 1 dropped its bombs 3 miles from the target when the pilot was hit in the eye by a perspex splinter. **Summary** Sorties: 3257 (including 668 from North Africa, and 39 airlifting 884 liberated POW's from Europe to England). Operational/Non-operational Flying Hours: 17,478/7106. Bombs dropped: 8776 tons. Victories: Aircraft: 9 destroyed, 5 damaged, in 60 engagements. Casualties: Operational: 52 aircraft (including 12 in the Mediterranean); 313 aircrew, of whom 37 were killed, 252 missing (of whom 16 proved safe), 14 POW, 10 injured. Non-operational: 4 aircraft; 9 personnel killed, 5 injured.
Honours and Awards 1 DSO, 1 bar to DFC, 49 DFC's, 1 CGM, 11 DFM's, 1 MiD. **Battle Honours** English Channel and North Sea 1943-1945. Baltic 1944-1945. Fortress Europe 1943-1944. France and Germany 1944-1945: *Biscay Ports 1943-1944, Ruhr 1943-1945, Berlin 1944, German Ports 1943-1945, Normandy 1944, Rhine.* Biscay 1943-1944. Sicily 1943. Italy 1943: *Salerno.*

---

[1]Ground crew by sea 16-27 May, aircraft 5-7 June. See also No. 331 (Medium Bomber) Wing.
[2]Flew the squadron's 2000th four-engine sortie, on 21 March 1945 (F/L J.F. Thomas and crew).

256 Handley Page Halifax B.Mk.III, coded QB-O, of No. 424 (B) Squadron at Skipton-on-Swale, Yorks. on 13 November 1944.

257 Avro Lancaster Mk.I NG347, aircraft "P" of No. 424 (B) Squadron climbing out from its base at Skipton-on-Swale, Yorks.

258 Avro Lancaster B. Mk.III, coded QB-V, of No. 424 (B) Squadron.

No. 425 Squadron

Badge A lark volant
Motto Je te plumerai (I shall pluck you)
Authority King George VI, January 1945
The badge is derived from the squadron nickname 'Alouette', and the motto comes from the refrain of the French Canadian folk song. The lark is shown in hovering position indicative of a bomber over the target about to strike the enemy.

Formed at Dishforth, Yorkshire, England on 25 June 1942 as the RCAF's 22nd — fifth Bomber — squadron formed overseas, the unit flew Wellington and Halifax aircraft on strategic and tactical bombing operations. It was a unique formation within the RCAF in that the organization order designated it "French-Canadian" and Bomber Command was combed for French-Canadian air and ground crew to fill its ranks. From June to October 1943, the squadron flew tropicalized Wellington aircraft in North Africa in support of the invasions of Sicily and Italy. After hostilities in Europe it was selected as part of "Tiger Force" for duty in the Pacific, converted to Canadian-built Lancaster aircraft, and returned to Canada for reorganization and training. The sudden end of the war in the Far East resulted in the squadron being disbanded at Debert, Nova Scotia on 5 September 1945.

Brief Chronology Formed at Dishforth, Yorks., Eng. 25 Jun 42. Disbanded at Debert, N.S. 5 Sep 45.
Title or Nickname "Alouette" (October 1942)
Adoption La Presse Newspaper Auxiliary, Montreal, and City of Quebec (March 1945).
Commanders
W/C J.M.W. St. Pierre 25 Jun 42 - 30 Sep 43 *OTE*.
W/C J.A.D.B. Richer, DFC 1 Oct 43 - 3 Apr 44 *OTE*.
W/C A.R. McLernon, DFC 4 Apr 44 - 23 May 44.
W/C J.H.L. Lecomte 24 May 44 - 20 Aug 44.
W/C H.C. Ledoux 21 Aug 44 - 10 Jun 45.
S/L L.P.J. Dupuis, DFC 15 Jun 45 - 5 Sep 45.
Higher Formations and Squadron Locations
Bomber Command:
No. 4 Group,
Dishforth, Yorks. 25 Jun 42 - 31 Dec 42.
No. 6 (RCAF) Group,
Dishforth, Yorks. 1 Jan 43 - 15 May 43.
En route to North Africa[1] 16 May 43 - 22 Jun 43.
Mediterranean Air Command:
No. 205 Group,
No. 331 (RCAF) Wing,
Kairouan/Zina, Tunisia 23 Jun 43 - 29 Sep 43.
Hani East Landing Ground, Tunisia 30 Sep 43 - 17 Oct 43.
En route to England (minus aircraft) 26 Oct 43 - 5 Nov 43.
Bomber Command:
No. 6 (RCAF) Group,
No. 61 (RCAF) Base,
Dishforth, Yorks. 6 Nov 43 - 9 Dec 43.
No. 62 (RCAF) Base,
Tholthorpe, Yorks. 10 Dec 43 - 12 Jun 45.
En route to Canada 13 Jun 45 - 14 Jun 45.
RAF "Tiger Force" (for ops),

RCAF Eastern Air Command (for training):
No. 6 (RCAF) Group,
No. 663 (RCAF) Wing,
Debert, N.S. 15 Jun 45 - 5 Sep 45.
Representative Aircraft (Unit Code KW)
Vickers Wellington B.Mk.III (Aug 42 - Apr 43)
X3364 J   X3648 R   X3763 E   X3803 H   X3872 A
Z1603 L   Z1729 T   Z1742 C   BJ605 A   BJ657 G
BJ644 Q   BJ755 D   BJ783 F   BJ894 K   BJ918 F
BK308 C   BK332 O   BK333 N   BK334 V   BK337 P
BK344 B   BK401 B   BK557 S   BK593 S
Vickers Wellington B.Mk.X (England, Apr - May 43)
HE423 A   HE475 E   HE486 T   HE491 U   HE500 C
HE592 Q   HE955 O   HE655 B   HE733 S   HE865 K
HE901 V   HE903 W   HZ277 F   HE355 D   LN409 I
MS493 Q
Vickers Wellington B.Mk.X (North Africa, May - Oct 43)
HE260 P   HE261 R   HE267 I   HE269 R   HE270 K
HE329 O   HE330 C   HE516 S   HE521 V   HE522 B
HE551 T   HE900 E   HE930 A   HE931 L   HE970 C
HE976 H   HE977 I   HE978 X   HE979 Y   HZ514 Q
HZ809 A   LN436 D   LN440 W   MS492 F
Handley Page Halifax B.Mk.III (Dec 43 - May 45)
LL594 X   LL595 K   LL596 U   LW390 J   LW391 N
LW412 Q   LW413 Q   LW414 E   LW415 K   LW424 T
LW428 C   LW715 Q   MZ418 C   MZ419 E   MZ454 S
MZ602 T   MZ815 C   MZ954 M   NA201 S   NA518 G
NP999 W   NR136 P   NR176 R   NR271 N
Avro Lancaster B.Mk.X (May - Sep 45, not on operations)
KB875 U   KB876 L   KB894 A   KB899 X   KB903 R
KB912 Q   KB915 H   KB916 C   KB917 E   KB918 P
KB924 T   KB926 F   KB930 N   KB931 S   KB932 O
KB934 I   KB936 G   KB944 K   KB954 V   KB962 D
Operational History: First Mission 5/6 October 1942, 5 Wellington III's from Dishforth despatched to bomb Aachen; 4 bombed the primary target, the fifth what was believed to be Aachen. **Last Mission** 25 April 1945, 18 Halifax III's from Tholthorpe bombed gun positions on the Island of Wangerooge. **Summary** Missions/Sorties: 328/3694 (including 88/741 from North Africa.) Operational/Non-operational Flying Hours: 20,231/7653. Bombs dropped: 9152 tons. Victories: Aircraft: 7 destroyed, 1 damaged. Casualties: Operational: 55 aircraft; 345 aircrew, of whom 37 were killed, 153 presumed dead, 91 POW (of whom 1 escaped), 7 interned, 53 evaded capture or proved safe, 4 missing. Non-operational: 11 aircraft; 73 personnel, of whom 64 were killed, 8 injured, 1 died of natural causes. **Honours and Awards** 2 MBE's, 4 bars to DFC, 63 DFC's, 2 GM's, 18 DFM's, 1 DFC (USA), 4 MiD's. **Battle Honours** English Channel and North Sea 1942-1943. Fortress Europe 1942-1944. France and Germany 1944-1945: *Biscay Ports 1943-1944, Ruhr 1942-1945, Berlin 1944, German Ports 1942-1945, Normandy 1944, Rhine.* Biscay 1942-1943. Sicily 1943. Italy 1943: *Salerno.*

---

[1] Ground crew by sea, 16-27 May; aircraft took off 5 June. One Wellington (HE268 KW-K) was shot down over the Bay of Biscay and the 7-man crew interned in Portugal. See also No. 331 (Medium Bomber) Wing.

259   Handley Page Halifax Mk.III MZ454 aircraft "S" of No. 425 (B) Squadron in its dispersal at Tholthorpe, Yorks.

260   Avro Lancaster Mk.X KB936 coded KW-G and displaying the No. 425 (B) Squadron lark. The inscription in French is taken from the unit's badge ("I shall pluck you").

No. 426 Squadron

Badge A thunderbird
Motto On wings of fire
Authority King George VI, October 1943

The thunderbird is a mythical bird, the sight of which is supposed to cause havoc and death to those who perceive it. It was the name given by some Indians to the first aeroplane they saw. The thunderbird signified disaster to those on the ground who have incurred its displeasure.

Formed at Dishforth, Yorkshire, England on 15 October 1942 as the RCAF's 24th — seventh Bomber — squadron formed overseas, the unit flew Wellington, Lancaster and Halifax aircraft on strategic and tactical bombing operations. After hostilities in Europe, to meet a need for long-range transport units to support the proposed "Tiger" Force for duty in the Pacific, it was redesignated Transport on 25 May 1945 and converted to Liberator aircraft. Between October and December 1945, it airlifted British troops to Egypt and Indian troops from Egypt to India, and British troops back to England. The squadron was disbanded at Tempsford, Bedfordshire on 1 January 1946.[1]

Brief Chronology Formed as No. 426 (B) Sqn, Dishforth, Yorks., Eng. 15 Oct 42. Redesignated No. 426 (T) Sqn, Driffield, Yorks. 25 May 45. Disbanded at Tempsford, Beds. 1 Jan 46.

Title or Nickname "Thunderbird"
Adoption: City of Regina Auxiliary, Sask.

Commanders
W/C S.S. Blanchard 15 Oct 42 - 14 Feb 43 *KIA*.
W/C L. Crooks (RAF), DSO, DFC 15 Feb 43 - 17 Aug 43 *KIA*.
W/C W.H. Swetman, DSO, DFC 18 Aug 43 - 4 Apr 44 *OTE*.
W/C E.C. Hamber, DFC 8 Apr 44 - 10 Jul 44 *OTE*.
W/C C.W. Burgess 11 Jul 44 - 2 Jan 45 *OTE*.
W/C F.C. Carling-Kelly 3 Jan 45 - 28 Jan 45 *POW*.
W/C C.M. Black, DFC 29 Jan 45 - 24 May 45.
W/C D.R. Miller, AFC 25 May 45 - 11 Dec 45.
W/C J.F. Green, DFC 12 Dec 45 - 1 Jan 46.

Higher Formations and Squadron Locations
Bomber Command:
No. 4 Group,
Dishforth, Yorks. 15 Oct 42 - 31 Dec 42.
No. 6 (RCAF) Group,
No. 61 (RCAF) Base,
Dishforth, Yorks. 1 Jan 43 - 17 Jun 43.
No. 62 (RCAF) Base,
Linton-on-Ouse, Yorks. 18 Jun 43 - 24 May 45.
Transport Command:
No. 47 Group,
Driffield, Yorks. 25 May 45 - 24 Jun 45.
Tempsford, Beds. 25 Jun 45 - 1 Jan 46.

Representative Aircraft (Unit Code OW)
Vickers Wellington Mk.III (Oct 42 - Apr 43) X1599 E
Vickers Wellington B.Mk.X (Mar - Jun 43)
Avro Lancaster B.Mk.II (Jul 43 - May 44) DS621 N
DS647 R DS686 D DS687 L DS711 B DS741 T
DS757 D DS759 K DS760 M DS763 O DS771 P
DS775 W DS789 A DS829 J DS838 J DS840 C
DS841 Q DS852 P LL621 N LL634 F LL687 L
LL688 R LL700 X
Handley Page Halifax B.Mk.III (Apr - Jun 44) LK796 M
LK871 K LK879 P LK880 C LK883 E LW377 G
LW382 Q LW598 J LW682 M MZ598 J MZ600 D
MZ645 N MZ750 S MZ682 N MZ690 U NA510 E
Handley Page Halifax B.Mk.VII (Jun 44 - May 45)
LW199 C LW200 N LW201 D LW203 F LW204 K
LW205 Y LW206 Q LW207 W LW209 E LW210 B
LW510 D LW775 G NP684 O NP709 A NP737 Z
NP770 G NP779 C NP797 G NP799 J NP800 S
NP811 G NP818 M PN238 L RG452 V
During Dec 44 - Mar 45 the squadron also had 5 B.Mk.III's: NA202 A NA204 N NR134 Z NR144 V RG350 K
Consolidated Liberator C.Mk.VI&VIII (Jul - Dec 45, Unit Code OLW) KG918 X KG983 Y KH179 HW
KH181 LW KH224 NW KH329 Z KH333 JW
KH381 W KK255 Q KK267 O KK340 R
KK374 T KL618 K KL619 H KL621 D
KL625 V KL639 F KL650 J KL663 C KL670 G
KN833 OW

Operational History: First Mission 14/15 January 1943, 7 Wellington III's from Dishforth bombed Lorient, France. **Last Mission - Bombing** 25 April 1945, 20 Halifax VII's from Linton-on-Ouse bombed gun positions on the Island of Wangerooge; 1 aircraft failed to return. **First Mission, Transport** 30 September 1945, 2 Liberators departed Tempsford for India each carrying 26 passengers and both returned on 7 October with 26 passengers. Aircraft and captains were: KL641 OLW-A (F/L K.W. Warner to India, F/O W.A. Craig return to England), KK341 OLW-P (F/O H.E. Miskiman to India, F/L L. Greenburgh return to England). **Last Mission** 20 December 1945, 4 Liberators departed for India; the last (KK265 OLW-O) returned to England on 29 December. **Summary** Bomber Missions/Sorties: 268 (including 242 bombing, 19 sea-mining, 7 sea searches)/3233. Operational/Non-operational Flying Hours: 11,184/3327. Bombs dropped: 8997 tons. Victories: Aircraft: 4 destroyed, 1 probably destroyed, 5 damaged. Casualties: Operational: 88 aircraft; 557 aircrew, of whom 87 were killed, 337 presumed dead, 97 POW, 36 evaded capture, were interned or proved safe. Non-operational: 4 personnel killed. Transport Sorties: 242. Operational/Non-operational Flying Hours: 5631/3716. Airlifted: 5500 passengers. Casualties: nil.
Honours and Awards 2 DSO's, 2 bars to DFC, 130 DFC's, 1 CGM, 25 DFM's, 2 BEM's, 1 DFC (USA), 13 MiD's.
Battle Honours English Channel and North Sea 1943. Baltic 1943. Fortress Europe 1943-1944. France and Germany 1944-1945: *Biscay Ports 1943-1944, Ruhr 1943-1945, Berlin 1943-1944. German Ports 1943-1945, Normandy 1944, Rhine.* Biscay 1943.

---

[1] Personnel volunteering for further service were assigned along with the squadron's aircraft to an RAF Liberator squadron at Tempsford.

261 Vickers Wellington Mk.III X1599, coded OW-E, of No. 426 (B) Squadron in the fall of 1942.

262 Handley Page Halifax Mk.VII NP818, coded OW-M, of No. 426 (B) Squadron based at Linton-on-Ouse, Yorks.

263 Avro Lancaster Mk.II DS840, coded OW-C, of No. 426 (T) Squadron running up on the chocks.

264 Consolidated Liberator C.Mk.VIII KL658, coded W-M, of No. 426 (T) Squadron at Driffield, Yorks.

No. 427 Squadron

Badge In front of a maple leaf a lion rampant
Motto Ferte manus certas (Strike sure)
Authority King George VI, October 1943
The combination of a lion representing England and a maple leaf for Canada indicates the formation of this squadron in England.
Formed at Croft, Yorkshire, England on 7 November 1942 as the RCAF's 25th — eighth Bomber — squadron formed overseas, the unit flew Wellington, Halifax and Lancaster aircraft on strategic and tactical bombing operations. After hostilities in Europe, it remained in England as part of Bomber Command's strike force, under which it airlifted Allied prisoners of war, and British troops from Italy, back to England. The squadron was disbanded at Leeming, Yorkshire on 1 June 1946.
Brief Chronology Formed at Croft, Yorks., Eng. 7 Nov 42. Disbanded at Leeming, Yorks. 1 Jun 46.
Title or Nickname "Lion"
Adoption Metro-Goldwyn-Mayer Studios, Hollywood, USA, May 1943 (MGM also symbolized by lion).
Commanders
W/C D.H. Burnside (RAF), DFC and Bar 7 Nov 42 - 5 Sep 43 OTE.
W/C R.S. Turnbull, DFC, AFC, DFM 6 Sep 43 - 13 Jun 44 2OTE.
W/C C.J. Cribb (RAF), DFC 14 Jun 44 - 27 Aug 44 OTE.
W/C E.M. Bryson 28 Aug 44 - 26 Sep 44 OTE.
W/C V.F. Ganderton, DSO, DFC 27 Sep 44 - 8 May 45 2 OTE.
W/C E.M. Bryson, DFC 9 May 45 - 11 Jun 45.
W/C R.D.P. Blagrave 12 Jun 45 - 2 Oct 45.
W/C J.C.R. Brown, DFC and Bar 3 Oct 45 - 1 Jun 46.
Higher Formations and Squadron Locations
Bomber Command:
No. 4 Group,
Croft, Yorks. 7 Nov 42 - 31 Dec 42.
No. 6 (RCAF) Group,
Croft, Yorks. 1 Jan 43 - 4 May 43.
No. 63 (RCAF) Base,
Leeming, Yorks. 5 May 43 - 29 Aug 45.
No. 1 Group,
Leeming, Yorks. 30 Aug 45 - 1 Jun 46.
Representative Aircraft (Unit Code ZL)
Vickers Wellington B.Mk.III (Nov 42 - Mar 43)
Z1604 P   BK364 K   BK389 L
Vickers Wellington B.Mk.X (Feb - May 43)
Handley Page Halifax B.Mk.V (May 43 - Feb 44) EB241 A
EB243 U   EB246 S   EB247 P   DK146 Q   DK186 L
DK226 Y   LK627 K   LK633 G   LK637 V   LK643 T
LK644 C   LK658 H   LK684 R   LK799 N   LK923 B
LK965 W   LK972 F   LK974 J   LK975 E   LK976 Z
LL139 D   LL169 L   LL194 C
Handley Page Halifax B.Mk.III (Jan 44 - Mar 45)
MZ304 B   LK755 S   LV828 Q   LV829 D   LV830 H
LV831 P   LV883 W   LV922 B   LV986 V   LV987 K
LV994 J   LV995 Y   LV996 E   LW161 V   LW162 D
LW163 U   LW548 L   LW558 A   LW559 F   LW572 R
LW575 O   LW576 C   LW577 K   LW759 Z
Avro Lancaster B.Mk.I & III (Mar 45 - May 46)
ME393 D   ME426 C   ME498 K   ME501 T   NX548 J
NX549 U   NX550 V   NX551 G   NX552 S   NX553 H
NX554 F   NX555 R   PA260 Q   PA263 E   PA271 W
RA534 A   RA536 N   RA537 P   RA538 B   RA539 O
Operational History: First Mission 14 December 1942, 3 Wellington III's from Croft despatched to lay mines in the Frisian Islands area; 2 aborted owing to weather. **First Mission, Bombing** 15/16 January 1943, 6 Wellington III's from Croft despatched to bomb Lorient, France; 5 bombed the primary target, 1 failed to return. **Last Mission** 25 April 1945, 10 Lancasters from Leeming despatched to bomb gun positions on the Island of Wangerooge; 9 bombed the primary target, 1 bombed buildings on the Island of Spiekeroog. **Summary** Sorties: 3328. Operational/Non-operational Flying Hours: 18,512/11,271. Bombs dropped: 10,294 tons. Victories: Aircraft: 10 destroyed, 1 probably destroyed, 10 damaged. Casualties: Operational: 90 aircraft; 522 aircrew, of whom 35 were killed, 477 missing (of whom 10 POW, 11 proved safe), 10 injured. Non-operational: 6 aircraft; 22 personnel, of whom 15 were killed, 5 injured, 2 died of natural causes. **Honours and Awards** 4 DSO's, 6 bars to DFC, 147 DFC's, 1 AFC, 2 CGM's, 16 DFM's, 8 MiD's. **Battle Honours** English Channel and North Sea 1943-1945. Baltic 1944-1945. Fortress Europe 1943-1944. France and Germany 1944-1945: *Biscay Ports 1943-1944, Ruhr 1943-1945, Berlin 1943-1944, German Ports 1943-1945, Normandy 1944, Rhine. Biscay 1944.*

265 Handley Page Halifax B.Mk.III, "Gutsy Girty", of No. 427 (B) Squadron at Leeming, Yorks. on 3 November 1944, prior to another night operation.

No. 428 Squadron

Badge In a shroud, a death's head
Motto Usque ad finem (To the very end)
Authority King George VI, February 1945
The badge refers to the squadron 'Ghost', a designation earned through many hours of night bombing operations, and also the death and destruction which it carried to the enemy.
Formed at Dalton, Yorkshire, England on 7 November 1942 as the RCAF's 26th — ninth Bomber — squadron formed overseas, the unit flew Wellington, Halifax and Canadian-built Lancaster aircraft on strategic and tactical bombing operations. After hostilities in Europe, it was selected as part of "Tiger Force" for duty in the Pacific, and was the first squadron to return to Canada for reorganization and training. The sudden end of the war in the Far East resulted in the squadron being disbanded at Yarmouth, Nova Scotia on 5 September 1945.
Brief Chronology Formed at Dalton, Yorks., Eng. 7 Nov 42. Disbanded at Yarmouth, N.S. 5 Sep 45.
Title or Nickname "Ghost"
Adoption Imperial Order of the Daughters of the Empire, Toronto, Ont.
Commanders
W/C A. Earle (RAF) 7 Nov 42 - 20 Feb 43.
W/C D.W.M. Smith (Can/RAF), DFC 21 Feb 43 - 14 Sep 43 *POW*.
W/C W.R. Suggitt, DFC 15 Sep 43 - 30 Oct 43 *OTE*.
W/C D.T. French, DFC 31 Oct 43 - 8 May 44 *OTE*.
W/C W.A.G. McLeish, DFC 9 May 44 - 7 Aug 44 *OTE*.
W/C A.C. Hull, DFC 8 Aug 44 - 1 Jan 45 *OTE*.
W/C M.W. Gall 2 Jan 45 - 2 Jun 45.
W/C C.M. Black, DFC 26 Jul 45 - 5 Sep 45.
Higher Formations and Squadron Locations
Bomber Command:
No. 4 Group,
Dalton, Yorks. 7 Nov 42 - 31 Dec 42.
No. 6 (RCAF) Group,
Dalton, Yorks. 1 Jan 43 - 3 Jun 43.
No. 64 (RCAF) Base,
Middleton St. George, Yorks. 4 Jun 43 - 30 May 45.
En route to Canada 31 May 45 - 2 Jun 45.
RAF "Tiger Force" (for ops),
RCAF Eastern Air Command (for training):
No. 6 (RCAF) Group,
No. 661 (RCAF) Wing,
Yarmouth, N.S. 25 Jul 45 - 5 Sep 45.
Representative Aircraft (Unit Code NA)
Vickers Wellington B.Mk.III (Nov 42 - Apr 43) X3541 N
X3550 J   Z1719 P   Z1727 K   BK154 V   BK155 B
BK156 Q   BK337 D   BK562 E   BK563 C   BK564 R
DF635 I   DF668 W
Vickers Wellington B.Mk.X (Apr - Jun 43) HE158 L
HE174 T   HE175 U   HE176 F   HE177 G   HE288 P
HE319 Y   HE321 Z   HE322 J   HE367 V   HE505 X
HE656 A   HE727 K   HE728 B   HE738 O   HE750 R
HE751 S   HE864 D   HE873 H   HE911 I   HE917 W
HE918 C   LN424 E   MS481 Q
Handley Page Halifax B.Mk.V (Jun 43 - Jan 44)
W1271 P   DK196 Z   DK228 D   DK229 W   DK230 V
DK233 X   DK235 J   DK237 L   DK238 I   DK239 Q
EB209 C   DB210 E   EB211 F   EB212 U   EB213 G
EB214 S   EB215 T   EB216 R   LK901 O   LK913 N
LK914 K   LK927 Y   LK928 B   LK930 M
Handley Page Halifax B.Mk.II (Nov 43 - Jun 44)
HR857 K   HR988 C   HX183 E   JH971 H   JN271 M
JN955 L   JN967 X   JN968 I   JN969 V   JN973 U
JP113 A   JP122 J   JP124 N   JP127 T   JP130 Y
JP132 D   JP191 B   JP195 F   JP197 Q   JP198 G
JP199 O   JP201 S   JP279 P   LW285 Z
Avro Lancaster B.Mk.X (Jun 44 - Sep 45)[1] KB709 G
KB711 N   KB744 J   KB747 X   KB757 C   KB760 P
KB763 S   KB764 B   KB770 D   KB777 V   KB778 Y
KB780 T   KB781 U   KB782 H   KB791 A   KB792 I
KB793 E   KB795 Q   KB820 M   KB838 O   KB855 F
KB867 L   KB882 R   KB920 K
Operational History: First Mission 26/27 January 1943, 6 Wellington III's from Dalton despatched to bomb U-boat base at Lorient, France; 5 bombed the primary target, 1 aborted. **Last Mission** 25 April 1945, 15 Lancaster X's from Middleton St. George bombed gun positions on the Island of Wangerooge. **Summary** Missions/Sorties: 283/3467. (These figures include 66 minelaying missions,[2] and 45 sorties airlifting 1055 liberated POW's). Operational/Non-operational Flying Hours: 21,188/7089. Bombs dropped: 9378 tons. Victories: Aircraft: 3 destroyed, 1 probably destroyed, 3 damaged. Casualties: Operational: 84 aircraft, 463 aircrew, of whom 63 were killed, 377 missing, 13 POW, 2 evaded capture, 8 injured. Non-operational: 4 aircraft; 20 personnel, of whom 16 were killed, 3 injured, 1 died. **Honours and Awards** 2 DSO's, 71 DFC's, 2 CGM's, 6 DFM's. **Battle Honours** English Channel and North Sea 1943-1944. Baltic 1944. Fortress Europe 1943-1944. France and Germany 1944-1945: *Biscay Ports 1943-1944, Ruhr 1943-1945, Berlin 1943-1944, German Ports 1943-1945, Normandy 1944, Rhine.* Biscay 1943-1944.

---

[1] The squadron flew the RCAF's first Lancaster X mission on 14/15 July 1944 when seven aircraft bombed St Pol, France. The order of take-off was KB737 NA-R, KB704 NA-E, KB758 NA-Z, KB725 NA-L, KB742 NA-M, KB705 NA-F and KB739 NA-W.
[2] On 4/5 January 1944 six Halifax II's carried out Bomber Command's first high-level minelaying operation, dropping their mines by parachute into the inner harbour at Brest from heights of 14-15,000 feet. Prior to this, mines had been planted from heights below 6000 feet.

266 Vickers Wellington B.Mk.X HE158 aircraft "L" of No. 428 (B) Squadron illustrates the severity of battle damage a Wellington could sustain and still return to its base (Dalton, Yorks.).

267 Avro Lancaster Mk.X aircraft of No. 428 (B) Squadron depart for Canada on 31 May 1945.

No. 429 Squadron

Badge On a mount, a bison the head lowered
Motto Fortunae nihil (Nothing to chance)
Authority King George VI, June 1944
The bison is a fierce and powerful opponent indigenous to Canada.
Formed at East Moor, Yorkshire, England on 7 November 1942 as the RCAF's 27th — tenth Bomber — squadron formed overseas, the unit flew Wellington, Halifax and Lancaster aircraft on strategic and tactical bombing operations. After hostilities in Europe, it remained in England as part of Bomber Command's strike force, under which it airlifted Allied prisoners of war, and British troops from Italy, back to England. The squadron was disbanded at Leeming, Yorkshire on 1 June 1946.
Brief Chronology Formed at East Moor, Yorks., Eng. 7 Nov 42. Disbanded at Leeming, Yorks. 1 Jun 46.
Title or Nickname "Bison"
Adoptions City of Bradford, Yorks., Eng. (January 1943) City Council of Lethbridge, Alta. (1944)
Commanders
W/C J.A.P. Owen (RAF) 7 Nov 42 - 31 May 43.
W/C J.L. Savard, DFC 1 Jun 43 - 22 Jun 43 KIA.
W/C J.A. Piddington (Can/RAF) 28 Jun 43 - 27 Jul 43 KIA.
W/C J.D. Patterson, DFC 30 Jul 43 - 2 Mar 44.
W/C A.F. Avant, DFC 1 May 44 - 10 Oct 44.
W/C R.L. Bolduc 11 Oct 44 - 9 Apr 45.
W/C E.H. Evans 10 Apr 45 - 31 May 45.
W/C J. Comar, DFC and Bar 26 Sep 45 - 1 Jun 46.
Higher Formations and Squadron Locations
Bomber Command:
No. 4 Group,
East Moor, Yorks. 7 Nov 42 - 31 Mar 43.
No. 6 (RCAF) Group,
No. 62 (RCAF) Base,
East Moor, Yorks. 1 Apr 43 - 12 Aug 43.
No. 63 (RCAF) Base,
Leeming, Yorks. 13 Aug 43 - 29 Aug 45.
No. 1 Group,
Leeming, Yorks. 30 Aug 45 - 1 Jun 46.
Representative Aircraft (Unit Code AL)
Vickers Wellington B.Mk.III (Nov 42 - Aug 43)
Vickers Wellington B.Mk.X (Jan - Aug 43)
Handley Page Halifax B.Mk.II (Aug 43 - Jan 44)
JD164 K JD275 T JD318 F JD325 F JD333 W
JD327 E JD411 A LW285 X
Handley Page Halifax B.Mk.V (Nov 43 - Mar 44)
LK662 Q LK697 D LK734 C LK746 K LK947 R
LK974 Z LK993 J LK995 C LL168 M LL170 N
LL171 U LL178 V
Handley Page Halifax B.Mk.III (Mar 44 - Mar 45)
HX339 D MZ282 A MZ303 R MZ312 Y MZ318 V
MZ424 Z MZ755 O MZ824 G MZ825 E MZ880 X
MZ906 H NP954 P NR203 P LK800 N LK804 Q
LV860 T LV913 N LV914 V LV950 C LV967 R
LV989 R LW127 F LW136 Z LW137 K
Avro Lancaster B.Mk.I & III (Mar 45 - May 46)
ME534 O ME536 Q ME537 N ME538 E ME539 A
ME540 P ME543 B NG343 J NG344 U NG345 V
NP967 L NN701 T PA225 G PA226 H PA227 C

PA273 R PA274 F PD209 K RE153 V RE155 X
RE571 D RF207 S RF252 G RF257 W

Operational History: First Mission 21 January 1943, 3 Wellington X's from East Moor despatched to lay mines at Terschelling; 1 completed the mission, 1 aborted, 1 failed to return. **First Mission, Bombing** 26/27 January 1943, 10 Wellington X's from East Moor despatched to bomb Lorient, France; 6 bombed the primary target, 3 aborted, 1 failed to return. **Last Mission** 25 April 1945, 10 Lancaster I/III's from Leeming despatched to bomb gun positions on the Island of Wangerooge; 9 bombed the primary target, 1 aborted. **Summary** Sorties: 3221 (including 45 airlifting 1055 POW's back to England). Operational/Non-operational Flying Hours: 17,289/10,934. Bombs dropped: 9356 tons. Victories: Aircraft: 7 destroyed, 1 probably destroyed, 10 damaged. Casualties: Operational: 71 aircraft; 451 aircrew, of whom 82 were killed, 322 missing, 23 POW, 7 injured; 17 proved safe. Non-operational: 11 aircraft; 25 personnel, of whom 15 were killed, 10 injured. **Honours and Awards** 2 bars to DFC, 45 DFC's, 1 AFC, 1 CGM, 7 DFM's. **Battle Honours** English Channel and North Sea 1943-1945. Baltic 1943-1945. Fortress Europe 1943-1944. France and Germany 1944-1945: *Biscay Ports 1943-1944, Ruhr 1943-1945, Berlin 1943-1944, German Ports 1943-1945, Normandy 1944, Rhine. Biscay 1943-1944.*

268 *Handley Page Halifax Mk.II, with the inscription "Easy does it," is inspected on 22 November 1943 in the service of No. 429 (B) Squadron at Leeming, Yorks.*

269 *Avro Lancaster Mk.I PA226 aircraft "H" of No. 429 (B) Squadron showing battle damage to the starboard wing.*

No. 430 Squadron

Badge none
Formed as an Army Co-operation unit at Hartford Bridge, Hampshire, England on 1 January 1943 as the RCAF's 30th — third and last army co-operation — squadron formed overseas, and redesignated Fighter Reconnaissance on 28 June 1943, the unit flew Mustang and Spitfire aircraft on air intelligence work, carrying out photographic reconnaissance for Allied invasion planners, and before-and-after photographs of air attacks on German "Noball" (V-1 flying bomb) launching sites. After 6 June 1944, it provided tactical photographic reconnaissance for ground forces in North-West Europe. The squadron was disbanded at Luneburg, Germany on 7 August 1945.
Brief Chronology Formed as No. 430 (AC) Sqn, Hartford Bridge, Hants., Eng. 1 Jan 43. Redesignated No. 430 (FR) Sqn, Dunsfold, Surrey 28 Jun 43. Disbanded at Luneburg, Ger. 7 Aug 45.
Title or Nickname "City of Sudbury"
Adoption City of Sudbury, Ont. (July 1943).
Commanders
W/C E.H.G. Moncrieff, AFC 1 Jan 43 - 4 Jul 43.
S/L R.A. Ellis, DFC 15 Jul 43 - 13 Sep 43.
S/L F.H. Chesters 14 Sep 43 - 2 Oct 44 OTE.
S/L J. Watts 3 Oct 44 - 2 Mar 45 OTE.
S/L C.D. Bricker, DFC 3 Mar 45 - 11 May 45 OTE.
S/L H.W. Russell 12 May 45 - 7 Aug 45.
Higher Formations and Squadron Locations
Army Co-operation Command:
No. 39 (RCAF) Wing,
Hartford Bridge, Hants. 1 Jan 43 - 11 Jan 43.
Dunsfold, Surrey 12 Jan 43 - 31 May 43.
Armament Practice Camp, Weston Zoyland, Som. 25 Apr - 5 May 43.
Fighter Command:
No. 11 Group,
Dunsfold, Surrey 1 Jun 43 - 4 Jul 43.
Second Tactical Air Force:
No. 83 (Composite) Group,
No. 39 (RCAF) Sector (disbanded 1 Jul 44),
No. 129 (RCAF) Wing,
Ashford, Kent 13 Aug 43 - 14 Oct 43.
Gatwick, Surrey 15 Oct 43 - 31 Mar 44.
No. 15 Armament Practice Camp, Peterhead, Scot. 5-18 Jan. 44.
Exercise "Eagle", Clifton, Yorks. 9-25 Feb 44.
No. 128 (RCAF) Wing,
Odiham, Hants. 1 Apr 44 - 28 Jun 44.
B.(Base)8 Sommervieu, Fr. 29 Jun 44 - 1 Jul 44.
No. 39 (RCAF) Wing,
B.8 Sommervieu, Fr. 2 Jul 44 - 13 Aug 44.
B.21 Ste-Honorine-de-Ducy, Fr. 14 Aug 44 - 31 Aug 44.
B.34 Avrilly, Fr. 1 Sep 44 - 19 Sep 44.
B.66 Diest, Bel. 20 Sep 44 - 3 Oct 44.
B.78 Eindhoven, Neth. 4 Oct 44 - 6 Mar 45.
B.90 Petit-Brogel, Bel. 7 Mar 45 - 9 Apr 45.
B.108 Rheine, Ger. 10 Apr 45 - 15 Apr 45.
B.116 Wunstorf, Ger. 16 Apr 45 - 27 Apr 45.
B.154 Soltau, Ger. 28 Apr 45 - 7 May 45.
B.156 Luneburg, Ger. 8 May 45 - 7 Aug 45.
No. 17 Armament Practice Camp, Warmwell, Dorset., Eng. 23 Jul - 2 Aug 45.

Representative Aircraft (Unit Code G9)
Curtiss Tomahawk Mk.I & II (Jan - Feb 43, not on operations)
North American Mustang Mk.I (Jan 43 - Dec 44) AG227 H
AG349 A   AG377 D   AG433 Z   AG488 V   AG552 M
AG553 R   AG627 C   AG628 R   AL978 B   AL986 T
AL997 O   AM103 O   AM125 W   AM170 P   AM227 L
AM228 N   AM237 K   AP178 J   AP179 F   AP180 E
AP186 Y   AP188 O   AP235 X
Supermarine Spitfire F.R.Mk.XIV (Nov 44 - Aug 45)
RM180 A   RM817 V   RM818 C   RM820 M   RM821 B
RM822 P   RM847 K   RM848 S   RM850 D   RM851 F
RM852 J   RM853 N   RM856 R   RM857 E   RM860 O
RM866 T   RM974 L   RM876 H   RM910 Y   RM996 T
RN114 W   RN115 Q   RN116 C   RN202 C
Operational History: First Mission 1 June 1943, 2 Mustangs from Dunsfold (F/L N.S. Clarke and F/O T.M. Pethick) rhubarb, attacked a train just north of Sées and a second train between Voutré and Evron. Many hits were scored, no enemy aircraft or flak encountered. **Last Mission** 5 May 1945, 2 Spitfire XIV's from Soltau (W/C R.C.A. Waddell in RM824 "Z" and F/O C.W. Anderson in RM817 "V") — tactical reconnaissance over an enemy airfield at Kiel.
Summary Sorties: 4946. Operational/Non-operational Flying Hours: 5831/5838. Victories: Aircraft: 2 destroyed, 1 damaged. Ground: destroyed 31 locomotives, 13-plus electrical pylons, 4 armoured fighting vehicles, 4 motor vehicles, 3 boats. Casualties: Operational: 29 aircraft; 23 pilots, of whom 3 were killed, 13 presumed dead, 3 POW, 4 injured. Non-operational: 7 personnel, of whom 3 were killed, 2 wounded, 2 injured. **Honours and Awards** 9 DFC's, 1 Air Medal (USA), 1 Croix de Guerre (Fr). **Battle Honours** Fortress Europe 1943-1944. France and Germany 1944-1945: *Normandy 1944, Arnhem, Rhine*.

No. 431 Squadron

Badge An Iroquois Indian's head
Motto The hatiten ronteriios (Warriors of the air)
Authority King George VI, March 1945
The Iroquois Indian's head represents the squadron's nickname. The motto is in Iroquois.
Formed at Burn, Yorkshire, England on 11 November 1942 as the RCAF's 28th — 11th Bomber — squadron formed overseas, the unit flew Wellington, Halifax and Canadian-built Lancaster aircraft on strategic and tactical bombing operations. After hostilities in Europe, it was selected as part of "Tiger Force" for duty in the Pacific, and returned to Canada for reorganization and training. The sudden end of the war in the Far East resulted in the squadron being disbanded at Dartmouth, Nova Scotia on 5 September 1945.
Brief Chronology Formed at Burn, Yorks., Eng. 11 Nov 42. Disbanded at Dartmouth, N.S. 5 Sep 45.
Title or Nickname "Iroquois"
Adoption Town of Simcoe, Ont.
Commanders
W/C J. Coverdale (RAF) 1 Dec 42 - 21 Jun 43 *KIA*.
W/C W.F.M. Newson, DFC 26 Jun 43 - 10 May 44 *OTE*.
W/C H.R. Dow, DFC 14 May 44 - 25 Jul 44 *POW*.
W/C E.M. Mitchell 27 Jul 44 - 10 Jan 45 *OTE*.
W/C R.F. Davenport 14 Jan 45 - 11 Mar 45 *KIA*.
W/C W.F. McKinnon 18 Mar 45 - 15 Jun 45.
W/C E.M. Bryson, DFC 4 Aug 45 - 5 Sep 45.
Higher Formations and Squadron Locations
Bomber Command:
No. 4 Group,
Burn, Yorks. 11 Nov 42 - 14 Jul 43.
No. 6 (RCAF) Group,
Tholthorpe, Yorks. 15 Jul 43 - 9 Dec 43.
No. 64 (RCAF) Base,
Croft, Yorks. 10 Dec 43 - 6 Jun 45.
En route to Canada 7 Jun 45 - 11 Jun 45.
RAF "Tiger Force" (for ops),
RCAF Eastern Air Command (for training):
No. 6 (RCAF) Group,
No. 662 (RCAF) Wing,
Dartmouth, N.S. 12 Jun 45 - 5 Sep 45.
Representative Aircraft (Unit Code SE)
Vickers Wellington B.Mk.X (Dec 42 - Jul 43) HE182 A
HE183 J   HE184 M   HE197 G   HE198 D   HE199 R
HE200 P   HE201 T   HE202 Z   HE203 B   HE204 C
HE205 X   HE213 F   HE379 H   HE396 K   HE440 Y
HE443 O   HE503 S   LN282 L   LN283 P   LN248 Q
LN291 V   LN403 W   LN405 U
Handley Page Halifax B.Mk.V (Jul 43 - Apr 44) DK185 S
DK264 V   LK632 M   LK639 E   LK640 Q   LK649 G
LK657 K   LK701 L   LK705 X   LK897 P   LK898 O
LK905 D   LK918 F   LK963 H   LK964 T   LK991 U
LL150 N   LL172 B   LL173 C   LL174 Z   LL175 A
LL231 J   LL232 R   LL233 Y
Handley Page Halifax B.Mk.III (Mar - Oct 44)
LK828 S   LK833 R   LK837 H   LW412 F   LW462 B
LW576 O   MZ364 M   MZ509 C   MZ517 D   MZ529 E

MZ537 L MZ600 N MZ655 T MZ656 X MZ657 K
MZ859 A MZ860 P MZ861 Q MZ922 J NA498 G
NP999 W NR121 B NR122 Y NR123 V
Avro Lancaster B.Mk.X (Oct 44 - Sep 45) KB741 Y
KB774 D KB801 S KB802 V KB806 X KB807 B
KB808 U KB809 Q KB810 H KB811 T KB812 F
KB813 S KB815 K KB817 P KB818 G KB819 J
KB821 P KB822 W KB823 E KB827 M KB847 R
KB872 N KB888 O KB900 C

Operational History: First Mission 2/3 March 1943, 7 Wellington X's from Burn despatched to lay mines in the "Nectarines" (Frisian Islands) area: 5 were successful, 2 aborted. **First Mission, Bombing** 5/6 March 1943, 3 Wellington X's from Burn bombed Essen, Germany. **Last Mission** 15 April 1945, 15 Lancaster X's from Croft despatched to bomb gun positions on the Island of Wangerooge; 12 bombed the primary target; 1, unable to release its bombs over the aiming point, jettisoned the load over the centre of the smoke pall, hitting the airfield; 2 failed to return (believed to have collided). **Summary** Sorties: 2584 (including 11 airlifting 240 POW's back to England). Operational/Non-operational Flying Hours: 14,621/8986. Bombs dropped: 14,004 tons. Victories: Aircraft: 6 destroyed, 1 probably destroyed, 4 damaged. Casualties: Operational: 72 aircraft; 490 aircrew, of whom 313 were killed, 54 missing, 104 POW, 18 safe, 1 injured. Non-operational: 14 personnel killed. **Honours and Awards** 1 DSO, 63 DFC's, 10 DFM's, 2 CGM's, 1 MiD. **Battle Honours** English Channel and North Sea 1943-1944. Baltic 1943-1944. Fortress Europe 1943-1944. France and Germany 1944-1945: *Biscay Ports 1943-1944, Ruhr 1943-1945, Berlin 1943-1944, German Ports 1943-1945, Normandy 1944, Rhine.* Biscay 1943-1944.

270   Avro Lancaster Mk.X KB774, coded SE-D, of No. 431 (B) Squadron at its base at Croft, Yorks.

No. 432 Squadron

Badge In front of a full moon a cougar leaping down
Motto Saeviter ad lucem (Ferociously towards the light)
Authority King George VI, March 1945
The badge indicates the squadron's night bombing attacks on the enemy. The motto refers to the unit's fierce desire to fight for light against the darkness of oppression.
Formed at Skipton-on-Swale, Yorkshire, England on 1 May 1943 as the RCAF's 31st — 12th Bomber, first to be formed under No. 6 (RCAF) Group — squadron formed overseas, the unit flew Wellington, Lancaster and Halifax aircraft on strategic and tactical bombing operations. It was disbanded at East Moor, Yorkshire on 15 May 1945.
Brief Chronology Formed at Skipton-on-Swale, Yorks., Eng. 1 May 43. Disbanded at East Moor, Yorks. 15 May 45.
Title or Nickname "Leaside"
Adoption Town of Leaside, Ont.
Commanders
W/C H.W. Kerby 1 May 43 - 29 Jul 43 *KIA.*
W/C W.A. McKay 30 Jul 43 - 30 May 44 *OTE.*
W/C J.K.F. MacDonald 31 May 44 - 25 Jul 44 *MIA.*[1]
W/C A.D.R. Lowe 26 Jul 44 - 28 Sep 44 *OTE.*
W/C J.K.F. MacDonald 29 Sep 44 - 28 Jan 45 *OTE.*
S/L S.H. Minhinnick 29 Jan 45 - 27 Feb 45 *OTE.*
W/C K.A. France 28 Feb 45 - 15 May 45.
Higher Formations and Squadron Locations
Bomber Command:
No. 6 (RCAF) Group,
Skipton-on-Swale, Yorks. 1 May 43 - 18 Sep 43.
No. 62 (RCAF) Base,
East Moor, Yorks. 19 Sep 43 - 15 May 45.
Representative Aircraft (Unit Code QO)
Vickers Wellington B.Mk.X (May - Oct 43) HE222 E
HE348 P HE514 K HE555 A HE567 Z HE568 Y
HE638 F HE686 R HE688 V HE732 O HE800 A
HE818 Q HE906 H HE918 C HE456 D HF571 U
HF599 L HZ272 W HZ484 N JA118 G LN240 S
LN435 J LN454 B MS485 T
Avro Lancaster B.Mk.II (Oct 43 - Feb 44) DS739 Y
DS740 Z DS757 L DS788 E DS789 V DS792 U
DS794 W DS829 A DS830 S DS831 N DS832 K
DS834 O DS844 H DS848 R DS850 M DS851 D
DS852 C LL617 J LL618 F LL632 G LL636 B
LL686 F LL732 H LL724 N
Handley Page Halifax B.Mk.III (Feb - Jul 44) LK764 F
LK765 H LK766 V LK807 J LK811 N LW582 M
LW583 L LW584 Y LW592 A LW593 O LW594 G
LW595 Q LW596 D LW597 C LW598 K LW614 S
LW615 U LW616 R LW617 J MZ506 X MZ582 Z
MZ603 E MZ632 W MZ633 O
Handley Page Halifax B.Mk.VII (Jul 44 - May 45)
NP687 A NP688 X NP689 M NP690 G NP691 V
NP692 D NP693 Q NP694 R NP695 K NP697 F
NP698 U NP699 O NP701 G NP703 H NP704 L
NP705 Y NP707 W NP708 E NP736 B NP801 N
NP802 S NP803 I NP804 Q NP805 J NP807 P
PN208 G PN233 D PN454 P

Operational History: First Mission 23/24 May 1943, 15 Wellington X's from Skipton-on-Swale despatched to bomb Dortmund, Germany; 11 bombed the primary target, 4 aborted. **Last Mission** 25 April 1945, 19 Halifax VII's from East Moor bombed gun positions on the Island of Wangerooge. **Summary** Missions/Sorties: 246/3130 (including 53 sea mining). Operational/Non-operational Flying Hours: 16,607/6166. Bombs dropped: 8980 tons. Victories: Aircraft: 5 destroyed, 2 probably destroyed, 5 damaged. Casualties: Operational: 73 aircraft; 448 aircrew, of whom 30 were killed, 8 missing, 252 presumed dead, 123 POW (of whom 5 escaped); 35 evaded capture. Non-operational: 42 personnel, of whom 38 were killed, 3 injured, 1 died of natural causes. **Honours and Awards** 2 DSO's, 1 bar to DFC, 119 DFC's, 1 CGM, 20 DFM's, 1 Croix de Guerre (Fr). **Battle Honours** English Channel and North Sea 1943. Fortress Europe 1943-1944. France and Germany 1944-1945: *Biscay Ports 1944, Ruhr 1943-1945, Berlin 1943-1944, German Ports 1943-1945, Normandy 1944, Rhine.* Biscay 1943.

---

[1] Shot down 25 July 1944, evaded capture, and returned to England.

271 Avro Lancaster Mk.II DS848 awaits the completion of her crew's last cigarette before departing on another night flight over Germany with No. 432 (B) Squadron.

No. 433 Squadron

Badge In front of a hurt a porcupine
Motto Qui s'y frotte s'y pique (Who opposes it gets hurt)
Authority King George VI, December 1945
This squadron was adopted by the Porcupine District of Northern Ontario. The hurt or blue disc symbolizes the "hurt" done to the enemy and the sky through which the unit operates. The motto refers to the squadron and its nickname.

Formed at Skipton-on-Swale, Yorkshire, England on 25 September 1943 as the RCAF's 32nd — 14th and last Bomber[1] — squadron formed overseas, the unit flew Halifax and Lancaster aircraft on strategic and tactical bombing operations. After hostilities in Europe, it remained in England as part of Bomber Command's strike force, under which it lifted prisoners of war back to England. The squadron was disbanded on 15 October 1945.
Brief Chronology Formed at Skipton-on-Swale, Yorks., Eng. 25 Sep 43. Disbanded 15 Oct 45.
Title or Nickname "Porcupine"
Adoption Porcupine district of northern Ontario
Commanders
W/C C.B. Sinton, DFC 9 Nov 43 - 30 May 44 OTE.
W/C A.J. Lewington 31 May 44 - 5 Nov 44 OTE.
W/C G.A. Tambling 6 Nov 44 - 1 Aug 45.
W/C C.E. Lewis, DFC 25 Sep 45 - 15 Oct 45.
Higher Formations and Squadron Location
Bomber Command:
No. 6 (RCAF) Group,
No. 63 (RCAF) Base,
Skipton-on-Swale, Yorks. 25 Sep 43 - 30 Aug 45.
No. 1 Group,
Skipton-on-Swale, Yorks. 31 Aug 45 - 15 Oct 45.
Representative Aircraft (Unit Code BM)
Handley Page Halifax B.Mk.III (Nov 43 - Jan 45)
HX280 O   HX281 H   HX282 K   HX283 R   HX284 B
HX285 E   HX287 U   HX288 F   HX289 T   HX290 V
HX291 W   HX292 G   LV842 D   LV911 I   LW368 L
MZ807 C   MZ808 P   MZ857 N   MZ872 Q   MZ883 S
MZ905 J   MZ944 M   NR120 A   NR136 R
Avro Lancaster B.Mk.I & III (Jan - Oct 45) ME375 D
ME457 U   NG232 H   NG233 E   NG441 L   NG459 K
NG460 A   NG496 N   NG497 E   NG498 T   NN779 J
NP930 W   PA219 M   PA337 X   PB893 G   PB903 F
PB908 C   RA505 R   RA506 O   RA509 P   RA511 Q
RA512 S   RA513 Y   SW273 V
Operational History: First Mission 2/3 January 1944, 4 Halifax III's from Skipton-on-Swale despatched to lay mines in the "Nectarines" (Frisian Islands) area; 3 were successful, 1 aborted. **First Mission, Bombing** 18/19 January 1944, 8 Halifax III's from Skipton-on-Swale despatched to bomb Berlin; 6 bombed the primary target, 1 bombed Kiel, 1 returned early. **Last Mission** 25 April 1945, 10 Lancaster I/III's from Skipton-on-Swale bombed gun positions on the Island of Wangerooge. **Summary** Missions/Sorties: 209/2316 (54 missions were minelaying). Operational/Non-operational Flying Hours: 12,488/6059. Bombs dropped:

7486 tons. Victories: Aircraft: 5 destroyed, 3 probably destroyed, 2 damaged. Casualties: Operational: 38 aircraft; 241 aircrew, of whom 56 were killed, 96 presumed dead, 56 POW, 33 missing (of whom 9 captured or proved safe). Non-operational: 15 personnel killed. **Honours and Awards** 2 bars to DFC, 132 DFC's, 9 DFM's, 1 BEM, 14 MiD's, 1 Air Medal (USA). **Battle Honours** English Channel and North Sea 1944-1945. Baltic 1944-1945. Fortress Europe 1944. France and Germany 1944-1945: *Biscay Ports 1944, Ruhr 1944-1945, Berlin 1944, German Ports 1944-1945, Normandy 1944, Rhine.* Biscay 1944.

---

¹Although this was the last RCAF bomber squadron formed overseas, an existing torpedo bomber squadron (No. 415) was redesignated bomber on 12 July 1944.

No. 434 Squadron

Badge A representation of the schooner "Bluenose"
Motto In excelsis vincimus (We conquer in the heights)
Authority King George VI, October 1945.
The squadron was adopted by the Rotary Club of Halifax, Nova Scotia and took the nickname "Bluenose" in reference to the common nickname for Nova Scotians. The schooner 'Bluenose' is well known for its fine record.
Formed at Tholthorpe, Yorkshire, England on 13 June 1943 as the RCAF's 31st — 13th Bomber — squadron formed overseas, the unit flew Halifax and Canadian-built Lancaster aircraft on strategic and tactical bombing operations. After hostilities in Europe, it was selected as part of "Tiger Force" for duty in the Pacific, and returned to Canada for reorganization and training. The sudden end of the war in the Far East resulted in the squadron being disbanded at Dartmouth, Nova Scotia on 5 September 1945.
Brief Chronology Formed at Tholthorpe, Yorks., Eng. 13 Jun 43. Disbanded at Dartmouth, N.S. 5 Sep 45.
Title or Nickname "Bluenose"
Adoption Rotary Club of Halifax, N.S.
Commanders
W/C C.E. Harris (Can/RAF), DFC 15 Jun 43 - 6 Feb 44.
W/C C.S. Bartlett, DFC 7 Feb 44 - 12 Jun 44 *KIA*.
W/C F.H. Watkins, DFC 13 Jun 44 - 29 Aug 44.
W/C A.P. Blackburn, DFC 30 Aug 44 - 7 Apr 45.
W/C J.C Mulvihill, AFC 8 Apr 45 - 5 Sep 45.
Higher Formations and Squadron Locations
Bomber Command:
No. 6 (RCAF) Group,
Tholthorpe, Yorks. 13 Jun 43 - 10 Dec 43.
No. 64 (RCAF) Base,
Croft, Yorks. 11 Dec 43 - 9 Jun 45.
En route to Canada 10 Jun 45 - 14 Jun 45.
RAF "Tiger Force" (for ops),
RCAF Eastern Air Command (for training):
No. 6 (RCAF) Group,
No. 662 (RCAF) Wing,
Dartmouth, N.S. 15 Jun 45 - 5 Sep 45.
Representative Aircraft (Unit Code WL)
Handley Page Halifax B.Mk.V (Jun 43 - May 44) EB217 A
EB258 T   DK251 F   DK262 R   LL666 T   LK669 Z
LK708 L   LK709 E   LK801 D   LK907 M   LK945 O
LK972 N   LL136 E   LL137 D   LL243 U   LL288 P
Handley Page Halifax B.Mk.III (May - Dec 44) LW171 M
LW433 W   LW713 P   LW714 H   MZ405 F   MZ420 F
MZ421 A   MZ626 T   MZ276 Y   MZ878 E   MZ920 C
MZ921 Q   NA552 U   NP939 W   NR121 G   NR144 V
Avro Lancaster B.Mk.X (Dec 44 - Sep 45) KB789 V
KB814 S   KB816 G   KB824 E   KB825 A   KB826 K
KB830 D   KB832 F   KB833 B   KB834 Y   KB835 J
KB836 H   KB840 N   KB842 L   KB843 Q   KB844 W
KB846 P   KB849 T   KB850 O   KB852 R   KB863 P
KB973 G   KB880 L   KB885 Q

During Feb-Mar 45 six B.Mk.I & III's were flown by the squadron: NG343 U NG344 Z NG345 Q NG497 P PA225 O PA226 X
Operational History: First Mission 2 August 1943, 4 Halifax V's from Tholthorpe — sea search. **First Mission, Bombing** 12/13 August 1943, 10 Halifax V's from Tholthorpe despatched to bomb Milan, Italy; 9 bombed the primary target, 1 aborted. **Last Mission** 25 April 1945, 15 Lancaster X's from Croft bombed gun positions on the Island of Wangerooge. **Summary** Missions/Sorties: 198/2582 (missions included 179 bombing, 17 minelaying, 1 diversionary, 1 sea search; sorties include 45 on POW airlift). Operational/Non-operational Flying Hours: 14,622/6579. Bombs dropped: 10,358 tons (plus 225 mines). Victories: Aircraft: 7 destroyed, 2 probably destroyed, 4 damaged. Casualties: Operational: 75 aircraft; 484 aircrew, of whom 34 were killed, 313 presumed dead, 121 POW (of whom 2 died, 2 escaped), 16 evaded capture or proved safe. Non-operational: 9 personnel, of whom 8 killed, 1 died of natural causes. **Honours and Awards** 6 bars to DFC, 108 DFC's, 6 DFM's, 1 BEM, 7 MiD's. **Battle Honours** English Channel and North Sea 1943-1944. Baltic 1943-1944. Fortress Europe 1943-1944. France and Germany 1944-1945: *Biscay Ports 1944, Ruhr 1943-1945, Berlin 1943-1944, German Ports 1944-1945, Normandy 1944, Rhine.*

No. 435 Squadron

Badge A chinthe on a plinth
Motto Certi provehendi (Determined on delivery)
Authority King George VI, August 1946.
The chinthe is a legendary monster which guards the temples in Burma where this squadron operated. The motto refers to the unit's activities as a transport squadron.
Formed at Gujrat, Punjab, India on 1 November 1944 as the RCAF's 34th — third Transport, second in India — squadron formed overseas, the unit flew Dakota aircraft in support of the British Fourteenth Army in northern Burma. After hostilities in the Far East, the squadron moved to England and provided transport service to Canadian units on the Continent until disbanded at Down Ampney, Gloucestershire, England on 1 April 1946.
Brief Chronology Formed at Gujrat, Punjab, India 1 Nov 44. Disbanded at Down Ampney, Glos., Eng. 1 Apr 46.
Title or Nickname "Chinthe"
Commanders
W/C T.P. Hartnett 1 Nov 44 - 10 Sep 45.
W/C C.N. McVeigh, AFC 11 Sep 45 - 1 Apr 46.
Higher Formations and Squadron Locations
South East Asia Command:
No. 229 Group,
No. 341 Wing,
Gujrat, Punjab, India 1 Nov 44 - 17 Dec 44.
Tulihal, Manipur, India 18 Dec 44 - 31 Aug 45.
En route to England[1] 27 Aug 45 - 19 Sep 45.
Transport Command:
No. 46 Group,
No. 120 (RCAF) Wing,
Down Ampney, Glos. 29 Aug 45 - 1 Apr 46.
10 aircraft, Croydon, Surrey 9 Oct 45 - 16 Mar 46.
Representative aircraft
Douglas Dakota Mk.III & IV (India, Oct 44 - Aug 45, no Unit Code)[2] FD915 O KG891 B KJ821 J
KJ883 N KN563
Douglas Dakota Mk.III & IV (England, Aug 45 - Mar 46, Unit Code ODM) FZ658 Q KG317 A KG337 B
KG414 G KG416 K KG486 K KG557 U KG559 O
KG563 N KG580 O KG632 B KG659 W KG668 S
KG713 Y KK143 A KN413 F KN511 Z KN665 E
KN666 KW KP226 P
Operational History: First Mission 20 December 1944, 16 Dakotas from Tulihal airlifted supplies for the Fourteenth Army to a hurriedly constructed landing strip at Tamu in the Kabaw Valley, and air supply drops at Pinlebu, east of the Chindwin River. Note: One previous mission was flown before the squadron was declared operational. On 9 December it flew 35 sorties on an emergency move of No. 149 (F) Squadron RAF to Imphal. **Last Mission, India** 30 August 1945, 4 Dakotas from Tulihal, staging through Toungoo, dropped supplies to British guerrillas. **First Mission, England** 18 September 1945, 4 Dakotas from Down Ampney transported 16 passengers and 8557 pounds of freight to various destinations in Europe. **Last Mission** 14 March 1946, Dakota KG587 ODM-T from Down Ampney

with F/L H.E. Carling and crew — special flight to Rennes, France and return. **Summary, Burma** Sorties: 15,681. Operational/Non-operational Flying Hours: 28,792/2734. Airlifted: 27,460 tons of freight, 14,000 passengers, 851 casualties. Casualties: Operational: 4 aircraft; 21 aircrew, of whom 2 were killed, 14 missing, 5 wounded. Non-operational: nil. **England** Sorties: 1018. Operational/Non-operational Flying Hours: 4803/2434. Airlifted: 383.3 tons of freight, 62.8 mail; 9293 passengers. Casualties: Operational: 3 aircraft; 9 aircrew and 6 passengers killed, 2 aircrew and 13 passengers injured. Non-operational: nil. **Honours and Awards** 1 MBE, 1 DFC, 2 AFC's, 1 MiD's. **Battle Honours** Burma 1944-1945.

---

[1] The first three aircraft left India on 27 August and the last three arrived in England on 19 September. At this time, the squadron was completing its reorganization as most of the India personnel were repatriated and replaced by crews newly arrived from Canada.
[2] On leaving India the squadron exchanged Dakotas with its relief, No. 233 (T) Squadron RAF.

272   Douglas Dakota Mk.IV KK105 aircraft "A" of No. 435 (T) Squadron operating in Burma on supply-dropping duties.

No. 436 Squadron

Badge An elephant's head couped carrying a log
Motto Onus portamus (We carry the load)
Authority King George VI, May 1945
The squadron operated as a transport unit from a base in India and adopts this badge and motto to symbolize its functions.
Formed at Gujrat, Punjab, India on 20 August 1944[1] as the RCAF's 33rd — second Transport, first of two in India — squadron formed overseas, the unit flew Dakota aircraft in support of the British Fourteenth Army in northern Burma. After hostilities, the squadron moved to England and provided transport service to Canadian units on the Continent until disbanded at Odiham, Hampshire, England on 22 June 1946.
Brief Chronology Formed at Gujrat, Punjab, India 20 Aug 44. Disbanded at Odiham, Hants., Eng. 22 Jun 46.
Title or Nickname "Elephant"[2]
Commanders
W/C R.A Gordon, DSO, DFC 26 Oct 44 - 30 Jul 45 OTE.
W/C R.L. Denison, DFC 31 Jul 45 - 22 Jun 46.
Higher Formations and Squadron Locations
South East Asia Command:
No. 229 Group,
No. 341 Wing,
Gujrat, Punjab, India 20 Aug 44 - 13 Jan 45.
Kanglatongbi, Assam, India 14 Jan 45 - 16 Mar 45.
No. 232 Group,
No. 342 Wing,
Mawnubyin, Akyab, Burma 17 Mar 45 - 14 May 45.
Kyaukpyu, Ramree Island, Burma 15 May 45 - 9 Sep 45.
16 aircraft, Kinmagan, Burma 13-21 Aug 45.[3]
En route to England[4] 24 Aug 45 - 15 Sep 45.
Transport Command:
No. 46 Group,
No. 120 (RCAF) Wing,
Down Ampney, Glos. 29 Aug 45 - 3 Apr 46.
7 aircraft, Biggin Hill, Kent 9 Dec 45 - 14 Apr 46.
Odiham, Hants. 4 Apr 46 - 22 Jun 46.
Representative Aircraft
Douglas Dakota Mk.III & IV (India, Oct 44 - Sep 45[5], no Unit Code) KG755 Q   KG790 P   KG794 N   KG855 K   KJ763 L   KJ820 A   KJ821 U   KJ841 F   KF845 V   KJ858 H   KJ887 J   KJ956 G   KJ964 T   KK107 M   KK113 B   KK126 D
Douglas Dakota Mk.III & IV (England, Sep 45 - Jun 46, Unit Code ODN) FZ665 X   FZ678 K   KG320 C   KG400 T   KG403 M   KG448 F   KG455 G   KG580 J   KG635 P   KG659 Q
Operational History: First Mission 15 January 1945, 7 Dakotas from Kangla airlifted 59 tons of supplies for 33 Corps at Shwebo (17.5 para dropped, 3.7 free dropped, 37.8 landed at airstrip). Note: One previous mission was flown before the squadron became fully operational. On 10 December 1944, 16 Dakotas were employed in an emergency airlift of No. 177 (F) Squadron RAF from Bikramganj to a forward area at Hathazari. **Last Mission, Burma** 31 August 1945, 5 Dakotas from Kyaukpyu — 7 sorties to

airlift 14.7 tons of freight, 1 ton of mail and 7 passengers. (2 aircraft collected mail at Chittagong for Meiktila, Toungoo and Mingaladon; 2 aircraft on deliveries to Kinmagan, Comilla and Alipore; 1 aircraft with freight from Comilla for Mingaladon). **First Mission, England** 9 September 1945, 3 Dakotas from Down Ampney — airlift to Paris, France. **Last Mission** 16 June 1946, Dakota KG587 ODN-L with F/L Nicholls and crew — special flight from Brussels, Belgium to Odiham with 3 passengers and 1.5 tons of freight. **Summary, India/Burma** Sorties: 9806. Operational/Non-operational Flying Hours: 31,798/4561. Airlifted: 29,000 tons of supplies, 15,000 troops, passengers and casualties. Casualties: Operational: 3 aircraft; 4 aircrew killed. Non-operational: 1 aircraft; 1 killed, 2 died of natural causes. **England** Operational/Non-operational Flying Hours: 6983/2597. Airlifted: 296.9 tons of mail, 1486.2 freight, 13,063 passengers and 3156 casualties. Casualties: Operational: nil. Non-operational: 1 aircraft; 4 aircrew killed. **Honours and Awards** 1 DSO, 26 DFC's, 1 AFM, 3 BEM's, 11 MiD's. **Battle Honours** Burma 1945.

---

[1] Although it officially came into being on this date, the unit did not receive its personnel from Canada until 9 October.
[2] In Burma, a nickname, "Canucks Unlimited," was painted on the squadron's aircraft.
[3] These aircraft were involved in airlifting the headquarters and squadron personnel of No. 244 Group and No. 910 Wing to Rangoon.
[4] First aircraft left Burma on 24 August and the last arrived in England on 15 September. Personnel from India and Burma were repatriated and replaced by crews newly arrived from Canada.
[5] The squadron exchanged Dakotas with its relief, No. 48 Squadron RAF, and carried personnel of No. 529 Squadron RAF back to England.

273   Douglas Dakota Mk.III FD946 aircraft "Y" of No. 436 (T) Squadron in Burma. The overpainted bars of the USAAF behind the fuselage roundel indicate the origin of the aircraft.

No. 437 Squadron

Badge A husky's head affronte erased
Motto Omnia passim (Anything anywhere)
Authority King George VI, April 1945
Nicknamed the "Husky Squadron," this unit adopted as its badge a husky's head indicative of its function of glider towing and the transportation of essential freight.
Formed at Blakehill Farm, Wiltshire, England on 14 September 1944 as the RCAF's 32nd — first of three Transport —squadrons formed overseas, the unit flew Dakota aircraft on air transport duty. Three days after its formation, it airlifted part of the British airborne force to Arnhem. Four days later, it suffered its first casualties when four of ten aircraft despatched on a resupply mission to Arnhem failed to return. On 24 March 1945, the unit took part in the airborne crossing of the Rhine at Wesel (Operation "Varsity") towing 24 Horsa gliders containing 230 men of the 1st Royal Ulster Rifles, 13 jeeps with trailers and six jeeps with 6-pounder guns. After hostilities, the squadron provided air transport service to Canadian units on the Continent until disbanded at Odiham, Hampshire on 15 June 1946.
Brief Chronology Formed at Blakehill Farm, Wilts., Eng. 14 Sep 44. Disbanded at Odiham, Hants. 15 Jun 46.
Title or Nickname "Husky"
Adoption Hudson's Bay Company, Winnipeg, Man.
Commanders
W/C J.A. Sproule (Can/RAF), DFC[1] 14 Sep 44 - 14 Sep 45.
W/C A.R. Holmes 15 Sep 45 - 15 Jun 46.
Higher Formations and Squadron Locations
Transport Command:
No. 46 Group,
Blakehill Farm, Wilts. 14 Sep 44 - 6 May 45.
No. 11 Wing,
B.(Base) 75 Nivelles, Bel. 7 May 45 - 6 Jun 45.
B.58 Melsbroek, Bel. 7 Jun 45 - 14 Sep 45.
No. 88 Group, 6 aircraft, Oslo, Norway 17 Jul - 27 Nov 45.
No. 120 (RCAF) Wing, 6 aircraft, Odiham, Hants. 1 Aug - 14 Nov 45.
B.56 Evère, Bel. 15 Sep 45 - 14 Nov 45.
No. 120 (RCAF) Wing,
Odiham, Hants. 15 Nov 45 - 15 Jun 46.
14 aircraft, No. 111 Wing, B.56 Evère, Bel. 15 Nov 45 - 12 Mar 46.
Croydon, Surrey 13 Mar - 30 May 46.
Representative Aircraft (Unit Code 1944-45 Z2, 1945-46 ODO)
Douglas Dakota Mk.III & IV (Sep 44 - Jun 46) FZ669 Q*
FZ692 R   FZ694 P   FZ695 AW   KG312 G   KG330 E
KG345 V   KG354 H   KG389 B*   KG394 O   KG395 Y
KG600 C   KG634 T   KG808 M   KN256 L   KN262 V
KN278 W   KN281 N
*Used on first Arnhem mission
Operational History: First Mission 17 September 1944. 15 Dakotas from Blakehill Farm despatched on Operation "Market Garden", the airborne assault on Arnhem, 1

aborted. Twelve Horsa gliders were released, containing 146 men of the British 1st Airborne Division, plus 16 bicycles, 10 motorcycles, 5 jeeps, 6 trailers, 2 handcarts, 4 blitz buggies, 3 wireless sets. **Last Mission** 30 May 1946, Dakota KG577 ODO-K with P/O J.E. Thompson and crew — Brussels, Belgium to Odiham with 6 passengers, 300 pounds of mail and 3024 pounds of freight. **Summary** Sorties: 11,625. Operational/Non-operational Flying Hours: 32,355/2884. Airlifted (only post-war data available, Nov 45 - Jun 46): 152.3 tons of mail, 1415.3 freight; 25,269 passengers, 180 casualties. Casualties: Operational: 14 aircraft; 17 aircrew, of whom 2 were killed, 14 missing, 1 wounded. Non-operational: nil. **Honours and Awards** 1 OBE, 8 DFC's, 2 AFC's, 1 DFM, 1 MiD's, 3 King's Commendations. **Battle Honours** France and Germany 1944-1945: *Arnhem, Rhine.*

---

¹Transferred from the RAF to the RCAF on 4 October 1944.

274 Douglas Dakota Mk.III KG425, coded Z2-M, of No. 437 (T) Squadron. The inscription "Royal Canadian Air Force" on the fuselage side was a forecast of future marking practices.

No. 438 Squadron

Badge The head of a wild cat affronte
Motto Going down
Authority King George VI, October 1945
The wild cat represents the squadron's nickname. The motto is taken from the last instructions of the formation leader when Typhoon fighter bombers of this unit were about to attack.
Formed in Canada as No. 118 (Fighter) Squadron on 13 January 1941, the unit was the first of six home squadrons transferred overseas (complete in personnel but without aircraft) in preparation for the Allied invasion of Europe, and was redesignated No. 438 (Fighter Bomber) Squadron at Digby, Lincolnshire, England on 18 November 1943. It flew Typhoon aircraft in the pre-invasion softening up of the German defences and, after D-Day, gave close support to ground forces by dive-bombing and strafing enemy strongpoints, bridges, rail and road traffic. The squadron was disbanded at Flensburg, Germany on 26 August 1945.
Brief Chronology Formed as No. 118 (F) Sqn, Rockcliffe, Ont. 13 Jan 41. Redesignated No. 438 (FB)¹ Sqn, Digby, Lincs., Eng. 18 Nov 43. Disbanded at Flensburg, Ger. 26 Aug 45.
Title or Nickname "Wild Cat"
Adoption City of Montreal
Commanders
S/L F.G. Grant, DFC 18 Nov 43 - 28 Jul 44.
S/L J.R. Beirnes, DFC 29 Jul 44 - 13 Oct 44 *2OTE.*
S/L R.F. Reid 14 Oct 44 - 29 Dec 44 *OTE.*
S/L P. Wilson 30 Dec 44 - 1 Jan 45 *KIA.*
S/L J.E. Hogg, DFC 20 Jan 45 - 23 Mar 45 *KIA.*
S/L J.R. Beirnes, DFC 6 Apr 45 - 1 Jun 45 *KIFA.*
S/L P. Bissky 4 Jun 45 - 26 Aug 45.
Higher Formations and Squadron Locations
Air Defence Great Britain:
No. 12 Group,
Digby, Lincs. 20 Nov 43 - 18 Dec 43.
Wittering, Leics. 19 Dec 43 - 9 Jan 44.
Second Tactical Air Force:
No. 83 (Composite) Group,
No. 22 (RCAF) Sector (disbanded 13 Jul 44),
No. 143 (RCAF) Wing,
Ayr, Scot. 10 Jan 44 - 17 Mar 44.
Hurn, Hants. 18 Mar 44 - 2 Apr 44.
Funtington, Sussex 3 Apr 44 - 19 Apr 44.
Hurn, Hants. 20 Apr 44 - 26 Jun 44.
B.(Base) 9 Lantheuil, Fr. 27 Jun 44 - 30 Aug 44.
B.24 St Andre, Fr. 31 Aug 44 - 2 Sep 44.
B.48 Glisy, Fr. 3 Sep 44 - 5 Sep 44.
B.58 Melsbroek, Bel. 6 Sep 44 - 25 Sep 44.
B.78 Eindhoven, Neth. 26 Sep 44 - 18 Mar 45.
No. 17 Armament Practice Camp, Warmwell, Dorset., Eng. 19 Mar - 2 Apr 45.
B.100 Goch, Ger. 3 Apr 45 - 11 Apr 45.
B.110 Osnabruck, Ger. 12 Apr 45 - 20 Apr 45.
B.150 Hustedt, Ger. 21 Apr 45 - 28 May 45.
B.166 Flensburg, Ger. 29 May 45 - 26 Aug 45.

Representative Aircraft (Unit Code F3)
Hawker Hurricane Mk.IV (Nov 43 - May 44 — not on operations) LD 973
Hawker Typhoon Mk.IB (Jan 44 - Aug 45)
DN619 G   EK383 N   EK481 H   JR135 J   MM959 B
MN283 L   MN345 D   MN375 W   MN398 A   MN424 S
MN547 Q   MN626 A   MN758 M   RB391 Y   RB407 T
SW398 E   SW414 G

Operational History: First Mission 20 March 1944, 4 Typhoons from Hurn — fighter sweep Cherbourg-Alderney; strafed a German staff car and troops without observed results. **Last Mission** 4 May 1945, 8 Typhoons from Hustedt armed with two 500-pound bombs — anti-shipping strike in Kiel Bay, attacked a 1000-ton vessel. No bomb hits, but cannon fire hits were observed during strafing runs. **Summary** Sorties: 4022. Operational/Non-operational Flying Hours: 4515/3961. Victories: Aircraft: 2 destroyed, 1 damaged. Ground: dropped 2070 tons of bombs (2537 1000- and 3205 500-pound) and credited with 430 rail cuts; destroyed/damaged 184/169 motor vehicles, 12/3 tanks, 5/73 locomotives, 101/532 trains, 1/38 barges, 5/0 bridges. Casualties: Operational: 28 aircraft; 31 pilots, of whom 17 were killed, 5 missing, 6 POW; 3 proved safe. Non-operational: 3 personnel killed. **Honours and Awards** 5 DFC's. **Battle Honours** Fortress Europe 1944. France and Germany 1944-1945: *Normandy 1944, Arnhem, Rhine.*

---

[1] The abbreviation "FB" for "Flying Boat" had been discontinued at the outbreak of the Second World War. Here, and later, "FB" is the abbreviation for "Fighter Bomber."

276   *Hawker Typhoon Mk.IB RB407 aircraft "T" of No. 438 (FB) Squadron taxis into a wet dispersal during operations on the Continent.*

275   *Hawker Hurricane Mk.IV and a group of No. 438 (FB) Squadron pilots during their conversion to the Hawker Typhoon.*

No. 439 Squadron

Badge none
Formed in Canada as No. 123 (Army Co-operation Training) Squadron on 15 January 1942, the unit was the second of six home squadrons transferred overseas (complete in personnel but without aircraft) in preparation for the Allied invasion of Europe, and was redesignated No. 439 (Fighter Bomber) Squadron at Wellingore, Lincolnshire, England on 31 December 1943. It flew Typhoon aircraft in the pre-invasion softening up of the German defences and, after D-Day, gave close support to ground forces by dive-bombing and strafing enemy strongpoints, bridges, rail and road traffic. The squadron was disbanded at Flensburg, Germany on 26 August 1945.

Brief Chronology Formed as No. 123 (ACT) Sqn, Rockcliffe, Ont. 15 Jan 42. Redesignated No. 439 (FB) Sqn, Wellingore, Lincs., Eng. 31 Dec 43. Disbanded at Flensburg, Ger. 26 Aug 45.

Title or Nickname "Westmount"

Adoption City of Westmount, Que.

Commanders
S/L W.M. Smith 31 Dec 43 - 9 Mar 44.
S/L H.H. Norsworthy, DFC 10 Mar 44 - 12 Sep 44 *2 OTE*.
S/L K.J. Fiset, DFC 13 Sep 44 - 8 Dec 44 *OTE*.
S/L R.G. Crosby 9 Dec 44 - 22 Jan 45 *WIA*.
S/L J.H. Beatty, DFC 25 Jan 45 - 26 Aug 45.

Higher Formations and Squadron Locations
Air Defence Great Britain:
No. 12 Group,
Wellingore, Lincs. 31 Dec 43 - 7 Jan 44.
Second Tactical Air Force:
No. 83 (Composite) Group,
No. 22 (RCAF) Sector (disbanded 13 Jul 44),
No. 143 (RCAF) Wing,
Ayr, Scot. 8 Jan 44 - 17 Mar 44.
Hurn, Hants. 18 Mar 44 - 1 Apr 44.
Funtington, Sussex 2 Apr 44 - 18 Apr 44.
Hurn, Hants. 19 Apr 44 - 26 Jun 44.
No. 16 Armament Practice Camp, Hutton Cranswick, Yorks. 11-20 May 44.
B.(Base) 9 Lantheuil, Fr. 27 Jun 44 - 30 Aug 44.
B.24 St Andre, Fr. 31 Aug 44 - 2 Sep 44.
B.48 Glisy, Fr. 3 Sep 44 - 6 Sep 44.
B.58 Melsbroek, Bel. 7 Sep 44 - 24 Sep 44.
B.78 Eindhoven, Neth. 25 Sep 44 - 29 Mar 45.
B.100 Goch, Ger. 30 Mar 45 - 2 Apr 45.
No. 17 Armament Practice Camp. Warmwell, Dorset, Eng. 3-21 Apr 45.
B.150 Hustedt, Ger. 22 Apr 45 - 28 May 45.
B.166 Flensburg, Ger. 29 May 45 - 26 Aug 45.

Representative Aircraft (Unit Code 5V)
Hawker Hurricane Mk.IV (Jan - Apr 44, not on operations)
Hawker Typhoon Mk.IB (Jan 44 - Aug 45) R8926 B
R8977 D   EK219 X   JP401 P   JR299 S   JR326 F   JR444 J
JR506 X   MN427 Y   MN464 N   MN516 W   RB257 S
RB441 Z   SW423 J   SW460 D

Operational History: First Mission 27 March 1944, 9 Typhoons from Hurn — fighter sweep over the Cherbourg Peninsula. **Last Mission** 4 May 1945, 7 Typhoons from Hustedt, each with two 500-pound bombs — anti-shipping strike against vessels in Kiel Bay, claimed 2 probably destroyed, 4 damaged. **Summary** Sorties: 3996. Operational/Non-operational Flying Hours: 4206/3426. Victories: Aircraft: 11 destroyed, 1 probably destroyed, 9 damaged. Ground: dropped 2108 tons of bombs (2664 1000- and 3100 500-pound), credited with 360 rail cuts: destroyed/damaged 6/0 bridges, 237/321 motor vehicles, 17/13 tanks, 5/92 locomotives, 65/396 trains, 10/27 barges. Casualties: Operational: 41 aircraft; 37 pilots, of whom 19 were killed, 9 missing, 6 POW; 3 proved safe. Non-operational: 5 pilots killed. **Honours and Awards** 12 DFC's.

Battle Honours Fortress Europe 1944. France and Germany 1944-1945: *Normandy 1944, Arnhem, Rhine.*

277   *Hawker Typhoon Mk.IB JP401, coded 5V-P, of No. 439 (FB) Squadron landing on an "improved" runway in Germany.*

No. 440 Squadron

Badge: None

Formed in Canada as No. 111 (Fighter) Squadron on 1 November 1941, the unit was the third of six home squadrons transferred overseas (complete in personnel but without aircraft) in preparation for the Allied invasion of Europe, and was redesignated No. 440 (Fighter Bomber) Squadron at Ayr, Scotland on 8 February 1944. It flew Typhoon aircraft in the pre-invasion softening up of the German defences and, after D-Day, gave close support to ground forces by dive-bombing and strafing enemy strongpoints, bridges, rail and road traffic. The squadron was disbanded at Flensburg, Germany on 26 August 1945.

Brief Chronology Formed as No. 111 (F) Sqn, Rockcliffe, Ont. 1 Nov 41. Redesignated No. 440 (FB) Sqn, Ayr, Scot. 8 Feb 44. Disbanded at Flensburg, Ger. 26 Aug 45.

Title or Nickname "City of Ottawa," "Beaver"

Adoption City of Ottawa (December 1944)

Commanders
S/L W.H. Pentland, DFC 8 Feb 44 - 7 Oct 44 *KIA*.
S/L A.E. Monson, DFC 8 Oct 44 - 15 Dec 44 *OTE*.
S/L H.O. Gooding, DFC 16 Dec 44 - 10 Mar 45 *OTE*.
S/L R.E. Coffey, DFC 11 Mar 45 - 30 Jul 45 *died*.[1]
S/L A.E. Monson, DFC 4 Aug 45 - 26 Aug 45.

Higher Formations and Squadron Locations
Second Tactical Air Force:
No. 83 (Composite) Group,
No. 22 (RCAF) Sector (disbanded 13 Jul 44),
No. 143 (RCAF) Wing,
Ayr, Scot. 14 Feb 44 - 17 Mar 44.
Hurn, Hants. 18 Mar 44 - 2 Apr 44.
Funtington, Sussex 3 Apr 44 - 19 Apr 44.
Hurn, Hants. 20 Apr 44 - 27 Jun 44.
B.(Base) 9 Lantheuil, Fr. 28 Jun 44 - 30 Aug 44.
B.24 St Andre, Fr. 31 Aug 44 - 2 Sep 44.
B.48 Glisy, Fr. 3 Sep 44 - 5 Sep 44.
B.58 Melsbroek, Bel. 6 Sep 44 - 25 Sep 44.
B.78 Eindhoven, Bel. 26 Sep 44 - 29 Mar 45.
B.100 Goch, Ger. 30 Mar 45 - 10 Apr 45.
B.110 Osnabruck, Ger. 11 Apr 45 - 19 Apr 45.
B.150 Hustedt, Ger. 20 Apr 45 - 28 May 45.
No. 17 Armament Practice Camp, Warmwell, Dorset., Eng. 23 Apr - 7 May 45.
B.166 Flensburg, Ger. 29 May 45 - 26 Aug 45.

Representative Aircraft (Unit Code I8)
Hawker Hurricane Mk.IV (Feb - Mar 44, not on operations)
Hawker Typhoon Mk.IB (Mar 44 - Aug 45) JP149 P
JR432 A   JR530 Y   MN171 E   MN257 D   MN298 A
MN366 G   MN378 A   MN403 J   MN428 Z   MN457 R
MN535 K   MN547 Q   MN595 M   MN603 P   MN691 V
MN703 L   MN709 B   MN720 W   MN777 U   PB389 P
PD452 X   PD589 R   RB377 Z   RB427 A   RB495 T
SW428 S

Operational History: First Mission 30 March 1944, 10 Typhoons from Hurn — anti-shipping sweep over the Channel Islands. **Last Mission** 21 April 1945, 11 Typhoons from Hustedt, each with one 500-pound bomb — dive-bombed Aachen, Germany. **Summary** Sorties: 4213. Operational/Non-operational Flying Hours: 4820/2264. Victories: Aircraft: 1 destroyed, 1 probably destroyed, 1 damaged. Ground: dropped 2215 tons of bombs (2735 1000- and 2591 500-pound), credited with 420 rail cuts; destroyed/damaged 5/0 bridges, 309/262 motor vehicles, 9/7 tanks, 2/48 locomotives, 93/500 trains, 4/50 barges. Casualties: Operational: 38 aircraft; 32 pilots, of whom 23 were killed, 5 missing, 3 POW, 1 evaded capture. Non-operational: 8 personnel, of whom 4 were killed, 3 wounded, 1 died of natural causes. **Honours and Awards** 5 DFC's. **Battle Honours** Fortress Europe 1944. France and Germany 1944-1945: *Normandy 1944, Arnhem, Rhine. Aleutians 1942-1943.*[2]

---

[1] Severely injured in an auto accident on 30 July 1945 and died 1 August.
[2] Earned as No. 111 (F) Squadron

278 *Hawker Typhoon Mk.IB MP149, coded I8-P, of No. 440 (FB) Squadron taxis cautiously through the jerricans at a hastily prepared base on the Continent.*

279 Hawker Typhoon Mk.I of No. 440 (FB) Squadron being armed with two 500-pound bombs.

No. 441 Squadron

Badge A silver fox's mask
Motto Stalk and kill
Authority King George VI, December 1945
The silver fox, an animal indigenous to Canada, represents the squadron's nickname.
Formed in Canada as No. 125 (Fighter) Squadron on 20 April 1942, the unit was the fourth of six home squadrons transferred overseas (complete in personnel but without aircraft) in preparation for the Allied invasion of Europe, and was redesignated No. 441 (Fighter) Squadron at Digby, Lincolnshire, England on 8 February 1944. It flew Spitfire aircraft on offensive and defensive air operations and, after D-Day, gave close support to ground forces in North-West Europe. In May 1945, the unit was re-equipped with Mustang aircraft for long-range bomber escort duty, but was not operational when hostilities ceased in Europe. The squadron was disbanded at Molesworth, Huntingdonshire on 7 August 1945.
Brief Chronology Formed as No. 125 (F) Sqn, Sydney, N.S. 20 Apr 42. Renumbered No. 441 (F) Sqn, Digby, Lincs., Eng. 8 Feb 44. Disbanded at Molesworth, Hunts. 7 Aug 45.
Title or Nickname "Silver Fox"
Commanders
S/L G.U. Hill, DFC and 2 Bars 11 Mar 44 - 25 Apr 44 *POW.*
S/L J.D. Browne, DFC 26 Apr 44 - 30 Jun 44 *OTE.*
S/L T.A. Brannagan, DFC 1 Jul 44 - 15 Aug 44 *POW.*
S/L R.H. Walker 26 Aug 44 - 7 Aug 45.
Higher Formations and Squadron Locations
Air Defence Great Britain:
No. 12 Group,
Digby, Lincs. 8 Feb 44 - 17 Mar 44.
Second Tactical Air Force:
No. 83 (Composite) Group,
No. 22 (RCAF) Sector,
No. 17 (RCAF) Sector (21 Apr 44, disbanded 13 Jul 44),
No. 144 (RCAF) Wing,
Holmsley South, Hants. 18 Mar 44 - 31 Mar 44.
Westhampnett, Sussex 1 Apr 44 - 11 Apr 44.
No. 16 Armament Practice Camp, Hutton Cranswick, Yorks. 12 Apr 44 - 22 Apr 44.
Funtington, Sussex 23 Apr 44 - 13 May 44.
Ford, Sussex 14 May 44 - 14 Jun 44.
B(Base) 3 Ste Croix-sur-Mer, Fr. 15 Jun 44 - 14 Jul 44.
No. 125 (RAF) Wing,
B.11 Longues, Fr. 15 Jul 44 - 12 Aug 44.
B.19 Lingèvres, Fr. 13 Aug 44 - 1 Sep 44.
B.40 Beauvais, Fr. 2 Sep 44 - 4 Sep 44.
B.52 Douai, Fr. 5 Sep 44 - 16 Sep 44.
B.70 Antwerp, Bel. 17 Sep 44 - 30 Sep 44.
Fighter Command:
No. 11 Group,
Hawkinge, Kent 1 Oct 44 - 28 Dec 44.
No. 13 Group,
Skeabrae, Orkney Island 30 Dec 44 - 2 Apr 45.

No. 11 Group,
Hawkinge, Kent 3 Apr 45 - 28 Apr 45.
Hunsdon, Herts. 29 Apr 45 - 16 May 45.
No. 12 Group,
Digby, Lincs. 17 May 45 - 15 Jul 45.
Molesworth, Hunts. 16 Jul 45 - 7 Aug 45.
Representative Aircraft (Unit Code 9G)
Supermarine Spitfire Mk.VB (Feb - Mar 44, not on operations)
Supermarine Spitfire Mk.IXB (Mar 44 - Jan 45)
Supermarine Spitfire F., H.F., L.F.Mk.IX (Feb - May 45)
North American Mustang Mk.III (May - Aug 45, not on operations)
Operational History: First Mission 27 March 1944, 2 Spitfire IXB's from Holmsley South — immediate readiness, scrambled against an aircraft approaching the south coast of England; intercepted in mid-Channel, the intruder was identified as a USAAF Mustang. **First Offensive Mission** 28 March 1944, 12 Spitfire IXB's from Holmsley South, with 12 from No. 442 Squadron, all part of No. 144 Wing — supporting fighter sweep over Dreux aerodrome, west of Paris, for USAAF Fortresses bombing aerodromes in northern France. With No. 442 Squadron acting as top cover, the unit strafed Dreux and claimed 4 aircraft destroyed, 1 probably destroyed, and 3 damaged on the ground. **First Victories** 25 April 1944, 12 Spitfire IXB's from Funtington, with 12 from No. 443 Squadron, all part of No. 144 Wing — escort to USAAF Fortresses and Liberators bombing Dreux aerodrome. Wing engaged enemy fighters and credited with 6 Fw.190's destroyed — 2 by each of the squadrons and 2 by W/C J.E. Johnson. One of the squadron's victories was shared by F/O's J.W. Fleming and L.A. Plummer. The other was shared by S/L Hill and F/O R.H. Sparling, both of whom failed to return; S/L Hill was taken prisoner, and F/O Sparling was killed. **Triple Victory** 13 Jul 1944, 12 Spitfire IXB's from Ste Croix-sur-Mer — armed reconnaissance, engaged 12 Fw.190's west of Argentan, claimed 10 destroyed without loss. F/O W.J. Myers credited with 3 destroyed. **Last Mission** 12 May 1945, 10 Spitfire IX's from Hunsdon — air cover to naval forces removing German prisoners-of-war from Guernsey, in the Channel Islands. **Summary** Sorties: 3148.
Operational/Non-operational Flying Hours: 4635/4023.
Victories: Aircraft: 56 destroyed, 12 damaged. Ground: dropped 12 tons of bombs; credited with 500-plus vehicles, tanks, barges and locomotives destroyed or damaged.
Casualties: Operational: 16 pilots, of whom 2 were killed, 7 presumed dead, 3 POW, 2 evaded capture; 2 proved safe. Non-operational: 1 killed, 2 presumed dead.
Squadron Aces F/L D.H. Kimball, DFC 6-0-1. F/L G.E. Mott, DFC 5½-0-⅓. F/L T.A. Brannagan, DFC 5½-0-0.
Honours and Awards 9 DFC's, 1 Croix de Guerre (Fr), 3 MiD's. **Battle Honours** Defence of Britain 1945. Fortress Europe 1944. France and Germany 1944-1945: *Normandy 1944, Arnhem, Walcheren.*

280 North American Mustang Mk.III and pilots of No. 441 (F) Squadron at Digby, Lincs. following the end of the war in Europe.

281 Supermarine Spitfire Mk.IXB MK193, coded Y2-E, of No. 442 (F) Squadron being guided on a narrow taxi strip in France, 1944.

No. 442 Squadron

Badge none
Formed in Canada as No. 14 (Fighter) Squadron on 2 January 1942, the unit was the fifth of six home squadrons transferred overseas (complete in personnel but without aircraft) in preparation for the Allied invasion of Europe, and was redesignated No. 442 (Fighter) Squadron at Digby, Lincolnshire, England on 8 February 1944. It flew Spitfire aircraft on defensive and offensive air operations and, after D-Day, gave close support to ground forces in North-West Europe. In March 1945 the unit was re-equipped with Mustang aircraft and employed on long-range bomber escort duty. The squadron was disbanded at Molesworth, Huntingdonshire on 7 August 1945.

Brief Chronology Formed as No. 14 (F) Sqn, Rockcliffe, Ont. 2 Jan 42. Renumbered No. 442 (F) Sqn, Digby, Lincs., Eng. 8 Feb 44. Disbanded at Molesworth, Hunts. 7 Aug 45.

Title or Nickname "Caribou"

Commanders
S/L B.R. Walker, DFC 8 Feb 44 - 27 Apr 44.
S/L B.D. Russel, DFC 28 Apr 44 - 6 Jul 44.
S/L H.J. Dowding, DFC and Bar 7 Jul 44 - 22 Sep 44 *OTE*.
S/L W.A. Olmstead, DSO, DFC and Bar 23 Sep 44 - 13 Dec 44 *2 OTE*.
S/L M.E. Jowsey, DFC 14 Dec 44 - 22 Feb 45 *MIA*.[1]
S/L M. Johnston 22 Feb 45 - 7 Aug 45.

Higher Formations and Squadron Locations
Air Defence Great Britain:
No. 12 Group,
Digby, Lincs. 8 Feb 44 - 17 Mar 44.
Second Tactical Air Force:
No. 83 (Composite) Group,
No. 22 (RCAF) Sector,
No. 17 (RCAF) Sector (21 Apr 44, disbanded 13 Jul 44),
No. 144 (RCAF) Wing,
Holmsley South, Hants. 18 Mar 44 - 31 Mar 44.
Westhampnett, Sussex 1 Apr 44 - 22 Apr 44.
Funtington, Sussex 23 Apr 44 - 14 May 44.
No. 16 Armament Practice Camp, Hutton Cranswick, Yorks. 25 Apr - 1 May 44.
Ford, Sussex 15 May 44 - 14 Jun 44.
B.(Base) 3 Ste Croix-sur-Mer, Fr. 15 Jun 44 - 14 Jul 44.
No. 126 (RCAF) Wing,
B.4 Beny-sur-Mer, Fr. 15 Jul 44 - 7 Aug 44.
B.18 Cristot, Fr. 8 Aug 44 - 31 Aug 44.
B.24 St André, Fr. 1 Sep 44.
B.25 Illiers l'Evêque, Fr. 2 Sep 44.
B.44 Poix, Fr. 3 Sep 44 - 6 Sep 44.
B.56 Evère, Bel. 7 Sep 44 - 20 Sep 44.
B.58 Le Culot, Bel. 21 Sep 44 - 2 Oct 44.
B.84 Rips, Neth. 3 Oct 44 - 13 Oct 44.
B.80 Volkel, Neth. 14 Oct 44 - 5 Dec 44.
No. 17 Armament Practice Camp, Warmwell, Dorset., Eng. 14-25 Nov 44.
B.88 Heesch, Neth. 6 Dec 44 - 22 Mar 45.
Fighter Command:
No. 11 Group,
Hunsdon, Herts., 23 Mar 45 - 16 May 45.
No. 12 Group,
Digby, Lincs. 17 May 45 - 16 Jul 45.
Molesworth, Hunts. 17 Jul 45 - 7 Aug 45.

Representative Aircraft (Unit Code Y2)
Supermarine Spitfire Mk.VB (Feb - Mar 44, not on operations)
Supermarine Spitfire Mk.IXB (Mar - Sep 44) MH718 M
MJ368 S   MJ515 W   MJ520 R   MJ608 W   MJ829 H
MJ967 X   MK131 X   MK141 J   MK149 B   MK181 C
MK193 E   MK194 H   MK206 I   MK295 X   MK464 Y
MK777 Z   MK826 X   MK416 D   NH325 H   NH412 S
PL280 F
Supermarine Spitfire Mk.IXE (Sep 44 - Mar 45)
MH456 R   MH728 L   MJ425 T   MJ466 J   MK303 B
MK564 M   MK844 Y   ML324 G   NH369 F   PL207 B
PL213 W   PL260 D   PL330 K   PL344 P   PL423 V
PL436 F   PL493 E   PL495 M   PT402 I   PT644 S
PT883 A   PV148 K   PV190 A   RR196 Q
North American Mustang Mk.III (Mar - Aug 45)
KH661 C   KH665 V   KH680 B   KH694 P   KH700 S
KH709 J   KH711 N   KH729 A   KH735 W   KH737 D
KH747 Y   KH765 R   KM122 F   KM218 Q

Operational History: First Mission 28 March 1944, 12 Spitfire IXB's from Holmsley South, with 12 from No. 441 Squadron, all part of No. 144 Wing — supporting fighter sweep for USAAF Fortresses bombing aerodromes in northern France. (The squadron acted as top cover for No. 441 Squadron in the strafing of Dreux aerodrome, west of Paris.) **First Victory** 22 June 1944, 7 Spitfire IXB's from Ste Croix-sur-Mer — returning from a fighter sweep, attacked by 8 enemy aircraft just west of Argentan, claimed 3 destroyed without loss. P/O W.R. Weeks in Y2-D and P/O F.B. Young in Y2-E each credited with a Bf.109 destroyed; S/L B.D. Russel in Y2-A and F/L J.T. Marriott in Y2-U shared a Fw.190 destroyed. The squadron diary credited P/O Weeks with drawing first blood. **Last Mission** 9 May 1945, 13 Mustangs from Hunsdon with long-range tanks — patrol over the Channel Islands while an Allied task force landed and liberated the areas: 3 returned early (mechanical trouble).[2] **Summary** Sorties: 4954, (including 82 Mustang). Operational/Non-operational Flying Hours: 7186/2977. Victories: Aircraft: 56 destroyed, 5 probably destroyed, 25 damaged.[3] Ground: dropped 328 tons of bombs, credited with 91 rail cuts; destroyed/damaged 909 motor vehicles, 125 locomotives, 194 trains. Casualties: Operational: 16 pilots: 1 killed, 9 presumed dead, 4 POW (1 escaped), 1 wounded; 1 proved safe. Non-operational: 2 personnel killed. **Squadron Ace** F/L D.C. Gordon, DFC 5½-0-1. **Honours and Awards** 1 DSO, 3 bars to DFC, 10 DFC's, 1 Croix de Guerre (Fr). **Battle Honours** Fortress Europe 1944. France and Germany 1944-1945: *Normandy 1944, Arnhem, Rhine. Aleutians 1943*.[4]

---

[1] Evaded capture and returned safely to England.
[2] The squadron's last bomber escort mission was on 25 April 1945 when 12 Mustangs from Hunsdon (with 12 of No. 611 Squadron) escorted Lancasters in an attack on Hitler's refuge at Berchtesgaden.
[3] Includes 2 Fw.190's credited to Mustang pilots on 16 April 1945 — one shared by F/O's R.J. Robillard (KH668 Y2-T) and L.H. Wilson (KH647 Y2-H), and one to F/L W.V. Shank (KH659 Y2-I).
[4] Earned as No. 14 (F) Squadron

No. 443 Squadron

Badge A hornet affronte
Motto Our sting is death
Authority King George VI, February 1946.
This unit was known as the Hornet Squadron.
Formed in Canada as No. 127 (Fighter) Squadron on 1 July 1942, the unit was the last of six home squadrons transferred overseas (complete in personnel but without aircraft) in preparation for the Allied invasion of Europe, and was redesignated No. 443 (Fighter) Squadron at Digby, Lincolnshire, England on 8 February 1944. It flew Spitfire aircraft on defensive and offensive air operations and, after D-Day, gave close support to ground forces in North-West Europe. After hostilities in Europe, the squadron served as one of four day fighter units assigned to the British Air Forces of Occupation (Germany) until disbanded at Utersen, Germany on 21 March 1946.
Brief Chronology Formed as No. 127 (F) Sqn, Dartmouth, N.S. 1 Jul 42. Renumbered No. 443 (F) Sqn, Digby, Lincs., Eng. 8 Feb 44. Disbanded at Utersen, Ger. 21 Mar 46.
Title or Nickname "Hornet"
Commanders
S/L H.W McLeod, DSO, DFC and Bar 8 Feb 44 - 27 Sep 44 *KIA.*
S/L A.H. Sager, DFC 30 Sep 44 - 29 Mar 45 *2 OTE.*
S/L T.J. De Courcy, DFC 5 Apr 45 - 7 Jun 45 *died.*[1]
F/L H.R. Finley, DFC 8 Jun 45 - 17 Sep 45 *OTE.*
S/L C.D. Bricker, DFC 18 Sep 45 - 21 Mar 46.
Higher Formations and Squadron Locations
Air Defence Great Britain:
No. 12 Group,
Digby, Lincs. 8 Feb 44 - 17 Mar 44.
Second Tactical Air Force:
No. 83 (Composite) Group,
No. 22 (RCAF) Sector,
No. 17 (RCAF) Sector (21 Apr 44, disbanded 13 Jul 44),
No. 144 (RCAF) Wing,
Holmsley South, Hants. 18 Mar 44 - 26 Mar 44.
No. 16 Armament Practice Camp, Hutton Cranswick, Yorks. 27 Mar 44 - 7 Apr 44.
Westhampnett, Sussex 8 Apr 44 - 21 Apr 44.
Funtington, Sussex 22 Apr 44 - 14 May 44.
Ford, Sussex 15 May 44 - 14 Jun 44.
B.(Base) 3 Ste. Croix-sur-Mer, Fr. 15 Jun 44 - 14 Jul 44.
No. 127 (RCAF) Wing,
B.2 Bazenville, Fr. 15 Jul 44 - 27 Aug 44.
B.26 Illiers l'Evêque, Fr. 28 Aug 44 - 20 Sep 44.
B.68 Le Culot, Bel. 21 Sep 44 - 29 Sep 44.
B.82 Grave, Neth. 30 Sep 44 -21 Oct 44.
B.58 Melsbroek, Bel. 22 Oct 44 - 3 Nov 44.
B.56 Evère, Bel. 4 Nov 44 - 2 Mar 45.
No. 17 Armament Practice Camp, Warmwell, Dorset., Eng. 18 Dec 44 - 2 Jan 44.
B.90 Petit-Brogel, Bel. 3 Mar 45 - 30 Mar 45.
B.78 Eindhoven, Neth. 31 Mar 45 - 11 Apr 45.
B.100 Goch, Ger. 12 Apr 45.
B.114 Diepholz, Ger. 13 Apr 45 - 27 Apr 45.
B.154 Soltau, Ger. 28 Apr 45 - 1 Jul 45.
British Air Forces of Occupation (Germany):
No. 83 (Composite) Group,
No. 126 (RCAF) Wing,
B.152 Fassberg, Ger. 2 Jul 45 - 6 Jul 45.
B.174 Utersen, Ger. 7 Jul 45 - 21 Mar 46.
Representative Aircraft (Unit Code 2I)
Supermarine Spitfire Mk.VB (Feb - Mar 44, not on operations)
Supermarine Spitfire Mk.IXB (Mar 44 - Feb 45) MJ366 J
MK343 G   MK366 V   MK397 T   MK455 R   MK605 W
MK607 S   MK636 E   ML194 S   MH300 V
Supermarine Spitfire Mk.XIV (Jan 45 - Jan 46)
Supermarine Spitfire Mk.XIVE (Jan - Mar 46)
Operational History: First Mission 13 April 1944, 12 Spitfire IXB's from Westhampnett — top cover escort for Bostons and Mitchells bombing Dieppe. Following this attack, 5 of the squadron's aircraft carried out strafing attacks over Rouen. **First Victory** 19 April 1944, 6 Spitfire IXB's from Westhampnett — part of a withdrawal fighter sweep for 64 Marauders of the USAAF returning from an attack on Malines, Belgium. S/L McLeod credited with a Do.217 destroyed near Brussels, the only enemy aircraft engaged. **Last Mission** 8 May 1945, 6 Spitfire XVI's from Soltau — fighter escort for Dakotas bound for Copenhagen with relief supplies. **Summary** Sorties: 5850. Operational/Non-operational Flying Hours: 7660/5103. Victories: Aircraft: 42 destroyed, 2 probably destroyed, 29 damaged. Ground: dropped 57 tons of bombs, credited with 19 rail and 3 road cuts; destroyed or damaged 1077 motor vehicles, 35 horsedrawn transport, 25 locomotives, 66 railway cars, 8 tanks. Casualties: Operational: 20 pilots of whom 1 was killed, 8 presumed dead, 6 POW, 1 evaded capture; 4 proved safe. Non-operational: 7 personnel; 1 killed, 2 missing, 4 injured. **Squadron Ace** S/L H.W. McLeod, DSO, DFC and Bar 8-0-0. **Honours and Awards** 1 DSO, 6 DFC's. **Battle Honours** Fortress Europe 1944. France and Germany 1944-1945: *Normandy 1944, Arnhem, Rhine.*

---

[1] As a result of an automobile accident near Schneverdingen, Germany.

282 *Supermarine Spitfire Mk.XVIE, coded 21-D and -N, of No. 443 (F) Squadron illustrate the variation in wing plan between the clipped and standard configurations.*

Air Observation Post Squadrons

Late in 1944, three Air Observation Post (AOP) squadrons were formed overseas. Although the Auster aircraft were flown by members of the Royal Canadian Artillery (RCA) on spotting and ranging artillery fire, the squadrons were administered and serviced by the RCAF.

The first thought of forming Canadian AOP squadrons occurred in September 1941 when three RCA officers, Captains D.R. Ely, R.R. MacNeil and R.A. Donald, were sent on a nine-month AOP training course. After completing the course, the Canadian military authorities decided against the formation of Canadian AOP squadrons and the three trained officers were loaned to the British to fill vacancies in squadrons being formed in England. Three months before the entry of Canadian troops in the Mediterranean theatre, the three officers were withdrawn from the British squadrons and posted to the 1st Field Regiment, Royal Canadian Horse Artillery. During the attack on Ortona, in December 1943, Captain Donald was killed and Captain MacNeil captured.

As the Italian campaign continued, it became obvious to senior Canadian Army officers that AOP squadrons were a necessary part of a modern army and, in June 1944, the commander of the First Canadian Army recommended, and the War Cabinet authorized, the formation of Nos. 1, 2 and 3 Canadian AOP Squadrons RCA. Major Ely was given the responsibility of organizing these units. In September 1944 it was decided to follow British precedent in the command and control of AOP squadrons. On the recommendation of the army commander, the War Cabinet re-authorized the three units as squadrons of the RCAF — Nos. 664, 665 and 666 (AOP) Squadrons — the pilots to be drawn from the artillery. The squadrons were formed under RAF Fighter Command's No. 70 Group and trained at No. 43 Operational Training Unit to observe artillery fire from the air and co-ordinate correction orders by the manoeuvring of the aircraft. As the squadrons were declared operational, they were sent to the Continent and placed under the operational control of the First Canadian Army. Soon after the end of hostilities in May 1945, these squadrons ceased to be highly specialized units and were reduced to the level of a hire-taxi service until disbanded one by one, between July 1945 and June 1946.

No. 664 Squadron

Brief Chronology Formed at Andover, Hants., Eng. 9 Dec 44. Disbanded at Apeldoorn, Neth. 1 Jun 46.
Commanders
Maj D.R. Ely 16 Dec 44 - 21 Jan 45.
Maj D.W. Blyth 22 Jan 45 - 1 Jun 46.
Higher Formations and Squadron Locations
Fighter Command:
No. 70 Group,
No. 43 Operational Training Unit,
Andover, Hants. 9 Dec 44 - 1 Feb 45.
Penshurst, Kent 2 Feb 45 - 21 Mar 45.
First Canadian Army:
Tilburg, Neth. 23 Mar 45 - 31 Mar 45.
Breda, Neth. 1 Apr 45 - 21 Apr 45.
Meppen, Ger. 22 Apr 45 - 5 May 45.
Rostrup, Ger. 6 May 45 - 16 Jun 45.
Apeldoorn, Neth. 17 Jun 45 - 1 Jun 46.
Representative Aircraft
Auster A.O.P.Mk.IV & V (Jan 45 - May 46)
MS945   MS946   MT166   TJ418
Operational History: First Mission 29 March 1945, Auster V RT564 with Capt G.M. Henderson and LAC R.S. Laye — reconnaissance of the Maas-Heusden area. **Last Mission** 5 May 1945, Auster V RT515 with Capt D.G. Rouse and LAC M.L. Wright — reconnaissance of a new air landing ground. **Summary** Sorties: 619. Operational/Non-operational Flying Hours: 583/4362. Casualties: Operational: 1 aircraft; 2 killed. Non-operational: nil.

No. 665 Squadron

Brief Chronology Formed at Andover, Hants., Eng. 22 Jan 45. Disbanded at Apeldoorn, Neth. 20 Jul 45.
Commanders
Maj D.R. Ely 22 Jan 45 - 4 Mar 45.
Maj N.W. Reilander 5 Mar 45 - 12 Jun 45.
Capt W.K. Buchanan 13 Jun 45 - 10 Jul 45.
Higher Formations and Squadron Locations
Fighter Command:
No. 70 Group,
No. 43 Operational Training Unit,
Andover, Hants. 22 Jan 45 - 16 Mar 45.
Oatlands Hill, Wilts. 17 Mar 45 - 18 Apr 45.
First Canadian Army:
B.77 Gilze-Rijen, Neth. 21 Apr 45 - 26 May 45.
Borne, Neth. 27 May 45 - 6 Jun 45.
Apeldoorn, Neth. 7 Jun 45 - 10 Jul 45.
Representative Aircraft
Auster A.O.P.Mk.IV & V (Feb - Jul 45) TJ346   TJ366   TJ402   TJ484
Operational History: First Mission 27 April 1945, Auster V TJ342 with Capt B.R.H. Watch — shoot on enemy gun positions on Duiveland Island. **Last Mission** 7 May 1945, Auster V TJ399 with Capt W.G. Milliken — reconnaissance. **Summary** Sorties: 58. Operational/Non-operational Flying Hours: 24/2092. Casualties: Operational: nil. Non-operational: 3 aircraft; 1 pilot killed, 3 injured (2 in non-flying accidents).

No. 666 Squadron

Brief Chronology Formed at Andover, Hants., Eng. 5 Mar 45. Disbanded at Apeldoorn, Neth. 1 Nov 45.
Commanders
Maj D.R. Ely 5 Mar 45 - 11 Jun 45.
Maj A.B. Stewart 12 Jun 45 - 1 Nov 45.
Higher Formations and Squadron Locations
Fighter Command:
No. 70 Group,
No. 43 Operational Training Unit,
Andover, Hants. 5 Mar 45 - 17 Apr 45.
Friston, Sussex 18 Apr 45 - 26 May 45.
First Canadian Army:
B.77 Gilze-Rijen, Neth. 28 May 45 - 5 Jun 45.
Hilversum, Neth. 6 Jun 45 - 24 Jun 45.
Apeldoorn, Neth. 25 Jun 45 - 1 Nov 45.
Representative Aircraft
Auster A.O.P. Mk.IV & V (Mar - Oct 45)
Operational History Never employed on operations. Flying Hours: 3728. Casualties: nil.

The Post-War Years 1945-1968

Establishment and Organization

On 1 January 1944 the RCAF reached its peak wartime strength of 215,200 all ranks (including 15,153 women), of whom 104,000 were in the British Commonwealth Air Training Plan, 64,928 were Home War Establishment (HWE) and 46,272 were serving overseas. There were 78 squadrons in service: 35 overseas, and 43 at home (of which six had been ordered overseas). By 1 April 1945 the strength of the RCAF was reduced to 164,846 all ranks through the termination of the BCATP and a reduction of the HWE. With the formal cessation of hostilities on 2 September 1945, and a proposed peace time establishment for the RCAF of 16,000 all ranks, a two-year "Interim Period" was declared during which the emphasis was to be on demobilization of approximately 90 per cent of the wartime force. On 6 February 1946 the Cabinet approved a peacetime RCAF of four components — a Regular Force, an Auxiliary, a Reserve and the Royal Canadian Air Cadets — and on 30 September 1947, when this came into effect, Canada's armed forces, which had been on active service since 1 September 1939, were officially "stood down."

The Regular Force had an authorized establishment of 16,100 all ranks and eight squadrons, and was to constitute a highly trained nucleus for a wartime force if one was required. With an actual strength of five squadrons, the RCAF resumed its pre-war activities of aerial photography (involving Nos. 413 and 414 Squadrons), air transport (Nos. 426 and 435 Squadrons) and communications (No. 412 Squadron).[1] Picking up where No. 8 (General Purpose) Squadron had stopped its work on mobilization, the two photographic units began an intensive five-year programme of photographing the whole of Canada.

The Auxiliary was authorized an establishment of 4500 all ranks and 15 squadrons, and was to provide a ready reserve of units that could be mobilized with a minimum of delay. Assigned the role of air defence, the Auxiliary began to form flying squadrons in April 1946, and continued in the defence role until 1954, when Regular Force CF-100 all-weather interceptor squadrons became operational and took over this role.

Two geographical air commands were formed to exercise administrative, training and operational control of all air force units within their boundaries: Central Air Command, with headquarters at Trenton and No. 10 Group at Halifax; and North West Air Command, with headquarters at Edmonton, No. 11 Group at Winnipeg, and No. 12 Group at Vancouver. In addition were No. 9 (Transport) Group and Maintenance Command, both of which were functional rather than regional in their operation. Beginning in 1948, the RCAF began to reorganize its command structure along functional as opposed to regional lines, as No. 9 (Transport) Group was elevated to Air Transport Command and No. 1 Air Defence Group was formed. In 1949, Maintenance Command became Air Materiel Command; Central Air Command was renamed Training Command; and Nos. 10 and 11 Groups were redesignated Maritime and Tactical Group respectively. Increased tension in world affairs in the early 1950's resulted in a further expansion, with the formation of No. 1 Air Division Europe, No. 5 Air Division (formerly No. 12 Group) and No. 14 (Training) Group, while a number of groups were elevated to command status: Air Defence Command, Maritime Air Command and Tactical Air Command.

From a Regular Force of 11,569 officers and airmen and an Auxiliary of 655 on 31 December 1947, the RCAF was to show a steady growth as relations between the Western democracies and the Communist bloc deteriorated. In January 1955 the strength "ceiling" was lifted to 51,000, placing the RCAF, for the first time in history, higher than the army (then 49,000). The Auxiliary experienced a corresponding growth, from 1400 in mid-1949 to nearly 5900 by mid-1952. From five Regular Force squadrons in 1947, the RCAF reached a peak of 29 Regular and twelve Auxiliary flying squadrons in 1955. Commencing in 1962, it was gradually reduced as the CF-100's were withdrawn from operational service and replaced by fewer CF-101 Voodoos, and as CF-104 Starfighters replaced the aging Sabres of the Air Division in Europe. This resulted in a decrease to 18 Regular operational and four training squadrons, and six Auxiliary squadrons, in February 1968 when the RCAF was integrated into the Canadian Armed Forces and ceased to exist as a separate and independent force.

Air Transport Command

Since the earliest days of aviation in Canada, air transport has played a vital part in penetrating the vast and sparsely populated land areas. The Canadian Air Force, formed in 1920, was responsible for the charting of air routes, aerial photography, and transportation of government officials throughout Canada by air. The eventual growth of commercial and private aviation relieved the RCAF of air mail and passenger services but aerial photography remained one of its main responsibilities. With the expansion of the RCAF from 1936 onward, air transport was prominently featured as two of its eight squadrons, Nos. 7 and 8 (General Purpose), were assigned to aerial photography and air transport. When the Second World War began, however, concern with the defence of Canada's eastern seaboard was of prime importance, and squadrons were reorganized and augmented to fill this need; consequently, only a small Communications Flight was retained for communications and transport requirements.[2]

It quickly became apparent, however, that a transatlantic ferry route would have to be established and maintained between Canada and the United Kingdom in order to supply the hard-pressed Royal Air Force with replacement aircraft. Goose Bay, Labrador was selected as the western terminus, and in August 1942 work began on the construction of an airfield. The vast quantities of material required for this development could be brought in only by sea or by air, and as sea transportation was slow and the sea was frozen over part of the year, the burden of building up and sustaining Goose Bay fell chiefly upon the air forces of Canada and the United States. By the end of 1942, a great backlog of materiel had accumulated at various points in eastern Canada, all of it urgently required at Goose Bay so as not to delay its development. It was to meet this need that No. 164 (Transport) Squadron — the premier air transport unit of the RCAF — was formed at Moncton, New Brunswick on 23 January 1943.

By the summer of 1943, demands on the RCAF for air transport facilities had so increased that an administrative organization was needed to co-ordinate the activities of the various units. On 5 August 1943 the Directorate of Air Transport Command was formed, consisting of a Transport Wing (Nos. 164 and 165 Squadrons), a Ferry Wing (Nos. 124 and 170 Squadrons), a Communications Wing (No. 12 Squadron), and a Training Establishment to provide crews for both home and overseas transport squadrons. By early 1945, control and administrative problems resulted in the reorganization of the Directorate into No. 9 (Transport) Group, and as the volume and importance of air transport continued to grow in the post-war period, it was renamed Air Transport Command on 1 April 1948.

Although the primary role of the Command has always been military air transport, its secondary roles have altered from time to time. During the war, the Directorate of Air Transport Command was also responsible for ferrying all types of training and operational aircraft within Canada. After the war, as No. 9 (Transport) Group, it was assigned the high-priority, but still secondary, role of completing the aerial photography of Canada, and when this was finished in 1950, the emphasis shifted to the introduction of new long-range transport aircraft to serve the needs arising from this country's involvement in the North Atlantic Treaty Organization and North American Air Defence Command.

In 1968, when Air Transport Command was absorbed into the Canadian Armed Forces, its organization was not affected, but its operational role placed more emphasis on troop-carrying, in keeping with the reorganization of the ground forces into highly mobile and air-transportable battle groups.

Maritime Air Command
Under the initial post-war establishment, there were no firm plans for the use of RCAF units in the defence of Canada's seaboard. Coastal defence was mainly the responsibility of the Royal Canadian Navy, with its anti-submarine destroyers and one aircraft carrier. The RCAF maintained only a small headquarters at Halifax. (No. 11 Group, of Central Air Command). By the late 1940's, however, the growing strength of the Russian submarine fleet, and its threat to the North Atlantic sealanes, resulted in the RCN's anti-submarine force being augmented by a major air force organization.

On 1 April 1949 No. 10 Group was redesignated Maritime Group, and on 1 November formed No. 2 (Maritime) Operational Training Unit at Greenwood, Nova Scotia for the specialized training of maritime aircrew. On 31 May 1950 No. 405 Squadron was formed at Greenwood, followed by No. 404 Squadron on 30 April 1951. On 1 April 1952 Maritime Group became an integral part of the newly formed Allied Command Atlantic of the North Atlantic Treaty Organization (NATO); and on 1 July it extended its responsibilities on the Pacific coast with the formation of No. 407 Squadron at Comox, British Columbia. The increasing size and complexity of the group, whose commander was also the air commander of NATO's Canadian Atlantic Sub-Area, prompted an elevation in status — to Maritime Air Command — on 1 June 1953. The Command reached an authorized strength of four squadrons on 1 May 1962, with the formation of No. 415 Squadron at Summerside, Prince Edward Island.

On 17 January 1966, almost two years before unification was complete, Maritime Air Command was absorbed into the Canadian Armed Forces' Maritime Command.

Air Defence Command
Under initial post-war plans, the air defence of major cities was the responsibility of Auxiliary flying squadrons supplemented by mobile Auxiliary radar units, but there was no separate organization for air defence. Accordingly, eight of the ten Auxiliary squadrons formed in 1946-48 were designated day fighter and equipped with either Vampire jet interceptor or Mustang piston-engined aircraft; no Regular squadrons were involved. In the United States, on the other hand, defence planning called for a permanent continental radar wall constantly on guard against air attack from any quarter, backed by a force of all-weather jet interceptors. The deteriorating international situation in the 1940's, plus the build-up of the Soviet long-range bomber force, turned the thinking of the Canadian government and its military advisers towards the American concept. This resulted in the RCAF forming No. 1 Air Defence Group (later Air Defence Command) at St Hubert, Quebec on 1 December 1948, together with No. 1 (Fighter) Operational Training Unit and No. 410 (Fighter) Squadron, the first such Regular units on its post-war establishment.

Although Canada was unable to provide fighter squadrons to the United Nations during the Korean war of June 1950 to July 1953 (22 fighter pilots did serve with American units), this conflict, along with other political developments, did have an immediate and direct affect on the growth of Canadian air defence forces. At a NATO ministerial meeting in December 1950, particular stress was placed on the need for increased air power as a deterrent to Communist aggression in Western Europe, and Canada undertook to station in Europe by the fall of 1954 an air division of 12 Sabre-equipped fighter squadrons. Further, in the defence of North America, Canada committed itself to additional measures to make this continent safe from air attack.

In February 1951 Canada and the United States jointly announced an agreement to co-operate in the air defence of North America through a closely integrated radar system.[3] Canada had already decided to improve its air defence forces by developing and building an all-weather jet interceptor, the Avro CF-100 Canuck. First flown on 10 January 1950, it was delivered to No. 3 All-Weather (Fighter) Operational Training Unit in November 1952 and to the newly-formed No. 445 Squadron in April 1953. Over the next 18 months, a further eight squadrons were formed, so that by 1955 Air Defence Command had reached its allotted strength of 19 fighter squadrons — nine Regular with CF-100's, and ten Auxiliary with Vampires and Mustangs.[4]

North American Air Defence Command
During the construction of the radar warning lines and the build-up of the RCAF's defence forces, Canadian and American officials considered how best to combine their air defences. The concept of an international radar system gradually evolved into one of a single command embracing all air defence facilities and forces of both countries. On 1 August 1957 an agreement to integrate the Canadian and US air defences into a single command, named North American Air Defence Command (NORAD), was announced. The agreement provided, among other things, that NORAD would be maintained for a period of ten years or such shorter period as might be agreed to by both

parties. It was further agreed that the Commander-in-Chief was to be an American and that the Deputy Commander-in-Chief was to be a Canadian. Inasmuch as the C-in-C was to be responsible to both the US and the Canadian Chiefs of Staff, his appointment was subject to his acceptability to the Canadian government. By the same token the deputy, who in the absence of the C-in-C, reported also to the US Joint Chiefs, had to be acceptable to the US authorities.[5]

In September 1958 the Canadian government outlined three innovations to be made in the air defence system: the introduction of the Bomarc supersonic surface-to-air missile, designed for atomic warheads; the installation of additional large radar units and gap fillers to bolster the Pinetree Line; and the construction of a combat control centre under a system known as "semi-automatic ground environmental" (SAGE), combining all the capabilities of high-speed computers and high-performance radar. The combat control centre would be housed underground at NORAD Regional Headquarters, situated at North Bay.

The rapidly changing technology of air defence had far-reaching effects on the strength of Air Defence Command. The Auxiliary Force radar units were disbanded, and the flying squadrons reassigned to light transport and emergency rescue duties. The Regular Force squadrons were reduced from a peak strength of nine CF-100 units in 1955 to five CF-101B's in 1961, and then to three in 1964. Nevertheless, they remained an active part of NORAD in the 1970's.[6]

While its responsibility was broadening from the defence of major cities in Canada to a greater role in maritime defence and partnership with the United States in continental air defence, the RCAF also became involved in regional air defence overseas.

North Atlantic Treaty Organization

After the Second World War the Western democracies rapidly dismantled their armed forces and placed their trust in the newly-created United Nations Organization and the great powers forming its Security Council, to maintain peace. Their hopes, however, were soon dashed by the expansionism of the Soviet Union and its casting of 30 vetoes in the Security Council.

In March 1948 the recent Communist coup in Czechoslovakia and the increased threat to the security of the Western European nations led to the Treaty of Brussels, which joined Belgium, Luxembourg, The Netherlands and the United Kingdom into a defensive grouping. In September this developed into the Western Union Defence Organization. On 4 April 1949 a new security pact was signed, enlarging Western Union defence into the North Atlantic area. Canada, Belgium, Denmark, France, Iceland, Italy, Luxembourg, The Netherlands, Norway, Portugal, the United Kingdom and the United States signed the North Atlantic Treaty, which declared that an armed attack against one or more of the signatories, whether in Europe or North America, would be regarded as an attack against them all, and bound them to take action to restore and maintain the security of the North Atlantic area. On ratification, the North Atlantic Treaty Organization (NATO) came into effect on 24 August 1949.[7] As relations between NATO and the Communist-bloc nations further deteriorated, and as a direct result of the Communist-inspired invasion of South Korea in June 1950, the United States proposed the creation of an integrated military force in Western Europe under a supreme commander. In December 1950 the military commitments of the member nations were established and in April 1951 Supreme Headquarters Allied Powers in Europe (SHAPE) was set up.

Militarily, NATO was divided into three major commands — Allied Command Europe, Allied Command Atlantic and Allied Command Channel. Allied Command Europe, under the Supreme Allied Commander Europe (SACEUR), stretched from Norway to Turkey and was divided into four regions — Northern, Central, Southern and Mediterranean. The area of responsibility of Allied Forces Central Europe extended from Denmark southward to the Austro-German border. Early in April 1951, Allied Air Forces Central Europe (AIRCENT) was created to maintain nuclear retaliatory forces ready to assist, immediately and by all possible means, in the defence of adjacent NATO territory. A year later, to exercise effective operational control of its forces, AIRCENT formed two multinational Allied Tactical Air Forces. In the northern sector was Second ATAF with Belgian, British, German and Netherlands air force units; and in the southern sector Fourth ATAF with American, Canadian, French[8] and, later, German units.

No. 1 Air Division Europe

The RCAF component of Fourth ATAF, known as No. 1 Air Division Europe, was to consist of four day fighter wings, each of three squadrons of Sabre aircraft.[9] In view of the war damage to European air bases, it was decided to build four new airfields: two in France (Grostenquin and Marville), and two in Germany (Baden-Soellingen and Zweibrucken). Until these were ready, one RCAF fighter wing was to be based in England.

No. 1 (Fighter) Wing was formed at North Luffenham on 1 November 1951 and became operational on the 15th with the arrival of its first Sabre squadron, No. 410. This squadron had left Halifax on 31 October, the aircraft being shipped to Scotland aboard the Royal Canadian Navy's carrier *Magnificent*, and on 15 November the pilots flew their Sabres from Renfrew to North Luffenham. On 12 February 1952 No. 441 Squadron reached North Luffenham in the same manner. The third squadron arrived in June 1952 in a way that made RCAF history and established the pattern for the transfer of the air division's remaining nine squadrons. Between 30 May and 15 June, 21 Sabres of No. 439 Squadron were flown from Uplands to England via Greenland and Iceland as Operation "Leapfrog I". Despite bad weather all the way, every aircraft made the 3560-mile trip without mishap.

On 1 October 1952 No. 1 Air Division Europe, although far from complete, became an operational component of Fourth ATAF. Its build-up continued with the arrival of No. 2 Wing at Grostenquin on 11 October 1952 (the first RCAF formation to be stationed on the Continent since March 1946), No. 3 at Zweibrucken on 6 April 1953, and No. 4 at Baden-Soellingen on 4 September. No. 1 Wing moved from North Luffenham to Marville in March 1955.

Between 1956 and August 1957, in view of a shortage of all-weather interceptor aircraft in NATO, the air division began to replace one Sabre day-fighter squadron in each of the four wings with a CF-100 all-weather fighter squadron from Canada. In July 1959 the government announced that in the fall of 1962 it would begin to re-equip the eight Sabre squadrons with Canadian-built versions of the American F-104 Starfighter and disband the four CF-100 squadrons. This re-equipping of the air division resulted in a change of role from all-weather night and day interception to tactical nuclear strike support of ground forces.

In 1962 there arose a conflict of interest between France and the United States over the storing of nuclear weapons on French soil without full French control. Consequently, all NATO nuclear weapons were withdrawn from France. Because No. 1 Wing was assigned to a strike reconnaissance role and possessed no such weapons, it remained at Marville for the time being while No. 2 Wing, at Grostenquin, was disbanded early in 1964 and its two squadrons were redeployed to the two German bases. In March 1966 the French government announced that it was going to withdraw its own forces from NATO operational control and bring them home, and requested that all NATO units either leave the country by 1 April 1967 or else accept French operational control. In February 1967 the Canadian government stated that France had agreed to turn over the French base at Lahr, Germany in exchange for the Canadian base at Marville.

Earlier, in 1966, the Canadian government had announced that it had decided against the procurement of replacement Starfighters needed to offset anticipated losses associated with normal attrition. The consequence of this decision in the years ahead was that there would be a progressive lowering in the number of Starfighter squadrons in the face of attrition. As an immediate consequence, the strength of the air division was cut from eight to six Starfighter squadrons in 1967, with a further decrease in 1968 when, as an austerity measure, No. 3 Wing at Zweibrucken was disbanded and the six Starfighter squadrons were redistributed between Nos. 1 and 4 Wings. By early 1970, Canada had decided upon a major reduction of all its forces in Europe, and on 1 July, No. 1 Air Division Europe was reduced to an air group of three squadrons employing conventional weapons for close support of ground operations.

Throughout this post-war period, Canada's air industry had grown enormously, not only in support of the RCAF. Canadian-built aircraft were also supplied under Mutual Aid to NATO allies. Mutual Aid has been examined in an earlier publication sponsored by the Canadian War Museum (John Griffin, *Canadian Military Aircraft; Serials and Photographs* (Queen's Printer, 1969) while the triumphs of the air industry in Canada — including the ill-fated Avro-Arrow — will be dealt with in a forthcoming museum publication (John Griffin, "Canadian Military Aircraft; Form and Function").

The Auxiliary

In recognition of the valuable contribution made by the Auxiliary to the wartime RCAF — it had represented one-third of the establishment in August 1939 and provided the first two of three squadrons to go overseas — post war plans gave the Auxiliary an even more important role than before the war. It was to be developed as a national guard, almost on a par with the Regular Force, and for the next decade it would enjoy a high priority in defence measures.[10]

The RCAF peace establishment, prescribed on February 1946, called for an Auxiliary of fifteen flying squadrons. On 15 April authority was granted for the formation of seven squadrons at specified locations: No. 400 (Toronto), Nos. 401 and 438 (Montreal), No. 402 (Winnipeg), No. 418 (Edmonton), No. 424 (Hamilton), and No. 442 (Vancouver).[11] In the autumn of 1946 each squadron received Harvard aircraft and commenced service flying training, in the east as fighter interceptor and in the west as fighter bomber. On 1 April 1947 an eighth squadron was authorized — No. 406, at Saskatoon.

In January 1948 the RCAF received its first jet interceptor aircraft, de Havilland Vampires, and by March had issued them to Nos. 400, 401, 438 and 442 Squadrons, while Mustang aircraft were issued to No. 424 in a fighter role and to No. 402 as fighter bomber. Nos. 406 and 418 Squadrons were equipped with twin-engine Mitchell light bombers.[12] Also in 1948, two more squadrons were formed — No. 420 in London and No. 403 in Calgary — both with Mustang aircraft. No. 411 Squadron was formed at Toronto in 1950 and the twelfth and last squadron, No. 443, at Vancouver in 1951; both were equipped with Vampire aircraft. In 1955 Silver Star jet trainers were issued to the Auxiliary squadrons in preparation for their re-equipment with Sabre aircraft being returned from the air division in Europe. On 28 October 1956 the first of the Sabre Mark 5's were officially handed over to No. 401 Squadron and the re-equipping of Nos. 400, 411, 438, 442 and 443 began.

In addition to the flying squadrons, the Auxiliary formed specialized units to back up those squadrons and support the Regular Force as well. On 15 December 1948, the first of eleven Auxiliary radar squadrons — later redesignated Aircraft Control and Warning — was formed at Montreal. These units initially manned mobile radar units and later, permanent sites in the Pinetree Line, and provided the Auxiliary flying units with precise control for intercepting unidentified aircraft. There were also medical, technical training and intelligence units. To administer the growing number of units, a wing headquarters was formed in late 1950 in cities where two or more Auxiliary establishments were located.

In 1958 the government announced a major change in the future employment of naval reserves, the Militia and the RCAF Auxiliary. Changing technology had made it probable that war, if it came, would be fought with immediately available forces, and would be over before reserves could play their traditional part. Accordingly, the new role of reserves was to be civil defence and rescue work.

On 26 March 1958 the Air Council announced that the Auxiliary flying units would be employed as light transport and search-and-rescue units, their fighter and bomber aircraft to be replaced with twin-engine Expeditor aircraft and, in 1960, with single-engine Otters. During 1961, the introduction of stronger radar in the Pinetree Line resulted in the disbandment of the Auxiliary aircraft control and warning units. On 1 April, control of the Auxiliary passed from Air Defence Command to Air Transport Command, the latter being more in line with the new role of the flying units. Exactly three years later it was reduced to four wing headquarters and six flying squadrons: Nos. 401 and 438, both at Montreal; Nos. 400 and 411, Toronto; No. 402, Winnipeg; and No. 418 Squadron at Edmonton.

On 1 February 1968 the RCAF Auxiliary ceased to exist when it was integrated into the Canadian Forces Reserve to provide a pool of trained air personnel to meet the emergency requirements of civil defence.

The Unification of the Forces 1964-1968

In March 1964 a White Paper was tabled in the House of Commons to review Canada's defence policy and forecast the future development and use of the armed forces. The introduction defined the objectives of defence policy, which could not be separated from foreign policy, as the preservation of peace by supporting collective defence measures; the deterrence of military aggression; and the protection and surveillance of Canadian territory, air space and coastal waters. The White Paper then went on to propose that the three separate military services be unified in a single organization, the Canadian Armed Forces,[13] the fundamental considerations being more effective control, the streamlining of procedures — particularly decision-making — and the reduction of overhead costs.

The White paper pointed out that a measure of integration had already been effected as early as 1946, with the abolition of separate defence ministries "for Naval Services" and "for Air" (introduced in 1940) and a return to a single Minister of National Defence. On 1 April 1947 the Defence Research Board had been established to provide all three services with scientific advice. A further step towards integration was the appointment, in 1951, of a Chairman of the Chiefs of Staff Committee to co-ordinate the training and operations of the three services. Other examples of integration were the Royal Military Colleges (actually tri-service); common legal, medical and chaplain services; inter-service logistics, for which a standard catalogue of all material had been devised; and food procurement, dental and postal services, provided by the army for all three services.

As a first step towards total unification, on 13 April 1964 the government introduced Bill C-90, "Integration of the Headquarters Staff", whereby a single Chief of the Defence Staff (CDS) was to replace the individual service chiefs. On 16 July the Bill was given Royal assent, and on 1 August Air Chief Marshal F.R. Miller was appointed as the first CDS, along with the heads of new functional branches of Canadian Forces Headquarters (CFHQ). This meant that the former three services were no longer independent entities for the purpose of control and administration.

Following the appointment of the CDS and the reorganization of CFHQ, attention was turned to the command and control of integrated units. It was decided in January 1966 to establish six functional commands to replace the eleven mainly regional service commands; every Regular establishment in Canada was to be reallocated to the appropriate command by 1 April 1966.

Mobile Command was formed to maintain combat-ready land and tactical air forces capable of rapid deployment, in circumstances ranging from NATO service in Europe to United Nations and other peace-keeping operations. Its Tactical Air Group would consist of CF-5 tactical ground-support, Buffalo transport aircraft, heavy and light helicopters.

Maritime Command, embodying all sea and air maritime forces on both the Atlantic and Pacific seaboards, was primarily responsible for anti-submarine defence, but was to become increasingly capable of such other tasks as patrolling the Arctic region.

Air Transport Command would provide the forces with "strategic airlift capability", the emphasis being on troop-carrying operations.

Air Defence Command was to contribute squadrons of CF-101 Voodoo interceptors, and surveillance and control radar, to North American Air Defence Command.

Training Command was responsible for all individual training, including flying and trades training.

Materiel Command would provide the necessary supply and maintenance support to the other functional commands.

In addition there were: Reserve and National Survival; Communications System (elevated to command status in July 1970); and Canadian Forces Europe (consisting of the 4th Mechanized Combat Battle Group and No. 1 Canadian Air Group) as independent organizations reporting directly to CFHQ.

Although the detailed structure of the Canadian Armed Forces has undergone changes as priorities have altered, the basic aim remains the forging of a highly compact and mobile force that can be deployed to meet future needs.

On 4 November 1966, Bill C-243, "The Canadian Forces Reorganization Act," was introduced to amend the National Defence Act; the Canadian Army, the Royal Canadian Navy and the Royal Canadian Air Force, previously separate and independent services, would become one. Following debate in the House of Commons and further examination in the Defence Committee, the Bill was given third and final reading in April 1967, clearing the way for unification.

The Canadian Forces Reorganization Act came into effect on 1 February 1968. With that, the identity of the RCAF, its records and its achievements, were laid to rest in the pages of Canadian aviation and military history.

---

[1]The RCAF announced early in 1947 that it was going to perpetuate the wartime overseas squadrons by re-adopting "400-block" numbers. The four squadrons then on strength were re-numbered accordingly; No. 12 became No. 412, No. 13, No. 413, and No. 14 No. 414, while No. 164 was reorganized into two squadrons, Nos. 426 and 435.

[2]No. 8 Squadron went to Sydney, Nova Scotia on coastal work and No. 7 was reduced to its communications flight, the remainder of its personnel going to other operational squadrons.

[3]Three radar chains were eventually built. The Pinetree Line, along the border, was a joint undertaking; the Mid-Canada Line, running along the 55th parallel, was a completely Canadian project, largely automatic in operation; and the Distant Early Warning (DEW) Line was built north of the Arctic Circle as a joint undertaking. First announced in 1955, the DEW Line was declared operational in July 1957.

The Pinetree Line, unlike the other two, was not solely a warning line but a command and control system for the identification and interception of unknown aircraft. It was an integral part of a system of continuous radar-cover for the whole of Continental USA and part of Canada — eventually reaching roughly the 60th parallel of latitude. To describe it as a "line", therefore, conveys a wrong impression, the "Pinetree System" would be more correct.

[4]Between November 1956 and August 1957, four of the CF-100 squadrons were sent to No. 1 Air Division Europe, replacing four of the Sabre squadrons which were then reactivated in Canada with CF-100's. To maintain Air Defence Command's strength during the changeover, the CF-100 units in Canada earmarked for duty in Europe were maintained over-strength; and when the main body departed for Europe the surplus became the nucleus of the new CF-100 squadron.

[5]NORAD Command now consists of the US Army Air Defense Command, USAF Air Defense Command, Canadian Forces (formerly RCAF) Air Defence Command, and certain units of the US Navy. These commands train and equip units for the commander-in-chief, who directs, controls and co-ordinates their operational activities. The area encompassed is over ten and a half million square miles, and in order that the commander-in-chief can manage such widely deployed forces, the continent is divided into Regions and these in turn are sub-divided into Divisions. Where a Region or Division overlaps the international boundary, the principle has been that the commander is provided by the "host" country and his deputy by the other.

[6]In August 1971 Canada announced that it intended to dispense with the two Bomarc installations and return the missiles and their atomic warheads to the United States. The two squadrons, Nos. 446 and 447, were disbanded on 1 September 1972.

[7]Greece and Turkey joined NATO on 18 February 1952, and West Germany on 9 May 1955.

[8]Until France's eventual withdrawal from NATO operational control, the First French Tactical Air Corps was a component of the Fourth ATAF.

[9]Canada also undertook to provide facilities for the training in this country of aircrew from Belgium, Denmark, France, Italy, The Netherlands, Norway, Portugal and the United Kingdom. This arrangement was later extended to include Greece, Turkey and West Germany when they joined NATO. The cost of the training was borne by Canada under the Mutual Aid Plan. Between May 1951 and July 1957, 4600 pilots and navigators were trained. In April 1957 an independent, three-year extension had provided for the training of further Danish, Dutch and Norwegian aircrew (155 in all) annually with those countries paying a token sum of $5000 per student; and in December 1958 a similar agreement was undertaken with West Germany to train 360 aircrew.

[10]Between 1 October 1950 and 1 September 1951 the Auxiliary was known as the Reserve.

[11]The localization of three of the Auxiliary squadrons were based on prewar Auxiliary service: No. 400, formerly No. 110 at Toronto; No. 402, formerly No. 112 at Winnipeg; and No. 401, perpetuating both No. 115 at Montreal and No. 1 Squadron of the Regular Force. Except for No. 442 Vancouver which traced its West Coast service to No. 14, the remainder had been wartime adoptions by the cities involved.

[12]No. 418 Squadron's role was changed to light bomber on 1 January 1948, while No. 406 was formed in that role.

[13]The terms Canadian Armed Forces (CAF) and Canadian Forces (CF) are both official. As service aircraft came to be marked "CAF", this abbreviation is preferred by the authors; the shorter term, in full or abbreviated, is used only in special cases such as the occasional quotation.

UNIT HISTORIES
POST-WAR SQUADRONS

No. 400 "City of Toronto" Squadron (Auxiliary)

Badge *see Overseas Squadrons, No. 400 Squadron*
Formed at Toronto, Ontario on 15 April 1946, the squadron flew Vampire and Sabre aircraft in a fighter role until October 1958 when it was reassigned to a light transport and emergency rescue role and re-equipped with Expeditor and Otter aircraft. On 10 June 1961 it became the first RCAF unit to receive a Squadron Standard for 25 years service as No. 110 and 400 Squadron. On 1 February 1968 the squadron was integrated into the Canadian Armed Forces as No. 400 "City of Toronto" Air Reserve Squadron.

Brief Chronology Formed as No. 400 (FB) Sqn (Aux), Toronto, Ont. 15 Apr 46. Redesignated No. 400 (F) Sqn (Aux) 1 Feb 47. Titled No. 400 "City of Toronto" (F) Sqn (Aux) 6 Nov 52. Redesignated No. 400 "City of Toronto" Sqn (Aux) 1 Oct 58. Integrated into CAF as No. 400 "City of Toronto" Air Res Sqn 1 Feb 68.

Title "City of Toronto"

Commanders
W/C G.W. Gooderham, AFC 8 Feb 47 - 31 Oct 48 *ret.*
W/C C.A. James, AFC 1 Nov 48 - 7 May 51.
W/C A.E. Fleming 8 May 51 - 8 Mar 52 *KIFA.*
W/C R.H. Rohmer, DFC 9 Mar 52 - 5 Sep 52.
W/C W.N. Stowe, DFC 6 Sep 52 - 6 May 54 *ret.*
S/L C.I.M. Ettles 7 May 54 - 30 Jun 55.
S/L W.A. Curtis, Jr. 1 Jul 55 - 2 May 57 *ret.*
W/C H.B. Davis 3 May 57 - 31 Dec 59.
W/C B.A. Howard, CD 1 Jan 60 - 2 Oct 63.
W/C G.E. Gilroy, CD 3 Oct 63 - 31 Dec 66.
W/C G.M. Georgas, CD 1 Jan 67 - 31 Jan 68.

Higher Formations and Squadron Locations
Training Command,
Air Defence Command (1 Aug 51),
No. 2 Group (Auxiliary) (15 Jan 51 - 1 Mar 57),
Air Transport Command (1 Oct 58):
No. 14 Wing (Auxiliary) (1 Oct 50),
Malton, Ont. 15 Apr 46 - 17 Oct 47.
Downsview,[1] Ont. 19 Oct 46 - 31 Jan 68.

Representative Aircraft (Unit Code 1947-51 AA, 1951-58 GW)
North American Harvard Mk.II (Jul 46 - Sep 58)
3125 AA-F   3165 AA-H   3329 AA-M
de Havilland Vampire Mk.III (Mar 48 - May 55)
17002   17020   17036   17051 AA-P
Canadair Silver Star Mk.3 (Apr 55 - Oct 58)
21318   21445   21467   21505
Canadair Sabre Mk.5 (Oct 56 - Oct 58)
23104   23227   23312   23364
Beechcraft Expeditor Mk.3 (Oct 58 - Oct 66)
1442   1516   1559   2343
de Havilland Otter (Oct 60 - Jan 68)
3663   3685   3745   9416

Honours and Awards Squadron Standard (10 June 1961)

---

[1] Known as De Havilland Aerodrome until February 1947.

283 North American Harvard Mk.II 3154, radio call sign VC-AAG, of No. 400 Squadron (Aux) at Downsview, Ont.

284 De Havilland Vampire Mk.III 17051, VC-AAP, of No. 400 Squadron (Aux). Nose and boom stripes were blue and white, applied to all the squadron aircraft in this period.

285 Canadair Silver Star Mk.3 21463 bearing the unit identifier GW of No. 400 Squadron (Aux).

287 Beechcraft Expeditor Mk.3N 1472 of No. 400 Squadron (Aux) on 31 October 1966, the day the Expeditors were removed from squadron service.

288 De Havilland Otter 3673 of No. 400 Squadron (Aux) at Downsview, Ont. on 16 April 1967.

286 Canadair Sabre Mk.5 23204 on the No. 400 Squadron (Aux) flight-line at Downsview, Ont.

No. 401 "City of Westmount" Squadron (Auxiliary)

Badge *see Overseas Squadrons, No. 401 Squadron*
Formed at Montreal, Quebec on 15 April 1946, the squadron flew Vampire and Sabre aircraft in a fighter role until October 1958 when it was reassigned to a light transport and emergency rescue role and re-equipped with Expeditor and Otter aircraft. On 5 May 1962 it received a Squadron Standard for 25 years service as No. 1 and 401 Squadron. On 1 February 1968 the squadron was integrated into the Canadian Armed Forces as No. 401 "City of Westmount" Air Reserve Squadron.

Brief Chronology Formed as No. 401 (F) Sqn (Aux), Montreal, Que. 15 Apr 46. Titled No. 401 "City of Westmount" (F) Sqn (Aux) 4 Sep 52. Redesignated No. 401 "City of Westmount" Sqn (Aux) 1 Nov 58. Integrated into CAF as No. 401 "City of Westmount" Air Res Sqn 1 Feb 68.

Title "City of Westmount"
Nickname "Ram"
Commanders
W/C F.G. Grant, DSO, DFC 10 Oct 47 - 23 Aug 48 *ret.*
W/C J.W. Reid, DFC 24 Aug 48 - 14 Dec 51.
W/C H.J. Everard, DFC, CD 15 Dec 51 - 13 Jan 55.
W/C J.M. Grisdale 15 Sep 55 - 15 Sep 57 *ret.*
W/C A.J. Edwards, CD 1 Apr 58 - 31 Mar 61 *ret.*
G/C H.J. Everard, DFC, CD 1 Apr 61 - 31 Sep 61.
W/C T. Bowie, CD 1 Nov 61 - 31 Sep 65 *ret.*
W/C G.D. Adkins, CD 1 Oct 65 - 31 Jan 68.

Higher Formations and Squadron Location
Training Command,
Air Defence Command (1 Jun 51),
No. 1 Group (Auxiliary),
Air Transport Command (1 Apr 61):
No. 11 Wing (Auxiliary) (1 Sep 61),
St Hubert, Que. 15 Apr 46 - 31 Jan 68.

Representative Aircraft (Unit Code 1947-51 AB, 1951-58 QT)
North American Harvard Mk.II (Sep 46 - Sep 58) 2680 3046 3300 3832
de Havilland Vampire Mk.III (Mar 48 - Feb 56) 17061 17017 17024 17039
Canadair Silver Star Mk.3 (Dec 54 - Oct 58) 21437 21439
Canadair Sabre Mk.5 (Oct 56 - Oct 58) 23152 23340 23346
Beechcraft Expeditor Mk.3 (Sep 58 - Oct 66) HB112 1386 1418 2331
de Havilland Otter (Oct 60-Jan 68) 9409 9426 9689 9692

Honours and Awards Squadron Standard (5 May 1962)

289 De Havilland Vampire Mk.III VC-ABA of No. 401 Squadron (Aux) based at St Hubert, Que.

290 North American Harvard Mk.IIB 3288 VC-ABG of No. 401 Squadron (Aux) with blue cowling and rudder.

291 Canadair Silver Star Mk.3 21530, bearing the ram's head of No. 401 Squadron (Aux), at St Hubert, Que. on 29 March 1958.

292 Canadair Sabre Mk.5 23315, coded AB and bearing the No. 401 Squadron (Aux) ram's head, at St Hubert, Que. on 29 March 1958.

293 Beechcraft Expeditor Mk.3NM 2325 of No. 401 Squadron (Aux) with the ram's head unit insignia on the nose.

No. 402 "City of Winnipeg" Squadron (Auxiliary)

Badge A standing grizzly bear totem of the North Pacific Coast Indians
Motto We stand on guard
Authority Queen Elizabeth II, November 1956

The grizzly bear holds a prominent place in Indian mythology and is believed to have supernatural powers.
Formed at Winnipeg, Manitoba on 15 April 1946, the squadron flew Mustang aircraft in a fighter and fighter bomber role until January 1957 when it was redesignated transport, but employed on navigation training, and re-equipped with Expeditor aircraft. In March 1958 the unit was reassigned to a light transport and emergency rescue role and received Otter aircraft. On 26 November 1961 it received a Squadron Standard for 25 years service as No. 112 and 402 Squadron. On 1 February 1968 the squadron was integrated into the Canadian Armed Forces as No. 402 "City of Winnipeg" Air Reserve Squadron.

Brief Chronology Formed as No. 402 (FB) Sqn (Aux), Winnipeg, Man. 15 Apr 46.[1] Redesignated No. 402 (F) Sqn (Aux) 1 Mar 47. Redesignated No. 402 (FB) Sqn (Aux) 1 Apr 49. Titled No. 402 "City of Winnipeg" (FB) Sqn (Aux) 18 Sep 50. Redesignated No. 402 "City of Winnipeg" (F) Sqn (Aux) 10 Nov 53. Redesignated No. 402 "City of Winnipeg" (T) Sqn (Aux) 25 Jan 57.[2] Redesignated No. 402 "City of Winnipeg" Sqn (Aux) 1 Apr 58. Integrated into CAF as No. 402 "City of Winnipeg" Air Res Sqn 1 Feb 68.

Title "City of Winnipeg"
Nickname "Winnipeg Bears"
Commanders
W/C R.J. Clement, DFC 2 Aug 46 - 21 Dec 48 *ret.*
W/C L.M. Cameron, DFC 22 Dec 48 - 27 Feb 50 *ret.*
W/C W.B. Breckon 28 Feb 50 - 7 Nov 51 *ret.*
W/C D.W. Rathwell, DFC 8 Nov 51 - 7 Oct 53 *ret.*
W/C J.M. Reid, DFC 8 Oct 53 - 14 May 56 *ret.*
W/C D.M. Gray, CD 15 May 56 - 17 Jul 60.
W/C J.T. Patterson, CD 18 Jul 60 - 15 Oct 62.
W/C D.R. Scott, CD 16 Oct 62 - 15 Oct 65.
W/C J.A. Brown, CD 21 Oct 65 - 31 Jan 68.

Higher Formations and Squadron Location
North West Air Command,
Tactical Air Command (1 Aug 51),
Air Defence Command (16 Nov 53),
Training Command (25 Jan 57),
Air Transport Command (1 Apr 61):
No. 17 Wing (Auxiliary) (1 Apr 61),
Stevenson Field, Man. 15 Apr 46 - 31 Jan 68.

Representative Aircraft (Unit Code 1947-51 AC, 1951-58 SV)
North American Harvard Mk.II (Jul 46 - Mar 57)
2552 2962 3070 3343
de Havilland Vampire Mk.III (Apr 48 - Nov 50)
17023 17059 AC-P 17077 17081
North American Mustang Mk.IV (Nov 50 - Sep 56)
9563 AC-U 9564 AC-F 9569 AC-S 9570 AC-B
9571 AC-O
Canadair Silver Star Mk.3 (Nov 54 - Aug 56) 21297 21445

Beechcraft Expeditor Mk.3 (Aug 56 - Mar 64)
1508   1556   2328   2329
de Havilland Otter (May 60 - Jan 68)
9403   9405   9415   9416
Honours and Awards Squadron Standard (26 November 1961)

---

[1] Was to have been formed as No. 402 (Fighter Reconnaissance) Squadron (Auxiliary) on 1 May 1946, but this order was amended.
[2] Although designated Transport, it became in fact a navigational training unit.

296   North American Mustang Mk.IV 9258 of No. 402 Squadron (Aux) lined up with six of the squadron's aircraft at Stevenson Field, Winnipeg, Man.

294   North American Harvard Mk.IIB 3316 of No. 402 Squadron (Aux) at Rivers, Man. in 1956. The rudder and cowling bear the blue stripes of the unit.

297   Canadair Silver Star Mk.3 21297 at Trenton, Ont. on 11 August 1965, immediately following service with No. 402 Squadron (Aux).

295   De Havilland Vampire Mk.III 17023 of No. 402 Squadron (Aux) displaying the crest of the City of Winnipeg and the scroll indicating No. 402 Squadron.

298   Beechcraft Expeditor Mk.3N 1457 of No. 402 Squadron (Aux) photographed at Rivers, Man. in July 1956.

299   De Havilland Otter 9415 of No. 402 Squadron (Aux) visiting Mt. Hope, Ont. on 21 January 1968.

No. 403 "City of Calgary" Squadron (Auxiliary)

Badge *see Overseas Squadrons, No. 403 Squadron*
Formed at Calgary, Alberta on 15 October 1948, the squadron flew Mustang aircraft in a fighter bomber and fighter role until January 1957 when it was redesignated a transport unit and re-equipped with Expeditor aircraft. In March 1958 it was reassigned to a light transport and emergency rescue role and received Otter aircraft. A reduction of the Auxiliary Force resulted in the squadron being disbanded on 1 April 1964.
Brief Chronology Formed as No. 403 (FB) Sqn (Aux), Calgary, Alta. 15 Oct 48.[1] Titled No. 403 "City of Calgary" (FB) Sqn (Aux) 3 Sep 52. Redesignated No. 403 "City of Calgary" (F) Sqn (Aux) 16 Nov 53. Redesignated No. 403 "City of Calgary" (T) Sqn (Aux) 25 Jan 57. Redesignated No. 403 "City of Calgary" Sqn (Aux) 1 Apr 58. Disbanded 1 Apr 64.
Title "City of Calgary"
Nickname "Wolf"
Commanders
W/C W.A. Mostyn-Brown, AFC 15 Jan 50 - 31 Jan 52.
W/C D.B. Freeman, DFC 1 Feb 52 - 29 Nov 53 *died*.
W/C A.R. Cruickshank 30 Nov 53 - 29 Nov 56 *ret*.
W/C G.M. Kelly, CD 30 Nov 56 - 28 Feb 59.
S/L H.T. Johnstone 1 Mar 59 - 25 Nov 59 *ret*.
S/L W.H. Huston, CD 26 Nov 59 - 7 Sep 60.
W/C J.G. McLaws 8 Sep 60 - 31 Jan 63 *ret*.
W/C W.H. Huston, CD 1 Feb 63 - 1 Apr 64.
Higher Formations and Squadron Location
North West Air Command,
Tactical Air Command (1 Aug 51),
Air Defence Command (16 Nov 53),
Training Command (25 Jan 57),
Air Transport Command (1 Apr 61):
No. 30 Wing (Auxiliary) (1 Aug 54),
Calgary,[2] Alta. 15 Oct 48 - 1 Apr 64.
Representative Aircraft (Unit Code 1948-51 AD, 1951-58 PR)
North American Harvard Mk.II (Aug 49 - Mar 57)
2653 2654 3131 3286
North American Mustang Mk.IV (Nov 50 - Oct 56)
9244 9291 9590 9594
Canadair Silver Star Mk.3 (Nov 55 - Feb 57) 21438 21460
Beechcraft Expeditor Mk.3 (Aug 56 - Mar 64)
1417 1513 1594 2354
de Havilland Otter (Oct 56 - Mar 64) 3668

300 North American Mustang Mk.IV 9263 of No. 403 Squadron (Aux) at Calgary, Alta.

---

[1]Was to have been formed as No. 403 (Fighter Reconnaissance) Squadron (Auxiliary) on 15 September 1948, but the order was amended.
[2]Renamed Lincoln Park on 1 September 1961

No. 404 Squadron

Badge *see Overseas Squadrons, No. 404 Squadron*
Formed as a Maritime Reconnaissance unit at Greenwood, Nova Scotia on 30 April 1951, and redesignated Maritime Patrol on 17 July 1956, the squadron was the second of four formed in Maritime Air Command, and flew Lancaster, Neptune and Argus aircraft on East Coast maritime duty. On 1 February 1968 the squadron was integrated into the Canadian Armed Forces.

Brief Chronology Formed as No. 404 (MR) Sqn, Greenwood, N.S. 30 Apr 51. Redesignated No. 404 (MP) Sqn 17 Jul 56. Integrated into CAF 1 Feb 68.

Nickname "Buffalo"

Commanders
W/C D.E. Galloway, MBE, CD 18 Aug 51 - 12 Jun 54.
W/C B.H. Moffit, DFC, AFC, CD 13 Jul 54 - 9 Feb 57.
W/C I.A.H. MacFarlane, CD 10 Feb 57 - 31 Jul 60.
W/C A.M. Halkett, DFM, CD 1 Aug 60 - 30 Jun 62.
W/C A.J. Ireland, CD 1 Jul 62 - 20 Jan 64.
W/C H.A. Carswell, CD 21 Jan 64 - 2 Sep 65.
W/C D.W. Souchen, CD 3 Jan 66 - 31 Jan 68.

Higher Formations and Squadron Location
Maritime Air Command,
Canadian Forces Maritime Command (17 Jan 66):
Greenwood, N.S. 30 Apr 52 - 31 Jan 68.

Representative Aircraft (Unit Code 1947-51 AF, 1951-58 SP)
Avro Lancaster Mk.X (Apr 51 - Sep 55) FM102
FM220 AF-M   KB355 AF-A   KB959
Lockheed Neptune (Mar 55 - 1960) 24108   24110
24122   24124
Canadair CP-107 Argus Mk.1 & 2 (Apr 59 - Jan 68) 20714
20725   20735   20742

301   Avro Lancaster Mk.X KB959 VC-AFA of No. 404 (MP) Squadron going on sea patrol from its base at Greenwood, N.S.

302   Lockheed Neptune 24110 of No. 404 (MP) Squadron in the early midnight blue finish common to this aircraft-type at the time of its being taken into RCAF service.

303   Canadair Argus Mk.2 20725 on Atlantic patrol in the service of No. 404 (MP) Squadron.

No. 405 Squadron

Badge see Overseas Squadrons, No 405 Squadron
Formed as a Maritime Reconnaissance unit at Greenwood, Nova Scotia on 31 March 1950, and redesignated Maritime Patrol on 17 July 1956, the squadron was the first of four formed in Maritime Air Command, and the first to fly Lancaster, Neptune and Argus aircraft on East Coast maritime duty. On 1 February 1968 the squadron was integrated into the Canadian Armed Forces.

Brief Chronology Formed as No. 405 (MR) Sqn, Greenwood, N.S. 31 Mar 50.[1] Redesignated No. 405 (MP) Sqn 17 Jul 56. Integrated into CAF 1 Feb 68.

Nickname "Eagle"[2]

Commanders
W/C D.T. French, DFC 31 Mar 50 - 31 Jul 50 *KIFA*.[3]
W/C A.G. Kenyon, CD 18 Aug 51 - 2 Oct 51.
W/C W.P. Pleasance, DFC and Bar, CD 28 Jan 52 - 29 Jun 52.
W/C J.E. Creeper, DFC, CD 30 Jun 52 - 14 Aug 54.
W/C E.J. Smith, CD 15 Aug 54 - 14 Jan 56.
W/C J.C. Mulvihill, AFC, CD 21 Jan 56 - 16 Jul 58.
W/C C. Torontow, AFC, CD 17 Jul 58 - 14 Jul 61.
W/C J.F. Drake, AFC, CD 15 Jul 61 - 8 Sep 64.
W/C A. Lehn, CD 2 Dec 64 - 4 Aug 66.
W/C W.M. Houser, CD 5 Aug 66 - 31 Jan 68.

Higher Formations and Squadron Location
Maritime Air Command,
Canadian Forces Maritime Command (17 Jan 66):
Greenwood, N.S. 31 Mar 50 - 31 Jan 68.

Representative Aircraft (Unit Code 1946-51 AG, 1951-58 VN)
Avro Lancaster Mk.X (Apr 50 - Nov 55)
FM173 AF-F   FM213 AG-J   FM223 AF-O   FM229 AG-R
KB857 AG-N   KB868 AG-S   KB920 AF-A   KB929 AG-B
KB946 AG-D   KB950 AG-L   KB964 AG-H   KB997 AG-M
Lockheed Neptune (Mar 55 - 1958) 24112   24119
24121   24123
Canadair CP-107 Argus Mk.1 & 2 (Aug 58 - Jan 68)
20713   20717   20728   20741

---

[1] An order for the formation of a No. 405 (Bomber Reconnaissance) Squadron was issued on 1 April 1947, then amended to 1 October 1947, then cancelled.
[2] Also adopted by No. 423 AW(F) Squadron.
[3] Killed while on a resupply mission to Alert, Northwest Territories, in Lancaster KB965.

304   Avro Lancaster Mk.X KB925 VC-AGA at Dartmouth, N.S. on 24 October 1958 while serving No. 405 (MP) Squadron.

305   Lockheed Neptune 24111 wearing the unit identifier VN of No. 405 (MP) Squadron.

306   Canadair Argus Mk.1 20710 of No. 405 (MP) Squadron during flight trials on 28 March 1957.

No. 406 "City of Saskatoon" Squadron (Auxiliary)

Badge *see Overseas Squadrons, No. 406 Squadron*
Formed at Saskatoon, Saskatchewan on 1 April 1947, the squadron flew Mitchell aircraft in a light bomber role until March 1958 when it was reassigned to a light transport and emergency rescue role and re-equipped with Expeditor and Otter aircraft. A reduction of the Auxiliary Force resulted in the squadron being disbanded on 1 April 1964.
Brief Chronology Formed as No. 406 (TacB) Sqn (Aux), Saskatoon, Sask. 1 Apr 47. Redesignated No. 406 (LB) Sqn (Aux) 1 Apr 49. Titled No. 406 "City of Saskatoon" (LB) Sqn (Aux) 3 Sep 52. Redesignated No. 406 "City of Saskatoon" Sqn (Aux) 1 Apr 58. Disbanded 1 Apr 64.
Title "City of Saskatoon"
Nickname "Lynx"
Commanders
W/C J. Baillie 1 Jul 47 - 31 Dec 51 *ret.*
W/C A.A. Meyers 1 Jan 52 - 31 Dec 54 *ret.*
W/C R.J. Henry, DFC, CD 1 Jan 55 - 16 Jul 56 *ret.*
W/C W.B. Maloney, CD 17 Jul 56 - 31 Dec 59 *ret.*
W/C J.P. Coggins 1 Jan 60 - 31 Mar 62 *ret.*
W/C T. Jasieniuk, CD 1 Apr 62 - 1 Apr 64.
Higher Formations and Squadron Location
North West Air Command,
Tactical Air Command (1 Aug 51),
Training Command (1 Oct 58),
Air Transport Command (1 Apr 61):
No. 23 Wing (Auxiliary) (1 Jan 55),
Saskatoon, Sask. 1 Apr 47 - 1 Apr 64.
Representative Aircraft (Unit Code 1947-51 AH, 1951-58 XK & QP)
North American Harvard Mk.II (May 47 - Oct 51) 2723 3021 3177 3240
North American Mitchell Mk.III (Jun 48 - Jun 58) FW251 HD331 HD334 KL144
Canadair Silver Star Mk.3 (Sep 56 - Jun 58) 21151 21159
Beechcraft Expeditor Mk.3 (Feb 58 - Oct 63) HB139 HB207 2320
de Havilland Otter (Feb 61 - Mar 64)

307 *North American Mitchell Mk.III 5244 at Saskatoon, Sask. with No. 406 Squadron (Aux).*

308 *Canadair Silver Star Mk.3 21159 of No. 406 Squadron (Aux) based at Saskatoon, Sask.*

309 *Beechcraft Expeditor Mk.3N 2362 of No. 406 Squadron (Aux) at Saskatoon, Sask. on 24 October 1958.*

No. 407 Squadron

Badge see *Overseas Squadrons, No. 407 Squadron*
Formed as a Maritime Reconnaissance unit at Comox, British Columbia on 1 July 1952, and redesignated Maritime Patrol on 17 July 1956, the squadron was the third of four — the only West Coast — unit formed in Maritime Command and flew Lancaster and Neptune aircraft. On 1 February 1968 the squadron was integrated into the Canadian Armed Forces.

Brief Chronology Formed as No. 407 (MR) Sqn, Comox, B.C. 1 Jul 52. Redesignated No. 407 (MP) Sqn 17 Jul 56. Integrated into CAF 1 Feb 68.

Nickname "Demon"

Commanders
W/C C.W. McNeill, CD 1 Jul 52 - 8 Sep 54.
W/C W. McLeod, CD 9 Sep 54 - 12 Jul 55.
W/C W.D. Foster, DFC, CD 13 Jul 55 - 21 May 59.
W/C J.C. McCarthy, DSO, DFC, CD 22 May 59 - 4 Jul 62.
W/C L.H. Croft, CD 5 Jul 62 - 22 Jun 64.
W/C K.O. Moore, DSO, CD 23 Jun 64 - 22 Dec 65.
W/C H.E Smale, CD 23 Dec 65 - 31 Jan 68.

Higher Formations and Squadron Location
Maritime Command (Pacific),
Canadian Forces Maritime Command (Pacific) 17 Jan 66):
Comox, B.C. 1 Jul 52 - 31 Jan 68.

Representative Aircraft (Unit Code 1951-58 RX)[1]
Avro Lancaster Mk.X (Jul 52 - May 59) FM136
FM210   KB975   KB996
Lockheed Neptune (May 58 - Jan 68) 24102   24111
24123   24125

---

[1]Neptune aircraft received from Greenwood often carried the previous codes of Nos. 404 and 405 Squadrons (SP and VN) until repainted.

311   Lockheed Neptune 24102, with jet packs installed, as used by No. 407 (MR) Squadron.

310   Avro Lancaster Mk.X KB894 with unit identifier RX of No. 407 (MR) Squadron, based at Comox, B.C. and photographed at Saskatoon, Sask.

No. 408 Squadron

Badge *see Overseas Squadrons, No. 408 Squadron*
Formed as a Photographic unit at Rockcliffe (Ottawa), Ontario on 10 January 1949, the squadron flew modified Lancaster aircraft equipped with shoran[1] for geodetic control in aerial photography in the far north and Arctic regions of Canada, and Canso, Norseman, Otter and Dakota aircraft to fly in and maintain the ground stations associated with shoran. When this programme was completed, the unit was redesignated Reconnaissance on 18 July 1957 and flew Lancaster aircraft on Arctic surveillance patrols. In April 1964 it moved to Rivers, Manitoba, was redesignated Transport Support and Area Reconnaissance, and re-equipped with C-119 and Silver Star aircraft. On 1 February 1968 the squadron was integrated into the Canadian Armed Forces.

Brief Chronology Formed as No. 408 (P) Sqn, Rockcliffe, Ont. 10 Jan 49. Redesignated No. 408 (R) Sqn 18 Jul 57. Redesignated No. 408 (TS & AR) Sqn, Rivers, Man. 1 May 64. Redesignated No. 408 (TacS & AR) Sqn 15 Mar 66. Integrated into CAF 1 Feb 68.

Nickname "Goose"

Commanders
W/C C.L. Olsson, DFC 10 Jan 49 - 19 Nov 50.
W/C D.J.G. Jackson, CD 20 Nov 50 - 22 Sep 51.
W/C H.M. Smith, DFC, CD 23 Sep 51 - 2 Apr 54.
W/ J.G. Showler, AFC, CD 3 Apr 54 - 14 Jul 57.
W/C J.F. Mitchell, DFC, AFC, CD 15 Jul 57 - 20 Aug 58.
W/C F.W.H. MacDonnell, CD 21 Aug 58 - 2 Oct 61.
W/C S.F. Cowan, CD 3 Oct 61 - 25 Jul 63.
W/C R.G. Orpen, CD 26 Jul 63 - 22 Aug 65.
S/L P. Bissky, CD 23 Aug 65 - 1 May 66.
W/C H.A. McKay, DFC, CD 2 May 66 - 31 Jan 68.

Higher Formations and Squadron Locations
Air Transport Command:
Rockcliffe, Ont. 10 Jan 49 - 31 Mar 64.
Rivers, Man. 1 Apr 64 - 14 Mar 66.
Canadian Forces Mobile Command:
No. 10 Tactical Air Group,
Rivers, Man. 15 Mar 66 - 31 Jan 68.

Representative Aircraft (Unit Code 1949-51 AK, 1951-58 MN)
Avro Lancaster Mk.X (Jan 49 - Mar 64) FM120   FM199   FM207   KB961
Consolidated Canso A (Jan 49 - Sep 59) 9815   11091   11093   11094
Noorduyn Norseman Mk.VI (May 51 - Sep 53) 366   368
de Havilland Otter (Feb 53 - Sep 57) 3661   3666   3672   3684
Avro Lancaster Mk.X (Dec 53 - Mar 64) KB882   KB976
Douglas Dakota Mk.III & IV (Jan 55 - Nov 66) KG256   KG345   KG389   KG423
Fairchild C-119G Flying Boxcar (Apr 64 - May 65) 22115   22117   22119   22132
Canadair Silver Star Mk.3 (Apr 64 - Jan 68)
Armament 21118   21555   21580   21582
Photo 21556   21557   21565   21633
Trainer 21566
Lockheed C-130B Hercules (Apr 65 - Mar 66) 10301   10302   10303   10304

312   *Noorduyn Norseman Mk.IV 787 VC-AKF of No. 408 (P) Squadron being tied-up at the dock at Golden Lake, Ont. on 14 May 1953.*

313   *Avro Lancaster Mk.X FM122 with unit identifier MN, flown by No. 408 (P) Squadron while based at Rockcliffe, Ont.*

314   *Consolidated Canso A 11056, with modified nose and blisters, at Rivers, Man. in July 1956 with No. 408 (P) Squadron.*

---

[1] *A short-range navigational device developed during the war.*

315 Douglas Dakota Mk.III KG345 in service with No. 408 (R) Squadron.

318 De Havilland Otter 3663 flown by No. 408 (P) Squadron.

316 Fairchild Flying Boxcar 22115 of No. 408 (TS & AR) Squadron.

319 Canadair Silver Star Mk.3 21137, in the markings of No. 408 (TS & AR) Squadron, visits Downsview, Ont. on 2 February 1968.

317 Lockheed C-130B Hercules 10303 of No. 408 (TS & AR) Squadron based at Rivers, Man., 1965.

No. 409 Squadron

Badge *see Overseas Squadrons, No. 409 Squadron*
Formed as an All-Weather (Fighter) unit at Comox, British Columbia on 1 November 1954,[1] the squadron flew CF-100 and CF-101 aircraft on North American (West Coast) air defence duty, and on 1 February 1968 was integrated into the Canadian Armed Forces.

Brief Chronology Formed at Comox, B.C. 1 Nov 54. Integrated into CAF 1 Feb 68.

Nickname "Nighthawk"

Commanders
S/L F.E. Haley, CD 1 Nov 54 - 4 Mar 56.
W/C T.J. Evans, CD 5 Mar 62 - 21 Jul 58.
W/C H.E. Bridges, DFC, CD 22 Jul 58 - 11 Aug 61.
W/C E.G. Ireland, DFC, CD 12 Aug 61 - 1 Jul 62.
W/C G. Inglis, CD 10 Oct 62 - 21 Oct 65.
W/C W.H. Vincent, CD 22 Oct 65 - 1 Aug 67.
W/C G.W. Patterson, CD 2 Aug 67 - 31 Jan 68.

Higher Formations and Squadron Locations
North American Air Defence Command (12 Sep 57),
RCAF Air Defence Command:
Comox, B.C. 1 Nov 54 - 31 Jan 68.
Namao, Alta (conversion to DF-101 under No. 425 AW(F) Sqn) 5 Feb — 13 Mar 62.

Representative Aircraft (Unit Code LP)
Avro CF-100 Canuck Mk.4A (Feb - Jun 55) 18247
18253   18269   18315
Avro CF-100 Canuck Mk.4B (May 55 - Sep 57) 18363
18370   18399   18402
Avro CF-100 Canuck Mk.5 (Feb 57 - Mar 62) 18487
18580   18608   18631
McDonnell CF-101B Voodoo (Mar 62 - Feb 68)

---

[1] An order had been issued on 1 January 1947 for the formation of a No. 409 (Fighter) Squadron (Auxiliary) at Victoria, British Columbia, but cancelled shortly afterwards.

320   Avro CF-100 Mk.3 18178 of No. 409 AW(F) Squadron provides the background for a well-chained young cougar in March 1958.

321   Avro CF-100 Mk.4A 18247 and 18315 of No. 409 AW(F) Squadron in formation over the Rocky Mountains in March 1955.

322   Avro CF-100 Mk.4B 18402 at Vancouver, B.C. in June 1956 with No. 409 AW(F) Squadron.

323   Avro CF-100 Mk.5 18631, with wing tip stores and the Air Defence Command rudder marking, in the service of No. 409 AW(F) Squadron.

324   McDonnell Voodoo 17396 with No. 409 AW(F) Squadron at Comox, B.C.

No. 410 Squadron

Badge see Overseas Squadrons, No. 410 Squadron
Formed in a Fighter role at St Hubert (Montreal), Quebec on 1 December 1948, the squadron was the first post-war Regular Force fighter unit, the first to fly Vampire and Sabre aircraft, and the first to join No. 1 (Fighter) Wing of No. 1 Air Division Europe. In 1956, it was decided to replace one Sabre squadron in each of the Air Division's four wings with an all-weather fighter unit. When No. 445 AW(F) Squadron arrived from Canada, No. 410 was deactivated at Marville, France on 1 October 1956 and reactivated as All-Weather (Fighter) at Uplands (Ottawa), Ontario on 1 November. The squadron flew CF-100 and CF-101 aircraft on North American air defence until disbanded on 1 April 1964.

Brief Chronology Formed as No. 410 (F) Sqn, St Hubert, Que. 1 Dec 48. Deactivated at Marville, Fr. 1 Oct 56. Reactivated as No. 410 AW(F) Sqn, Uplands, Ont. 1 Nov 56. Disbanded 1 Apr 64.

Nickname "Cougar"

Commanders
S/L R.A. Kipp, DSO, DFC and Bar 4 Jan 49 - 25 Jul 49 KIFA.[1]
S/L L.A. Hall 24 Dec 49 - 15 Apr 52.
S/L D. Warren, DFC 27 May 52 - 3 Aug 54.[2]
S/L A.W. Fisher, CD 15 Aug 54 - 1 Oct 56.
Squadron inactive
S/L H.E. Bodien, DSO, DFC 1 Nov 56 - 14 Jun 58.
S/L E.W. Garrett, DFC, CD 15 Jun 58 - 17 Mar 59.
S/L J.W. Whiteley, CD 18 Mar 59 - 31 Aug 59.
W/C J.E. McClure, DFC, CD 1 Sep 59 - 30 May 61.
W/C M.F. Doyle, CD 1 Jun 61 - 13 Jul 61.
W/C K.W. MacDonald, CD 14 Jul 61 - 1 Apr 64.

Higher Formations and Squadron Locations
Air Defence Command:
St Hubert, Que. 1 Dec 48 - 28 Oct 51.
En route overseas aboard HMCS *Magnificent* 19 Oct 51 - 13 Nov 51.
No. 1 Air Division:
No. 1 (Fighter) Wing,[3]
North Luffenham, Notts., Eng. 15 Nov 51 - 14 Nov 54.
No. 4 (Fighter) Wing,[4]
Baden-Soellingen, Ger. 15 Nov 54 - 18 Apr 55.
No. 1 (Fighter) Wing,
Marville, Fr. 19 Apr 55 - 1 Oct 56.
Squadron inactive
North American Air Defence Command,
RCAF Air Defence Command:
Uplands, Ont. 1 Nov 56 - 1 Apr 64.
Namao, Alta. (conversion to CF-101 under No. 425 AW(F) Sqn) 11 Nov - 21 Dec 61.

Representative Aircraft (Unit Code 1948-56 AM 1956-58 AN)
de Havilland Vampire Mk.III (Dec 48 - May 51) 17050 17067   17076   17084
Canadair Sabre Mk.2 (May 51 - Nov 54) 19102 AM-N 19128 AM-H 19129 AM-Q 19144 AM-T
Canadair Sabre Mk.5 (Nov 54 - Sep 56)

Avro CF-100 Canuck Mk.5 (Nov 56 - Nov 61) 18533
18591   18607   18748
McDonnell CF-101B Voodoo (Nov 61 - Mar 64) 17395
17398   17402   17450

---

[1] Killed during an aerobatic display at the Canadian National Exhibition in Toronto flying Vampire 17084.
[2] Served in Korea on exchange duty with the USAF from July to December 1953.
[3] While in England, under the operational control of RAF Fighter Command through its No. 11 Group.
[4] A temporary arrangement while No. 1 (F) Wing was moving from North Luffenham to Marville.

326   Canadair Sabre Mk.2 19102 VC-AMN, the first production Sabre from Canadair and first employed by No. 410 (F) Squadron, working up from its base at St Hubert, Que.

325   De Havilland Vampire Mk.III 17063   VC-AMP of No. 410 (F) Squadron, during Exercise "Sweetbriar," at Whitehorse, N.W.T. on 19 February 1950.

327   Avro CF-100 Mk.5 18483 with unit identifier AN, denoting No. 410 AW(F) Squadron, in May 1959.

328   McDonnell Voodoo 17463 and the remainder of No. 410 AW(F) Squadron at their Uplands, Ont. base.

No. 411 "County of York" Squadron (Auxiliary)

Badge see Overseas Squadrons, No. 411 Squadron
Formed at Toronto, Ontario on 1 October 1950, the squadron flew Vampire and Sabre aircraft in a fighter role until October 1958 when it was reassigned to a light transport and emergency rescue role and re-equipped with Expeditor and Otter aircraft. On 1 February 1968 the squadron was integrated into the Canadian Armed Forces as No. 411 "County of York" Air Reserve Squadron.

Brief Chronology Formed as No. 411 (F) Sqn (Aux), Toronto, Ont. 1 Oct 50.[1] Titled No. 411 "County of York" (F) Sqn (Aux) 2 Jun 52. Rdesignated No. 411 "County of York" Sqn (Aux) 1 Oct 58. Integrated into CAF as No. 411 "County of York" Air Res Sqn 1 Feb 68.

Title "County of York"

Nickname "Grizzly Bear"

Commanders
W/C R.I.A. Smith, DFC and Bar 1 Oct 50 - 10 Jul 52.
W/C R.H. Rohmer, DFC 2 Sep 52 - 10 Mar 53.
W/C C. Darrow, DFC 11 Mar 53 - 3 Sep 54.
S/L M.F. Cliff 8 Oct 54 - 15 Aug 55.
W/C J.W.P. Draper, DFC 7 Nov 55 - 6 Nov 58.
W/C F.J. Mills 7 Nov 58 - 13 Sep 61.
W/C E.M. Lane, CD 14 Sep 61 - 30 Nov 64.
W/C J.R.R. Neroutsos, CD 1 Dec 64 - 17 Mar 66.
W/C T.H. Ussher, CD 18 Mar 66 - 31 Jan 68.

Higher Formations and Squadron Location
Training Command,
Air Defence Command (1 Aug 51),
No. 2 Group (Auxiliary) (15 Jan 51 - 1 Mar 57),
Air Transport Command (1 Oct 58):
No. 14 Wing (Auxiliary) (1 Oct 50),
Downsview, Ont. 1 Oct 50 - 31 Jan 68.

Representative Aircraft (Unit Code 1950-51 AN, 1951-58 KH)
North American Harvard Mk.II (Feb 51 - Sep 58) 2903 2918 3784
de Havilland Vampire Mk.III (Jul 51 - Nov 55) 17017 17035 17045 17067
Canadair Silver Star Mk.3 (Oct 55 - Oct 58) 21463
Canadair Sabre Mk.5 (Oct 56 - Sep 58) 23104 23208 23306 23308
Beechcraft Expeditor Mk.3 (Sep 58 - Oct 66) 1442 1472 1559 1561
de Havilland Otter (Feb 62 - Jan 68) 9411 9413 9420 9422

---

[1] Formation date was to have been 1 February 1947, but only the Radar Convoy Section was formed on 18 April 1948.

329 North American Harvard Mk.II 2918 of No. 411 Squadron (Aux) visits Trenton, Ont. in September 1958.

330 De Havilland Vampire Mk.III 17042 VC-ANT of No. 411 Squadron (Aux).

331 Canadair Silver Star Mk.3 21463 of No. 411 Squadron (Aux) at Downsview, Ont.

332 Canadair Sabre Mk.5 23312 of No. 411 Squadron (Aux) tops the overcast above its base at Downsview, Ont.

335 Consolidated Liberator G.R.Mk.VIT 574 in the post-war markings of No. 412 (K) Squadron at Rockcliffe, Ont. June 1947.

333 Beechcraft Expeditor Mk.3N 1442 at Downsview, Ont. in the service of No. 411 Squadron (Aux).

336 Grumman Goose Mk.II 386 VC-AOQ of No. 412 (T) Squadron on 14 May 1953, on the shore of the Ottawa River.

334 De Havilland Otter 9420 of No. 411 Squadron (Aux) visiting Mt. Hope, Ont., on 13 May 1967.

337 Beechcraft Expeditor Mk.3TM 1534 VC-AON of No. 412 (T) Squadron.

No. 412 Squadron

Badge see Overseas Squadrons, No. 412 Squadron
Formed as No. 12 (Communications) Squadron at Rockcliffe (Ottawa), Ontario on 30 August 1940,[1] the unit was redesignated No. 412 (Composite)[2] Squadron on 1 April 1947, then No. 412 (Transport) Squadron on 1 April 1949. Throughout its history it was known as the "VIP" squadron because its chief responsibility was the transportation of "Very Important Persons", including British royalty and Canadian prime ministers. On 25 September 1964 the unit received its Squadron Standard for 25 years service as No. 12 and 412 Squadron. On 1 February 1968 the squadron was integrated into the Canadian Armed Forces.

Brief Chronology Formed as No. 12 (Comm) Sqn, Rockcliffe, Ont. 30 Aug 40. Redesignated No. 412 (K)[2] Sqn 1 Apr 47. Redesignated No. 412 (T) Sqn 1 Apr 49. Integrated into CAF 1 Feb 68.

Nickname "Falcon"

Ancestry No. 7 (General Purpose) Squadron

Commanders
W/C W.H. Swetman, DSO, DFC 1 Apr 47 - 12 Aug 49.
W/C B.H. Moffitt, DFC, AFC 13 Aug 49 - 26 Aug 50.
W/C R.I. Trickett, DFC 6 Sep 50 - 11 Feb 52.
W/C H.A. Morrison, DSO, DFC, AFC, CD 12 Feb 52 - 7 Jan 55.
W/C W.G.S. Miller, CD 8 Jan 55 - 2 Mar 58.
W/C W.K. Carr, DFC, CD 3 Mar 58 - 3 Aug 60.
W/C J.W. Borden, DFC, CD 4 Aug 60 - 22 Aug 63.
W/C M.G. Bryan, CD 23 Aug 63 - 14 Jun 66.
W/C D.C. MacKenzie, CD 15 Jan 66 - 31 Jan 68.

Higher Formation and Squadron Locations
Air Transport Command:
Rockcliffe, Ont. 1 Apr 47 - 31 Aug 55.
Uplands, Ont. 1 Sep 55 - 31 Jan 68.

Representative Aircraft (Unit Code AO)
Consolidated Liberator G.R.Mk.VIT (Apr - Nov 47)
570   to   574   3729   11101
Douglas Dakota Mk.III & IV (Apr 47 - Feb 68)
350   827   981   1000
Beechcraft Expeditor Mk.3 (Apr 47 - 1968) 1534 AO-N
Canadair North Star (Apr 49 - Apr 62) 17518   17520
Canadair C-5 (Aug 50 - Apr 66) 10000
Grumman Goose Mk.II (1953) 386
de Havilland Comet Mk.1A (May 53 - Jul 65) 5301   5302
North American Mitchell Mk.III (Sep 56 - Nov 60)
5220   5248
Canadair CC-109 Cosmopolitan (Jul 59 - Feb 68)
11151 to 11153   11157 to 11163
Canadair CC-106 Yukon (1962 - Feb 68) 15929   15932

Honours and Awards Squadron Standard (25 September 1964).

---

[1] From AFHQ Communications Flight, itself formed from the Communications Flight of No. 7 (GP) Squadron, which had been disbanded on 10 September 1939.
[2] Officially abbreviated "K", this term covered air transport and providing practice flying for staff personnel. The latter ceased to be a role of No. 412 Squadron in December 1947, when an AFHQ Practice Flight was formed.

338   *Douglas Dakota Mk.III KJ936 VC-AOH of No. 412 (T) Squadron at Rockcliffe, Ont.*

339   *North American Mitchell B-25J 5220, a VIP light transport conversion operated by No. 412 (T) Squadron. This photograph records a visit to St Hubert, Que. on 23 January 1957.*

340   *Canadair North Star Mk.M1 17518 with No. 412 (T) Squadron at Uplands, Ont.*

341 Canadair C-5 17524 operated by No. 412 (T) Squadron from its base at Uplands, Ont. This Pratt and Whitney powered aircraft was a one-of-a-kind purchase by the RCAF.

344 Canadair Cosmopolitan 11157 visits Downsview, Ont. in June 1960 in the service of No. 412 (T) Squadron.

342 De Havilland Comet Mk.1 5301, one of two Comets operated by No. 412 (T) Squadron. The RCAF holds claim to being the first air force to operate purely jet-powered transport aircraft.

343 Canadair Yukon 15929 of No. 412 (T) Squadron visits Downsview, Ont. in January 1968.

No. 413 Squadron

Badge see Overseas Squadrons, No. 413 Squadron
Unofficially formed as No. 13 (Photographic) Squadron at Rockcliffe (Ottawa), Ontario on 20 May 1944, and officially renumbered as No. 413 Squadron on 1 April 1947,[1] the unit flew Mitchell and Lancaster aircraft on aerial photography. On 1 April 1949 it was redesignated Survey Transport and flew Dakota, Canso and Norseman aircraft in logistical support and the transportation of survey parties in the far north. The squadron was disbanded on 1 November 1950.
Brief Chronology Unofficially formed as No. 13 (P) Sqn, Rockcliffe, Ont 20 May 44. Renumbered No. 413 (P) Sqn 1 Apr 47. Redesignated No. 413 (ST) Sqn 1 Apr 49. Disbanded 1 Nov 50.
Nickname "Tusker"
Commanders
S/L J.A. Wiseman, AFC 1 Apr 47 - 4 Jan 48.
S/L C.S. Olsen, CD 5 Jan 48 - 24 Aug 49.
W/C A.P. Blackburn, DFC 25 Apr 49 - 22 Jan 50.
W/C D.J.G. Jackson 3 Jan 50 - 1 Nov 50.
Higher Formations and Squadron Location
Air Transport Command:
No. 22 (Photographic) Wing,
Rockcliffe, Ont. 1 Apr 47 - 1 Nov 50.
Representative Aircraft (Unit Code AP)
Noorduyn Norseman Mk.VI (Apr 47 - Oct 50)
371 AP-N 787 AP-P 2495 AP-P 2496 AP-Q
Consolidated Canso A (Apr 47 - Oct 50
11003 AP-K 11018 AP-J 11047 AP-W
11075 AP-K 11079 AP-M
North American Mitchell Mk.II (Apr 47 - Oct 48)
KL145 891 892 894
Avro Lancaster Mk.X (Jan - Apr 48) FM207
FM212 AP-A FM216 AP-O FM218 AP-U
Douglas Dakota Mk.IV (Apr 49 - Oct 50)
KN427 AP-R 635
Honours and Awards 2 AFC's,[2] 1 King's Commendation.[3]

---

[1] See No. 13 Squadron for details
[2] Awarded to F/L J.F. Drake and F/O J.E. Goldsmith for their part in Operation "Polco," the locating of the North Magnetic Pole, in July 1950.
[3] To Sgt A.B. Hillman for his work in Operation "Polco."

Formed as a Fighter unit at Bagotville, Quebec on 1 August 1951 with Vampire and Sabre aircraft, the squadron joined No. 3 (Fighter) Wing at Zweibrucken, Germany in April 1953. In 1956, it was decided to replace one Sabre squadron in each of No. 1 Air Division Europe's four wings with an all-weather fighter unit. When No. 440 AW(F) Squadron arrived from Canada, No. 413 was deactivated on 7 April 1957 and reactivated as All-Weather (Fighter) at Bagotville on 1 May. The squadron flew CF-100 aircraft on North American air defence until disbanded on 30 December 1961.[1]
Brief Chronology Formed as No. 413 (F) Sqn, Bagotville, Que. 1 Aug 51.[2] Deactivated at Zweibrucken, Ger. 7 Apr 57. Reactivated as No. 413 AW(F) Sqn, Bagotville, Que. 1 May 57. Disbanded 30 Dec 61.
Nickname "Tusker"
Commanders
S/L J.D. Lindsay, DFC 1 Aug 51 - 6 Mar 53.
S/L W.I. Gordon 7 Mar 53 - 1 Sep 53.
S/L E.P. Wood, DFC 8 Jun 54 - 10 Jul 50.
W/C H.C. Stewart, AFC, CD 27 Aug 56 - 7 Apr 57.
Squadron inactive
S/L C. Allison, CD 1 Jul 57 - 14 Jul 59.
W/C J.R.D. Schultz, DFC and Bar, CD 24 Aug 59 - 20 Aug 60.
S/L E.G. Smith, DFC, CD 21 Aug 60 - 28 Jan 61.
W/C E.D. Kelly, CD 2 Feb 61 - 30 Dec 61.
Higher Formations and Squadron Locations
Air Defence Command:
Bagotville, Que. 1 Aug 51 - 6 Mar 53.
En route overseas (Operation "Leapfrog III") 7 Mar 53 - 6 Apr 53.
No. 1 Air Division Europe:
No. 3 (Fighter) Wing,
Zweibrucken, Ger. 7 Apr 53 - 7 Apr 57.
Squadron inactive
North American Air Defence Command,
RCAF Air Defence Command:
Bagotville, Que. 1 May 57 - 30 Dec 61.
Representative Aircraft (Unit Code AP)
de Havilland Vampire Mk.III (Aug - Dec 51)
Canadair Sabre Mk.2 (Nov 51 - Jun 54) 19213
19222 19236 19242
Canadair Sabre Mk.5 (Jun 54 - Sep 55) 23083
23145 23155 23180
Canadair Sabre Mk.6 (Sep 55 - Apr 57)
Avro CF-100 Canuck Mk.5 (May 57 - Dec 61) 18519
18555 18640 18668

---

[1] In its 18 years and 7 months of active service, the squadron had the distinction of being the first RCAF squadron to have operated on every continent except South America.
[2] Originally to have been formed as an all-weather fighter unit, but the order was amended on 8 November 1951 to day fighter.

345 *Avro Lancaster Mk.X FM218 VC-APU on photographic operations with No. 413 (P) Squadron in 1948.*

346   Consolidated Canso A 11003 VC-APK, bearing the No. 413 (P) Squadron crest under the cockpit, on the flight-line at Rockcliffe, Ont.

347   Noorduyn Norseman Mk.VI 366 VC-APV, operating on floats, in northern Ontario with No. 413 (P) Squadron.

348   North American Mitchell Mk.II 891, with modified nose panels to accommodate an oblique camera installation, serving with No. 413 (P) Squadron.

349   Canadair Sabre Mk.2 19224 of No. 413 (F) Squadron undergoing maintenance checks at Zweibrucken, Germany.

350   Canadair Sabre Mk.5 23083 bearing unit identifier AP and the yellow "X" wing and tail markings of Exercise "Carte Blanche." Photographed at Metz, France during an engine check in the service of No. 413 (F) Squadron.

351   Avro CF-100 Mk.5 18651 displaying the unit identifier AP and the elephant crest of No. 413 AW(F) Squadron, based at Bagotville, Que.

No. 414 Squadron

Badge none
Unofficially formed as No. 14 (Photographic) Squadron at Rockcliffe (Ottawa), Ontario on 12 June 1944,[1] and officially renumbered No. 414 on 1 April 1947, the squadron flew Dakota aircraft on vertical photographic duty until disbanded on 1 November 1950.
Brief Chronology Unofficially formed as No. 14 (P) Sqn, Rockcliffe, Ont. 12 Jun 44. Renumbered No. 414 (P) Sqn 1 Apr 47. Disbanded 1 Nov 50.
Commanders
S/L R.F. Milne, DFC 1 Apr 47 - 19 Dec 48.
W/C W.B.M. Millar 20 Dec 48 - 1 Nov 50.
Higher Formations and Squadron Location
Air Transport Command:
No. 22 (Photographic) Wing,
Rockcliffe, Ont. 1 Apr 47 - 1 Nov 50.
Representative Aircraft (Unit Code AQ)
Douglas Dakota Mk.IIIP & IVP (Apr 47 - Oct 50)
KG635   KN278   984   989

---

[1] See No. 14 (Photographic) Squadron for details.

Badge Over a cloud a knight on a charger
Motto Totis viribus (With all our might)
Authority Queen Elizabeth II, June 1960
The squadron became unofficially known as the Black Knight Squadron. The knight connotes fair but mortal combat. The cloud indicates that the fighting is in the air. Black is the customary garb of the night fighter, coupled with the white horse and red trimmings it gives the squadron tricolours.
Formed as a Fighter unit at Bagotville, Quebec on 1 November 1952 with Sabre aircraft, the squadron joined No. 4 (Fighter) Wing at Baden-Soellingen, Germany in August 1953. In 1956, it was decided to replace one Sabre squadron in each of No. 1 Air Division Europe's four wings with an all-weather fighter unit. When No. 419 AW(F) Squadron arrived from Canada, No. 414 was deactivated on 14 July 1957 and reactivated as All-Weather (Fighter) at North Bay, Ontario on 5 August. The squadron flew CF-100 and CF-101 aircraft on North American air defence until disbanded on 30 June 1964.
Brief Chronology Formed as No. 414 (F) Sqn, Bagotville, Que. 1 Nov 52. Deactivated, Baden-Soellingen, Ger. 14 Jul 57. Reactivated as No. 414 AW(F) Sqn, North Bay, Ont. 5 Aug 57. Disbanded 30 Jun 64.
Nickname "Black Knight"
Commanders
W/C J.F. Allan, CD 12 Nov 52 - 1 Feb 54.
S/L J.R. Ritch, CD 15 May 54 - 8 Apr 56.
S/L L.J. Liggett, CD 15 May 56 - 14 Jul 57.
Squadron inactive
W/C E. Wilson, CD 7 Apr 58 - 30 Jun 59.
W/C C.D.L. Hare, DFC, AFC, CD 1 Jul 59 - 14 Nov 60 KIFA.
W/C H.A. McKay, DFC, CD 1 Feb 61 - 31 Jul 61.
W/C E.G. Smith, DFC, CD 1 Aug 61 - 11 Oct 62.
W/C K.F. Lowe, CD 12 Oct 62 - 30 Jun 64.
Higher Formations and Squadron Locations
Air Defence Command:
Bagotville, Que. 1 Nov 52 - 23 Aug 53.
En route overseas (Operation "Leapfrog IV") 24 Aug 53 - 3 Sep 53.
No. 1 Air Division Europe:
No. 4 (Fighter) Wing,
Baden-Soellingen, Ger. 4 Sep 53 - 14 Jul 57.
Squadron inactive.
North American Air Defence Command,
RCAF Air Defence Command:
North Bay, Ont. 5 Aug 57 - 30 Jun 64.
Namao, Alta. (conversion to CF-101 under No. 425 AW(F) Sqn) 19 Mar - 19 Apr 62.
Uplands, Ont. (North Bay runways under repair for CF-101) 28 Apr - 4 Oct 62.
Representative Aircraft (Unit Code AQ)
Canadair Sabre Mk.4 (Nov 52 - Nov 53)[1]
19453   19585   19606   19675
Canadair Sabre Mk.5 (Nov 53 - Aug 55) 23008
23031   23249   23345
Canadair Sabre Mk.6 (Jul 55 - Jul 57) 23416
23441   23468   23498
Avro CF-100 Canuck MK. 5 (Aug 57 - Feb 62)
18515   18645   18665   18770
McDonnell CF-101B Voodoo (Feb 62 - Jun 64)
17393   17403   17407

---

[1] Turned over to the Royal Air Force under the Mutual Aid Plan.

Formed as the Electronic Warfare Unit at St Hubert (Montreal), Quebec on 1 April 1959 with CF-100 aircraft, and redesignated No. 414 (Electronic Warfare) Squadron on 15 September 1967, the unit provided Air Defence Command ground control radar personnel and airborne interceptor crews with training and experiences in combatting radar jamming. On 1 February 1968 the squadron was integrated into the Canadian Armed Forces.
Brief Chronology Formed as EWU at St Hubert, Que. 1 Apr 59. Redesignated No. 414 (EW) Sqn 15 Sep 67. Integrated into CAF 1 Feb 68.
Nickname "Black Knight"
Commanders
W/C P.B. St. Louis, MBE, CD 15 Sep 67 - 31 Jan 68.
Higher Formation and Squadron Location
Air Defence Command:
St Hubert, Que. 15 Sep 67 - 31 Jan 68.
Representative Aircraft
Avro CF-100 Canuck Mk.5 (Sep 67 - Jan 68)

352 *Douglas Dakotas of No. 414 (P) Squadron on 12 April 1949.*

355 *McDonnell Voodoo 17403 of No. 414 AW(F) Squadron.*

353 *Canadair Sabre Mk.6 23407 of No. 414 (F) Squadron at Rabat, Morocco on 15 July 1956.*

356 *Avro CF-100 Mk.5 18789 of No. 414 (EW) Squadron at Calgary, Alberta.*

354 *Avro CF-100 Mk.5 18636 with the knight's head crest of No. 414 AW(F) Squadron on the engine intake side panels.*

357 *Avro CF-100 Mk.5 18556 of the Electronic Warfare Unit at St Hubert, Que. on 1 May 1964.*

No. 415 Squadron

Badge *see Overseas Squadrons, No. 415 Squadron*
Formed as a Maritime Patrol unit at Summerside, Prince Edward Island on 1 May 1961 as the fourth and last squadron formed in Maritime Air Command, the unit flew Argus aircraft on the East Coast, and on 1 February 1968 was integrated into the Canadian Armed Forces.
Brief Chronology Formed at Summerside, P.E.I. 1 May 61. Integrated into CAF 1 Feb 68.
Nickname "Swordfish"
Commanders
W/C S.S. Mitchell, CD 1 May 61 - 4 Sep 64 *ret*.
W/C W.B. Asbury, CD 5 Sep 64 - 4 Oct 65.
W/C J.V. Pierpoint, CD 5 Oct 65 - 31 Jan 68.
Higher Formations and Squadron Location
Maritime Air Command,
Canadian Forces Maritime Command (17 Jan 66):
Summerside, P.E.I. 1 May 61 - 31 Jan 68.
Representative Aircraft
Canadair CP-107 Argus (May 61 - Jan 68) 20720 20733 20736 20737

358 Canadair Argus Mk.1 20720 displaying the red swordfish on its fin to denote that it serves with No. 415 (MP) Squadron.

No. 416 Squadron

Badge *see Overseas Squadrons, No. 416 Squadron*
Formed as a Fighter unit at Uplands (Ottawa), Ontario on 8 January 1951 with Mustang and, later, Sabre aircraft, the squadron joined No. 2 (Fighter) Wing at Grostenquin, France in September 1952. In 1956, it was decided to replace one Sabre squadron in each of No. 1 Air Division Europe's four wings with an all-weather fighter unit. When No. 423 AW(F) Squadron arrived from Canada, No. 416 was deactivated on 31 January 1957 and reactivated as All-Weather (Fighter) at St Hubert (Montreal), Quebec on 1 February, and flew CF-100 aircraft on North American air defence. Pending re-equipment with CF-101 aircraft, the unit was again deactivated on 1 September 1961. Reactivated at Bagotville, Quebec on 1 January 1962, it subsequently moved to Chatham, New Brunswick in November. On 1 February 1968 the squadron was integrated into the Canadian Armed Forces.
Brief Chronology Formed as No. 416 (F) Sqn, Uplands, Ont. 8 Jan 51. Deactivated at Grostenquin, Fr. 31 Jan 57. Reactivated as No. 416 AW (F) Sqn, St Hubert, Que. 1 Feb 57. Deactivated 1 Sep 61. Reactivated at Bagotville, Que. 1 Jan 62. Integrated into CAF at Chatham, N.B. 1 Feb 68.
Nickname "Black Lynx"
Commanders
S/L D.C. Laubman, DFC and Bar 8 Jan 51 - 17 Mar 52.
S/L J. MacKay, DFC and Bar 18 Mar 52 - 25 Aug 54.[1]
S/L K.C. Lett 22 Oct 54 - 25 Mar 56.
S/L J.L.A. Roussell, DFC 26 Mar 56 -31 Jan 57.
Squadron inactive.
W/C W.L. Drake, CD 7 Mar 57 - 19 Aug 59.
W/C R.B. West, DFC, AFC, CD 20 Aug 59 - 30 Nov 59.
W/C W.J. Buzza, CD 17 Feb 60 - 12 Dec 60.
S/L D.J. Gagnon 13 Dec 60 - 1 Sep 61.
Squadron inactive.
W/C E.D. Kelly, CD 1 Jan 62 - 9 Oct 64.
W/C J.C. Henry, CD 13 Oct 64 - 29 Dec 67.
W/C S.A. Miller, CD 30 Dec 67 - 31 Jan 68.
Higher Formations and Squadron Locations
Air Defence Command:
Uplands, Ont. 8 Jan 51 - 27 Sep 52.
En route overseas (Operation "Leapfrog II") 28 Sep 52 - 10 Oct 52.
No. 1 Air Division:
No. 2 (Fighter) Wing,
Grostenquin, Fr. 11 Oct 52 - 31 Jan 57.
North American Air Defence Command,
RCAF Air Defence Command:
St Hubert, Que. 1 Feb 57 - 1 Sep 61.
Squadron inactive
Bagotville, Que. 1 Jan. 62 - 4 Feb 62.
Namao, Alta. (conversion to CF-101 under No. 425 AW(F) Sqn) 2-31 Jan 62.
Uplands, Ont. 5 Feb 62 - 3 Jul 62.
Bagotville, Que. 4 Jul 62 - 15 Nov 62.
Chatham, N.B. 16 Nov 62 - 31 Jan 68.

Representative Aircraft (Unit Code AS)
North American Mustang Mk.IV (Jan 51 - Mar 52)
9240 to 9245 9247 to 9251
Canadair Sabre Mk.2 (Mar 52 - Mar 54)
19226 19317 19338 19382
Canadair Sabre Mk.5 (Mar 54 - May 55)
23077 23082 23099 23237
Canadair Sabre Mk.6 (Apr 55 - Jan 57)
Avro CF-100 Canuck Mk.5 (Feb 57 - Aug 61)
18567 18586 18606 18750
McDonnell CF-101B Voodoo (Feb 62 - Jan 68)
17470 17477

---

[1] Served in Korea on exchange duty with the USAF from March to September 1953.

359   North American Mustang Mk.IV 9246 leading a formation of No. 416 (F) Squadron aircraft including one coded aircraft, AS-N, in the number four position, on 24 January 1952.

360   Canadair Sabre Mk.2 19250 with No. 416 (F) Squadron, No. 1 Air Division, Grostenquin, France.

361   Canadair Sabre Mk.5 23168 serving No. 416 (F) Squadron.

362   Avro CF-100 Mk.3D 18145 at St Hubert, Que. in August 1957, while serving No. 416 AW(F) Squadron.

363   Avro CF-100 Mk.5 18616 in the service of No. 416 AW(F) Squadron on 29 May 1957, near St Hubert, Que.

364 McDonnell Voodoo 17445 of No. 416 AW(F) Squadron and bearing the rudder stripes of Uplands, Ont.

No. 417 Squadron

Badge *see Overseas Squadrons, No. 417 Squadron*
Formed as a Fighter Reconnaissance unit at Rivers, Manitoba on 1 June 1947, the squadron flew Mustang aircraft in close support training with army units until disbanded on 1 August 1948.
Brief Chronology Formed at Rivers, Man. 1 Jan 47. Disbanded 1 Aug 48.
Commanders
W/C J.D. Mitchner, DFC and Bar 11 Jun 47 - 1 Aug 48.
Higher Formation and Squadron Location
North West Air Command:
Rivers, Man. 1 Jun 47 - 1 Aug 48.
Representative Aircraft (Unit Code AT)
North American Harvard Mk.II (Jul 47 - Jul 48)
3063 AT-W.
North American Mustang Mk.IV (Jul 47 - Jul 48)
9567 AT-F   9573   9577   9580 AT-B

365 North American Harvard Mk.IIB 3063 VC-ATW in service with No. 417 (FR) Squadron.

366 North American Mustang Mk.IV 9580 VC-ATB of No. 417 (FR) Squadron while based at Rivers, Man.

No. 418 "City of Edmonton" Squadron (Auxiliary)

Badge An Eskimo on an ice-floe holding a harpoon
Motto Piyautailili (Defend even unto death)
Authority Queen Elizabeth II, June 1958
The Eskimo holding a harpoon to symbolize the function of the unit which stands on guard on Canada's northern frontiers. The ice with its reflections is to symbolize the northland.
Formed at Edmonton, Alberta on 15 April 1946, the squadron flew Mitchell aircraft in a light bomber role until March 1958 when it was reassigned to a light transport and emergency rescue role and re-equipped with Expeditor and Otter aircraft. On 27 May 1967 it received a Squadron Standard for 25 years service. On 1 February 1968 the squadron was integrated into the Canadian Armed Forces as No. 418 "City of Edmonton" Air Reserve Squadron.
Brief Chronology Formed as No. 418 (FB) Sqn (Aux), Edmonton, Alta. 15 Apr 46. Redesignated No. 418 (LB) Sqn (Aux) 1 Jan 47. Redesignated No. 418 (TacB) Sqn (Aux) 1 Apr 47. Redesignated No. 418 (LB) Sqn (Aux) 1 Apr 49. Titled No. 418 "City of Edmonton" (LB) Sqn (Aux) 3 Sep 52. Redesignated No. 418 "City of Edmonton" Sqn (Aux) 31 Mar 58. Integrated into CAF as No. 418 "City of Edmonton" Air Res Sqn 1 Feb 68.
Title "City of Edmonton"
Commanders
W/C D.R. Jacox, AFC 1 May 46 - 13 Mar 50.
W/C A.W. Speed 14 Mar 50 - 17 Jun 52 *ret*.
W/C A.D.R. Lowe, DFC, AFC 18 Jun 52 - 16 Sep 54.
W/C J.M. Flint, CD 17 Sep 54 - 18 Sep 55.
W/C J.K. Campbell, CD 19 Sep 55 - 30 Jan 59.
W/C O.W. Cornish, CD 31 Jan 59 - 14 Oct 61 *ret*.
W/C F.T. Guest, CD 15 Oct 61 - 31 Dec 62.
W/C R. Klesko, CD 1 Jan 63 - 1 Oct 66.
W/C S.E.W.J. Wood, CD 2 Oct 66 - 31 Jan 68.
Higher Formations and Squadron Locations
Tactical Air Command,
Training Command (1 Jan 59),
Air Transport Command (1 Apr 61):
No. 18 Wing (Auxiliary) (1 Oct 50),
Edmonton, Alta. 15 Apr 46 - 30 Sep 55.
Namao, Alta. 1 Oct 55 - 31 Jan 68.
Representative Aircraft (Unit Code 1947-51 AU, 1951-55 HO)
North American Harvard Mk.II (Oct 46 - Jan 53)
2610 3060 3100 3287
North American Mitchell Mk.II & III (Jan 47 - Mar 58)[1]
HD313 FW278 KL143 5233
Beechcraft Expeditor Mk.3 (Apr 48 - Nov 66)
1498 1524 1547 2377
Canadair Silver Star Mk.3 (Aug 56 - May 58) 21907
de Havilland Otter (Oct 60 - Jan 68)
9404 9407 9408 9418
Honours and Awards Squadron Standard (27 May 1967).

367 *North American Mitchell Mk.II KL149 with engine covers removed, and displaying on the nose panel an Al Capp character approved for use by No. 418 Squadron (Aux).*

368 *North American Harvard Mk.II 3030 with the forward panels removed while at Trenton, Ont. in service with No. 418 Squadron (Aux).*

---

[1] During the Second World War, the squadron had used characters from the comic strip "Li'l Abner" to name and decorate its aircraft. In August 1950 permission was obtained from the creator, Al Capp, for the post-war squadron to do likewise.

369 Beechcraft Expeditor Mk.3N 1488 serving No. 418 Squadron (Aux) at Trenton, Ont. on 27 July 1962.

370 De Havilland Otter 9417 with fluorescent rescue markings while serving No. 418 Squadron (Aux).

No. 419 Squadron

Badge see *Overseas Squadrons, No. 419 Squadron*
Formed as an All-Weather (Fighter) unit at North Bay, Ontario on 15 March 1954, the squadron flew CF-100 aircraft on North American air defence until August 1957 when it then joined No. 1 Air Division Europe to replace No. 414 (Fighter) Squadron in No. 4 (Fighter) Wing at Baden-Soellingen, Germany. On the withdrawal of CF-100 aircraft from operational service, the squadron was disbanded on 31 December 1962.
Brief Chronology Formed at North Bay, Ont. 15 Mar 54. Disbanded at Baden-Soellingen, Ger. 31 Dec 62.
Nickname "Moose"
Commanders
W/C E.G. Ireland, DFC 8 Jan 55 - 1 Aug 56.
W/C R.E. MacBride, DFC, CD 20 Sep 56 - 2 Dec 58.
W/C O.C. Brown, CD 15 Jun 59 - 20 Jun 60.
W/C P.E. Etienne, DFC, CD 27 Jun 60 - 31 Dec 62.
Higher Formations and Squadron Locations
Air Defence Command:
North Bay, Ont. 15 Mar 54 - 4 Aug 57.
En route overseas (Operation "Nimble Bat IV") 5 Aug 57.
No. 1 Air Division Europe:
No. 4 (Fighter) Wing,
Baden-Soellingen, Ger. 6 Aug 57 - 31 Dec 62.
No. 3 (Fighter) Wing,
Zweibrucken, Ger. 15 Jun - 9 Oct 59.
Representative Aircraft (Unit Code UD)
Avro CF-100 Canuck Mk.4A (Mar 54 - May 55)
18200   18206   18213   18220
Avro CF-100 Canuck Mk.4B (May 55 - Dec 62)
18339   18378   18400   18419

371 Avro CF-100 Mk.4B 18351, bearing the moose crest of No. 419 AW(F) Squadron, at Baden-Soellingen, Germany.

372 Avro CF-100 Mk.4A 18204, coded UD, of No. 419 AW(F) Squadron with its wing tip stores removed.

No. 420 "City of London" Squadron (Auxiliary)

Badge *see Overseas Squadrons, No. 420 Squadron*
Formed at London, Ontario on 15 September 1948, the squadron flew Mustang aircraft in a fighter role until disbanded on 1 September 1956.
Brief Chronology Formed as No. 420 (F) Sqn (Aux), London, Ont. 15 Sep 48.[1] Titled No. 420 "City of London" (F) Sqn (Aux) 4 Sep 52. Disbanded 1 Sep 56.
Title "City of London"
Nickname "Snowy Owl"
Commanders
W/C A.D. Haylett, AFC 15 Sep 48 - 14 Dec 53.
W/C G.M. Burns, DFC 15 Dec 53 - 1 Sep 56.
Higher Formations and Squadron Location
Training Command,
Air Defence Command (1 Aug 51):
No. 22 Wing (Auxiliary) (15 Dec 53),
Crumlin airport (London), Ont. 15 Sep 48 - 1 Sep 56.
Representative Aircraft (Unit Code AW)
North American Harvard Mk.II (Aug 49 - Aug 56)
2942 AW-F   3176 AW-G   3337 AW-B   3785
North American Mustang Mk.IV (Dec 50 - Aug 56)
9223 AW-J   9224 AW-N   9228 AW-K   9229 AW-L
9256 AW-X
Canadair Silver Star Mk.3 (Oct 54 - Aug 56) 21447

374 North American Mustang Mk.IV 9224 VC-AWN takes part in an exercise with No. 420 Squadron (Aux).

[1] Formation had been authorized on 1 May 1947, but on 1 October was postponed to September 1948.

373 North American Harvard Mk.IIB 3337 VC-AWB of No. 420 Squadron (Aux) based at London, Ont.

No. 421 Squadron

Badge *see Overseas Squadrons, No. 421 Squadron*
Formed as a Fighter unit at Chatham, New Brunswick on 15 September 1949, the squadron flew Vampire aircraft and, during 1951, was stationed in the United Kingdom for operational training with the Royal Air Force. In December 1951 it was re-equipped with Sabre aircraft and in October 1952 joined No. 2 (Fighter) Wing at Grostenquin, France.[1] Selected as one of eight Sabre squadrons in No. 1 Air Division Europe to be re-equipped with CF-104 Starfighter aircraft for a nuclear strike role, the squadron was deactivated on 1 August 1963 and was reactivated as Strike Attack on 21 December. When No. 2 Wing was disbanded in February 1964, the squadron joined No. 4 Wing at Baden-Soellingen, Germany. On 1 February 1968 the squadron was integrated into the Canadian Armed Forces.

Brief Chronology Formed as No. 421 (F) Sqn, Chatham, N.B. 15 Sep 49.[2] Deactivated at Grostenquin, Fr. 1 Aug 63. Reactivated as No. 421 (ST/A) Sqn 2 Dec 63. Integrated into CAF at Baden-Soellingen, Ger. 1 Feb 68.

Nickname "Red Indian"

Commanders
S/L R.T.P. Davidson, DFC, CD 15 Sep 49 - 25 Nov 52.
S/L R.G. Middlemiss, DFC, CD 29 Nov 51 - 25 Sep 53.
S/L E.P. Wood, DFC, CD 23 Oct 53 - 17 May 54.
S/L J.R.F. Johnson, DFC, AFC, CD 18 May 54 - 26 Apr 56.
S/L C.D.A. Bourque 27 May 56 - 6 Jul 58.
S/L L.D. Allatt, CD 7 Jul 58 - 29 Jun 59.
W/C R. Van Adel, CD 30 Jun 59 - 10 Jul 62.
W/C A.J. Bauer, CD 11 Jul 62 - 1 Aug 63.
Squadron inactive
W/C J.B. Lawrence, CD 29 Feb 64 - 5 May 67.
W/C R.H. Annis, CD 28 Apr 67 - 31 Jan 68.

Higher Formations and Squadron Locations
Air Defence Command:
Chatham, N.B. 15 Sep 49 - 15 Jan 51.
En route overseas (minus aircraft) 16 Jan 51 - 18 Jan 51.
RAF Fighter Command:
No. 11 Group,
Odiham, Hants., Eng. 19 Jan 51 - 12 Nov 51.
En route to Canada (minus aircraft) 13 Nov 51 - 14 Dec 51.
Air Defence Command:
St Hubert, Que. 15 Dec 51 - 27 Sep 52.
En route overseas (Operation "Leapfrog II") 28 Sep 52 - 10 Oct 52.
No. 1 Air Division Europe:
No. 2 (Fighter) Wing,
Grostenquin, Fr. 11 Oct 53 - 1 Aug 63.
Squadron inactive
No. 2 Wing, Grostenquin, Fr. 2 Dec 63 - 13 Feb 64.
No. 4 Wing, Baden-Soellingen, Ger. 14 Feb 64 - 31 Jan 68.

Representative Aircraft (Unit Code AX)
de Havilland Vampire Mk.III (Sept 49 - Dec 50)
17001 17045 17060 17065
de Havilland Vampire Mk.5 (England, Jan - Oct 50, no code) VZ261 VZ264 VZ339 VZ343
Gloster Meteor T.Mk.7 (Jan - Oct 50) WA740 WA742
Canadair Sabre Mk. 2 (Dec 51 - Mar 54)
19287 19350 19365 19389
Canadair Sabre Mk.5 (Mar 54 - Jun 56) 23153 23166
Canadair Sabre Mk.6 (Jun 56 - Jul 63) 23595 23742 23654 23684
Canadair CF-104 Starfighter (Dec 63 - Feb 68)
12856 12858 12867 12894

---

[1] It thus became the only RCAF squadron to log three tours of overseas duty.
[2] The squadron was to have formed on 15 February 1949, but the order was amended to 1 June 1949 with Greenwood, Nova Scotia as the location, and further amended 15 September 1949 to Chatham.

375 *De Havilland Vampire Mk.III 17030 VC-AXC of No. 421 (F) Squadron at Chatham, N.B. on 27 October 1950.*

376 *De Havilland Vampire Mk.5 VZ339 of the Royal Air Force on assignment to No. 421 (F) Squadron, based at Odiham, Hants. in 1951.*

377 Gloster Meteor T.Mk.7 WA742 and WA740 with a line-up of Vampires, all used by No. 421 (F) Squadron while based at Odiham, Hants. under RAF Fighter Command in the spring of 1951.

379 Canadair Sabre Mk.5 23062 of No. 421 (F) Squadron displays the markings of 2 (F) Wing's aerobatic team in November 1955.

378 Canadair Sabre Mk.2 line-up at Grostenquin, France includes the aircraft of No. 421 (F) Squadron on the right. Typical of its kind is 19239, bearing the unit identifier AX with red and white nose and tail bands.

380 Canadair Sabre Mk.6 23684 is rolled out on the flight line for use by No. 421 (F) Squadron.

381 Canadair CF-104 12807, in the service of No. 421 (ST/A) Squadron, landing with drogue 'chute deployed while participating in the Fifth Tactical Weapons Meet on 20 June 1966.

No. 422 Squadron

Badge *see Overseas Squadrons, No. 422 Squadron*
Formed as a Fighter unit at Uplands (Ottawa), Ontario on 1 January 1953 with Sabre aircraft, the squadron joined No. 4 (Fighter) Wing at Baden-Soellingen, Germany in August. Selected as one of eight Sabre units in No. 1 Air Division Europe to be re-equipped with CF-104 Starfighter aircraft for a nuclear strike role, it was deactivated on 15 April 1963 and reactivated as Strike Attack on 15 July. On 1 February 1968 the squadron was integrated into the Canadian Armed Forces.

Brief Chronology Formed as No. 422 (F) Sqn, Uplands, Ont. 1 Jan 53. Deactivated at Baden-Soellingen, Ger. 15 Apr 63. Reactivated as No. 422 (ST/A) Sqn 15 Jul 63. Integrated into CAF 1 Feb 68.

Nickname "Tomahawk"

Commanders
S/L W.J. Buzza, CD 1 Jan 53 - 7 Jun 55.
W/C C.C. Lee, CD 14 Jul 55 - 5 Jan 56.
S/L C.C. Magee, DFC, CD 6 Jan 56 - 9 Dec 56.
W/C G.G. Wright, AFC, CD 10 Dec 56 - 15 Jul 59.
S/L R.G. Murray, AFC, CD 16 Jul 59 - 31 Aug 60.
W/C F.J. Kaufman, CD 1 Sep 60 - 28 Feb 63.
W/C P.J.S. Higgs, CD 1 Mar 63 - 15 Apr 63.
Squadron inactive
W/C W.H.F. Bliss, CD 15 Jul 63 - 31 Jul 66.
W/C W.J. Stacey, CD 1 Aug 66 - 31 Jul 67.
W/C R.K. Scott, CD 1 Aug 67 - 31 Jan 68.

Higher Formations and Squadron Locations
Air Defence Command:
Uplands, Ont. 1 Jan 53 - 26 Aug 53.
En route overseas (Operation "Leapfrog IV") 27 Aug 53 - 3 Sep 53.
No. 1 Air Division Europe:
No. 4 (Fighter) Wing,
Baden-Soellingen, Ger. 4 Sep 53 - 15 Apr 63.
Squadron inactive
No. 4 Wing,
Baden-Soellingen, Ger. 15 Jul 63 - 31 Jan 68.

Representative Aircraft (Unit Code TF)
Canadair Sabre Mk.2 (Jan - May 53) 19210
Canadair Sabre Mk.4 (Apr 53 - Dec 54)[1]
19601   19642   19664   19686
Canadair Sabre Mk.5 (Jan 55 - Sep 55)
23043   23179   23233   23300
Canadair Sabre Mk.6 (Sep 55 - Apr 63)
23383   23431   23493   23500
Canadair CF-104 Starfighter (Jul 63 - Jan 68)
12844   12879   12884   12885

---

[1]Turned over to the Royal Air Force under the Mutual Aid Plan.

382 Canadair Sabre Mk.4 19801, bearing the national markings of the Royal Air Force with the colourful unit markings of No. 422 (F) Squadron, serving the Air Division in Europe. The Royal Air Force serial XB727 appears on the aft fuselage.

383 Canadair Sabre Mk.5 23068, in the markings of No. 422 (F) Squadron, at Baden-Soellingen, Germany in the spring of 1955.

384 Canadair Sabre Mk.6 23742 of No. 422 (F) Squadron dispersed at Baden-Soellingen, Germany in 1962.

No. 423 Squadron

Badge see *Overseas Squadrons, No. 423 Squadron*
Formed as an All-Weather (Fighter) unit at St Hubert (Montreal), Quebec on 1 June 1953, the squadron flew CF-100 aircraft on North American air defence until February 1957 when it then joined No. 1 Air Division Europe to replace No. 416 (Fighter) Squadron in No. 2 (Fighter) Wing at Grostenquin, France. On the withdrawal of CF-100 aircraft from operational service, the squadron was disbanded on 31 December 1962.

Brief Chronology Formed at St Hubert, Que. 1 Jul 53. Disbanded at Grostenquin, Fr. 31 Dec 62.

Nickname "Eagle"[1]

Commanders
W/C R.J. Lawlor, DFC, CD 1 Jul 53 - 31 Oct 54.
S/L L.P.S. Bing, DFC, CD 1 Nov 54 - 14 Aug 55.
W/C J.H.L. Lecomte, DFC, CD 3 Sep 55 - 8 Nov 56.
W/C K.B. Handley, CD 9 Nov 56 - 10 Mar 58.
W/C R.B. Murray, CD 13 Apr 58 - 4 Oct 59.
W/C J.D.W. Campbell, DFC, CD 5 Oct 59 - 16 Jul 61.
W/C W.J. Buzza, CD 2 Aug 61 - 30 May 62.
W/C R.D. Sloat, CD 1 Jun 62 - 31 Dec 62.

Higher Formations and Squadron Locations
Air Defence Command:
St Hubert, Que. 1 Jul 53 - 11 Feb 57.
En route overseas (Operation "Nimble Bat II") 12 Feb 57 - 15 Feb 57.
No. 1 Air Division Europe:
No. 2 (Fighter) Wing,
Grostenquin, Fr. 16 Feb 57 - 31 Dec 62.

Representative Aircraft (Unit Code NQ)
Avro CF-100 Canuck Mk.3B (Jul 53 - May 55)
18149   18160   18170   18177
Avro CF-100 Canuck Mk.4B (Feb 55 - Dec 62)
18321   18369   18383   18431
From April to December 1956 the squadron flew Mk.5's while their Mk.4B's were sent to Malton for overhaul and camouflaging by Avro prior to departure for Europe.

---

[1] Also adopted by No. 405 (MP) Squadron.

385   Avro CF-100 Mk.3B 18161 being prepared to start up with No. 423 AW(F) Squadron at St Hubert, Que. in March 1954.

386   Avro CF-100 Mk.4B 18389, in the markings of No. 423 AW(F) Squadron, displays its belly armament pack and wing tip tanks.

No. 424 "City of Hamilton" Squadron (Auxiliary)

Badge see Overseas Squadrons, No. 424 Squadron
Formed at Hamilton, Ontario on 15 April 1946, the squadron flew Mustang aircraft in a fighter role until 1 September 1957 when it was then reassigned to a light transport and emergency rescue role and re-equipped with Expeditor and Otter aircraft. On 21 October 1961 the unit received its Squadron Standard for 25 years service as No. 119 and 424 Squadron. A reduction of the Auxiliary Force resulted in the squadron being disbanded on 1 April 1964.
Brief Chronology Formed as No. 424 (LB) Sqn (Aux), Hamilton, Ont. 15 Apr 46.[1] Redesignated No. 424 (F) Sqn (Aux) 1 Apr 47. Titled No. 424 "City of Hamilton" (F) Sqn (Aux) 19 Sep 52. Redesignated No. 424 "City of Hamilton" (T) Sqn (Aux) 1 Sep 57. Redesignated No. 424 "City of Hamilton" Sqn (Aux) 1 Apr 58. Disbanded 1 Apr 64.
Title "City of Hamilton"
Nickname "Hamilton Tigers"
Commanders
W/C D.H. Wigle 1 May 46 - 31 Aug 48 *ret*.
W/C D.B. Annan, DFC, AFC, CD 1 Sep 48 - 30 Sep 50.
W/C G.C. Frostad, CD 1 Oct 50 - 30 Nov 53.
W/C M.G. Marshall, DFC, CD 1 Dec 53 - 14 Nov 57.
W/C R.C. Small 15 Nov 57 - 28 Feb 59 *ret*.
W/C C.H. Forsyth, CD 1 Mar 59 - 28 Feb 62.
W/C G.W. Johnson, DFC, CD 1 Mar 62 - 1 Apr 64.
Higher Formations and Squadron Location
Training Command,
Air Defence Command (1 Aug 51),
Training Command (1 Sep 57),
Air Transport Command (1 Apr 60):
No. 16 Wing (Auxiliary) (1 Oct 50),
Mount Hope airport (Hamilton), Ont. 15 Apr 46 - 1 Apr 64.
Representative Aircraft (Unit Code BA)
North American Harvard Mk.II (Nov 47 - Dec 58)
3047   3140   3278   3766 BA-P
North American Mustang Mk.IV (Nov 50 - Sep 56)
9255 BA-U   9275   9577   9590
Canadair Silver Star Mk.3 (Oct 54 - Oct 56) 21469
Beechcraft Expeditor Mk. 3 (May 56 - Mar 64)
FR944   1444   2290   2292
de Havilland Otter (Oct 61 - Mar 64) 9422
Honours and Awards Squadron Standard (21 October 1961).

---

[1]Was to have been formed as a bomber unit on 1 May 1946, but the order was amended.

387   North American Harvard Mk.II 3766 VC-BAP of No. 424 Squadron (Aux) at Brantford, Ont. in June 1950.

388   North American Mustang Mk.IV 9253 VC-BAS of No. 424 Squadron (Aux), based at Mt. Hope, Ont.

389   Canadair Silver Star Mk.3 21317 of No. 400 Squadron (Aux) in formation with 21469 of No. 424 Squadron (Aux) over Hamilton, Ont. in 1955.

390 Beechcraft Expeditor Mk.3TM 1570 at Mt. Hope, Ont. on 21 January 1964, with No. 424 Squadron (Aux).

391 De Havilland Otter 9422, on wheel-skis, at Mt. Hope, Ont. on 3 February 1963, with No. 424 Squadron (Aux).

No. 425 Squadron

Badge *see Overseas Squadrons, No. 425 Squadron*
Formed as an All-Weather (Fighter) unit at St Hubert (Montreal), Quebec on 1 October 1954, the squadron flew CF-100 aircraft on North American air defence. Selected as one of five units to be re-equipped with CF-101 aircraft, it was deactivated on 1 May 1961 pending delivery of the aircraft. Reactivated at Namao (Edmonton), Alberta on 15 October 1961, the squadron initially received the trainer version of the CF-101 and served as a training unit to convert the remaining four squadrons to this aircraft. It afterwards moved to Bagotville, Quebec, in July 1962, and was declared operational on 1 October when No. 3 All-Weather (Fighter) Operational Training Unit assumed responsibility for all future CF-101 training. On 1 February 1968 the squadron was integrated into the Canadian Armed Forces.

Brief Chronology Formed at St Hubert, Que. 1 Oct 54. Deactivated 1 May 61. Reactivated at Namao, Alta. 15 Oct 61. Integrated into CAF at Bagotville, Que. 1 Feb 68.

Nickname "Alouette"

Commanders
W/C D.L.S. MacWilliam, AFC, CD 1 Nov 54 - 12 Apr 55.
W/C F.W. Hillock 13 Apr 55 - 5 Aug 57.
W/C W.M. Middleton, DFC, CD 7 Aug 57 - 15 Jul 69.
W/C J.E.Goldsmith, DFC, AFC, CD 16 Jul 60 - 1 May 61.
Squadron inactive
W/C J.R.D. Schultz, DFC and Bar, CD 20 Oct 61 - 21 Jul 62.
W/C G. H. Nichols, CD 22 Jul 62 - 28 Sep 64.
W/C M.J. Dooher, CD 29 Sep 64 - 15 Jul 66.
W/C W.J. Marsh, CD 15 Aug 66 - 4 Jul 67.
W/C H.I. Pike, CD 16 Aug 67 - 31 Jan 68.

Higher Formations and Squadron Locations
North American Air Defence Command,
RCAF Air Defence Command:
St Hubert, Que. 1 Nov 54 - 1 May 61.
Squadron inactive
Namao, Alta. (non-operational) 15 Oct 61 - 20 Jul 62.
Bagotville, Que. 21 Jul 62 - 31 Jan 68.
North Bay, Ont. (Bagotville runways being resurfaced) 15 May - 15 Jul 66.

Representative Aircraft (Unit Code BB)
Avro CF-100 Canuck Mk.4A & 4B (Oct 54 - Apr 58)
18270   18288   18304   18502
Avro CF-100 Canuck Mk.5 (Mar 56 - Apr 61)
18520   18555   18617   18758
McDonnell CF-101F Voodoo (Oct 61 - Oct 62)[1]
17393   17337   17349   17372
McDonnell CF-101B Voodoo (Oct 61 - Jan 68)
17391   17436   17453   17482

---

[1] These were the trainer version and were transferred to No. 3 AW(F) OTU at Bagotville.

392 Avro CF-100 Mk.5 18532 in the markings of No. 425 AW(F) Squadron. The lark on the cowling represents the squadron's "Alouette" badge.

393 McDonnell Voodoo 17391 on patrol with No. 425 AW(F) Squadron.

394 Canadair North Star Mk.1 17511 in the midst of typical ground handling equipment while serving No. 426 (T) Squadron during the Korean war.

No. 426 Squadron

Badge *see Overseas Squadrons, No. 426 Squadron*
Formed as a Transport unit at Dartmouth, Nova Scotia on 1 August 1946 from the Dartmouth portion of No. 164 (Transport) Squadron, the squadron moved to Dorval (Montreal), Quebec in March 1947 and was re-equipped from Dakota to four-engine North Star aircraft for long-range transport duty. From July 1950 to June 1954 it was employed on the Korean airlift (Operation "Hawk") and made 600 round trips across the North Pacific between Vancouver and Tokyo, logging 34,000 flying hours and carrying 13,000 personnel and 7,000,000 pounds of freight and mail without mishap. In 1956 it airlifted United Nations Emergency Force personnel and equipment to the Middle East and, in 1960-62, to the Belgian Congo. The unit moved to Trenton, Ontario in September 1959, and in January 1962 to St Hubert (Montreal) Quebec. The squadron was disbanded on 1 September 1962.[1].

Brief Chronology Formed at Dartmouth, N.S. 1 Aug 46. Disbanded at St Hubert, Que. 1 Sep 62.
Nickname "Thunderbird"
Ancestry No. 164 (Transport) Squadron
Commanders
W/C C.A. Willis, DFC 1 Aug 46 - 28 Feb 47.
W/C C.G.W. Chapman, DSO 1 Mar 47 - 17 Feb 49.
W/C C.H. Mussells, OBE, DSO, DFC 18 Feb 49 - 31 Mar 51.
W/C J.K.F. MacDonald, DFC, CD 1 Jun 51 - 28 Aug 52.
W/C H.W. Lupton, AFC, CD 8 Sep 52 - 20 Jul 55.
W/C A.J. Mackie, DFC, CD 21 Jul 55 - 3 Nov 59.
W/C J.O. Maitland, CD 4 Nov 59 - 18 Oct 61.
W/C A.J. Mackie, DFC, CD 20 Oct 61 - 1 Sep 62.
Higher Formation and Squadron Locations
Air Transport Command:
Dartmouth, N.S. 1 Aug 46 - 9 Mar 47.
Dorval, Que. 10 Mar 47 - 31 Aug 59.
Korean Airlift (Operation "Hawk") McChord AFB, Tacoma, Wash., USA 25 Jul 50 - 15 Jun 51.
Trenton, Ont. 1 Sep 59 - 31 Dec 61.
St Hubert, Que. 1 Jan 62 - 1 Sep 62.
Representative Aircraft
Douglas Dakota Mk.IV (Aug 46 - Jun 48)
Canadair North Star (Sep 47 - Aug 62)
17502 to 17505   17509 to 17512
Honours and Awards 1 OBE (W/C Mussells), 1 MBE, 4 AFC's, 2 AFM's, 13 Queen's Commendations.

---

[1] The squadron was replaced in the long-range transport role by No. 437 (T) Squadron, equipped with Yukon aircraft.

No. 427 Squadron

Badge *see Overseas Squadrons, No. 427 Squadron*
Formed as a Fighter unit at St Hubert (Montreal), Quebec on 1 August 1952 with Sabre aircraft, the squadron joined No. 3 (Fighter) Wing at Zweibrucken, Germany in March 1953. Selected as the first of eight squadrons in No. 1 Air Division Europe to be re-equipped with CF-104 Starfighter aircraft for a nuclear strike role, it was deactivated on 15 December 1962 and reactivated as Strike Attack on the 17th. On 1 February 1968 the squadron was integrated into the Canadian Armed Forces.

Brief Chronology Formed as No. 427 (F) Sqn. St Hubert, Que. 1 Aug 52. Deactivated at Grostenquin, Fr. 15 Dec 62. Reactivated as No. 427 (ST/A) Sqn, Zweibrucken, Ger. 17 Dec 62. Integrated into CAF 1 Feb 68.

Nickname "Lion"

Commander
S/L C.L.V. Gervais 1 Aug 52 - 30 Aug 53.
S/L D.K. Burke, CD 2 Jun 54 - 17 Sep 55.
W/C D. Laidler, CD 18 Sep 55 - 13 Jul 56.
W/C W.R. Tew, DFC, CD 1 Oct 56 - 31 Mar 58.
W/C H.R. Knight, CD 1 May 58 - 30 May 60.
W/C P.B. St. Louis, MBE, CD 15 Dec 60 - 15 Dec 62.
Squadron inactive
W/C R.G. Middlemiss, DFC 17 Dec 62 - 10 Feb 64.
W/C W.R. Knight, CD 11 Feb 64 - 6 Jul 65.
W/C J.F. Dunlop, DFC, CD 7 Jul 65 - 15 Sep 65.
W/C P.J.S. Higgs, CD 17 May 66 - 1 Sep 67.
W/C R.E. Carruthers, CD 2 Sep 67 - 31 Jan 68.

Higher Formations and Squadron Locations
Air Defence Command:
St. Hubert, Que. 1 Aug 52 - 6 Mar 53.
En route overseas (Operation "Leapfrog III") 7 Mar 53 - 6 Apr 53.
No. 1 Air Division:
No. 3 (Fighter) Wing,
Zweibrucken, Ger. 7 Apr 53 - 15 Jun 62.
No. 1 (Fighter) Wing,
Grostenquin, Fr. (withdrawal of Sabre aircraft) 16 Jun 62 - 15 Dec 62.
Squadron inactive
No. 3 Wing,
Zweibrucken, Ger. 16 Dec 62 - 31 Jan 68.
No. 4 Wing,
Baden-Soellingen, Ger. (Zweibrucken runways
being resurfaced) 1 Jul - 15 Sep 66.

Representative Aircraft (Unit Code 1947-51 BD, 1951-58 BB)
Canadair Sabre Mk. 2 (Sep 52 - Jun 53)
19409   19427   19433   19444.
Canadair Sabre Mk.5 (May 53 - Sep 55)
Canadair Sabre Mk.6 (Sep 55 - Dec 62) 23393   23522
Canadair CF-104 Starfighter (Dec 62 - Jan 68)
12627   12794   12802   12823

395   *Canadair Sabre Mk.2 19451 and 19415 of No. 427 (F) Squadron fly the wingmen positions to the No. 434 (F) Squadron leader, while a No. 413 (F) Squadron Sabre flies the slot position.*

396   *Canadair Sabre Mk.6 23458 of No. 427 (F) Squadron, displaying both the unit identifier BB and the lion fin marking.*

No. 428 Squadron

Badge *see Overseas Squadrons, No. 428 Squadron*
Formed as an All-Weather (Fighter) unit at Uplands (Ottawa), Ontario on 21 June 1954, the squadron flew CF-100 aircraft on North American air defence until disbanded on 1 June 1961.
Brief Chronology Formed at Uplands, Ont., 21 Jun 54. Disbanded, 1 Jun 61.
Nickname "Ghost"
Commanders
W/C E.W. Smith, DSO, CD 4 Jan 55 - 26 Apr 57.
S/L P.F. Greenway, CD 27 Apr 57 - 23 Apr 59.
W/C M.F. Doyle, CD 24 Apr 59 - 1 Jun 61.
Higher Formations and Squadron Location
North American Air Defence Command,
RCAF Air Defence Command:
Uplands, Ont. 21 Jun 54 - 1 Jun 61
Representative Aircraft (Unit Code HG),
Avro CF-100 Canuck Mk.4A (Jun 54 - Apr 55)
18211   18223   18224   18225
Avro CF-100 Canuck Mk.4B (Feb 55 - Aug 56)
18331   18356   18359   18380
Avro CF-100 Canuck Mk.5 (Mar 56 - May 61)
18464   18542   18614   18768

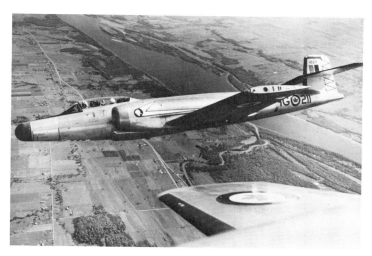

397   *Avro CF-100 Mk.4A 18211, with wing tip tanks removed, holds formation with another No. 428 AW(F) Squadron camera aircraft on 25 November 1954 near Uplands, Ont.*

398   *Avro CF-100 Mk.4B 18362 of No. 428 AW(F) Squadron.*

399   *Avro CF-100 Mk.5 18525 in a line-up of No. 428 AW(F) Squadron aircraft at Uplands, Ont., 16 October 1957.*

No. 429 Squadron

Badge *see Overseas Squadrons, No. 429 Squadron*
Formed as a Tactical Transport unit at St Hubert (Montreal), Quebec on 21 August 1967, the squadron flew Buffalo aircraft for the Canadian Forces Mobile Command, and on 1 February 1968 was integrated into the Canadian Armed Forces.
Brief Chronology Formed at St Hubert, Que. 21 Aug 67. Integrated into CAF 1 Feb 68.
Nickname "Bison"
Commanders
W/C J.W. Fitzsimmons, CD 21 Aug 67 - 31 Jan 68.
Higher Formations and Squadron Location
Canadian Forces Mobile Command:
No. 10 Tactical Air Group,
St Hubert, Que. 21 Aug 67 - 31 Jan 68.
4 aircraft, Namao, Alta. 21 Aug 67 - 31 Jan 68.
Representative Aircraft
de Havilland CC-115 Buffalo (Sep 67 - Jan 68)
9454 to 9458   9462   9463

400 De Havilland Buffalo 9452 serving with No. 429 (Tac T) Squadron, based at St Hubert, Que.

No. 430 Squadron

Badge In front of a sun in splendour a Gyrfalcon's head erased
Motto *Celeriter certoque* (Swiftly and surely)
Authority Queen Elizabeth II, August 1955
The gyrfalcon, found in Northern Canada, is noted for its exceptional and alert fighting ability. The sun in splendour is used to depict the squadron's original role of fighter reconnaissance; (the bringing of light or information to the troops).
Formed as a Fighter unit at North Bay, Ontario on 1 November 1951 with Sabre aircraft, the squadron joined No. 2 (Fighter) Wing at Grostenquin, France in September 1952. Selected as one of eight squadrons in No. 1 Air Division Europe to be re-equipped with CF-104 aircraft for a nuclear strike role, it was deactivated on 1 June 1963 and reactivated as Strike Attack on 30 September. When No. 2 Wing was disbanded on 24 February 1964, the squadron moved to No. 3 Wing at Zweibrucken, Germany. On 1 February 1968 the squadron was integrated into the Canadian Armed Forces.
Brief Chronology Formed as No. 430 (F) Sqn, North Bay, Ont. 1 Nov 51. Deactivated at Grostenquin, Fr. 1 Jun 63. Reactivated as No. 430 (ST/A) Sqn 30 Sep 63. Integrated into CAF at Zweibrucken, Ger. 1 Feb 68.
Nickname "Silver Falcon"
Commanders
W/C J.F. Edwards, DFC and Bar, DFM 1 Nov 51 - 26 Jan 53.
S/L P.L. Gibbs, DFC, AFC, CD 27 Jan 53 - 17 Jan 55.
W/C C.D. Barnett, DFC, CD 18 Jan 55 - 17 Mar 55.
S/L W.S. Harvey, CD 18 Mar 55 - 28 Oct 55.
S/L C.W. White, CD 29 Oct 55 - 2 Mar 58.
S/L R.V. Virr, CD 3 Mar 58 - 18 Dec 60.
W/C J.T. Mullen, CD 19 Dec 60 - 1 Jun 63.
Squadron inactive
W/C H.R. Knight, CD 30 Sep 63 - 31 Jan 64.
W/C A.J. Bauer, CD 24 Feb 64 - 2 Jul 66.
W/C W.G. Paisley 3 Jul 66 - 11 Oct 66.
W/C J.W. Whitley, CD 15 Nov 66 - 31 Jan 68.
Higher Formations and Squadron Locations
Air Defence Command:
North Bay, Ont., 1 Nov 51 - 27 Sep 52.
En route overseas (Operation "Leapfrog II") 28 Sep 52 - 10 Oct 52.
No. 1 Air Division Europe:
No. 2 (Fighter) Wing,
Grostenquin, Fr. 11 Oct 52 - 31 May 63.
Squadron inactive
No. 2 Wing,
Grostenquin, Fr. 30 Sep 63 - 23 Feb 64.
No. 3 Wing,
Zweibrucken, Ger. 24 Feb 64 - 31 Jan 68.
No. 1 Wing, Marville, Fr. (Zweibrucken runways being resurfaced) 1 Jul - 15 Sep 66.

Representative Aircraft (Unit Code 1951-58 BH)
Canadair Sabre Mk.2 (Nov 51 - Jun 53)
19318   19373   19397   19406
Canadair Sabre Mk.5 (Jun 53 - Jun 57)
23139   23201   23310
Canadair Sabre Mk.6 (May 57 - May 63)   23586
Canadair CF-104 Starfighter (Sept 63 - Jan 68)   12853

401   *Canadair Sabre Mk.2 19353 of No. 430 (F) Squadron at North Luffenham, England in October 1952.*

402   *Canadair Sabre Mk.6 23733 of No. 430 (F) Squadron at Grostenquin, France in the summer of 1957.*

403   *Canadair Sabre Mk.5 23156, with the unit identifier BH of No. 430 (F) Squadron, in a formation of No. 2 Wing aircraft on 18 November 1955.*

No. 432 Squadron

Badge see Overseas Squadrons, No. 432 Squadron
Formed as an All-Weather (Fighter) unit at Bagotville, Quebec on 1 October 1954, the squadron flew CF-100 aircraft on North American air defence until disbanded on 15 October 1961.
Brief Chronology Formed at Bagotville, Que. 1 Oct 54. Disbanded 15 Oct 61.
Nickname "Black Cougar"
Commanders
W/C A.G. Lawrence, DFC, AFC, CD 18 Oct 54 - 24 Jun 55.
W/C K.B. Handley, CD 2 Sep 55 - 7 Nov 56.
W/C J.H.L. Lecomte, DFC, CD 21 Nov 56 - 7 Mar 57.
W/C H.R. Norris, AFC, CD 8 Mar 57 - 5 Sep 57 KIFA.
W/C J.R.D. Braham, DSO and 2 Bars, DFC and 2 Bars, AFC, CD 27 Oct 57 - 29 Jul 60.
W/C A.B. Hammond, DFC, CD 1 Aug 60 - 15 Oct 61.
Higher Formations and Squadron Location
North American Air Defence Command,
RCAF Air Defence Command:
Bagotville, Que. 1 Oct 54 - 15 Oct 61.
Representative Aircraft (Unit Code DL)
Avro CF-100 Canuck Mk.4A (Dec 54 - May 55)
18289   18297   18305   18318
Avro CF-100 Canuck Mk.4B (Apr 55 - Aug 56)
18324   18349   18352   18361
Avro CF-100 Canuck Mk.5 (Jun 56 - Oct 61)
18539   18593   18600   18771

404   *Avro CF-100 Mk.4B 18343 with the black cougar and identifier DL, denoting No. 432 AW(F) Squadron.*

405   *Avro CF-100 Mk.5 18741 with the markings of No. 432 AW(F) Squadron at Bagotville, Que.*

No. 433 Squadron

Badge *see Overseas Squadrons, No. 433 Squadron*
Formed as an All-Weather (Fighter) unit at Cold Lake, Alberta on 15 November 1954, and moved to North Bay, Ontario in October 1955, the squadron flew CF-100 aircraft on North American air defence until disbanded on 1 August 1961.
Brief Chronology Formed at Cold Lake, Alta. 15 Nov 54. Disbanded at North Bay, Ont. 1 Aug 61.
Nickname "Porcupine"
Commanders
W/C J.C. Hovey, DFC, CD 28 Jun 56 - 25 Mar 59.
W/C E. Wilson, CD 29 Jul 59 - 19 Nov 60.
W/C G.H. Nichols 20 Nov 60 - 1 Aug 61.
Higher Formations and Squadron Locations
North American Air Defence Command,
RCAF Air Defence Command:
Cold Lake, Alta. 15 Nov 54 - 11 Oct 55.
North Bay, Ont. 17 Oct 55 - 1 Aug 61.
Representative Aircraft (Unit Code FG)
Avro CF-100 Canuck Mk.4 (Feb 55 - Sep 57)
18325   18329   18333   18336
Avro CF-100 Canuck Mk.5 (Sep 56 - Jul 61)
18560   18567   18656   18665

406   *Avro CF-100 Mk.4B 18433 based at North Bay, Ont. in the service of No. 433 AW(F) Squadron in September 1956.*

407   *Avro CF-100 Mk.5 18643 of No. 433 AW(F) Squadron based at North Bay, Ont.*

No. 434 Squadron

Badge *see Overseas Squadrons, No. 434 Squadron*
Formed as a Fighter unit at Uplands (Ottawa), Ontario on 1 July 1952 with Sabre aircraft, the squadron joined No. 3 (Fighter) Wing at Zweibrucken, Germany in March 1953.[1] Selected as one of eight squadrons in No. 1 Air Division Europe to be re-equipped with CF-104 Starfighter aircraft for a nuclear strike role, it was deactivated on 15 January 1963 and reactivated as Strike Attack on 8 April. When the Air Division was reduced to six squadrons, the squadron was again deactivated on 1 March 1967.[2]
Brief Chronology Formed as No. 434 (F) Sqn, Uplands, Ont. 1 Jul 52. Deactivated at Marville, Fr. 15 Jan 63. Reactivated as No. 434 (ST/A) Sqn, Zweibrucken, Ger. 8 Apr 63. Deactivated 1 Mar 67.
Nickname "Bluenose"
Commanders
W/C J.D. Mitchner, DFC and Bar 2 Jul 52 - 20 Feb 53.
S/L J.W. Fiander, CD 18 May 53 - 28 Nov 54.
S/L A.L. Sinclair, DFC, CD 29 Nov 54 - 29 Apr 56.
S/L E.R.B. Gray, CD 30 Apr 56 - 7 Apr 57.
W/C H.C. Stewart, AFC, CD 8 Apr 57 - 31 May 57.
S/L J.F. Dunlop, DFC, CD 11 Jun 57 - 25 May 59.
W/C H.F. Wenz, CD 26 May 59 - 12 Jul 62.
S/L J. Ursulak, CD 13 Jul 62 - 23 Sep 62.
S/L K.E. Lewis, CD 24 Sep 62 - 15 Jan 63.
Squadron inactive
W/C O.B. Philp, DFC, CD 8 Apr 63 - 3 Dec 65.
W/C J.A.G.F. Villeneuve, AFC, CD 4 Dec 65 - 1 Mar 67.
Higher Formations and Squadron Locations
Air Defence Command:
Uplands, Ont. 1 Jul 52 - 6 Mar 53.
En route overseas (Operation "Leapfrog III") 7 Mar 53 - 6 Apr 53.
No. 1 Air Division Europe:
No. 3 (Fighter) Wing,
Zweibrucken, Ger. 7 Apr 53 - 15 Jun 62.
No. 1 (Fighter) Wing,
Marville, Fr. (for withdrawal of Sabre aircraft) 16 Jun 62 - 15 Jan 63.
Squadron inactive
No. 3 Wing, Zweibrucken, Ger. 8 Apr 63 - 1 Mar 67.
No. 4 Wing, Baden-Soellingen, Ger. (Zweibrucken runways being resurfaced) 1 Jul - 15 Sep 66.
Representative Aircraft (Unit Code 1952-58 BR)
Canadair Sabre Mk.2 (July 52 - Nov 53)
19117   19307   19434   19449
Canadair Sabre Mk.5 (Oct 53 - Jan 57)
23053   23065   23085
Canadair Sabre Mk.6 (Jan 57 - Jan 63)
23480   23642   23688   23707
Canadair CF-104 Starfighter (Apr 63 - Feb 67)
12811   12815   12820   12824

---

[1] In May 1945 its predecessor had flown Canadian-built Lancaster B.X's from England to Canada. Thus No. 434 became the first squadron of the RCAF to fly the Atlantic in both directions.
[2] The squadron was still inactive when the RCAF was integrated into the Canadian Armed Forces, but was reactivated two weeks later (15 February 1968) as an operational training squadron with CF-5 aircraft.

408 Canadair Sabre Mk.2 19307 of No. 434 (F) Squadron at Zweibrucken, Germany on 7 April 1953.

409 Canadair Sabre Mk.6 aircraft 23624, 23688, 23674 and 23480, in diamond formation, flown by No. 434 (F) Squadron in Europe.

No. 435 Squadron

Badge *see Overseas Squadrons, No. 435 Squadron*
Formed as a Transport unit at Edmonton, Alberta on 1 August 1946 from the Edmonton portion of No. 164 (Transport) Squadron, the squadron flew Dakota and C-119 aircraft on western Canada transport duty and parachute training at Rivers, Manitoba. From November 1956 to January 1957, it airlifted members of the United Nations Emergency Force from Italy to Egypt. In 1960, it was re-equipped with C-130 Hercules aircraft and, during the 1962-63 conversion of No. 1 Air Division Europe squadrons from Sabre to Starfighter aircraft, ferried 137 CF-104 aircraft from Canada to Europe. On 1 February 1968 the squadron was integrated into the Canadian Armed Forces.

Brief Chronology Formed at Edmonton, Alta. 1 Aug 46. Integrated into CAF 1 Feb 68.
Nickname "Chinthe"
Commanders
W/C P.J. Grant 16 Aug 46 - 30 Aug 48.
W/C W.F. Parks, DFC 31 Aug 48 - 10 Jul 50.
W/C M.E. Pollard, DSO, DFC, AFC, CD 11 Jul 50 - 15 Aug 52.
W/C G.J.J. Edwards, DFC, CD 16 Aug 52 - 26 Mar 54.
W/C W.C. Klassen, DFC, CD 27 Mar 54 - 7 Jul 57.
W/C C.C.W. Marshall, DFC, CD 23 Sep 57 - 11 May 61.
W/C E.E. Hurlburt, DFC, CD 4 Aug 61 - 16 Jul 65.
W/C J.P. O'Callaghan, CD 26 Jul 65 - 17 Jul 67.
W/C K.C. Lee, CD 18 Jul 67 - 31 Jan 68.
Higher Formation and Squadron Locations
Air Transport Command:
Edmonton, Alta. 1 Aug 46 - 13 Sep 55.
Namao, Alta. 14 Sep 55 - 31 Jan 68.
United Nations Emergency Force: No. 114 (RCAF) Air Transport Unit, Naples (Capodichino), Italy 20 Nov 56 - 22 Jan 57. (4 aircraft remained until the spring of 1958).
Representative Aircraft (Unit Code 1947-51 BN, 1951-58 ET)
Douglas Dakota Mk.IV (Aug 46 - Oct 60)
KG936   KN278   983   991
Fairchild C-119G Flying Boxcar (Sep 52 - Jun 65)
22101 to 22111   22123   22135
Lockheed C-130B Hercules (1960 - 1966)   10301   10317
Lockheed C-130E Hercules (1966 - 1968)
10307   10312   10313   10314   10316   10318   10320

410 Fairchild Flying Boxcar 22101, showing its trim lines, in the service of No. 435 (T) Squadron.

411 Lockheed C-130E Hercules 10307 being loaded at Downsview, Ont. by No. 435 (T) Squadron.

No. 436 Squadron

Badge see *Overseas Squadrons, No. 436 Squadron*
Formed as a Transport unit at Dorval (Montreal), Quebec on 1 April 1949, the squadron flew C-119 aircraft on transport duty. In January 1955, it helped move No. 1 (Fighter) Wing from North Luffenham, England to Marville, France, and in June, during a British railway strike, carried out an emergency airlift of supplies to units of No. 1 Air Division in Europe. On 1 July 1956 the squadron moved to Downsview (Toronto), Ontario; between November 1956 and January 1957 it airlifted members of the United Nations Emergency Force from Italy to Egypt. It moved to Uplands (Ottawa), Ontario in August 1964, and in 1965 was re-equipped with C-130 Hercules aircraft. On 1 February 1968 the squadron was integrated into the Canadian Armed Forces.

Brief Chronology Formed at Dorval, Que. 1 Apr 53. Integrated into CAF at Uplands, Ont. 1 Feb 68.
Nickname "Elephant"
Commanders
S/L K.C.M. Dobbin 1 Apr 53 - 19 Aug 53.
S/L R.K. Trumley, CD 20 Aug 53 - 1 Jul 56.
W/C J.T. McCutcheon, DFC, AFC, CD 2 Jul 56 - 15 Sep 60.
W/C J.S. Miller, CD 16 Sep 60 - 1 Mar 64 *ret*.
W/C W.R. Lloyd, CD 11 May 64 - 11 Apr 66.
W/C L.W. Hussey, CD 25 Jul 66 - 31 Jan 68.
Higher Formation and Squadron Locations
Air Transport Command:
Dorval, Que. 1 Apr 53 - 1 Jul 56.
Downsview, Ont. 2 Jul 56 - 31 July 56.
United Nations Emergency Force; No. 114 (RCAF) Air Transport Unit, Naples (Capodichino), Italy 20 Nov 56 - 31 Jan 57.
Uplands, Ont. 1 Aug 64 - 31 Jan 68.
Representative Aircraft
Fairchild C-119G Flying Boxcar (Apr 53 - Jul 65)
22116 22121 22126 22135
Lockheed C-130E Hercules (1965 - 1968)
10305 10312 10321 10327

412 Fairchild Flying Boxcar 22131 of No. 436 (T) Squadron based at Downsview, Ont.

413 Lockheed C-130E Hercules 10305 serving No. 436 (T) Squadron.

No. 437 Squadron

Badge *see Overseas Squadrons, No. 437 Squadron*
Formed as a Transport unit at Trenton, Ontario on 1 October 1961, the squadron flew Yukon aircraft on transatlantic service to Canadian military bases in Europe. On 1 February 1968 the squadron was integrated into the Canadian Armed Forces.
Brief Chronology Formed at Trenton, Ont. 1 Oct. 61. Integrated into CAF 1 Feb 68.
Nicknames "Husky", "Yukon"
Commanders
W/C J.O. Maitland, CD 19 Oct 61 - 25 May 64.
W/C D.R. Adamson, CD 26 May 64 - 8 Aug 66.
W/C R.G. Husch, CD 9 Aug 66 - 31 Jan 68.
Higher Formation and Squadron Location
Air Transport Command:
Trenton, Ont. 1 Oct 61 - 31 Jan 68.
Representative Aircraft
Canadair CC-106 Yukon (Oct 61 - Jan 68).
Passenger   15921   15923 to 15925   15928
Freight   15922   159326   15927   15930   15931
VIP   15929   15932

414 Canadair Yukon 15928 on transatlantic service with No. 437 (T) Squadron.

No. 438 "City of Montreal" Squadron (Auxiliary)

Badge see Overseas Squadrons, No. 438 Squadron
Formed at Montreal, Quebec on 15 April 1946, the squadron flew Vampire and Sabre aircraft in a fighter role until November 1958 when it was then reassigned to a light transport and emergency rescue role and re-equipped with Expeditor and Otter aircraft. On 5 May 1961 the unit received a Squadron Standard for 25 years service as No. 115, 118 and 438 Squadron. On 1 February 1968 the squadron was integrated into the Canadian Armed Forces as No. 438 "City of Montreal" Air Reserve Squadron.

Brief Chronology Formed as No. 438 (FB) Sqn (Aux), Montreal, Que. 15 Apr 46. Redesignated No. 438 (F) Sqn (Aux) 1 Apr 47. Titled No. 438 "City of Montreal" (F) Sqn (Aux) 1 May 50. Redesignated No. 438 "City of Montreal" Sqn (Aux) 1 Nov 58. Integrated into CAF as No. 438 "City of Montreal" Air Res Sqn 1 Feb 68.

Title "City of Montreal"
Nickname "Wild Cat"
Commanders
W/C R.J.C. Hebert, DFC 1 May 46 - 14 May 50 *ret*.
W/C A.R. Morrissette, AFC 15 May 50 - 14 Feb 52 *ret*.
W/C J.A.G.P.F. Valois 15 Feb 52 - 14 Feb 55 *ret*.
W/C J.E.M.M. Gauthier, CD 15 Feb 55 - 14 Nov 60.
W/C J.L.Y. Gagne, CD 15 Nov 60 - 31 Oct 63 *ret*.
W/C J.D. Fisher, CD 1 Nov 63 - 30 Aug 67.
W/C A.E. Gamble, CD 31 Aug 67 - 31 Jan 68.

Higher Formations and Squadron Location
Training Command,
Air Defence Command (15 Nov 49),
No. 1 Group (Auxiliary) (15 Jan 51 - 15 Apr 57),
Air Transport Command (1 Apr 61):
No. 11 Wing (Auxiliary) (1 Oct 50 - 30 Apr 59 and 1 Sep 61 - 1 Feb 68),
St Hubert, Que. 15 Apr 46 - 31 Jan 68.

Representative Aircraft (Unit Code 1947-58 BQ)
North American Harvard Mk.II (Nov 46 - Mar 47)
2760 3213 3300 3343
de Havilland Vampire Mk.III (Apr 48 - Sep 53)
17014 17022 17033 17078
Canadair Silver Star Mk.3 (Nov 54 - Sep 58) 21469
Canadair Sabre Mk.5 (Oct 56 - Nov 58)
23052 23110 23313 23356
Beechcraft Expeditor Mk.3 (Nov 58 - Mar 64)
1534 1570 2352 2356
de Havilland Otter (Sep 60 - Jan 68)
9406 9409 9423 9427

Honours and Awards Squadron Standard (5 May 1961).

415 North American Harvard Mk.II 3213, coded BQ, in the service of No. 438 Squadron (Aux).

416 De Havilland Vampire Mk.III 17067 in formation with 17002, 17065 and 17078 of No. 438 Squadron (Aux).

417 Canadair Silver Star Mk.3 21326 of No. 438 Squadron (Aux), with an extensive bilingual identification on the nose panel, on 28 September 1957 at St Hubert, Que.

418 Canadair Sabre Mk.5 23356 at St Hubert, Que. on 29 March 1958 in the service of No. 438 Squadron (Aux).

419 De Havilland Otter 9423 of No. 438 Squadron (Aux) visiting Trenton, Ont. on 11 June 1966.

No. 439 Squadron

Badge In front of a fountain a sabre-toothed tiger's head erased
Motto Fangs of death
Authority Queen Elizabeth II, May 1954
This unit adopted the sabre-toothed tiger's head and became known as the sabre-toothed squadron. The fountain is indicative of the fact that the unit was the first squadron to fly Sabre aircraft direct from Canada to the United Kingdom.

Formed as a Fighter unit at Uplands (Ottawa), Ontario on 1 September 1951 with Sabre aircraft, the squadron joined No. 1 (Fighter) Wing at North Luffenham, Nottinghamshire, England in June 1952. In doing so, it intitiated the first of four Operations "Leapfrog" — mass transatlantic flights by Sabre-equipped units to No. 1 Air Division Europe. Despite bad weather throughout the move, the 21 aircraft made the 3560-mile trip without mishap, to complete the formation of No. 1 Wing. Early in 1955, the unit moved with the wing to its French base at Marville. Selected as one of eight squadrons of No. 1 Air Division Europe to be re-equipped with CF-104 Starfighter aircraft, it was deactivated on 1 November 1963 and reactivated as Strike Reconnaissance on 2 March 1964. On 1 February 1968 the squadron was integrated into the Canadian Armed Forces.

Brief Chronology Formed as No. 439 (F) Sqn, Uplands, Ont. 1 Sep 51. Deactivated at Marville, Fr. 1 Nov 63. Reactivated as No. 439 (ST/R) Sqn 1 Mar 64. Integrated into CAF at Lahr, Ger. 1 Feb 68.
Nickname "Sabre-Toothed Tiger"
Commanders
S/L C.D. Bricker, Jr., DFC, CD 1 Sep 51 - 12 Jul 53.
S/L K.J.H.M. Belleau, CD 13 Jul 53 - 23 Apr 56.
S/L D.J. Bullock, CD 27 Apr 56 - 30 Jan 57.
S/L A.W. Fisher, CD 31 Jan 57 - 30 Jun 57.
S/L R.Y. Cannon, CD 1 Jul 57 - 31 May 59.
S/L C.J. Day, CD 1 Jun 59 - 17 Jul 59 *died.*
W/C J. MacKay, DFC and Bar, CD 19 Oct 59 - 30 Jun 60.
W/C J.P. Bell, CD 1 Jul 60 - 11 Dec 62.
W/C P.B. St. Louis, MBE, CD 3 Jan 63 - 1 Nov 63.
Squadron inactive
S/L J.L. Frazer, CD 2 Mar 64 - 8 Nov 66.
W/C S.P. Gulyus, CD 9 Nov 66 - 31 Jan 68.
Higher Formations and Squadron Locations
Air Defence Command:
Uplands, Ont. 1 Sep 51 - 29 May 52.
En route overseas (Operation "Leapfrog I") 30 May 52 - 14 Jun 52.
No. 1 Air Division Europe:
No. 1 (Fighter) Wing,
North Luffenham, Notts., Eng.[1] 15 Jun 52 - 31 Mar 55.
Marville, Fr. 1 Apr 55 - 30 Nov 63.
Squadron inactive
No. 1 Wing,
Marville, Fr. 2 Mar 64 - 31 Mar 67.
Lahr, Ger. 1 Apr 67 - 31 Jan 68.
Representative Aircraft (Unit Code 1951-58 IG)
Canadair Sabre Mk.2 (Nov 51 - Feb 55) 19112 19140
19192 19208

Canadair Sabre Mk.5 (Mar 55 - Jul 56) 23336 23357 23360
Canadair Sabre Mk.6 (Jul 56 - Oct 63) 23581 23549
Canadair CF-104 Starfighter (Mar 64 - Jan 68)

---

¹Under the operational control of RAF Fighter Command through its No. 11 Group.

420 *Canadair Sabre Mk.2 19114 lined up with the remainder of No. 439 (F) Squadron at North Luffenham, England.*

421 *Canadair Sabre Mk.5 23360 leads in echelon left formation with No. 439 (F) Squadron prior to the application of Air Division markings.*

422 *Canadair Sabre Mk.6 23661 with No. 439 (F) Squadron based at Marville, France.*

No. 440 Squadron

Badge A bat in front of clouds
Motto Ka ganawaitak saguenay (He who protects the Saguenay)
Authority Queen Elizabeth II, October 1955
The bat suggestive of night flying and use of radar, and the clouds conditions of poor visibility.
Formed as an All-Weather (Fighter) unit at Bagotville, Quebec on 1 October 1953, the squadron flew CF-100 aircraft on North American air defence until May 1957 when it then joined No. 1 Air Division Europe to replace No. 413 (Fighter) Squadron in No. 3 Wing at Zweibrucken, Germany. On the withdrawal of CF-100 aircraft from operational service, the squadron was disbanded on 31 December 1962.
Brief Chronology Formed at Bagotville, Que. 1 Oct 53. Disbanded at Zweibrucken, Ger. 31 Dec 62.
Nickname "Bat"
Commanders
W/C A.G. Lawrence, DFC, AFC 1 Oct 53 - 17 Oct 54.
W/C C.V.B. Carson, CD 18 Oct 54 - 11 Jan 56.
W/C J.M. Sutherland, CD 9 Feb 56 - 28 Mar 57.
W/C W.A.G. McLeish, DFC, AFC, CD 29 Mar 57 - 20 Jun 57 KIFA.
W/C S.F. Cowan, CD 27 May 58 - 2 Aug 59.
W/C J.F. Corrigan, DFC, CD 3 Aug 59 - 24 Jul 61.
W/C R.F. Herbert, CD 31 July 61 - 31 Dec 62.
Higher Formations and Squadron Locations
Air Defence Command:
Bagotville, Que. 1 Oct 53 - 10 May 57.
En route overseas (Operation "Nimble Bat III") 11 May 57.
No. 1 Air Division Europe:
No. 3 (Fighter) Wing,
Zweibrucken, Ger. 12 May 57 - 15 Jun 62.
No. 4 (Fighter) Wing,
Baden-Soellingen, Ger. (for withdrawal of CF-100 aircraft) 16 Jun 62 - 31 Dec 62.
Representative Aircraft (Unit Code KE)
Avro CF-100 Canuck Mk.3B (Oct 53 - Sep 55)
18126 18134 18137 18150
Avro CF-100 Canuck Mk.4B (Feb 55 - Dec 62)
18320 18363 18402 18416

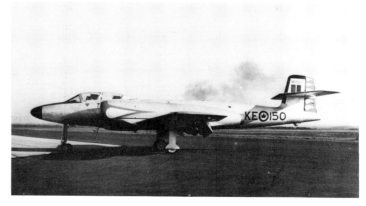

423 *Avro CF-100 Mk.3B 18150 in the markings of No. 440 AW(F) Squadron.*

424 Avro CF-100 Mk.4B 18332, with the unit identifier KE of No. 440 AW(F) Squadron, over Bagotville, Que. on 23 April 1957.

425 Avro CF-100 Mk.5 18498, with the crocodile tears of No. 440 AW(F) Squadron, on the engine cowls and rudder, tucks up on take-off from Bagotville, Que. on 1 August 1956.

No. 441 Squadron

Badge see Overseas Squadrons, No. 441 Squadron
Formed as a Fighter unit at St Hubert (Montreal), Quebec on 1 March 1951 with Vampire and later with Sabre aircraft, the squadron joined No. 1 (Fighter) Wing at North Luffenham, Nottinghamshire, England in March 1952; its personnel and Sabre aircraft were brought to England by the Royal Canadian Navy's carrier *Magnificent*. Early in 1955 the unit moved with the wing to its French base at Marville. Selected as one of eight squadrons in No. 1 Air Division Europe to be re-equipped with CF-104 Starfighter aircraft, it was deactivated on 1 September 1963 and reactivated as Strike Reconnaissance on the 15th. On 1 February 1968 the squadron was integrated into the Canadian Armed Forces.

Brief Chronology Formed as No. 441 (F) Sqn, St Hubert, Que. 1 Mar 51. Deactivated at Marville, Fr. 1 Sep 63. Reactivated as No. 441 (ST/R) Sqn 15 Sep 63. Integrated into CAF at Lahr, Ger. 1 Feb 68.

Nickname "Silver Fox"

Commanders
S/L A.R. MacKenzie, DFC 1 Mar 51 - 6 Nov 52[1].
S/L W.T.H. Gill, DFC, CD 1 Feb 53 - 1 Mar 54.
S/L D.R. Cuthbertson, AFC, CD 2 Mar 54 - 11 Sep 56.
S/L A.W. Fisher, CD 1 Oct 56 - 3 Dec 56.
S/L L.J. Hill, CD 4 Dec 56 - 13 Aug 59.
W/C H. McLachlan, DFC, CD 14 Aug 59 - 18 Jul 61.
W/C D.A.B. Smiley, DFC 19 Jul 61 - 1 Jun 63.
S/L E.L. Arnold, CD 2 Jun 63 - 25 Aug 63.
S/L L.C. Price 26 Aug 63 - 1 Sep 63.
Squadron inactive
W/C R.M. Edwards, AFC, CD 15 Sep 63 - 11 Oct 65.
W/C J.F. Dunlop, DFC, CD 12 Oct 65 - 31 Jan 68.

Higher Formations and Squadron Locations
Air Defence Command:
St Hubert, Que. 1 Mar 51 - 12 Feb 52.
Dorval, Que. (St Hubert runways under repair) 1 Jun - 15 Aug 51.
En route overseas (aboard HMCS *Magnificent*) 13 Feb 52 - 28 Feb 52.
No. 1 Air Division Europe:
No. 1 (Fighter) Wing,
North Luffenham, Notts., Eng.[2] 1 Mar 52 - 20 Dec 54.
No. 3 (Fighter) Wing,
Zweibrucken, Ger.[3] 21 Dec 54 - 31 Mar 55.
No. 1 (Fighter) Wing,
Marville, Fr. 1 Apr 55 - 1 Sep 63.
Squadron inactive
No. 1 Wing,
Marville, Fr. 15 Sep 63 - 31 Mar 67.
Lahr, Ger. 1 Apr 67 - 31 Jan 68.

Representative Aircraft (Unit Code 1951-58 BT)
de Havilland Vampire Mk.III (Mar - Jun 51) 17004
17067 BT-O
Canadair Sabre Mk.2 (Jun 51 - Jun 53) 19134 BT-B
19142 BT-M   19149 BT-W   19150 BT-A   19152 BT-C
19160 BT-D   19162 BT-H   19163 BT-G

Canadair Sabre Mk.5 (Jan 55 - Aug 56)   23206
Canadair Sabre Mk.6 (Aug 56 - Sep 63)
Canadair CF-104 Starfighter (Sep 63 - Jan 68)

[1]Posted to Korea, S/L MacKenzie was shot down on 5 December 1952 and taken prisoner (see Appendix 3).
[2]Under the operational control of RAF Fighter Command through its No. 11 Group.
[3]A temporary location while No. 1 Wing was moving to Marville.

428   Canadair Sabre Mk.6 23495 on patrol with No. 441 (F) Squadron from Marville, France.

426   De Havilland Vampire Mk.III 17067 VC-BTO with No. 441 (F) Squadron, St Hubert, Que.

429   Canadair CF-104 12854 serving No. 441 (F) Squadron on 29 October 1966 with No. 1 Wing in Europe.

427   Canadair Sabre Mk.5 23017 of No. 441 (F) Squadron at Marville, France on 27 February 1956.

No. 442 "City of Vancouver" Squadron (Auxiliary)

Badge A representation of Haietlik or Lightning Snake
Motto Un dieu, une reine, un coeur (One God, one Queen, one heart)
Authority Queen Elizabeth II, March 1957
According to legend of the Nootka Tribe Indians, Haietlik was dispatched to kill its enemies.

Formed at Vancouver, British Columbia on 15 April 1946, the squadron flew Vampire, Mustang and Sabre aircraft in a fighter role until October 1958 when it was reassigned to a light transport and emergency rescue role and re-equipped with Expeditor and Otter aircraft. A reduction of the Auxiliary Force resulted in the squadron being disbanded on 1 April 1964.

Brief Chronology Formed as No. 442 (F) Sqn (Aux), Vancouver, B.C. 15 Apr 46. Titled No. 442 "City of Vancouver" (F) Sqn (Aux) 3 Sep 52. Redesignated No. 442 "City of Vancouver" Sqn (Aux) 1 Sep 58. Disbanded 1 Apr 64.
Title "City of Vancouver"
Nickname "Caribou"
Commanders
W/C R.F. Begg 1 May 46 - 14 Jul 47 *ret.*
W/C J.W. Reid, DFC 15 Jul 47 - 23 Oct 47.
W/C D.C. MacDonald, DFC, CD 8 Dec 47 - 2 Dec 48 *ret.*
W/C G.W. Northcott, DSO, DFC and Bar 3 Dec 48 - 3 Jan 51.
W/C E.A. Alexander, AFC 4 Jan 51 - 31 Dec 53.
W/C G.M. Smith, DFC 1 Jan 54 - 31 Dec 54 *ret.*
W/C D.C. Cameron, CD 1 Jan 55 - 7 May 58.
W/C D.F.R. Aitken, CD 8 May 58 - 8 May 61 *ret.*
W/C J.L.T. Edwards, CD 9 May 61 - 1 Apr 64.
Higher Formations and Squadron Locations
North West Air Command,
Air Defence Command (1 Jul 51),
Air Transport Command (1 Apr 61):
No. 19 Wing (Auxiliary) (1 Oct 50),
Vancouver, B.C. 15 Apr 46 - 8 Nov 48.
Sea Island, B.C. 9 Nov 48 - 1 Apr 64.
Representative Aircraft (Unit Codes BU, SL)
North American Harvard Mk.II (Jul 46 - Sep 58)
2639 3040 3132 3332
de Havilland Vampire Mk.III (Apr 48 - Oct 56)
17074 BU-X 17075 BU-Y 17076 BU-W
North American Mustang Mk.IV (Nov 50 - Oct 56)
9266 9277 9595 9598
Canadair Sabre Mk.5 (Aug 56 - Aug 58) 23088
Beechcraft Expeditor Mk.3 (Oct 58 - Mar 64)
1401 1411 1477 1541
de Havilland Otter (Oct 60 - Mar 64) 9410 9420

430   North American Harvard Mk.IIB 3062 VC-BUC, with red rudder and cowling, in the service of No. 442 Squadron (Aux) at Vancouver, B.C.

431   De Havilland Vampire Mk.III 17012 of No. 442 Squadron (Aux) at Sea Island, B.C. in July 1956.

432   North American Sabre Mk.5 23222 of No. 442 Squadron (Aux) tops the mountains from its Vancouver, B.C. base.

433 De Havilland Otter 9420 at Trenton, Ont. on its return from No. 442 Squadron (Aux).

No. 443 "City of New Westminster" Squadron (Auxiliary)

Badge *see Overseas Squadrons, No. 443 Squadron*
Formed at Vancouver, British Columbia on 15 September 1951, the squadron flew Vampire, Mustang and Sabre aircraft in a fighter role until October 1958 when it was reassigned to a light transport and emergency rescue role and re-equipped with Expeditor and Otter aircraft. A reduction of the Auxiliary Force resulted in the squadron being disbanded on 1 April 1964.

Brief Chronology Formed as No. 443 (F) Sqn (Aux), Vancouver, B.C. 1 Sep 51. Titled No. 443 "City of New Westminster" (F) Sqn (Aux) 3 Sep 52. Redesignated No 443 "City of New Westminster" Sqn (Aux) 1 Sep 58. Disbanded 1 Apr 64.

Title "City of New Westminster"
Nickname "Hornet"
Commanders
W/C R.B. Barker, DFC, CD 1 Dec 51 - 31 Aug 56.
W/C R.O. Hetherington 1 Sep 56 - 31 Aug 58 *ret.*
W/C J.D. Fisher 1 Sep 58 - 31 Aug 61 *ret.*
W/C C.O.P. Smith, DFC, CD 1 Sep 61 - 1 Apr 64.
Higher Formations and Squadron Location
Air Defence Command,
Air Transport Command (1 Apr 61):
No. 19 Wing (Auxiliary),
Sea Island, B.C. 1 Sep 51 - 1 Apr 64.
Representative Aircraft (Unit Code PF)
North American Harvard Mk.II (Dec 51 - Sep 58)
2569   2831
North American Mustang Mk.IV (Nov 52 - Oct 56)
9226   9275   9293   9297
Canadair Sabre Mk.5 (Aug 56 - Aug 58)
23115   23167   23311
Beechcraft Expeditor Mk.3 (Aug 58 - Mar 64)
HB143   HB210   1396   1488
de Havilland Otter (Oct 60 - Mar 64)

434 North American Harvard Mk.II 3137 at Vancouver, B.C. in 1955 with No. 443 Squadron (Aux).

435 North American Mustang Mk.IV 9268 of No. 443 Squadron (Aux) based at New Westminster, B.C.

436 Canadair Sabre Mk.5 23310 with No. 443 Squadron (Aux) at Vancouver, B.C. in June 1957.

No. 444 Squadron

Badge none

Formed as an Air Observation Post unit at the Canadian Joint Air Training Centre at Rivers, Manitoba on 1 October 1947, the squadron flew Chipmunk and Auster aircraft in training army pilots in ranging and directing artillery fire. When this function was transferred to the Light Aircraft School, the squadron was disbanded on 1 April 1949.

Brief Chronology Formed at CJATC, Rivers, Man. 1 Oct 47. Disbanded 1 Apr 49.

Commanders
Maj N.W. Reilander 1 Oct 47 - 1 Apr 49.

Higher Formations and Squadron Location
North West Air Command:
Canadian Joint Air Training Centre,
Rivers, Man. 1 Oct 47 - 1 Apr 49.

Representative Aircraft
de Havilland Chipmunk (May 48 - Mar 49)
Auster A.O.P. Mk.VI (May 48 - Mar 49)   16652 to 16656   16661   16665

437 De Havilland Chipmunk Mk.I 18001 VC-BVY, in aluminum finish with maple leaf roundels without white fields, used by No. 444 (AOP) Squadron.

438 Auster A.O.P.Mk.VI 16662 at Winnipeg, Man. in 1948 with No. 444 (AOP) Squadron.

No. 444 Squadron, Re-formed

Badge A Cobra ready to strike
Motto Strike swift, strike sure
Authority Queen Elizabeth II, November 1954

This unit adopted the cobra in a position to strike as being indicative of the role of a fighter squadron.

Formed as a Fighter unit at St Hubert (Montreal), Quebec on 1 March 1953 with Sabre aircraft — the last of twelve units for No. 1 Air Division Europe — the squadron joined No. 4 (Fighter) Wing at Baden-Soellingen, Germany in September. Selected as one of eight squadrons of the Air Division to be re-equipped with CF-104 Starfighter aircraft for a nuclear strike role, it was deactivated on 1 March 1963 and reactivated as Strike Attack on 27 May. When the Air Division was reduced to six squadrons, the squadron was once more deactivated on 1 April 1967.[1]

Brief Chronology Formed as No. 444 (F) Sqn, St Hubert, Que. 1 Mar 53. Deactivated at Baden-Soellingen, Ger. 1 Mar 63. Reactivated as No. 444 (ST/A) Sqn 27 May 63. Deactivated 1 Apr 67.

Nickname "Cobra"

Commanders
S/L E.R. Heggtveit, CD 2 Mar 53 - 22 Jul 53.
S/L J. MacKay, DFC and Bar 23 Jul 53 - 8 Sep 53.
W/C H.F. Darragh, AFC 16 Jan 54 - 15 Mar 54 *KIFA*.
S/L J.B. Lawrence, CD 15 May 54 - 30 Jun 57.
S/L D.F. Archer, CD 1 Jul 57 - 4 Jan 59.
S/L E. Garry, CD 5 Jan 59 - 5 Sep 60.
W/C R.V. Smith, CD 6 Sep 60 - 23 Feb 61 *died*.
W/C J.L.A. Roussell, DFC, CD 10 Aug 61 - 1 Mar 63.
Squadron inactive
W/C K.J. Thorneycroft, CD 27 May 63 - 20 Jan 66.
W/C R.H. Annis, CD 1 Aug 66 - 1 Apr 67.

Higher Formations and Squadron Locations
Air Defence Command:
St Hubert, Que. 1 Mar 53 - 26 Aug 53.
En route overseas (Operation "Leapfrog IV") 27 Aug 53 - 3 Sep 53.
No. 1 Air Division Europe:
No. 4 (Fighter) Wing,
Baden-Soellingen, Ger. 4 Sep 53 - 1 Mar 63.
Squadron inactive
No. 4 Wing, Baden-Soellingen, Ger. 27 May 63 - 1 Apr 67.

Representative Aircraft (Unit Code 1953-58 VH)
Canadair Sabre Mk.4 (Mar 53 - Mar 54)[2] 19575 19677 19691 19702
Canadair Sabre Mk.5 (Feb 54 - Jan 57) 23020 23078 23109 23150
Canadair Sabre Mk.6 (Apr 55 - Feb 63) 23390 23455 23479 23868
Canadair CF-104 Starfighter (May 63 - Mar 67) 12717 12839 12876 12890

439   Canadair Sabre Mk.4 19702 serving No. 444 (F) Squadron in September 1953.

440   Canadair Sabre Mk.6 23625 being serviced at Baden-Soellingen, Germany while operating with No. 444 Squadron.

---

[1] Reactivated as a Tactical Helicopter unit in 1970.
[2] Turned over to the Royal Air Force under the Mutual Aid Plan.

No. 445 Squadron

Badge A wolverine rampant holding in its dexter paw a lightning flash
Motto Strike as lightning
Authority Queen Elizabeth II, April 1956

The wolverine, like the squadron, ventures forth to travel and hunt both day and night in all weather. It is indigenous to Canada and is ingenious, fearless, and of great strength. The flash of lightning is suggestive of the speed with which modern aircraft strike.

Formed as an All-Weather (Fighter) unit at North Bay, Ontario on 1 April 1953, the squadron was the first unit to fly CF-100 aircraft on North Americn air defence. In November 1956 it joined No. 1 Air Division Europe, replacing No. 410 (Fighter) Squadron in No. 1 Wing at Marville, France. On the withdrawal of CF-100 aircraft from operational service, the squadron was disbanded on 31 December 1962.

Brief Chronology Formed at North Bay, Ont. 1 Apr 53. Disbanded at Marville, Fr. 31 Dec 62.
Nickname "Wolverine"
Commanders
W/C G.E. Nickerson, DFC, CD 1 Apr 53 - 9 Nov 53.
W/C R.F. Hatton, DFC, CD 3 Feb 54 - 1 Aug 56.
W/C E.G. Ireland, DFC, CD 2 Aug 56 - 20 Jun 58.
W/C G. Sutherland, CD 1 Jul 58 - 31 Jul 59.
W/C K.W. MacDonald, CD 1 Aug 59 - 30 Jun 60.
W/C E.J. Trotter, DFC, DFM, CD 1 Jul 60 - Dec 62.
Higher Formations and Squadron Locations
Air Defence Command:
North Bay, Ont. 1 Apr 53 - 31 Aug 53.
Uplands, Ont. 1 Sep 53 - 31 Oct 56.
En route overseas (Operation "Nimble Bat I") 1 Nov 56.
No. 1 Air Division Europe:
No. 1 (Fighter) Wing,
Marville, Fr. 2 Nov 56 - 31 Dec 62.
Representative Aircraft (Unit Code SA)
Avro CF-100 Canuck Mk.3B (May 53 - Jun 54)
18133 18137 18140 18147
Avro CF-100 Canuck Mk.4B (Jun 54 - Dec 62)
18333 18385 18403 18445

441 Avro CF-100 Mk.3 18138 of No. 445 AW(F) Squadron over North Bay, Ont. in 1953.

442 Avro CF-100 Mk.4B 18386 en route to Marville, France in the service of No. 445 AW(F) Squadron.

No. 446 Squadron

Badge In front of two swords in saltire a Griffin segreant
Motto Vigilance — Swiftness — Strength
Authority Queen Elizabeth II, September 1965
The Griffin, a fabulous creature with the body of a lion and head and wings of an eagle, is a combination of two of the most powerful beasts. Its function is reputedly that of a guardian of treasure and, therefore, it is symbolic of watchfulness, alluding in this way to the unit's role in the defence of the North American Continent. The two-handed swords refer to the weight of the unit's armament.

Formed as a Surface-to-Air Missile unit at North Bay, Ontario on 28 December 1961, the squadron was equipped with nuclear-armed Bomarc surface-to-air missiles for North American air defence. On 1 February 1968 the squadron was integrated into the Canadian Armed Forces.
Brief Chronology Formed at North Bay, Ont. 28 Dec 61. Integrated into CAF 1 Feb 68.
Commanders
W/C A.G. Lawrence, DFC, AFC, CD 28 Dec 61 - 14 Jul 64.
W/C F.G. Fellows, CD 15 Jul 64 - 31 Jan 68.
Higher Formations and Squadron Location
North American Air Defence Command,
RCAF Air Defence Command:
North Bay, Ont. 28 Dec 61 - 31 Jan 68.
Representative Weapon
Lockheed CIM-10B Bomarc

443 Bomarc missile readied for launching at a typical Canadian site operated by Nos. 446 and 447 (SAM) Squadrons.

No. 447 Squadron

Badge In front of a red maple leaf a dagger per pale held by a dexter hand
Motto Monjak ecowi (Algonquin: may be translated "Toujours prêt" or "Always ready")
Authority Queen Elizabeth II, July 1966
The hand holding the dagger, the hilt of which is embellished by a fleur-de-lis, indicates the unit's location and its state of readiness. The maple leaf alludes to Canada.
Formed as a Surface-to-Air Missile unit at La Macaza, Quebec on 15 September 1962, the squadron was equipped with nuclear-armed Bomarc surface-to-air missiles for North American air defence. On 1 February 1968 the squadron was integrated into the Canadian Armed Forces.
Brief Chronology Formed at La Macaza, Que. 15 Sep 62. Integrated into CAF 1 Feb 68.
Commanders
W/C J.E.A. Laflamme, DFC 20 Sep 62 - 24 May 63.
W/C J.L.A. Roussell, DFC, CD 25 May 63 - 25 Jul 66.
W/C P.J. Roy, DFC, CD 26 Jul 66 - 31 Jan 68.
Higher Formations and Squadron Location
North American Air Defence Command,
RCAF Air Defence Command:
La Macaza, Que. 15 Sep 62 - 31 Jan 68.
Representative Weapon Lockheed CIM-10B Bomarc

No. 448 Squadron

Badge none
Formed as a Test unit at Cold Lake, Alberta on 20 June 1967, the squadron was responsible for the technical evaluation of new or modified aircraft armament and general systems. On 1 February 1968 the squadron was integrated into the Canadian Armed Forces.
Brief Chronology Formed at Cold Lake, Alta. 20 Jun 67. Integrated into CAF 1 Feb 68.
Commanders
W/C E.K. Fallis, CD 20 Jun 67 - 31 Jan 68.
Higher Formation and Squadron Location
Central Experimental and Proving Establishment:
Cold Lake, Alta. 20 Jun 67 - 31 Jan 68.
Representative Aircraft
Douglas Dakota Mk.1VM   10917
Sikorsky H-34A   9635
Canadair Silver Star Mk.3   21315   21404   21450
Canadair CF-104 Starfighter   12625   12702   12704   12705

444 Canadair Silver Star Mk.3 21315 at Malton, Ont. on 19 July 1966. This aircraft was attached to No. 448 (Test) Squadron during the latter half of 1967. The Central Experimental and Proving Establishment fin marking continued in use during this period.

445 Sikorsky H-34A helicopter 9633 of No. 448 (Test) Squadron based at Cold Lake, Alta.

446 Douglas Dakota Mk.IVM 10917 of No. 448 (Test) Squadron at Calgary, Alta. on 22 September 1967.

447 Canadair CF-104 12701 and a neat numerical sequence of No. 448 (Test) Squadron aircraft in formation from their Cold Lake, Alta. base.

Part Four
HIGHER FORMATIONS AND ANCILLARY UNITS
1939-1968

HEADQUARTERS

Air Force Headquarters

Badge In front of a circle inscribed with the motto PER ARDUA AD ASTRA and ensigned with the Imperial Crown, an eagle volant affronte the head to the sinister. Beneath the whole upon a scroll the words "Royal Canadian Air Force"
Motto Per ardua ad astra (Through adversity to the stars)
Authority King George VI, January 1943
The badge for the RCAF was based on that of the RAF and is very similar except for the head of the eagle to the sinister and a scroll below with the words 'Royal Canadian Air Force'. *The motto PER ARDUA AD ASTRA can be assigned to no author but is said to be derived from a line of Seneca AD ASTRA NULLA EST MOLLIS A TERRA VIA (There is no easy road from earth to heaven). The motto has been given several similar translations, the most preferable being "THROUGH ADVERSITY TO THE STARS" as it leaves one to imagine in English that the "stars" may stand for "glory," "heights," or "success."*
Formed as Canadian Air Force Headquarters at Ottawa, Ontario on 18 February 1920; redesignated Royal Canadian Air Force Headquarters on 1 April 1924; redesignated Air Force Headquarters 31 August 1939; ceased to exist on the formation of Canadian Forces Headquarters on 1 August 1964.[1]

Directors, CAF
A/C A.K. Tylee, OBE 17 May 20 - 21 Mar 21.
W/C R.F. Redpath 22 Mar 21 - 12 Jul 21.
W/C J.S. Scott, MC. AFC 13 Jul 21 - 30 Jun 22.
W/C J.L. Gordon, DFC 1 Jul 22 - 31 Mar 24.

Directors, RCAF
W/C W.G. Barker, VC, DSO, MC 1 Apr 24 - 18 May 24.
G/C J.S. Scott, MC, AFC 19 May 24 - 14 Feb 28.
W/C L.S. Breadner, DSC 15 Feb 28 - 29 Apr 32.
S/L A.A.L. Cuffe 30 Apr 32 - 31 Oct 32.

Senior Air Officers
G/C J.L. Gordon, DFC 1 Nov 32 - 31 May 33.
W/C G.O. Johnson, MC 1 Jun 33 -31 Dec 33.
A/V/M G.M. Croil, AFC 1 Jan 34 - 14 Dec 38.

Chiefs of the Air Staff[2]
A/V/M G.M. Croil, AFC 15 Dec 38 - 28 May 40.
A/M L.S. Breadner, CB, DSC 29 May 40 - 31 Dec 43.
A/M R. Leckie, CB, DSO, DSC, DFC 1 Jan 44 - 31 Aug 47.
A/M W.A. Curtis, CB, CBE, DSC, ED 1 Sep 47 - 31 Jan 53.
A/M C.R. Slemon, CB, CBE, CD 1 Feb 53 - 31 Aug 57.
A/M H. Campbell, CBE, CD 1 Sep 57 - 14 Sep 62.
A/M C.R. Dunlap, CBE, CD 15 Sep 62 - 31 Jul 64.

---

[1] The RCAF itself continued to exist until 1 February 1968 when the Canadian Forces Reorganization Act came into effect.
[2] Responsible directly to the Minister of National Defence — not, as in the case of the Senior Air Officer, through the Chief of the General Staff (Army).

RCAF Overseas Headquarters

Badge In front of a hurt an eagle volant carrying in the claw a sprig of maple
Motto Omni caelo (In every sky)
Authority King George VI, June 1944
The hurt, or blue disk, represents the sea or sky over which the eagle is carrying the maple of Canada.
Formed at London, England on 1 January 1940, the headquarters exercised administrative control over all RCAF personnel and units overseas and acted on behalf of the RCAF on all matters pertaining to their employment; disbanded on 22 July 1946.

Commanders[1]
W/C F.V. Heakes 1 Jan 40 - 6 Mar 40.
A/C G.V. Walsh, MBE 7 Mar 40 - 15 Oct 40.
A/C L.F. Stevenson 16 Oct 40 - 23 Nov 41.
A/M H. Edwards, CB 24 Nov 41 - 31 Dec 43.
A/M L.S. Breadner, CB, DSC 1 Jan 44 - 31 Mar 45 *ret*.
A/M G.O. Johnson, CB, MC 1 Apr 45 - 22 Jul 46.

---

[1] Originally "Officer Commanding, RCAF in Great Britain," the title underwent the following changes: 4 Jun 40: Air Officer Commanding, RCAF in Great Britain; 7 Nov 41: Air Officer in Chief, RCAF Overseas; 16 Jul 42: Air Officer Commanding in Chief, RCAF Overseas; 5 Feb 43: Air Officer Commanding in Chief, Headquarters, RCAF Overseas.

COMMANDS

Air Defence Command

Badge In front of two rays of lightning a long-tailed jaeger volant
Motto Detegere et destruere (To detect and to destroy)
Authority Queen Elizabeth II, November 1954
The jaeger is indicative of defending aircraft and the rays of lightnng represent radar and other electronics of the Command.
Formed as No. 1 Air Defence Group at Air Force Headquarters, Ottawa, Ontario on 1 December 1948; moved to St Hubert, Quebec on 1 November 1949 and declared operational on the 23rd; elevated to Air Defence Command on 1 June 1951; integrated into North American Air Defence (NORAD) Command on 12 September 1957; moved to North Bay, Ontario in August 1966; integrated into the Canadian Armed Forces on 1 February 1968.
Commanders
No. 1 Air Defence Group:
G/C W.R. MacBrien, OBE, CD 1 Dec 48 - 31 May 51.
Air Defence Command:
A/V/M C.R. Dunlap, CBE, CD 1 Jun 51 - 31 Aug 51.
A/V/M A.L. James, CBE, CD 1 Sep 51 - 31 Aug 54.
A/V/M L.E. Wray, OBE, AFC, CD 17 Jan 55 - 21 Aug 58.
A/V/M W.R. MacBrien, OBE, CD 22 Aug 58 - 31 Aug 62.
A/V/M M.M. Hendrick, OBE, CD 1 Sep 62 - 30 Aug 64 *ret.*
A/V/M M.D. Lister, CD 31 Aug 64 - 31 Mar 66.
A/V/M M.E. Pollard, DSO, DFC, AFC, CD 1 Apr 66 - 31 Jan 68.

448 *A typical radar site on the Canadian Pine Tree system located at Foymount, Ont. These stations were the most southern radar defences erected in the early 1950's.*

Air Materiel Command

Badge In front of a Terrestrial Globe supported by a gauntleted hand an Astral Crown
Motto Sustinemus (We support)
Authority Queen Elizabeth II, October 1964
The Terrestrial Globe supported by a gauntleted hand is to represent the support given by the Command to the RCAF across the world. The Astral Crown is to indicate Command status.
Formed as Maintenance Command at Uplands (Ottawa), Ontario on 1 October 1945; moved to No. 8 Temporary Building, Ottawa, Ontario on 15 March 1947; moved to Rockcliffe (Ottawa), Ontario and renamed Air Materiel Command on 1 April 1949; disbanded on 1 August 1965 on the formation of the Canadian Forces Materiel Command.
Commanders
A/V/M R.E. McBurney, CBE 1 Oct 45 - 23 Jun 46.
A/C F.R. Miller, CBE 24 Jun 46 - 4 Aug 48.
A/V/M R.E. McBurney, CBE 5 Aug 48 - 31 Dec 51 *ret.*
A/V/M H.B. Godwin, CBE, CD 1 Jan 52 - 19 Jul 55.
A/V/M J.L. Plant, CBE, AFC, CD 20 Jul 55 - 31 Aug 56 *ret.*
A/V/M R.C. Ripley, OBE, CD 1 Sep 56 - 1 Mar 57 *KIFA.*
A/V/M C.A. Cook, OBE, CD 2 Mar 57 - 24 Jul 58 *died.*
A/V/M C.L. Annis, OBE, CD 1 Jan 59 - 9 Sep 62.
A/V/M J.B. Millward, DFC, CD 10 Sep 62 - 1 Aug 65.

Air Training Command

Formed as Training Group at Camp Borden, Ontario on 1 April 1935; moved to Trenton, Ontario on 1 September 1937; elevated to Air Training Command on 15 September 1938; moved to Toronto, Ontario on 1 October 1938; disbanded on 1 January 1940 on the formation of No. 1 Training Command of the British Commonwealth Air Training Plan.
Commanders
G/C A.E. Godfrey, MC, AFC, VD 17 Oct 38 - 15 Dec 38.
A/C A.A.L. Cuffe 16 Dec 38 - 1 Jan 40.

Air Transport Command

Badge An albatross flying in front of a Terrestrial Globe
Motto Versatile and ready
Authority Queen Elizabeth II, December 1959
To symbolize the widespread operations of the Command over land and sea.
Formed within Air Force Headquarters as the Directorate of Air Transport Command at Ottawa, Ontario on 5 August 1943; redesignated No. 9 (Transport) Group and moved to Rockcliffe (Ottawa), Ontario on 5 February 1945 as an autonomous headquarters; elevated to Air Transport Command on 1 April 1948; moved to Lachine (Montreal), Quebec on 9 August 1951, then to Trenton, Ontario on 1 September 1959; integrated into the Canadian Armed Forces on 1 February 1968.
Commanders
Directorate of Air Transport Command:
G/C Z.L. Leigh, OBE 5 Aug 43 - 4 Feb 45.
No. 9 (Transport) Group:
G/C Z.L. Leigh, OBE 5 Feb 45 - 31 May 45.
A/C J.L. Plant, CBE 1 Jun 45 - 4 Feb 46.
A/C L.E. Wray, OBE, AFC 5 Feb 46 - 31 Mar 48.
Air Transport Command:
A/C L.E. Wray, OBE, AFC 1 Apr 48 - 15 Aug 48.
A/C A.D. Ross, GC, CBE, CD 16 Aug 48 - 3 Aug 51.
A/C R.C. Ripley, OBE, CD 4 Aug 51 - 17 Nov 53.
A/C H.M. Carscallen, DFC, CD 26 Jul 54 - 5 Aug 56.
A/C F.S. Carpenter, AFC, CD 6 Aug 56 - 26 Jun 61.
A/C R.J. Lane, DSO, DFC and Bar, CD 27 Jun 61 - 27 Dec 65.
A/C G.G. Diamond, AFC, CD 28 Dec 61 - 14 Apr 67 ret.
A/C A.C. Hull, DFC, CD 15 Apr 67 - 31 Jan 68.

449 Goose Bay, Labrador was opened during the Second World War to provide a mainland base for ferry flights. Spread over a high plateau, it offered long runways and unobstructed approaches to the multi-engine transports and bomber aircraft using the North Atlantic route to Europe.

Central Air Command

Formed at Trenton, Ontario on 1 March 1947 on the disbandment of No. 1 Air Command; was one of the two geographical air commands of the peacetime establishment, and controlled No. 10 Group (formerly Eastern Air Command) at Halifax, Nova Scotia; disbanded on 1 April 1949 on the formation of Training Command.
Commanders
A/V/M E.E. Middleton, CBE 1 Mar 47 - 1 Apr 49.

Eastern Air Command

Formed at Halifax, Nova Scotia on 15 November 1938; during the Second World War, it operated both as a command and as an operational group headquarters, as well as controlling No. 1 Group at St. John's, Newfoundland and No. 5 (Gulf) Group at Gaspé, Quebec; it also administered No. 12 (Operational Training) Group at Halifax; disbanded on 1 March 1947 on the formation of No. 10 Group of Central Air Command.
Commanders
A/V/M N.R. Anderson 17 Dec 38 - 17 Feb 42.
A/V/M A.A.L. Cuffe 18 Feb 42 - 13 Jan 43.
A/V/M G.O. Johnson, CB, MC 14 Jan 43 - 19 Mar 45.
A/V/M A.L. Morfee, CB, CBE 20 Mar 45 - 1 Mar 47.

Maritime Air Command

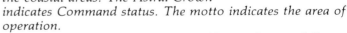

Badge Two Osprey heads conjoined and issuant from an Astral Crown
Motto Over the seas
Authority Queen Elizabeth II, June 1964
The Osprey like the Command ranges over a wide area including both coasts of Canada. The Osprey is a powerful bird with keen eyesight. It seldom misses its prey. The two heads shown issuant from an Astral Crown indicates watchfulness over the coastal areas. The Astral Crown indicates Command status. The motto indicates the area of operation.
Formed at Halifax, Nova Scotia as No. 10 Group of Central Air Command on 1 March 1947 on the disbandment of Eastern Air Command; redesignated Maritime Group on 1 April 1949 and became autonomous on 15 January 1951; elevated to Maritime Air Command on 1 June 1953; disbanded on 16 January 1966 on the formation of Canadian Forces Maritime Command.
Commanders
No. 10 Group, Central Air Command:
A/C F.G. Wait, CBE 31 May 47 - 31 Mar 49.
Maritime Group:
A/C F.G. Wait, CBE 1 Apr 49 - 12 Dec 49.
A/C R.C. Gordon, CBE, CD 24 Jan 50 - 8 Aug 51.
A/C A.D. Ross, GC, CBE, CD 9 Aug 51 - 31 May 53.
Maritime Air Command:
A/C A.D. Ross, GC, CBE, CD 1 Jun 53 - 6 Aug 54.
A/C M. Costello, CBE, CD 30 Aug 54 - 10 Jul 58.

A/C W.I. Clements, OBE, CD 11 Jul 58 - 31 Jul 63 *ret.*
A/C F.S. Carpenter, AFC, CD 1 Aug 63 - 30 Aug 65.
A/C R.A. Gordon, DSC, DFC, CD 13 Sep 65 - 16 Jan 66.

Maintenance Command
see Air Materiel Command

North West Air Command

Formed as the North West Staging Route at Edmonton, Alberta on 15 October 1942 under No. 4 Training Command of the British Commonwealth Air Training Plan; redesignated No. 2 Wing of Western Air Command on 1 January 1944; elevated to Command status and renamed North West Air Command on 1 June 1944; retained in the peacetime organization as one of the two geographical air commands, and controlled No. 11 Group at Winnipeg, Manitoba and No. 12 Group at Vancouver, British Columbia; was disbanded on 1 August 1951, when it was absorbed by Tactical Group (formerly No. 11 Group) to form Tactical Air Group (later Tactical Air Command).
Commanders
North West Staging Route:
W/C C.M.G. Farrell, DFC 15 Oct 42 - 31 Dec 42.
W/C W.J. McFarlane, 1 Jan 43 - 31 Dec 43.
G/C V.H. Petriarche, AFC 1 Jan 44 - 25 Feb 44.
No. 2 Wing:
G/C V.H. Petriarche, AFC 26 Feb 44 - 31 Mar 44.
North West Air Command:
A/V/M T.A. Lawrence, CB 1 Jun 44 - 15 Sep 46.
A/C R.C. Gordon, CBE 16 Sep 46 - 12 Mar 47.
A/V/M K.M. Guthrie, CB, CBE 13 Mar 47 - 31 Mar 49 *ret.*
A/V/M H.L. Campbell, CBE 1 Apr 49 - 3 Nov 49.
A/V/M C.R. Dunlap, CBE, CD 4 Nov 49 - 30 May 51.
A/C W.E. Bennett, CD 31 May 51 - 1 Aug 51.

Tactical Air Command

Badge A demi-polar bear issuant from an Astral Crown.
Motto Custos Borealis (Guardians of the North)
Authority Queen Elizabeth II, October 1954
The Astral Crown denotes a Command and the demi-polar bear is suggestive of its operational role within the northern boundaries of Canada.
Formed as No. 11 Group of North West Air Command at Winnipeg, Manitoba on 1 March 1947; redesignated Tactical Group on 1 April 1949; became autonomous on 15 January 1951; moved to Edmonton, Alberta and merged with North West Air Command on 1 August 1951 to form Tactical Air Group; elevated to Tactical Air Command on 1 June 1953; disbanded on 1 January 1959.
Commanders
No. 11 Group (later Tactical Group):
A/C M. Costello, CBE, CD 8 Mar 47 - 23 May 51.
Tactical Air Group:
A/C W.E. Bennett, CD 1 Aug 51 - 27 Nov 51.

A/C J.G. Kerr, CBE, AFC, CD 28 Feb 52 - 17 Dec 52.
A/C S.W. Coleman, CD 18 Dec 52 - 31 May 53.
Tactical Air Command:
A/C S.W. Coleman, CD 1 Jun 53 - 8 Nov 55.
G/C H.G. Richards, OBE, CD 20 Nov 55 - 15 Nov 58.

Training Command

Badge Issuant from an Astral Crown two torches in saltire
Motto Exercendum usque ad optimum (One must train up to the highest standard)
Authority Queen Elizabeth II, July 1962
The torches are introduced to suggest learning and the Astral Crown leadership and success.
Formed at Trenton, Ontario on 1 April 1949 on the disbandment of Central Air Command; controlled No. 14 (Training) Group at Winnipeg, Manitoba; moved to Winnipeg on 11 September 1958 and absorbed No. 14 Group; disbanded on 15 January 1966 on the formation of Canadian Forces Training Command.
Commanders
A/V/M E.E. Middleton, CBE 1 Apr 49 - 31 Aug 49 *ret.*
A/V/M C.R. Slemon, CB, CBE, CD 1 Sep 49 - 11 Jan 53.
A/V/M J.G. Kerr, CBE, AFC, CD 12 Jan 53 - 31 Aug 55.
A/V/M J.G. Bryans, CBE, CD 1 Sep 55 - 18 Nov 60 *ret.*
A/V/M H.M. Carscallen, DFC, CD 19 Nov 60 - 25 Aug 63 *ret.*
A/V/M C.H. Greenway, OBE, CD 26 Aug 63 - 26 Sep 65.
A/C C.W. Burgess, DFC, CD 27 Sep 65 - 15 Jan 66.

Western Air Command

Formed at Vancouver, British Columbia on 1 March 1938; moved to Victoria, British Columbia on 25 November 1939 and returned to Vancouver on 1 January 1943; during the Second World War, operated as both a command and an operational group headquarters, as well as controlling No. 4 Group at Prince Rupert, British Columbia and, for a short time, No. 2 Group at Victoria; disbanded on 1 March 1947 on the formation of No. 12 Group of North West Air Command.
Commanders
A/C G.O. Johnson, MC 5 Apr 38 - 20 Oct 39.
A/C A.E. Godfrey, MC, AFC 21 Oct 39 - 31 Dec 41.
A/V/M L.F. Stevenson, CB 1 Jan 42 - 9 Jun 44.
A/V/M F.V. Heakes, CB 10 Jun 44 - 13 Feb 46 *ret.*
A/V/M J.L. Plant, CBE, AFC 14 Feb 46 - 1 Mar 47.

No. 1 Air Command

Formed at Trenton, Ontario on 15 January 1945 by the merger of No. 1 and No. 3 Training Command of the British Commonwealth Air Training Plan, as a temporary organization to cover the closing down of the BCATP and the establishment of the peacetime command organization; disbanded on 1 March 1947 on the formation of Central Air Command.

Commanders
A/V/M A. Raymond, CBE 15 Jan 45 - 3 Jul 45 *ret*.
A/V/M E.E. Middleton, CBE 4 Jul 45 - 1 Mar 47.

No. 2 Air Command

Badge Between two wheat sheaves a bison's head caboshed
Motto Servituri patriae (To serve our country)
Authority King George VI, May 1947
The badge is symbolic of the prairie provinces over which units of this Command operated.
Formed at Winnipeg, Manitoba on 1 December 1944 by the merger of No. 2 and No. 4 Training Command of the British Commonwealth Air Training Plan as a temporary organization to cover the closing down of the BCATP and the establishment of the peacetime command organization; disbanded on 1 March 1947 on the formation of No. 11 Group of North West Air Command.

Commanders
A/V/M K.M. Guthrie, CB, CBE 1 Dec 44 - 1 Mar 47.

AIR DIVISIONS

No. 1 Air Division Europe

Badge In front of four maple leaves with stems joined to form a cross two sabres in saltire
Motto Ad custodiendam Europam (For the defence of Europe)
Authority Queen Elizabeth II, June 1962
The Division's four wings are represented by the maple leaves and the sabres are to signify the Division's fighting power and also have reference to the Sabre aircraft with which the wings of the unit were originally equipped.
Formally constituted at Paris, France on 1 October 1952 as an operational command of the North Atlantic Treaty Organization's Allied Command Europe; moved to Metz, France on 13 April 1953, then to Lahr, Germany on 1 April 1967; integrated into Canadian Armed Forces on 1 February 1968.

Commanders
A/V/M H.L. Campbell, CBE, CD 11 Dec 52 - 3 Aug 55.
A/V/M H.B. Godwin, CBE, CD 4 Aug 55 - 5 Aug 58.
A/V/M L.E. Wray, OBE, AFC, CD 4 Sep 58 - 17 Jul 63 *ret*.
A/V/M D.A.R. Bradshaw, DFC, CD 18 Jul 63 - 8 Jul 66.
A/V/M R.J. Lane, DSO, DFC and Bar, CD 9 Aug 66 - 31 Jan 68.

Higher Formations and Air Division Locations
Air Force Headquarters (for adm),
Allied Command Europe (for ops),
Allied Air Forces Central Europe:
Fourth Allied Tactical Air Force,
Paris, Fr. 1 Oct 52 - 12 Apr 53.
Metz, Fr. 13 Apr 53 - 31 Mar 67.
Lahr, Ger. 1 Apr 67 - 31 Jan 68.

No. 5 Air Division

Formed as No. 12 Group of North West Air Command at Comox, British Columbia on 1 March 1947; redesignated No. 12 Air Defence Group and reassigned to Air Defence Command on 1 July 1951; redesignated No. 5 Air Division on 1 September 1955; transferred to Maritime Air Command (Pacific) on 15 October 1962; disbanded on 31 December 1963.

Commanders
No. 12 Group:
A/C J.L. Plant, CBE, AFC 1 Mar 47 - 5 Nov 47.
W/C E.H. Evans 6 Nov 47 - 26 Mar 48.
G/C Z.L. Leigh, OBE, ED 27 Mar 48 - 1 Sep 49.
A/C J.A. Easton, OBE, CD 2 Sep 49 - 31 Jun 51.
No. 12 Air Defence Group:
A/C J.A. Easton, OBE, CD 1 July 51 - 16 Aug 51.
A/C R.C. Gordon, CBE, CD 17 Aug 51 - 21 Jan 53.
A/C W.A. Orr, CBE, CD 22 Jan 53 - 31 Aug 55.
No. 5 Air Division:
A/C W.A. Orr, CBE, CD 1 Sep 55 - 16 Aug 56.
A/C A.D. Ross, GC, CBE, CD 17 Aug 56 - 31 Mar 60 *ret*.
A/C G.G. Truscott, OBE, CD 1 Jul 60 - 18 Aug 63.
A/C R.C. Weston, CD 19 Aug 63 - 31 Dec 63.

Higher Formations and Air Division Location
North West Air Command,
Air Defence Command (1 Jul 51),
Maritime Air Command (Pacific) (15 Oct 62):
Comox, B.C. 1 Mar 47 - 31 Dec 63.

GROUPS

No. 1 Group
Eastern Air Command

Formed at St. John's, Newfoundland on 10 July 1941 to control anti-submarine air operations in the Northwest Atlantic area and the air defence of Newfoundland and Labrador; disbanded on 30 June 1945.

Commanders
A/C C.M. McEwen, MC, DFC 15 Aug 41 - 21 Dec 42.
A/V/M F.V. Heakes 22 Dec 42 - 30 Mar 44.
A/V/M A.L. Morfee, CBE 31 Mar 44 - 26 Mar 45.

A/C F.G. Wait 27 Mar 45 - 30 Jun 45.
Higher Formation and Group Location
Eastern Air Command:
St. John's, Nfld., 10 Jul 41 - 30 Jun 45.

No. 1 Air Defence Group
see Air Defence Command

No. 1 Group (Auxiliary)
Montreal, Quebec

Formed as RCAF (Reserve) Group Montreal on 15 January 1951; redesignated No. 1 Group (Reserve) on 1 June 1951, and No. 1 Group (Auxiliary) on 1 September; performed the functions of No. 11 (Operational) Wing (Auxiliary) and No. 12 (Technical Training) Wing (Auxiliary), both of which had been authorized but not formed; disbanded on 15 April 1957.
Commanders
G/C L.G.G.J. Archambault, AFC, CD 1 Feb 51 - 16 Aug 53.
G/C F.R.C. Carling-Kelly, AFC, CD 17 Aug 53 - 1 Sep 55.
G/C J.A.D.B. Richer, DFC, CD 2 Sep 55 - 15 Mar 57.
Higher Formation and Group Location
Air Defence Command:
Montreal, Que. 15 Jan 51 - 15 Apr 57.

No. 2 Group
Western Air Command

Formed at Victoria, British Columbia on 1 January 1943 as a temporary formation to cover the move of Western Air Command Headquarters from Victoria to Vancouver; disbanded on 15 March 1943.
Commanders
A/C E.L. McLeod 1 Jan 43 - 15 Mar 43.
Higher Formation and Group Location
Western Air Command:
Victoria, B.C. 1 Jan 43 - 15 Mar 43.

No. 2 Group (Auxiliary)
Toronto, Ontario

Formed as RCAF (Reserve) Group Toronto on 15 January 1951; redesignated No. 2 Group (Reserve) on 1 August 1951, then No. 2 Group (Auxiliary) on 1 September; controlled No. 14 (Operational) Wing (Auxiliary) and No. 15 (Technical Training) Wing (Auxiliary); disbanded on 1 March 1957.
Commanders
G/C G.A. Hiltz, AFC, CD 22 Jan 51 - 17 Jan 53.
G/C V.H. Patriarche, OBE, AFC, ED 18 Jan 53 - 6 Sep 54.
G/C Z.L. Leigh, OBE, ED 7 Sep 54 - 12 Feb 57.
Higher Formations and Group Location
Training Command,
Air Defence Command (1 Aug 51):
Toronto, Ont. 15 Jan 51 - 1 Mar 57.

No. 4 Group
Western Air Command

Formed at Prince Rupert, British Columbia on 16 June 1942 to provide administrative and operational control of RCAF units in northern British Columbia and the Yukon; disbanded on 1 April 1944.
Commanders
G/C R.C. Gordon 16 Jun 42 - 11 Jun 43.
G/C R.H. Foss 12 Jun 43 - 1 Apr 44.
Higher Formation and Group Location
Western Air Command:
Prince Rupert, B.C. 16 Jun 42 - 1 Apr 44.

No. 5 (Gulf) Group
Eastern Air Command

Formed at Gaspé, Quebec on 1 May 1943 to integrate and control air operations over the Gulf of St. Lawrence during the shipping season; inactive from 15 November 1943 to 30 April 1944, with headquarters located at Eastern Air Command Headquarters, Halifax, Nova Scotia; disbanded on 15 November 1944.
Commanders
G/C W.A. Orr, OBE 5 May 43 - 31 Oct 43.
W/C F.J. Ewart 1 Nov 43 - 15 Nov 44.
Higher Formation and Group Locations
Eastern Air Command:
Gaspé, Que. 1 May 43 - 14 Nov 43.
Halifax, N.S. (Group inactive) 15 Nov 43 - 30 Apr 44.
Gaspé, Que. 1 May 44 - 15 Nov 44.

No. 6 (RCAF) Group
RAF Bomber Command

Badge A maple leaf superimposed on a York rose
Motto Sollertia et ingenium (Initiative and skill)
Authority King George VI, October 1946
The York rose symbolizes the association with Yorkshire where the group was formed.
Formed at Linton-on-Ouse, Yorkshire, England on 25 October 1942 and moved to permanent quarters at Allerton Park, east of Knaresborough, on 6 December; declared operational at 0001 hours on 1 January 1943; transferred to RCAF's Eastern Air Command on 14 July 1945 to reorganize and train for service in the Pacific as part of RAF "Tiger Force"; disbanded on 1 September 1945.
Commanders
A/V/M G.E. Brookes, OBE 25 Oct 42 - 28 Feb 44.
A/V/M C.M. McEwen, CB, MC, DFC 29 Feb 44 - 13 Jul 45.
Main Headquarters (Halifax, N.S.):
A/C J.G. Kerr, AFC 14 Jul 45 - 1 Sep 45*.
Rear Headquarters (Allerton Park):
A/C J.L. Hurley 14 Jul 45 - 1 Sep 45.

Higher Formations and Group Locations
RAF Bomber Command:
Linton-on-Ouse, Yorks 25 Oct 42 - 5 Dec 42.
Allerton Park, Yorks 6 Dec 42 - 13 Jul 45.
(Rear Headquarters until 1 Sep 45)
RAF "Tiger Force",
RCAF Eastern Air Command (for training):
Halifax, N.S. 14 Jul 45 - 1 Sep 45.
Operational History: First Mission 3/4 January 1943, 6 Wellingtons of No. 427 (B) Squadron laid mines off the Frisian Islands. **First Bombing Mission** 13/14 January 1943, 14 Wellingtons despatched to bomb Lorient, France; 11 bombed the primary target, 2 returned early, 1 failed to return. **Maximum Mission** 6/7 October 1944, 293 Lancasters and Halifaxes despatched to bomb Dortmund, Germany; 273 bombed the primary target, 3 bombed the alternative, 15 dropped no bombs, 2 failed to return. **Last Mission** 25 April 1945, 102 Lancasters and 92 Halifaxes, with 160 aircraft from No. 4 Group, bombed gun positions on the Island of Wangerooge; 1 Lancaster and 2 Halifaxes failed to return. **Summary** Sorties: 40,822. Operational Flying Hours: 271,981. Bombs dropped: 126,122 tons (including mines). Victories: 116 aircraft destroyed, 24 probably destroyed, 92 damaged. Casualties: 814 aircraft; 3500-plus aircrew killed or presumed dead.

*Although the headquarters was being set up at Halifax, A/C Kerr operated from Ottawa during this period.

450 No. 6 (RCAF) Group Headquarters at Allerton Park, Yorks., England. Operations were planned at Allerton Hall from 1 January 1943.

No. 9 (Transport) Group
see Air Transport Command

No. 10 Group
see Maritime Air Command

No. 11 Group
see Tactical Air Command

No. 12 Group
see No. 5 Air Division

No. 12 (Operational Training) Group
Eastern Air Command

Formed at Halifax, Nova Scotia on 22 July 1942 to administer and control operational training units in Eastern Air Command; disbanded on 14 January 1945.
Commanders
A/C L.L. MacLean 22 Jul 42 - 4 May 43.
A/C W.J. Seward, CBE 5 May 43 - 30 Sep 44.
G/C G.T. Richardson 1 Oct 44 - 14 Jan 45.
Higher Formation and Group Location
Eastern Air Command:
Halifax, N.S. 22 Jul 42 - 14 Jan 45.

No. 14 (Training) Group
Training Command

Formed at Winnipeg, Manitoba on 1 August 1951 to administer and control training facilities reactivated in the Prairie Provinces to train North Atlantic Treaty Organization aircrew; absorbed into Training Command Headquarters when the latter moved to Winnipeg on 11 September 1959.
Commanders
A/C J.G. Bryans, CBE, CD 1 Aug 51 - 31 Aug 55.
A/C H.H.C. Rutledge, OBE, CD 1 Sep 55 - 11 Sep 59.
Higher Formation and Group Location
Training Command:
Winnipeg, Man. 1 Aug 51 - 11 Sep 59.

SECTORS

No. 17 Sector
RAF Second Tactical Air Force

Formed as No. 17 (Fighter) Wing at Headcorn, Kent, England on 4 July 1943 to administer and control No. 126 and No. 127 Airfield, plus No. 144 Airfield as of 21 April 1944; redesignated No. 17 Sector at Kenley, Surrey on 15 May 1944 when airfields became wings; disbanded at Crépon, France on 13 July 1944 on the reorganization of Second Tactical Air Force.
Commanders
G/C W.R. MacBrien 4 Jul 43 - 13 Jul 44.
Higher Formations and Sector Locations
Tactical Air Force Fighter Command, renamed Second Tactical Air Force (15 Nov 43):
No. 83 (Composite) Group,
Headcorn, Kent 4 Jul 43 - 13 Oct 43.
Kenley, Surrey 14 Oct 43 - 5 Jun 44.
En route to Europe, 6 Jun 44 - 13 Jun 44.
B.(Base) 2 Crépon, Fr. 14 Jun 44 - 13 Jul 44.

No. 22 Sector
RAF Second Tactical Air Force

Formed as No. 22 (Fighter) Wing at Ayr, Scotland on 9 January 1944 to administer and control No. 143 and No. 144 Airfield; reorganized as a Fighter Bomber formation on 16 April 1944, when it absorbed No. 16 (RAF) Wing along with No. 121 and No. 124 (RAF) Airfield, (No. 144 Airfield was transferred to No. 17 (Fighter) Wing); redesignated No. 22 Sector at Hurn, Hampshire, England on 15 May 1944 when airfields became wings; disbanded at Camilly, France on 13 July 1944 on the reorganization of Second Tactical Air Force.
Commanders
G/C P.Y. Davoud, DSO, DFC 9 Jan 44 - 13 Jul 44.
Higher Formations and Sector Locations
Air Defence Great Britain:
No. 12 Group,
Ayr, Scot. 9 Jan 44 - 20 Feb 44.
Digby, Lincs. 21 Feb 44 - 16 Mar 44.
Hurn, Hants. 17 Mar 44.
Second Tactical Air Force:
No. 83 (Composite) Group,
Hurn, Hants. 18 Mar 44 - 25 Mar 44.
Westhampnett, Sussex 26 Mar 44 - 15 Apr 44.
Hurn, Hants. 16 Apr 44 - 5 Jun 44.
En route to Europe 6 Jun 44 - 15 Jun 44.
B.(Base) 5 Le Fresne, Fr. 16 Jun 44 - 13 Jul 44.

BASES

No. 61 (Training) Base
No. 6 (RCAF) Group

Formed as Topcliffe Operational Base with headquarters at Topcliffe, Yorkshire, England on 1 March 1943, controlling RCAF Stations Topcliffe, Dishforth and Dalton; redesignated No. 6 (RCAF) Group Training Base on 30 April 1943, then No. 61 (Training) Base on 16 September, when it added RCAF Station Wombleton; transferred to No. 7 (Training) Group and renumbered No. 76 (RCAF) Training Base on 9 November 1944; disbanded on 1 September 1945.
Commanders
A/C C.M. McEwen, MC, DFC 5 Apr 43 - 25 Jun 43.
A/C B.F. Johnson 26 Jun 43 - 16 Feb 44.
A/C R.E. McBurney, AFC 17 Feb 44 - 15 May 44.
A/C F.G. Wait 16 May 44 - 7 Aug 44.
A/C J.L. Hurley 1 Sep 44 - 18 Sep 44.
A/C F.R. Miller 19 Sep 44 - 12 Jan 45.
A/C J.G. Kerr, AFC 13 Jan 45 - 30 May 45.
A/C N.W. Timmerman, DSO, DFC 1 Aug 45 - 1 Sep 45.
Higher Formations and Base Location
Bomber Command:
No. 6 (RCAF) Group,
No. 7 (Training) Group (9 Nov 44),
Topcliffe, Yorks. 1 Mar 43 - 1 Sep 45.

No. 62 "Beaver" (Operational) Base
No. 6 (RCAF) Group

Badge On a rock a beaver gnawing a log
Motto Ad opus diligenter (To the task with diligence)
Authority King George VI, September 1944

Formed as Linton-on-Ouse Operational Base with headquarters at Linton-on-Ouse, Yorkshire, England on 1 June 1943, controlling RCAF Stations Linton-on-Ouse, East Moor and Tholthorpe; redesignated No. 62 "Beaver" (Operational) Base on 6 October 1943; disbanded on 15 July 1945 when its units and establishment were transferred to RCAF Eastern Air Command as part of RAF "Tiger" Force for duty in the Pacific.
Commanders
A/C C.M. McEwen, MC, DFC 18 Jun 43 - 28 Feb 44.
A/C A.D. Ross, GC 29 Feb 44 - 27 Jun 44.
A/C J.E. Fauquier, DSO and 2 Bars, DFC 28 Jun 44 - 18 Sep 44.
A/C J.L. Hurley 19 Sep 44 - 30 May 45.
A/C J.G. Kerr, AFC 31 May 45 - 15 Jul 45.
Higher Formation and Base Location
Bomber Command:
No. 6 (RCAF) Group,
Linton-on-Ouse, Yorks. 1 Jun 43 - 15 Jul 45.

No. 63 (Operational) Base
No. 6 (RCAF) Group

Formed with headquarters at Leeming, Yorkshire, England on 1 May 1944, controlling RCAF Stations Leeming and Skipton-on-Swale; disbanded on 31 August 1945.
Commanders
A/C J.G. Bryans 1 May 44 - 12 Jan 45.
A/C F.R. Miller 13 Jan 45 - 25 May 45.
A/C J.L. Hurley 30 May 45 - 13 Jul 45.
Higher Formation and Base Location
Bomber Command:
No. 6 (RCAF) Group,
Leeming, Yorks. 1 May 44 - 31 Aug 45.

No. 64 (Operational) Base
No. 6 (RCAF) Group

Formed with headquarters at Middleton St. George, Durham, England on 1 May 1944, controlling RCAF Stations Middleton St. George and Croft; disbanded on 15 June 1945, when its units and establishment were transferred to RCAF Eastern Air Command as part of RAF "Tiger" Force for duty in the Pacific.
Commanders
A/C R.E. McBurney 1 May 44 - 28 Dec 44.
A/C C.R. Dunlap, CBE 22 Jan 45 - 24 Apr 45.
A/C H.B. Godwin 25 Apr 45 - 29 May 45.
A/C H.T. Miles 30 May 45 - 15 Jun 45.

Higher Formations and Base Location
Bomber Command:
No. 6 (RCAF) Group,
Middleton St. George, Durham 1 May 44 - 15 Jun 45.

WINGS

Canadian Digby Wing
RAF Fighter Command

Formed at Digby, Lincolnshire, England on 14 April 1941; inactive from 1 May 1942 to 18 April 1943; disbanded on 30 April 1944 when its squadrons were transferred to Second Tactical Air Force.
Wing Commanders Flying
W/C G.R. McGregor, DSO, DFC 14 Apr 41 - 31 Aug 41.
W/C H.P. Blatchford (Can/RAF), DFC 8 Sep 41 - 30 Apr 42.
Wing inactive
W/C L.S. Ford, DFC and Bar 19 Apr 43 - 4 Jun 43 *KIA*.
W/C L.V. Chadburn, DSO and Bar, DFC 5 Jun 43 - 30 Dec 43.
W/C N.H. Bretz, DFC 31 Dec 43 - 30 Apr 44.
Higher Formations and Wing Location
Fighter Command renamed,
Air Defence Great Britain (15 Nov 43):
No. 12 Group, Digby, Lincs. *(Satellite airfield at Wellingore)* 14 Apr 41 - 30 Apr 42 and 19 Apr 43 - 30 Apr 44.
Operational History: First Mission 15 April 1941, 12 Hurricane I's of No. 402 Squadron led by W/C McGregor, and supported by the Wittering Wing (Spitfires of No. 65 and No. 266 Squadron RAF) — fighter sweep over the Boulogne area. **First Victory** 26 June 1941, 2 Hurricane IIB's of No. 402 Squadron — convoy patrol; Sgt G.D. Robertson credited with a Ju.88 damaged. This was also the squadron's first victory. 13 October 1941, 36 Spitfire IIA's from No. 411, No. 412 and No. 266 Squadron (RAF) — fighter sweep Boulogne to Hardelot; engaged a mixed gaggle of Bf.109's and Fw.190's. P/O R.W. McNair of No. 411 Squadron credited with a Bf.109 destroyed and a second probably destroyed. McNair himself was shot down but bailed out over the Channel and was rescued. Sgt. E.N. MacDonald of No. 412 Squadron credited with a Bf.109 destroyed. Each aircraft "destroyed" was the first confirmed victory for the respective squadron. **Last Mission** 31 March 1944, 12 Spitfire VB's from No. 402 Squadron and 13 from No. 64 Squadron (RAF) — "Roadstead No. 44", Part I, close escort to 36 Beaufighters of Coastal Command detailed to attack an enemy convoy of six motor vessels known to have left Den Helder; no shipping sighted. **Summary** Sorties: 7560. Operational/Non-operational Flying Hours: 9394/19,403. Victories: 38 aircraft destroyed, 18 probably destroyed, 35 damaged. Casualties (June 1943 — April 1944): Operational: 6 aircraft; 6 pilots killed.

Canadian Kenley Wing
RAF Fighter Command

Formed at Kenley, Surrey, England on 25 November 1942; disbanded on 4 July 1943 when its squadrons were transferred to Second Tactical Air Force.
Wing Commanders Flying
W/C J.C. Fee, DFC 25 Nov 42 - 17 Jan 43 *KIA*.
W/C K.L.B. Hodson, DFC 22 Jan 43 - 28 Feb 43 *OTE*.
W/C J.E. Johnson (RAF), DSO and Bar, DFC and Bar 21 Mar 43 - 4 Jul 43.
Higher Formations and Wing Location
Fighter Command,
Tactical Air Force Fighter Command (12 Jun 43):
No. 11 Group,
Kenley, Surrey *(satellite airfield at Redhill)* 25 Nov 42 - 4 Jul 43.
Operational History: First Mission 26 November 1942, 2 Spitfire VB's of No. 401 Squadron — patrol of Shoreham-Beachy Head. **First Victory** 4 December 1942, 24 Spitfire VB's from No. 401 and No. 402 Squadron, led by W/C Fee — fighter sweep Audruicq-Gravelines as withdrawal wing, engaged enemy aircraft in the Guines area (between Ambleteuse and Audruicq). No. 402 Squadron claimed 1 Fw.190 destroyed by P/O N.A. Keene, 1 probably destroyed and 1 damaged by F/O H.A. Simpson; 2 pilots missing, 1 wounded. No. 401 Squadron — no claims, no losses. **Last Mission** 4 July 1943, 24 Spitfires from No. 401 and No. 411 Squadron, led by S/L B.D. Russel — "Ramrod No. 124", Part I, close escort to 12 Mitchells bombing Amiens. Ten to fifteen miles south of Abbeville, Blue Section of No. 411 Squadron was bounced by five Bf.109's, and over the target Red Section was also bounced by Bf.109's. No. 411 Squadron claimed 1 Bf.109 damaged by F/O D.R. Matheson, reported 1 pilot missing. 24 Spitfires from No. 403 and No. 421 Squadron, led by S/L H.C. Godefroy — "Ramrod No. 124", Part II, second fighter sweep (Somme — Amiens Estuary — Hardelot); sighted enemy aircraft, no engagement. **Summary** Sorties: 5936. Operational/Non-operational Flying Hours: 7996/6217. Victories: 96 aircraft destroyed, 6 probably destroyed, 54 damaged. Casualties: Wing diary does not systematically record casualties of squadrons under command.

No. 1 Wing
No. 1 Air Division Europe

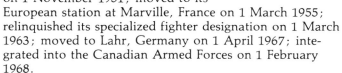

Badge A stone arrowhead point upwards in bend
Motto Pro pace armati (Armed for peace)
Authority Queen Elizabeth II, October 1954
The Indian arrowhead is to symbolize the unit's place of origin and its desire to be in forefront of any operation in which it might be able to participate.
Formed as No. 1 (Fighter) Wing at North Luffenham, Rutland, England on 1 November 1951; moved to its European station at Marville, France on 1 March 1955; relinquished its specialized fighter designation on 1 March 1963; moved to Lahr, Germany on 1 April 1967; integrated into the Canadian Armed Forces on 1 February 1968.
Commanders
G/C E.B. Hale, DFC, CD 1 Nov 51 - 19 Aug 53.
G/C J.D. Sommerville, DSO, DFC, CD 20 Aug 53 - 24 Jul 56.
G/C D.J. Williams, DSO, DFC, CD 25 Jul 56 - 18 Jun 59.
G/C D.P. Hall, CD 19 Jun 59 - 1 Aug 63.
G/C A.F. Avant, DSO, DFC, CD 2 Aug 63 - 9 Aug 66.
G/C R.G. Christie, CD 10 Sep 66 - 31 Jan 68.
Higher Formation and Wing Locations
No. 1 Air Division Europe:
North Luffenham, Rutland, Eng.[1] 1 Nov 51 - 28 Feb 55.
Marville, Fr. 1 Mar 55 - 31 Mar 67.
Lahr, Ger. 1 Apr 67 - 31 Jan 68.

[1] While in England, the wing was under the operational command of RAF Fighter Command through its No. 11 Group.

451 Canadair Sabre Mk.2 19188 photographed at Keflavik, Iceland en route to Europe as part of Operation "Leapfrog I" in June 1952.

452 Canadair Sabres cover the deck of the RCN aircraft carrier Magnificent *during the transfer of aircraft from Canada to Europe as part of the establishment of the Air Division.*

No. 2 Wing
Western Air Command
see North West Air Command

No. 2 Wing
No. 1 Air Division Europe

Badge: none
Formed as No. 2 (Fighter) Wing at Grostenquin, France on 1 October 1952; relinquished its specialized fighter designation on 1 March 1963; disbanded on 1 August 1964.
Commanders
G/C J.K.F. MacDonald, DFC, CD 1 Oct 52 - 27 Nov 52.
G/C M.E. Pollard, DSO, DFC, AFC, CD 28 Nov 52 - 25 May 53.
G/C W. Weiser, MBE, DFC, CD 30 Jun 53 - 28 Sep 55.
G/C W.F.M. Newson, DSO, DFC and Bar, CD 18 Oct 55 - 8 Jul 57.
G/C A.B. Searle, AFC, CD 9 Jul 57 - 22 Jul 61.
G/C R.E. MacBride, DFC, CD 23 Aug 61 - 29 Jul 63 *died.*
G/C E.R. Johnston, AFC, CD 25 Nov 63 - 1 Aug 64.
Higher Formation and Wing Location
No. 1 Air Division Europe:
Grostenquin, Fr. 1 Oct 52 - 1 Aug 64.

453 No. 2 (Fighter) Wing prepares to depart for Europe at Uplands, Ont. on 23 September 1952. Canadair Sabres of Nos. 416, 421 and 430 Squadrons line the taxi strips during the final inspection.

No. 3 Wing
No. 1 Air Division Europe

Badge The German eagle charged with three maple leaves
Motto Agmen primum libertatis (Freedom's vanguard)
Authority Queen Elizabeth II, April 1956
This wing was the first Canadian wing to be stationed in Germany in support of the North Atlantic Treaty Organization. The German eagle, charged with the maple leaves, is to symbolize this link with Germany after the Second World War.
Formed as No. 3 (Fighter) Wing at Zweibrucken, Germany on 2 February 1953; relinquished its specialized fighter designation on 1 March 1963; integrated into the Canadian Armed Forces on 1 February 1968.
Commanders
G/C A.C. Hull, DFC, CD 16 Feb 53 - 9 Jan 56.
G/C J.K.F. MacDonald, DFC, CD 2 May 56 - 6 Jul 60.
G/C V.L. Berg, CD 7 Jul 60 - 31 Jul 63.
G/C D.C. Laubman, DFC and Bar, CD 1 Aug 63 - 3 Aug 66.
G/C K.E. Lewis, CD 4 Aug 66 - 31 Jan 68.
Higher Formation and Wing Location
No. 1 Air Division Europe:
Zweibrucken, Ger. 2 Feb 53 - 31 Jan 68.

No. 4 Wing
No. 1 Air Division Europe

Badge Three lances with black and white pennants in front of an ogress with a Fesse wavy
Motto Auf wacht (On guard)
Authority Queen Elizabeth II, March 1961
The three lances represent the three squadrons which form the operational forces of the wing and symbolize the ability to carry out its duties by day and night. The ogress and Fesse wavy signify the Black Forest and the Rhine River area over which the unit is on guard.
Formed as No. 4 (Fighter) Wing at Baden-Soellingen, Germany on 1 July 1953; relinquished its specialized fighter designation on 1 March 1963; integrated into the Canadian Armed Forces on 1 February 1968.
Commanders
G/C R.S. Turnbull, DFC, AFC, DFM, CD 8 Jul 53 - 15 Apr 55.
G/C B.E. Christmas, CD 16 Apr 55 - 23 Jul 57.
G/C R.W. McNair, DSO, DFC and 2 Bars, CD 30 Aug 57 - 14 Sep 61.
G/C J.J. Jordan, AFC, CD 15 Sep 61 - 26 Jul 65.
G/C C. Allison, CD 18 Aug 65 - 31 Jan 68.
Higher Formation and Wing Location
No. 1 Air Division Europe:
Baden-Soellingen, Ger. 1 Jul 53 - 31 Jan 68.

454 Canadair Silver Star Mk.3 21614, in camouflage finish, with No. 4 Wing based at Baden-Soellingen, Germany.

455 Beechcraft Expeditor Mk.3TM 1533, at Odiham, Hants. on 6 September 1966, carries the identification "4 Wing" under the centre window.

456 Canadair CF-104 12751, with drag 'chute deployed on landing, carries the crest of No. 4 Wing under the cockpit, indicating service with either No. 421 or No. 444 (ST/A) Squadron.

No. 7 (Photographic) Wing
Air Transport Command

Formed at Rockcliffe (Ottawa), Ontario on 20 May 1944; renumbered No. 22 (Photographic) Wing on 1 April 1947; on completion of the major portion of the RCAF's postwar commitment to photograph all of Canada, was disbanded on 30 November 1949.
Commanders
No. 7 (P) Wing
W/C J.A.E. Schwartz 20 May 44 - 11 Jan 45.
W/C S Volk 12 Jan 45 - 31 Mar 47.
No. 22 (P) Wing
W/C S. Volk 1 Apr 47 - 3 May 47.
W/C R.I. Thomas, AFC 4 May 47 - 30 Nov 49.
Higher Formations and Wing Locations
Air Force Headquarters,
Air Transport Command (1 Apr 46):
Rockcliffe, Ont. 20 May 44 - 30 Nov 49.

No. 11 Wing (Auxiliary)
Montreal, Quebec

Authorized as RCAF (Reserve) Wing Montreal on 1 October 1950, its functions were performed by No. 1 Group (Auxiliary) until 15 April 1957, when that headquarters was disbanded, and subsequently by Air Defence Command staff; wing was formed on 1 September 1961 to administer and control No. 401 and No. 438 Squadrons; integrated into the Canadian Armed Forces on 1 February 1968 as No. 11 Air Reserve Wing.
Commanders
G/C H.J. Everard, DFC, CD 1 Sep 61 - 30 Sep 64 ret.
G/C J.E.M.M. Gauthier, CD 1 Oct 64 - 15 Nov 66 ret.
G/C J.D. Fisher, CD 1 Sep 67 - 31 Jan 68.
Higher Formation and Wing Location
Air Transport Command:
Montreal, Que. 1 Sep 61 - 31 Jan 68.

No. 12 (Technical Training) Wing (Auxiliary)
Montreal, Quebec

Authorized as RCAF (Reserve) Technical Training Wing Montreal on 15 March 1951, and redesignated No. 12 (Technical Training) Wing (Auxiliary) on 1 January 1955, its intended functions were performed by No. 1 Group (Auxiliary).

No. 14 Wing (Auxiliary)
Toronto, Ontario

Formed as RCAF (Reserve) Wing Toronto on 1 October 1950 to administer and control No. 400 and No. 411 Squadrons; redesignated RCAF (Reserve) Operational Wing (Toronto) on 15 January 1951, then No. 14 (Reserve) Operational Wing on 23 August 1951, No. 14 Operational Wing (Auxiliary) on 1 September 1951, and No. 14 Wing (Auxiliary) on 8 November 1957; integrated into the Canadian Armed Forces Reserve on 1 February 1968 as No. 14 Air Reserve Wing.
Commanders
G/C G.W. Gooderham, AFC 1 Oct 50 - 30 Jun 55 ret.
G/C R.C.A. Waddell, DSO, DFC, CD 1 Jul 55 - 23 Sep 59 ret.
G/C J.W.P. Draper, DFC, CD 29 Sep 59 - 31 Sep 64 ret.
G/C B.A. Howard, CD 1 Oct 64 - 31 Jan 68.
Higher Formations and Wing Location
Training Command,
Air Defence Command (1 Aug 51),
No. 2 Group (Auxiliary) (15 Jan 51 - 1 Mar 57),
Air Transport Command (1 Oct 58):
Toronto, Ont. 1 Oct 50 - 31 Jan 68.

No. 15 (Technical Training) Wing (Auxiliary)
Toronto, Ontario

Formed as RCAF (Reserve) Technical Training Wing Toronto on 1 April 1951; redesignated No. 15 (Reserve) Technical Training Wing on 1 August 1951, then No. 15 (Technical Training) Wing (Auxiliary) on 1 September 1951; disbanded on 10 January 1958.
Commanders
W/C O.B. Coumans, MBE, CD 1 Aug 51 - 15 Sep 57 ret.
W/C E.H. Mann 16 Sep 57 - 10 Jan 58.
Higher Formations and Wing Location
Training Command,
Air Defence Command (1 Aug 51):
No. 2 Group (Auxiliary) (1 Apr 51 - 1 Mar 57),
Toronto, Ont. 1 Apr 51 - 10 Jan 58.

No. 16 Wing (Auxiliary)
Hamilton, Ontario

Formed as RCAF (Reserve) Operational Wing Hamilton on 1 October 1950 to administer and control No. 424 Squadron; redesignated No. 16 (Reserve) Wing on 1

August 1951, then No. 16 Wing (Auxiliary) on 1 September 1951; disbanded on 1 April 1964.
Commanders
G/C D.B. Annan, DFC, AFC, CD 1 Oct 50 - 30 Sep 55 *ret.*
G/C D. Goldberg, DFC, CD 1 Oct 55 - 14 Mar 58 *ret.*
G/C G.C. Frostad, CD 15 Mar 58 - 14 Jan 60 *ret.*
G/C M.G. Marshall, DFC, CD 15 Jan 60 - 21 Sep 63 *ret.*
G/C P. Ardeline, DFC, CD 22 Sep 63 - 1 Apr 64.
Higher Formations and Wing Location
Training Command,
Air Defence Command (1 Aug 51),
Training Command (1 Sep 57),
Air Transport Command (12 Sep 59):
Hamilton, Ont. 1 Oct 50 - 1 Apr 64.

No. 17 (Fighter) Wing
Second Tactical Air Force
see No. 17 Sector

No. 17 Wing (Auxiliary)
Winnipeg, Manitoba

Formed as RCAF (Reserve) Wing Winnipeg on 1 October 1950 to administer and control No. 402 Squadron; redesignated No. 17 (Reserve) Wing on 1 August 1951, then No. 17 Wing (Auxiliary) on 1 September 1951; integrated into the Canadian Armed Forces Reserve on 1 February 1968 as No. 17 Air Reserve Wing.
Commanders
G/C G.H. Sellers, AFC 1 Oct 50 - 16 Mar 52 *ret.*
G/C W.R.D. Turner, AFC 17 Mar 52 - 31 Dec 54 *ret.*
G/C H.N. Scott, DFC, CD 1 Jan 55 - 31 Mar 59 *ret.*
G/C G.S. Varnam 1 Apr 59 - 15 Oct 62 *ret.*
G/C D.M. Gray, CD 16 Oct 62 - 31 Jan 68.
Higher Formations and Wing Location
North West Air Command,
Tactical Air Command (1 Aug 51),
Air Defence Command (1 Jan 55),
Training Command (25 Sep 57),
No. 14 (Training) Group (disbanded 11 Sep 59),
Air Transport Command (1 Apr 61):
Winnipeg, Man. 1 Oct 50 - 31 Jan 68.

No. 18 Wing (Auxiliary)
Edmonton, Alberta

Formed as RCAF (Reserve) Wing Edmonton on 1 October 1950 to administer and control No. 418 Squadron; redesignated No. 18 (Reserve) Wing on 1 August 1951, then No. 18 Wing (Auxiliary) on 1 September 1951; integrated into the Canadian Armed Forces Reserve on 1 February 1968 as No. 18 Air Reserve Wing.
Commanders
G/C D.R. Jacox, AFC 1 Nov 50 - 16 Apr 52 *ret.*
G/C G.K. Wynn 17 Apr 52 - 17 Apr 55 *ret.*
G/C A.D.R. Lowe, DFC, AFC 18 Apr 55 - 30 Sep 55 *ret.*
G/C K.M. Flint, CD 1 Oct 55 - 30 Jan 59 *ret.*
G/C J.K. Campbell, CD 31 Jan 59 - 31 Dec 62.
G/C F.T. Guest, CD 1 Jan 63 - 31 Jan 68.

Higher Formations and Wing Location
North West Air Command,
Tactical Air Command (1 Aug 51),
Training Command (1 Jan 59),
No. 14 (Training) Group (disbanded 11 Sep 59),
Air Transport Command (1 Apr 61):
Edmonton, Alta. 1 Nov 50 - 31 Jan 68.

No. 19 Wing (Auxiliary)
Vancouver, British Columbia

Badge A Sharp Shinned Hawk perched on a branch of Dogwood
Motto Vestigia nulla restrorsum (No retreat)
Authority Queen Elizabeth II, October 1962
The Sharp Shinned Hawk, which is one of the short winged birds of prey is indigenous to the area of British Columbia where the unit is located. Although it is not a very large bird, it is considered a force to be reckoned with. The Western Dogwood is plentiful in the country over which the wing operates.
Formed as RCAF (Reserve) Wing Vancouver on 1 October 1950 to administer and control No. 442 and No. 443 Squadron; redesignated No. 19 (Reserve) Wing on 1 August 1951, then No. 19 Wing (Auxiliary) on 1 September 1951; disbanded on 1 April 1964.
Commanders
G/C G.W. Northcott, DSO, DFC and Bar 1 Oct 50 - 7 May 55 *ret.*
G/C E.A. McNab, AFC, CD 8 May 55 - 7 May 58 *ret.*
G/C R.B. Barker, DFC, CD 8 May 58 - 31 Aug 61 *ret.*
G/C D.C. Cameron, CD 1 Sep 61 - 1 Apr 64.
Higher Formations and Wing Location
North West Air Command,
Air Defence Command (26 May 51),
Air Transport Command (1 Apr 61):
Vancouver, B.C. 1 Oct 50 - 1 Apr 64.

No. 22 (Fighter) Wing
Second Tactical Air Force
see No. 22 Sector

No. 22 (Photographic) Wing
Air Transport Command
see No. 7 (Photographic) Wing

No. 22 Wing (Auxiliary)
London, Ontario

Formed on 15 December 1953 to administer and control No. 420 Squadron; disbanded on 1 April 1957.
Commanders
G/C A.D. Haylett, AFC, CD 15 Dec 53 - 1 Apr 57.
Higher Formation and Wing Location
Air Defence Command:
London, Ont. 15 Dec 53 - 1 Apr 57.

No. 23 Wing (Auxiliary)
Saskatoon, Saskatchewan

Formed on 1 January 1955 to administer and control No. 406 Squadron; disbanded on 1 April 1964.
Commanders
G/C A.A. Myers, CD 1 Jan 55 - 31 Mar 57 *ret*.
G/C E.B. Van Slyck, DFC, CD 1 Apr 57 - 31 Mar 62 *ret*.
G/C D.J. Kelly, CD 1 Apr 62 - 1 Apr 64.
Higher Formations and Wing Location
Tactical Air Command,
Training Command (1 Oct 58),
No. 14 (Training) Group (disbanded 11 Sep 59),
Air Transport Command (1 Apr 61):
Saskatoon, Sask. 1 Jan 55 - 1 Apr 64.

No. 30 Wing (Auxiliary)
Calgary, Alberta

Formed on 1 August 1954 to administer and control No. 403 Squadron; disbanded on 1 April 1964.
Commanders
G/C W.A. Mostyn-Brown, AFC, CD 1 Aug 54 - 28 Feb 59 *ret*.
G/C G.M. Kelly, CD 1 Mar 59 - 31 Aug 61 *ret*.
G/C G.E. Sargenia, CD 1 Sep 61 - 1 Apr 64.
Higher Formations and Wing Location
Air Defence Command,
Training Command (25 Jan 57),
No. 14 (Training) Group (disbanded 11 Sep 59),
Air Transport Command (1 Apr 61):
Calgary, Alta. 1 Aug 54 - 1 Apr 64.

No. 39 (Reconnaissance) Wing
RAF Second Tactical Air Force

Formed as No. 39 (Army Co-operation) Wing at Leatherhead, Surrey, England on 12 September 1942 for reconnaissance duty under the operational control of First Canadian Army; transferred to Fighter Command on 1 June 1943 and relinquished its specialized "AC" designation; transferred to Second Tactical Air Force on 12 June 1943 and reorganized as No. 39 (Reconnaissance) Wing at Dunsfold, Surrey to administer and control No. 128 and No. 129 Airfield; redesignated No. 39 Sector at Odiham, Hampshire on 15 May 1944 when airfields became wings; reorganized as No. 39 (Reconnaissance) Wing at Sommervieu, France on 2 July 1944, absorbing No. 128 (Reconnaissance) Wing; disbanded at Luneberg, Germany on 7 August 1945.
Commanders
G/C D.M. Smith 12 Sep 42 - 9 Feb 44.
G/C E.H.G. Moncrieff, AFC 10 Feb 44 - 8 Feb 45.
G/C G.H. Sellers, AFC 9 Feb 45 - 15 May 45.
G/C R.C.A. Waddell, DSO, DFC 16 May 45 - 7 Aug 45.
Wing Commanders Flying
W/C J.H. Godfrey 2 Jul 44 - 5 Jul 44 *OTE*.
W/C R.C.A. Waddell, DSO, DFC[1] 16 Jul 44 - 15 May 45.
Higher Formations and Wing Locations
Army Co-operation Command:
No. 70 Group,
Leatherhead, Surrey 12 Sep 42 - 31 May 43.
Fighter Command:
No. 11 Group, Dunsfold, Surrey 1 Jun 43 - 11 Jun 43.
Tactical Air Force Fighter Command renamed Second Tactical Air Force (15 Nov 43):
No. 83 (Composite) Group,
Dunsfold, Surrey 12 Jun 43 - 31 Jul 43.
Redhill, Surrey 1 Aug 43 - 10 Aug 43.
Woodchurch, Kent 11 Aug 43 - 13 Oct 43.
Redhill, Surrey 14 Oct 43 - 31 Mar 44.
Odiham, Hants. 1 Apr 44 - 19 Jun 44.
En route to France 20 Jun 44 - 29 Jun 44.
B.(Base) 8 Sommervieu, Fr. 30 Jun 44 - 12 Aug 44.
B.21 Ste-Honorine-de- Ducy, Fr. 13 Aug 44 - 30 Aug 44.
B.34 Avrilly, Fr. 31 Aug 44 - 22 Sep 44.
B.64 Diest, Bel. 23 Sep 44 - 4 Oct 44.
B.78 Eindhoven, Neth. 5 Oct 44 - 7 Mar 45.
B.90 Petit-Brogel, Bel. 8 Mar 45 - 19 Mar 45.
B.104 Damme, Ger. 20 Mar 45 - 7 Apr 45.
B.108 Rheine, Ger. 8 Apr 45 - 14 Apr 45.
B.116 Wunstorf, Ger. 15 Apr 45 - 25 Apr 45.
B.154 Soltau, Ger. 26 Apr 45 - 8 May 45.
B.156 Luneburg, Ger. 9 May 45 - 7 Aug 45.
Operational History: First Mission 1 August 1943, Mustang I's of No. 430 Squadron from Redhill — standing patrols over a cable-laying vessel and its escort. **Last Mission** 8 May 1945, 4 Spitfire XI's of No. 400 Squadron — sea patrols. **Summary** Sorties: 13,526. Operational/Non-operational Flying Hours: 17,267/20,602. Victories: 40 aircraft destroyed, 5 probably destroyed, 25 damaged. Casualties: Operational: 42 aircraft; 27 pilots killed or missing.

---

[1] W/C Waddell flew a Spitfire F.R. Mk. XIV, RN114 RC-W.

No. 100 Wing (Auxiliary)
Vancouver, British Columbia

Authorized on 1 December 1938 to direct the organization and training of Auxiliary units in the Vancouver area; disbanded on 1 September 1939.
Commander
W/C A.D. Bell-Irving, MC 1 Dec 38 - 1 Sep 39.

No. 101 Wing (Auxiliary)
Toronto, Ontario

Authorized on 1 December 1938 to direct the organization and training of Auxiliary units in the Toronto area; disbanded on 1 September 1939.
Commander
W/C W.A. Curtis, DSC 1 Dec 38 - 1 Sep 39.

No. 102 Wing (Auxiliary)
Montreal, Quebec

Authorized on 1 December 1938 to direct the organization and training of Auxiliary units in the Montreal area; disbanded on 1 September 1939.
Commander
W/C J.A. Sully, AFC 1 Dec 38 - 1 Sep 39.

No. 120 (Transport) Wing
RAF Transport Command

Formed at RCAF Overseas Headquarters, London, England on 17 September 1945 to transport passengers, casualties, freight and mail in support of Canadian occupation forces in Germany; moved to Odiham, Hampshire on 7 October 1945; disbanded on 30 June 1946.
Commanders
G/C H.H.C. Rutledge, OBE 12 Sep 45 - 14 Jan 46.
G/C R.J. Lane, DSO, DFC and Bar 15 Jan 46 - 12 Jun 46.
Higher Formations and Wing Locations
Transport Command:
No. 46 Group,
London, Eng. 17 Sep 45 - 6 Oct 45.
Odiham, Hants. 7 Oct 45 - 30 Jun 46.
Summary of Operations Sorties: 1459. Operational/Non-operational Flying Hours: 17,056/1959. Airlifted: 50,961 passengers (including 3338 casualties), 511.9 tons of mail, 1978.9 tons of freight.

No. 126 (Fighter) Wing
RAF Second Tactical Air Force

Badge Four dragons' heads conjoined
Motto Fortitudo vincit (Courage wins)
Authority King George VI, May 1946
The four dragons' heads spitting fire represent the four Spitfire squadrons which composed the wing.
Formed as No. 126 Airfield at Redhill, Surrey, England on 4 July 1943 with Spitfire aircraft; redesignated No. 126 (Fighter) Wing at Tangmere, Sussex on 15 May 1944; transferred to the British Air Forces of Occupation (Germany) on 6 July 1945; disbanded at Utersen, Germany on 1 April 1946.
Commanders
W/C J.E. Walker, DFC and 2 Bars 9 Jul 43 - 26 Aug 43.
W/C K.L.B. Hodson, DFC and Bar 27 Aug 43 - 19 Jul 44.
G/C G.R. McGregor, OBE, DFC 20 Jul 44 - 27 Sep 45.
G/C W.E. Bennett 28 Sep 45 - 1 Apr 46.
Wing Commanders Flying
W/C B.D. Russel, DFC 9 Jul 43 - 16 Oct 43 *2 OTE.*
W/C R.W. McNair, DSO, DFC and 2 Bars 17 Oct 43 - 12 Apr 44.
W/C G.C. Keefer, DFC and Bar 17 Apr 44 - 7 Jul 44 *2 OTE.*
W/C B.D. Russel, DSO, DFC and Bar 8 Jul 44 - 26 Jan 45 *3 OTE.*
W/C G.W. Northcott, DSO, DFC and Bar 27 Jan 45 - 1 Apr 46.
Higher Formations and Wing Locations
Tactical Air Force Fighter Command, renamed Second Tactical Air Force (15 Nov 43):
No. 83 (Composite) Group,
No. 17 (RCAF) Sector (disbanded 13 Jul 44),
Redhill, Surrey 5 Jul 43 - 5 Aug 43.
Staplehurst, Kent 6 Aug 43 - 12 Oct 43.
Biggin Hill, Kent 13 Oct 43 - 14 Apr 44.
Tangmere, Sussex 15 Apr 44 - 8 Jun 44.
En route to France 9 Jun 44 - 15 Jun 44.
B.(Base) 3 Ste Croix-sur-Mer, Fr. 16 Jun 44 - 17 Jun 44.
B.4 Beny-sur-Mer, Fr. 18 Jun 44 - 7 Aug 44.
B.18 Cristot, Fr. 8 Aug 44 - 2 Sep 44.
B.44 Poix, Fr. 3 Sep 44 - 6 Sep 44.
B.56 Evere, Bel. 7 Sep 44 - 10 Sep 44.
B.68 Le Culot, Bel. 21 Sep 44 - 3 Oct 44.
B.84 Rips, Neth. 4 Oct 44 - 13 Oct 44.
B.80 Volkel, Neth. 14 Oct 44 - 6 Dec 44.
B.88 Heesch, Neth. 7 Dec 44 - 11 Apr 45.
B.108 Rheine, Ger. 12 Apr 45 - 14 Apr 45.
B.116 Wunstorf, Ger. 15 Apr 45 - 11 May 45.
B.152 Fassberg, Ger. 12 May 45 - 4 Jul 45.
B.174 Utersen, Ger. 5 Jul 45.
British Air Forces of Occupation (Germany):
No. 83 (Composite) Group,
B.174 Utersen, Ger. 6 Jul 45 - 1 Apr 46.
Operational History: First Mission 6 July 1943, 24 Spitfire VB's of No. 401 and No. 411 Squadron from Redhill, led by S/L B.D. Russel (commanding No. 411 Squadron) — "Rodeo No. 240", Part II, fighter sweep of the Gravelines-Berck area. **First Victory** 15 July 1943, 24 Spitfire VB's of No. 401 and No. 411 Squadron from Redhill, led by W/C B.D. Russel — "Ramrod No. 142", close support to 12 Bostons of No. 107 Squadron bombing Poix aerodrome. After leaving the target area, the formation was attacked by some 20 Fw.190's. S/L G.C. Semple, commanding No. 411 Squadron, credited with a Fw.190 damaged. The wing had no losses. 3 October 1953, 24 Spitfire IX's of No. 403 and No. 421 Squadron from Staplehurst — "Ramrod No. 257", Part I, top cover for 72 Marauders of the US Eighth Air Force bombing Woensdrecht aerodrome, attacked over the target by some 25 enemy aircraft. S/L R.W. McNair, commanding No. 421 Squadron, credited with a Fw.190 destroyed. The wing had no losses. **Last Mission** 5 May 1945, 8 Spitfire IXB's of No. 401 Squadron from Wunstorf — two 4-plane armed recces of the Hamburg area.
Summary Sorties: 29,631. Operational/Non-operational Flying Hours: 40,548/12,635. Victories: 355½ aircraft destroyed, 15 probably destroyed and 168 damaged, 1 V-1 destroyed. Casualties: Operational: 162 aircraft; 123 pilots killed or missing.

No. 127 (Fighter) Wing
RAF Second Tactical Air Force

Formed as No. 127 Airfield at Kenley, Surrey, England on 4 July 1943 with Spitfire aircraft; redesignated No. 127 (Fighter) Wing at Tangmere, Sussex on 15 May 1944; disbanded at Soltau, Germany on 7 July 1945.

Commanders
W/C M. Brown 11 Jul 43 - 18 Jul 44.
G/C W.R. MacBrien, OBE 19 Jul 44 - 11 Jan 45.
G/C P.S. Turner, DSO, DFC and Bar 12 Jan 45 - 7 Jul 45.

Wing Commanders Flying
W/C J.E. Johnson (RAF), DSO and Bar, DFC and Bar 5 Jul 43 - 18 Sep 43 *2OTE*.
W/C H.C. Godefroy, DSO, DFC and Bar 19 Sep 43 - 15 Apr 44 *2OTE*.
W/C L.V. Chadburn, DSO and Bar, DFC 16 Apr 44 - 12 Jun 44 *KIA*[1].
W/C R.A. Buckham, DFC and Bar 13 Jun 44 - 5 Jul 44 *OTE*.
W/C J.E. Johnson (RAF), DSO and 2 Bars, DFC and Bar 4 Jul 44 - 30 May 45.
W/C J.F. Edwards, DFC and Bar, DFM 6 Apr 45 - 7 Jul 45.

Higher Formations and Wing Locations
Tactical Air Force Fighter Command, renamed Second Tactical Air Force (15 Nov 43):
No. 83 (Composite) Group,
No. 17 (RCAF) Sector (disbanded 13 Jul 44),
Kenley, Surrey 5 Jul 43 - 5 Aug 43.
Lashenden, Kent 6 Aug 43 - 19 Aug 43.
Headcorn, Kent 20 Aug 43 - 13 Oct 43.
Kenley, Surrey 14 Oct 43 - 16 Apr 44.
Tangmere, Sussex 17 Apr 44 - 7 Jun 44.
En route to France 8 Jun 44 - 15 Jun 44.
B.(Base) 2 Crépon, Fr. 16 Jun 44 - 27 Aug 44.
B.26 Marcilly-la-Campagne, Fr. 28 Aug 44 - 17 Sep 44.
B.68 Le Culot, Bel. 18 Sep 44 - 26 Sep 44.
B.82 Grave, Neth. 27 Sep 44 - 21 Oct 44.
B.58 Melsbroek, Bel. 22 Oct 44 - 3 Nov 44.
B.56 Evère, Bel. 4 Nov 44 - 1 Mar 45.
B.90 Petit-Brogel, Bel. 2 Mar 45 - 30 Mar 45.
B.78 Eindhoven, Neth. 31 Mar 45 - 7 Apr 45.
B.100 Goch, Ger. 8 Apr 45 - 12 Apr 45.
B.114 Diepholz, Ger. 13 Apr 45 - 25 Apr 45.
B.154 Soltau, Ger. 26 Apr 45 - 7 Jul 45.

Operational History: First Mission and Victories 6 July 1943, 24 Spitfire IX's of No. 403 and No. 421 Squadron from Kenley, led by S/L H.C. Godefroy, commanding No. 403 Squadron — "Rodeo No. 240", Part I, second fighter sweep of Dieppe-Amiens-Doullens area, engaged two enemy formations. Near Amiens, 12 Bf.109's were sighted flying southeast and No. 403 Squadron was detailed to attack. The enemy aircraft immediately dived through cloud towards Abbeville except for their leader, who started to turn and was shot down by F/L H.D. Macdonald. At Doullens, 5 Bf.109's flying southeast were engaged by No. 421 Squadron, led by S/L R.W. McNair. Most of the enemy evaded combat by flying into the cloud, though not before S/L McNair had destroyed one Bf.109 and F/L A.H. Sager had damaged a second. Both engagements were without loss to the wing. **Last Mission** 7 May 1945, 52 Spitfire sorties from Soltau — escort to Dakotas, Stirlings and Halifaxes flying supplies to Copenhagen, Denmark.

Summary Sorties: 26,798. Operational/Non-operational Flying Hours: 38,003/9216. Victories: 245¼ aircraft destroyed, 14 probably destroyed, 155½ damaged. Casualties: Operational: 133 aircraft; 106 pilots killed or missing.

[1]Flying Spitfire IX MJ824 LV-C; mid-air collision over the Normandy beachhead.

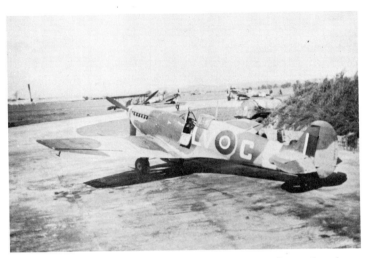

457 Supermarine Spitfire L.F.Mk.V EP548, flown by the leader of No. 127 Wing, W/C Lloyd Vernon Chadburn. Coded LV-C (his initials), the aircraft carried the Wing Commander's pennant on the forward left cowl, and under the cockpit the inscription "City of Oshawa" was displayed.

No. 128 (Reconnaissance) Wing
RAF Second Tactical Air Force

Formed as No. 128 Airfield at Dunsfold, Surrey, England on 4 July 1943 with Mustang aircraft; redesignated No. 128 (Reconnaissance) Wing at Odiham, Hampshire on 15 May 1944; disbanded on 2 July 1944 at Sommervieu, France, and absorbed into No. 39 Sector, which then became No. 39 (Reconnaissance) Wing.

Commanders
S/L J.D. Hall, DFC 20 Jul 43 - 3 Aug 43.
W/C J.M. Godfrey 4 Aug 43 - 2 Jul 44.

Higher Formations and Wing Locations
Tactical Air Force Fighter Command renamed Second Tactical Air Force (15 Nov 43):
No. 83 (Composite) Group,
No. 39 (RCAF) Sector,
Dunsfold, Surrey 4 Jul 43 - 27 Jul 43.
Woodchurch, Kent 28 Jul 43 - 14 Oct 43.
Redhill, Surrey 15 Oct 43 - 18 Feb 44.
Odiham, Hants. 19 Feb 44 - 30 Jun 44.
En route to France 1 Jul 44.
B.(Base) 8 Sommervieu, Fr. 2 Jul 44.

Operational History: First Mission 6 July 1943, 2 Mustang I's of No. 400 Squadron from Dunsfold — in search of ground targets in the Mont St Michel-Laval-Cabourg area, forced to turn back from Mont St Michel because of weather. **First Victory** 8 July 1943, 2 Mustang I's of No. 400 Squadron from Dunsfold — attacked railway targets in northern France. F/O J.M. Robb and F/O F.E.W. Hanton shared in destroying a Fieseler Storch light aircraft five miles from Cabourg and damaging 8 locomotives on the Rennes-Laval and Laval-Alençon lines. **Last Mission** 1 July 1944, 3 Mustang I's of No. 414 Squadron from Odiham — tactical reconnaissance of roads in the Ste Gauburge—Aube-sur-Rille-St Martin-Bubertré area, forced to return early on account of rain. **Summary** Sorties: 2088. Operational/Non-operational Flying Hours: 3264/5184. Victories: 12 aircraft destroyed, 2 probably destroyed, 6 damaged. Casualties: Operational: 4 aircraft; 4 pilots killed or missing.

No. 129 (Fighter Bomber) Wing
RAF Second Tactical Air Force

Formed as No. 129 Airfield at Gatwick, Surrey, England on 4 July 1943 for tactical reconnaissance with Mustang aircraft; assigned to fighter bomber role with Typhoon aircraft at Westhampnett, Sussex on 22 April 1944 and redesignated No. 129 (Fighter Bomber) Wing on 15 May;[1] disbanded at Plumetot, France on 13 July 1944 on the reorganization of Second Tactical Air Force.
Commanders
W/C E.H.G. Moncrieff, AFC 4 Jul 43 - 10 Feb 44.
W/C D.C.S. MacDonald, DFC 27 Feb 44 - 13 Jul 44.
Higher Formations and Wing Locations
Tactical Air Force Fighter Command, renamed Second Tactical Air Force (15 Nov 43):
No. 83 (Composite) Group,
No. 39 (RCAF) Sector,
Gatwick, Surrey 4 Jul 43 - 12 Aug 43.
Ashford, Kent 13 Aug 43 - 14 Oct 43.
Gatwick, Surrey 15 Oct 43 - 1 Apr 44.
Odiham, Hants 2 Apr 44 - 21 Apr 44.
No. 15 (RAF) Sector,
Westhampnett, Sussex 22 Apr 44 - 12 Jun 44.
En route to France 13 Jun 44 - 27 Jun 44.
B.(Base) 10 Plumetot, Fr. 28 Jun 44 - 13 Jul 44.
Operational History: First Mission 11 July 1943, two pairs of Mustang I's of No. 430 Squadron from Gatwick — attacked railway targets in northern France, all forced to return early because of weather. **Last Mission** 13 July 1944, Typhoons of No. 609 and No. 198 Squadron RAF from Plumetot — rocket attacks on ferries in the Seine River, good results observed.
Summary (These figures apply only to service under No. 39 Sector — not No. 15 (RAF) Sector.) Sorties: 797. Operational/Non-operational Flying Hours: 1077/4846. Victories: 8 aircraft destroyed, 2 damaged. Casualties: Operational: 9 aircraft; 8 pilots killed or missing.

---

[1]This change of role came about as a result of lack of units and individual pilots trained in tactical and photographic reconnaissance, and not enough time before the Allied invasion of Europe to carry out such specialized training.

No. 143 (Fighter Bomber) Wing
RAF Second Tactical Air Force

Formed as No. 143 Airfield at Ayr, Scotland on 10 January 1944 with Typhoon aircraft; redesignated No. 143 (Fighter Bomber) Wing at Hurn, Hampshire, England on 15 May 1944; disbanded at Flensburg, Germany on 7 September 1945.
Commanders
W/C F.W. Hillock 12 Jan 44 - 14 Jul 44.
G/C P.Y. Davoud, OBE, DSO, DFC 15 Jul 44 - 31 Dec 44.
W/C A.D. Nesbitt, DFC 1 Jan 45 - 7 Sep 45.
Wing Commanders Flying
W/C R. Marples (RAF), DFC 10 Jan 44 - 20 Jan 44.
W/C R.P.T. Davidson (RAF), DFC 21 Jan 44 - 8 May 44 MIA[1].
W/C M.T. Judd (RAF), DFC, AFC 14 May 44 - 13 Oct 44.
W/C F.G. Grant, DSO, DFC 14 Oct 44 - 7 Sep 45.
Higher Formations and Wing Locations
Air Defence Great Britain:
No. 12 Group,
No. 22 (RCAF) Sector,
Ayr, Scot. 10 Jan 44 - 17 Mar 44.
Second Tactical Air Force:
No. 83 (Composite) Group,
No. 22 (RCAF) Sector (disbanded 13 Jul 44),
Hurn, Hants. 18 Mar 44 - 1 Apr 44.
Funtington, Sussex 2 Apr 44 - 19 Apr 44.
Hurn, Hants. 20 Apr 44 - 17 Jun 44.
En route to France 18 Jun 44 - 27 Jun 44.
B.(Base) 9 Lantheuil, Fr. 28 Jun 44 - 29 Aug 44.
B.24 St André, Fr. 30 Aug 44 - 3 Sep 44.
B.48 Amiens, Fr. 4 Sep 44 - 24 Sep 44.
B.78 Eindhoven, Neth. 25 Sep 44 - 28 Mar 45.
B.100 Goch, Ger. 29 Mar 45 - 8 Apr 45.
B.108 Rheine, Ger. 9 Apr 45 - 11 Apr 45.
B.110 Achmer, Ger. 12 Apr 45 - 18 Apr 45.
B.150 Hustedt, Ger. 19 Apr 45 - 28 May 45.
B.166 Flensburg, Ger. 29 May 45 - 7 Aug 45.
Operational History: First Mission 20 March 1944, 4 Typhoon IB's of No. 438 Squadron from Hurn and staging through Kenley — "Rodeo No. 128", fighter sweep of the Cherbourg peninsula; attacked German staff car and troops on Montebourg-Carentan road. No results observed; all aircraft returned safely. **Last Mission** 4 May 1945, 15 Typhoon IB's of No. 438 and No. 439 Squadron from Hustedt — anti-shipping strike in Kiel Bay; claimed 2 small vessels probably destroyed and 4 damaged, without loss.
Summary Sorties: 12,926. Operational/Non-operational Flying Hours: 13,937/7825. Victories: Aircraft: 19 destroyed, 18 damaged. Ground (6 Jun 44 - 8 May 45): dropped 7936 1000- and 9896 500-pound bombs, fired 1,078,000 20-mm shells; credited with 1210 rail cuts; destroyed/damaged 736/841 motor vehicles, 38/24 tanks, 12/329 railway locomotives, 18/1697 gun positions, 6/18 barges. Casualties: Operational: 136 aircraft; 106 pilots killed or missing.

---

[1]Evaded capture and returned to England shortly after D-Day.

458 Hawker Typhoon Mk.IB flown by W/C R.T.P. Davidson while leading No. 143 Wing. The aircraft was coded RD (his initials), and displayed his unique victory flags, two Japanese, two Italian and two German aircraft confirmed.

No. 144 (Fighter) Wing
RAF Second Tactical Air Force

Formed as No. 144 Airfield at Digby, Lincolnshire, England on 1 February 1944 with Spitfire aircraft; redesignated No. 144 (Fighter) Wing at Ford, Sussex on 15 May 1944; disbanded at Ste Croix-sur-Mer, France on 13 July on the reorganization of Second Tactical Air Force.
Commanders
W/C J.E. Walker, DFC and 2 Bars 4 Mar 44 - 25 Apr 44 KIA.
W/C A.D. Nesbitt, DFC 1 May 44 - 13 Jul 44.
Wing Commanders Flying
W/C J.E. Johnson (RAF), DSO and 2 Bars, DFC and Bar 8 Mar 44 - 13 Jul 44.
Higher Formations and Wing Locations
Air Defence Great Britain:
No. 12 Group,
No. 22 (RCAF) Sector,
Digby, Lincs. 1 Feb 44 - 13 Mar 44.
Holmsley South, Hants. 14 Mar 44 - 17 Mar 44.
Second Tactical Air Force:
No. 83 (Composite) Group,
No. 22 (RCAF) Sector,
Holmsley South, Hants. 18 Mar 44 - 31 Mar 44.
Westhampnett, Sussex 1 Apr 44 - 20 Apr 44.
No. 17 (RCAF) Sector,
Funtington, Sussex 21 Apr 44 - 13 May 44.
Ford, Sussex 14 May 44 - 3 Jun 44.
En route to France 4 Jun 44 - 12 Jun 44.
B.(Base) 4 Beny-sur-Mer, Fr. 13 Jun 44.
B.3 Ste Croix- sur-Mer, Fr. 14 Jun 44 - 13 Jul 44.
Operational History: First Mission 28 March 1944, 24 Spitfire IX's of No. 441 and No. 442 Squadron from Holmsley South — "Rodeo No. 276", Part I. withdrawal cover of US Eighth Air Force Fortresses and Liberators returning from attacks on "Noball" (V-1) sites and aerodromes in France. The wing was led by W/C Johnson as far as the French coast (St Valery); there, because of radio trouble, he handed command over to W/C E.P. Wells (NZ/RAF). With No. 442 Squadron providing top cover, W/C Wells led No. 441 Squadron in a strafe of enemy aircraft on Dreux-Vernon aerodrome; the squadron claimed 4 destroyed, 1 probably destroyed and 3 damaged, without loss. **First Air Victory** 19 April 1944, 27 Spitfire IX's of No. 442 and No. 443 Squadron from Westhampnett and led by W/C Johnson — close escort to one of three formations of US Ninth Bomber Command Marauders attacking marshalling yards at Malines. The wing then swept the Venlo-Gilze-Rijen area and a Do.217, flying at 100 feet west of Diest, was destroyed by S/L H.W. McLeod, commanding No. 443 Squadron. The wing had no losses. **Last Mission** 13 July 1944, 48 Spitfire IX sorties flown from Ste Croix-sur-Mer — armed recces. The wing claimed 10 enemy aircraft destroyed, 1 probably destroyed and 1 damaged, plus 5 motor vehicles destroyed; 1 pilot reported missing. 12 Spitfires of No. 441 Squadron engaged 13 Fw.190's west of Argentan and claimed 10 destroyed, without loss. 12 Spitfires of No. 442 Squadron engaged Fw.190's and Bf.109's in the Cabourg area and claimed 1 Bf.109 probably destroyed with the loss of 1 pilot. 12 Spitfires of No. 442 Squadron claimed 5 motor vehicles destroyed without loss. 12 Spitfires of No. 443 Squadron engaged Fw.190's and Bf.109's in the Cabourg area and claimed 1 Bf.109 damaged without loss. **Summary** Sorties: 4126. Operational/Non-operational Flying Hours: 6153/3265. Victories: 60 aircraft destroyed, 2 probably destroyed, 17 damaged. Casualties: Operational: 16 aircraft; 11 pilots killed or missing.

No. 331 (Medium Bomber) Wing
North Africa

Formed at West Kirby, Cheshire, England on 7 May 1943 as part of the force despatched to North Africa for the invasions of Sicily and Italy; disbanded at Dishforth, Yorkshire, England on 18 December 1943.
Commanders
G/C C.R. Dunlap 1 May 43 - 16 Nov 43.
Higher Formations and Wing Locations
Bomber Command:
No. 6 (RCAF) Group,
West Kirby, Cheshire 7 May 43 - 15 May 43.
En route to North Africa by sea 16 May 43 - 26 May 43.
Allied Mediterranean Air Command:
Allied North West African Air Force,
Allied North West African Stategic Air Force,
No. 205 (RAF) Group,
Algiers, Algeria[1] 27 May 43 - 29 May 43.
Kairouan (Zina), Tunisia 30 May 43 - 29 Sep 43.
Hani East Landing Ground, Tunisia 30 Sep 43 - 16 Oct 43.
En route to England by sea 17 Oct 43 - 5 Nov 43.
Bomber Command:
No. 6 (RCAF) Group,
Dishforth, Yorks. 6 Nov 43 - 18 Dec 43.
Operational History: First Mission 26/27 June 1943, 5 Wellington X's of No. 420 Squadron and 10 of No. 425 Squadron despatched to bomb Sciacca, Sicily, where the 150th Wing of the Italian Air Force was stationed; 11 bombed the aerodrome, 1 bombed the town, 2 aborted; 1 aircraft of No. 425 Squadron failed to return. **Last Mission**

5 October 1943, 21 Wellington X's bombed the aerodrome at Grosseto, Italy. **Summary** Missions/Sorties: 87/2127.[2] Operational Flying Hours: 11,515. Bombs dropped: 3745 tons (plus 1881 mines). Casualties: Operational: 21 aircraft. Non-operational: 4 aircraft.

---

[1] The wing was waiting the completion of their two airfields on the plains of Tunisia, southwest of Kairouan (Zina and Pavillier). The groundcrew were in a transit camp near Algiers and the aircrew at Telergma, near Constantine, for two weeks.

[2] The wing was operational from 26 June to 5 October 1943, and its operations were divided into four phases: 26 June - 9 July, pre-invasion attacks on Sicily; 10 July - 17 August, support of ground operations in Sicily; 18 August - 2 September, pre-invasion attacks on Italy. 3 September - 5 October, support of ground operations in Italy.

No. 661 (Heavy Bomber) Wing
RAF "Tiger Force"

Formed at Yarmouth, Nova Scotia on 15 July 1945 for duty in the Pacific; disbanded on 5 September 1945.
Commander
W/C F.R. Sharp, DFC 15 Jul 45 - 5 Sep 45.[1]
Higher Formations and Wing Location
RAF "Tiger Force",
RCAF Eastern Air Command (for training):
No. 6 (RCAF) Group,
Yarmouth, N.S. 15 Jul 45 - 5 Sep 45.

---

[1] A/C J.G. Kerr, AFC, was to have commanded the wing but was acting commander of No. 6 Group's advanced headquarters at Ottawa.

No. 662 (Heavy Bomber) Wing
RAF "Tiger Force"

Formed at Dartmouth, Nova Scotia on 15 July 1945 for duty in the Pacific; disbanded on 5 September 1945.
Commander
G/C J.R. MacDonald, DFC 4 Aug 45 - 5 Sep 45.
Higher Formations and Wing Location
RAF "Tiger Force",
RCAF Eastern Air Command (for training):
No. 6 (RCAF) Group,
Dartmouth, N.S. 15 Jul 45 - 5 Sep 45.

No. 663 (Heavy Bomber) Wing
RAF "Tiger Force"

Formed at Debert, Nova Scotia on 1 August 1945 for duty in the Pacific; disbanded on 5 September 1945.
Commander
G/C J.H.L. Lecomte, DFC 4 Aug 45 - 5 Sep 45.
Higher Formations and Wing Location
RAF "Tiger Force",
RCAF Eastern Air Command (for training):
No. 6 (RCAF) Group,
Debert, N.S. 1 Aug 45 - 5 Sep 45.

No. 664 (Heavy Bomber) Wing
RAF "Tiger Force"

Formed at Greenwood, Nova Scotia on 1 August 1945 for duty in the Pacific; disbanded on 5 September 1945.
Commander
G/C W.A.G. McLeish, DFC 6 Aug 45 - 1 Sep 45.
Higher Formations and Wing Location
RAF "Tiger Force",
RCAF Eastern Air Command (for training):
No. 6 (RCAF) Group,
Greenwood, N.S. 1 Aug 45 - 5 Sep 45.

459 *Avro Lincoln B.Mk.XV FM300, manufactured in Canada and the first of many that would have equipped "Tiger Force" had the war in the Pacific continued. The end of the war closed down production and only FM300 was completed.*

RCAF ("X") Wing
Detached Operations, Anchorage, Alaska
Western Air Command

Formed at Elmendorf Field, Anchorage, Alaska on 2 June 1942 as part of the RCAF reinforcement to the United States Army Air Forces in Alaska, and employed on air defence and anti-shipping duty; disbanded on 15 September 1943 following the withdrawal of the Japanese from the Aleutian Islands.
Commanders
G/C G.R. McGregor, OBE, DFC 4 Jun 42 - 28 Feb 43.
W/C R.E.E. Morrow, DFC 6 Mar 43 - 15 Sep 43.[1]
Higher Formations and Wing Location
Western Air Command,
No. 4 Group (for adm),
US Alaskan Command:
11th Air Force Fighter Command (for ops),
Fort Richardson, Elmendorf Field, Anchorage, Alaska
2 Jun 42 - 15 Sep 43.
Operational History: First Offensive Mission 25 September 1942, 4 P-40E's of No. 111 (F) Squadron[2] from Adak Island, Aleutians — part of the fighter escort of 20 P-40E's for 9 B-24's and 12 P-39's bombing Japanese positions on Kiska Island. S/L K.A. Boomer credited with a Zero-type

float fighter destroyed. **First All-Canadian Offensive Mission** 26 April 1943, 8 P-40K-1's of No. 14 (F) Squadron from Adak Island in the Aleutians, each armed with one 300-pound demolition and six 20-pound fragmentation bombs, flew a total of three missions (24 sorties) against Japanese ground installations on Kiska. All aircraft returned safely. **Last Offensive Mission** 12 August 1943, 8 P-40K-1's of No. 14 (F) Squadron from Adak Island attacked Kiska. Pilots reported that the island appeared deserted, and it was later confirmed that the Japanese had indeed evacuated Kiska. **Summary** Offensive Missions/Sorties: 31/194. Victories: 1 aircraft destroyed. Casualties: Operational: nil. Non- operational: 6 aircraft; 4 pilots killed, 1 missing, 1 injured. **Honours and Awards** 1 OBE, 1 DFC, 11 Air Medals (USA).[3]

---

[1] Injured bailing out over Amchitka; temporarily replaced by W/C P.B. Pitcher from 19 May to 27 June 1943.
[2] The pilots were S/L K.A. Boomer, F/O's R. Lynch, J.G. Gohl and H.O. Gooding; the P-40E's were provided by the Americans.
[3] The OBE was awarded to G/C McGregor and the DFC to S/L Boomer.

RCAF ("Y") Wing
Detached Operations, Annette Island, Alaska
Western Air Command

Formed at Annette Island, Alaska on 14 June 1942[1] as part of the RCAF reinforcement to the United States Army Air Forces in Alaska, and employed in the defence of Prince Rupert, British Columbia, the main supply harbour for US forces in Alaska; disbanded on 18 November 1943 when the squadrons were redeployed to Prince Rupert.
Commanders
W/C A.D. Nesbitt, DFC 14 Jun 42 - 8 Oct 42.
W/C G.G. Diamond 9 Oct 42 - 5 Oct 43.
W/C R.F. Douglas 6 Oct 43 - 18 Nov 43.
Higher Formations and Wing Location
Western Air Command:
No. 4 Group,
Annette Island, Alaska 14 Jun 42 - 18 Nov 43.
Operational History: Employed only on defensive duties (details not recorded).

---

[1] This date is based on the arrival of W/C Nesbitt. One of the two squadrons was already on Annette Island, having arrived on 5 May 1942.

ANCILLARY UNITS
Coast Artillery Co-operation Detachments
1940-1944

During the Second World War, a few aircraft were stationed at each defended harbour for co-operation training with coast artillery batteries. Initially, squadrons were used, but they were found to be too large and the RCAF did not have enough squadrons to spare. It was therefore decided to form small detachments of one to four Lysander aircraft for this duty.
No. 1 (CAC) Detachment Formed at Saint John, N.B. from "B" Flight of No. 118 (CAC) Sqn 27 Sep 40. Disbanded 1 Apr 44.
No. 2 (CAC) Detachment Formed at Dartmouth, N.S. from "A" Flight of No. 118 (CAC) Sqn 27 Sep 40. Disbanded (absorbed by No. 121 (Composite) Sqn) 15 Aug 43.
No. 3 (CAC) Detachment Formed at Patricia Bay, B.C. on the disbandment of No. 111 (CAC) Sqn 1 Feb 41. Disbanded (absorbed by No. 122 (Composite) Sqn) 9 Jan 42.
No. 4 (CAC) Detachment Formed at Sydney, N.S. 1 Apr 41. Disbanded 1 Nov 43.
No. 5 (CAC) Detachment Formed at Sydney, N.S. 1 Dec 41. Moved to Torbay, Nfld. 23 Apr 42. Disbanded (absorbed by No. 121 (Composite) Sqn) 15 Jul 43.
No. 6 (CAC) Detachment Formed at Yarmouth, N.S. 15 Dec 41. Disbanded 1 Nov 43.

Operational Training Units
Canada (British Commonwealth Air Training Plan)

No. 1 Operational Training Unit Formed at Bagotville, Quebec on 6 June 1942 to train single-engine fighter pilots on Hurricane aircraft; graduated 940 pilots from an intake of 1012; disbanded on 31 January 1945.
No. 2 Operational Training Unit Formed at Pennfield Ridge, New Brunswick on 6 July 1942 to train transport crews on Dakota aircraft, but owing to a lack of aircraft was disbanded on 20 July 1942 and absorbed into No. 1 OTU, No. 34 (RAF) OTU and No. 164 (Transport) Squadron's Training Flight.
No. 3 Operational Training Unit To have formed at Shelburne, Nova Scotia on 3 August 1942 to train general reconnaissance crews on Catalina flying boats, but owing to unexpected delays in delivery the authority for formation was suspended on 30 July. The unit was formed at Patricia Bay, British Columbia on 9 November 1942 from No. 13 (Operational Training) Squadron, to train bomber reconnaissance crews on Stranraer and Canso A aircraft; by the end of 1944 had graduated 325 crews of from 7-9 members; disbanded on 3 August 1945.
No. 5 Operational Training Unit Formed at Boundary Bay, British Columbia on 1 April 1944 to train bomber crews on Liberator aircraft for operations in South East Asia (India and Burma); used Abbotsford as a satellite training field from 15 August 1944 to 1 October 1945; graduated 192 crews; disbanded 31 October 1945.
No. 6 Operational Training Unit No. 32 OTU was formed in West Kirby, Cheshire, England 7 August 1941 from No. 1 Torpedo Training School and arrived at Patricia Bay, British Columbia, on 22 August 1941 to train torpedo bomber crews on Hampden and Beaufort aircraft; converted to the training of transport crews on Dakota aircraft on 10 December 1943; moved to Comox, B.C. on 25 May 1944; graduated 314 crews; redesignated No. 6 (RCAF) OTU on 1 June 1944; graduated 228 crews to end of 1944; moved to Greenwood, Nova Scotia on 15 January 1946; disbanded on 31 March 1946.
No. 7 Operational Training Unit Formed as No. 31 OTU RAF in the United Kingdom on 25 April 1941 and arrived at Debert, Nova Scotia, on 22 May 1941 to train crews for Ferry Command and Coastal Command on Hudson air-

craft; training included 6542 flying hours on a total of 1404 anti-submarine sorties during which seven U-boats were sighted, two attacked, and one was damaged by Sgt. I.D. Wallace and crew on 4 July 1943 at 4337N 6403W; two aircraft and eight aircrew were lost; graduated 681 crews; redesignated No. 7 (RCAF) OTU on 1 July 1944 and converted to the training of Mosquito fighter-bomber crews with emphasis on intruder operations; graduated 98 crews to the end of 1944; disbanded on 20 July 1945.

No. 8 Operational Training Unit Formed as No. 36 OTU, RAF in the United Kingdom, arrived at Greenwood, Nova Scotia on 9 March 1942 to train general reconnaissance crews on Hudson aircraft; converted to training of Mosquito fighter-bomber crews on 3 July 1943; redesignated No. 8 (RCAF) OTU on 1 July 1944; graduated 588 crews by end of 1944; disbanded on 1 August 1945.

Overseas

No. 1659 Heavy Conversion Unit Formed at Leeming, Yorkshire, England on 6 October 1942 from No. 408 (Bomber) Squadron Conversion Flight to provide conversion training on Halifax aircraft; moved to Topcliffe, Yorkshire on 14 March 1943; converted to Lancaster B.Mk.X aircraft on 20 November 1944; disbanded on 10 September 1945.

No. 1664 Heavy Conversion Unit Formed at Croft, Yorkshire, England on 10 May 1943 to provide conversion training on Halifax aircraft; moved to Dishforth, Yorkshire on 7 December 1943; nicknamed "Caribou" on 21 June 1944; disbanded on 6 April 1945.

No. 1666 Heavy Conversion Unit Formed at Dalton, Yorkshire, England on 15 May 1943 to provide conversion training on Halifax aircraft; moved to Wombleton, Yorkshire on 21 October 1943; nicknamed "Mohawk" on 21 June 1944; converted to Lancaster B.Mk.X aircraft on 20 November 1944; disbanded on 3 August 1945.

No. 1679 Heavy Conversion Unit Formed at East Moor, Yorkshire, England on 20 May 1943 to provide conversion training on Lancaster B.Mk.II aircraft; moved to Wombleton, Yorkshire on 13 December 1943; disbanded on 27 January 1944.

No. 1691 (Bomber) Gunnery Flight Formed at Dalton, Yorkshire, England on 2 July 1943 to train air gunners; redesignated No. 1695 (Bomber) Defence Training Flight on 15 February 1944; disbanded on 28 July 1945.

No. 6 Group Aircrew School Formed at Dalton, Yorkshire, England on 6 August 1944 as a holding unit for aircrew passing from No. 22, No. 23 and No. 24 Operational Training Unit to heavy conversion units; provided physical conditioning, lectures on discipline, administration, escape and evasion, and instruction in synthetic trainers; disbanded on 3 August 1945.

Post-War

No. 1 (Fighter) Operational Training Unit Formed at St Hubert, Quebec on 1 September 1948 to train single-engine jet fighter pilots on Vampire aircraft; moved to Chatham, New Brunswick on 1 October 1949; converted to Sabre aircraft in 1950; redesignated Sabre Conversion Unit in April 1962 to provide conversion training for pilots destined to receive training on CF-104 aircraft at No. 6 (Strike and Reconnaissance) OTU.

No. 2 (Maritime) Operational Training Unit Formed at Greenwood, Nova Scotia on 1 November 1949 to train maritime crews on Lancaster and, later, Neptune and Argus aircraft; moved to Summerside, Prince Edward Island in November 1953; integrated into Canadian Armed Forces on 1 February 1968.

No. 3 All-Weather (Fighter) Operational Training Unit Formed at North Bay, Ontario on 3 November 1952 to train crews on CF-100 aircraft; moved to Cold Lake, Alberta on 22 May 1955, and to Bagotville, Quebec on 25 September 1961; converted to CF-101 aircraft in October 1962; integrated into Canadian Armed Forces on 1 February 1968.

No. 4 (Transport) Operational Training Unit Formed at Dorval (Montreal), Quebec on 1 March 1952 to train transport crews on Dakota, North Star, C-119 and other aircraft; moved to Trenton, Ontario on 23 January 1954 and added C-130, Otter, Yukon and Boeing 707 aircraft; integrated into Canadian Armed Forces on 1 February 1968.

No. 5 (Helicopter) Operational Training Unit Formed at Rockcliffe (Ottawa), Ontario on 1 June 1958 from No. 108 Communications Flight Training Section to train helicopter crews on the Sikorsky H-34; absorbed into No. 4 (T) OTU on 1 January 1959.

No. 6 (Strike and Reconnaissance) Operational Training Unit Formed at Cold Lake, Alberta on 1 September 1961 to train pilots on CF-104 aircraft for nuclear strike and reconnaissance with No. 1 Air Division Europe; integrated into Canadian Armed Forces on 1 February 1968.

460 Avro Lancaster Mk.III PB810 aircraft "P" of No. 1659 Heavy Conversion Unit, No. 6 (RCAF) Group.

Part Five
APPENDICES

Appendix 1

British Commonwealth Air Training Plan

Organization — January 1944

An asterisk denotes a double-size Elementary Flying Training School (Nos. 5, 19, 20 and 33). By 1944 schools in the 30-group were RAF units administered by the RCAF. Other 30-group schools had been renumbered — a former No. 33 EFTS as No. 18; No. 34 as No. 25; and No. 35 as No. 26.

No. 1 Training Command,
Headquarters Toronto, Ontario

Unit	Location	Aircraft Types
Manning Depots		
No. 1 MD	Toronto, Ont.	
Initial Training Schools		
No. 1 ITS	Toronto, Ont.	
No. 5 ITS	Bellevile, Ont.	
Elementary Flying Training Schools		
No. 1 EFTS	Malton, Ont.	Moth
No. 3 EFTS	London, Ont.	Finch
No. 7 EFTS	Windsor, Ont.	Finch
No. 9 EFTS	St. Catharines, Ont.	Moth
No. 10 EFTS	Hamilton, Ont.	Moth, Finch
No. 12 EFTS	Goderich, Ont.	Finch
No. 20 EFTS*	Oshawa, Ont.	Moth
Service Flying Training Schools		
No. 1 SFTS	Camp Borden, Ont.	Harvard
No. 5 SFTS	Brantford, Ont.	Anson
No. 6 SFTS	Dunnville, Ont.	Harvard, Yale
No. 9 SFTS	Centralia, Ont.	Anson, Harvard
No. 14 SFTS	Aylmer, Ont.	Anson, Harvard
No. 16 SFTS	Hagersville, Ont.	Anson, Harvard
No. 31 SFTS	Kingston, Ont.	Battle, Harvard
Air Observer Schools		
No. 1 AOS	Malton, Ont.	Anson
No. 4 AOS	London, Ont.	Anson
Bombing and Gunnery Schools		
No. 1 B&GS	Jarvis, Ont.	Anson, Battle, Bolingbroke, Lysander
No. 4 B&GS	London, Ont.	Anson, Battle, Bolingbroke, Lysander
No. 6 B&GS	Mountain View (Hamilton), Ont.	Anson, Battle, Bolingbroke, Lysander
No. 31 B&GS	Picton, Ont.	Anson, Battle, Bolingbroke, Lysander
Air Navigation Schools		
No. 3 ANS	Port Albert, Ont.	Anson
No. 33 ANS	Hamilton, Ont.	Anson
Wireless School		
No. 4 WS	Guelph, Ont.	Norseman, Moth
Central Flying School		
CFS	Trenton, Ont.	various

No. 2 Training Command
Headquarters Winnipeg, Manitoba

Unit	Location	Aircraft Types
Manning Depot		
No. 2 MD	Brandon, Man.	
Initial Training School		
No. 6 ITS	Saskatoon, Sask.	
Elementary Flying Training Schools		
No. 2 EFTS	Fort William, Ont.	Moth
No. 6 EFTS	Prince Albert, Sask.	Moth, Cornell
No. 14 EFTS	Portage la Prairie, Man.	Moth, Finch
No. 19 EFTS*	Virden, Man.	Moth, Cornell
No. 27 EFTS*	Neepawa, Man.	Moth, Cornell
Service Flying Training Schools		
No. 4 SFTS	Saskatoon, Sask.	Anson, Crane
No. 10 SFTS	Dauphin, Man.	Harvard
No. 11 SFTS	Yorkton, Sask.	Harvard, Crane, Anson
No. 12 SFTS	Brandon, Man.	Crane, Anson
No. 17 SFTS	Souris, Man.	Anson, Harvard
No. 18 SFTS	Gimli, Man.	Anson
No. 33 SFTS	Barberry, Man.	Anson
No. 35 SFTS	North Battleford, Sask.	Oxford
Air Observer Schools		
No. 5 AOS	Winnipeg, Man.	Anson
No. 6 AOS	Prince Albert, Sask.	Anson
No. 7 AOS	Portage la Prairie, Man.	Anson

Bombing and Gunnery Schools
No. 3 B&GS   Macdonald, Man.   Anson, Battle Bolingbroke, Lysander
No. 5 B&GS   Dafoe, Sask.   Anson, Battle, Bolingbroke, Lysander
No. 7 B&GS   Paulson, Man.   Anson, Battle, Bolingbroke, Lysander
Central Navigation School
No. 1 CNS   Rivers, Man.   Anson
Wireless School
No. 1 WS   Winnipeg, Man.   Moth, Stinson 105

No. 3 Training Command
Headquarters Montreal, Quebec

Initial Training School
No. 3 ITS   Victoriaville, Que.
Elementary Flying Training Schools
No. 4 EFTS   Windsor Mills, Que.   Finch, Moth
No. 11 EFTS   Cap de la Madeleine, Que.   Finch, Cornell
No. 13 EFTS   St. Eugene, Ont.   Finch
No. 17 EFTS   Stanley, N.S.   Finch, Moth
No. 21 EFTS   Chatham, N.B.   Finch
No. 22 EFTS   Quebec City, Que.   Finch
Service Flying Training Schools
No. 2 SFTS   Uplands (Ottawa), Ont.   Harvard, Yale
No. 8 SFTS   Moncton, N.B.   Anson
No. 13 SFTS   St. Hubert (Montreal), Que.   Harvard, Anson
Air Observer Schools
No. 8 AOS   Quebec City, Que.   Anson
No. 9 AOS   St. Jean, Que   Anson
No. 10 AOS   Chatham, N.B.   Anson
Bombing and Gunnery Schools
No. 9 B&GS   Mont-Joli, Que.   Anson, Battle, Bolingbroke, Lysander
No. 10 B&GS   Mount Pleasant, P.E.I.   Anson, Battle, Bolingbroke, Lysander
Wireless School
No. 3 WS   Montreal, Que.   Norseman, Moth
Flying Instructor School
No. 3 FIS   Arnprior, Ont.   various
Naval Air Gunners School
No. 1 NAGS   Yarmouth, N.S.   Swordfish
General Reconnaissance Schools
No. 1 GRS   Summerside, P.E.I.   Anson
No. 31 GRS   Charlottetown, P.E.I.   Anson

No. 4 Training Command
Headquarters Regina, Saskatchewan

Initial Training Schools
No. 2 ITS   Regina, Sask.
No. 4 ITS   Edmonton, Alta.
Elementary Flying Training Schools
No. 5 EFTS*   High River, Alta.   Moth, Cornell
No. 8 EFTS   Vancouver, B.C.   Moth
No. 15 EFTS   Regina, Sask.   Moth, Cornell
No. 16 EFTS   Edmonton, Alta.   Moth, Finch

No. 23 EFTS   Davidson, Sask.   Cornell
No. 24 EFTS   Abbotsford, B.C.   Cornell
No. 25 EFTS   Assiniboia, Sask.   Cornell
No. 31 EFTS   De Winton, Alta   Moth, Stearman, Cornell
No. 32 EFTS   Bowden, Alta.   Moth, Stearman, Cornell
No. 33 EFTS*   Caron, Sask.   Cornell
No. 36 EFTS   Pearce, Alta.   Moth, Stearman
Service Flying Training Schools
No. 3 SFTS   Calgary, Alta.   Anson, Crane
No. 7 SFTS   McLeod, Alta.   Anson
No. 15 SFTS   Claresholm, Alta.   Anson
No. 19 SFTS   Vulcan, Alta.   Anson
No. 32 SFTS   Moose Jaw, Sask.   Oxford
No. 34 SFTS   Medicine Hat, Alta.   Harvard, Oxford
No. 36 SFTS   Penhold, Alta.   Oxford
No. 37 SFTS   Calgary, Alta.   Oxford, Harvard, Anson
No. 38 SFTS   Estevan, Sask.   Anson
No. 39 SFTS   Swift Current, Sask.   Oxford
No. 41 SFTS   Weyburn, Sask.   Anson, Harvard
Air Observer Schools
No. 2 AOS   Edmonton, Alta.   Anson
No. 3 AOS   Regina, Sask.   Anson
Bombing and Gunnery Schools
No. 2 B&GS   Mossbank, Sask.   Anson, Battle Bolingbroke, Lysander
No. 8 B&GS   Lethbridge, Alta.   Anson, Battle, Bolingbroke, Lysander
Flying Instructor School
No. 2 FIS   Pearce, Alta.   various
Wireless School
No. 2 WS   Calgary, Alta.   Harvard, Fort

Output

Trade	RCAF	RAF	RAAF	RNZAF	Total
Pilot	25646	17796[1]	4045	2220	49707
Navigator	7280	6922	944	724	15870
Navigator B[2]	5154	3113	699	829	9795
Navigator W	421	3847		30	4298
Air Bomber	6659	7581	799	634	15673
Wireless Operator/ Air Gunner	12744	755	2975	2122	1896
Air Gunner	12917	2096[3]	244	443	15700
Flight Engineer	1913				1913
TOTAL	72734	42110	9706	7002	131552

Originally, only three types of aircrew were to be trained — pilots, air observers and wireless operators/air gunners — but the expansion of air operations, and the introduction of the heavy bomber with its large crew, made it necessary to train the eight categories listed above. Navigator B's received additional training as air bombers and Navigator W's as wireless operators.

---

[1] Includes 2629 Fleet Air Arm pilots.
[2] Includes air observers.
[3] Includes 704 naval air gunners.

Appendix 2

RAF "Tiger Force" (Pacific)

RCAF Element as proposed 23 November 1944

a. Headquarters: No. 6 (RCAF) Group.
b. 12 very long range heavy bomber squadrons, Nos. 405, 408, 415, 419, 420, 424, 425, 426, 427, 428, 429 and 431.
c. 6 long range fighter squadrons, Nos. 400, 401, 402, 403, 438 and 440.
d. 3 long range transport squadrons, Nos. 407, 422 and 423.
e. 1 air/sea rescue squadron, No. 404.
f. 6 heavy bomber stations, 2 squadrons per station, to be formed in Canada at Dartmouth, Debert, Pennfield, Greenwood, Yarmouth and Sydney.
g. 2 fighter stations, 3 squadrons per station, to be formed in Canada at Charlottetown and Chatham.
h. Transport wing to be formed in the United Kingdom.
i. Forward equipment depot to be formed in Canada at Moncton.
j. Forward repair depot to be formed in Canada at Scoudouc.

Proposed Composition — 10 July 1945

No. 5 (RAF) Group
No. 551 Wing: Nos. 83 and 97 (B) Squadrons
No. 552 Wing: Nos. 57 and 106 (B) Squadrons
No. 553 Wing: No. 9 (B) Squadron; No. 75 Squadron RNZAF
No. 554 Wing: Nos. 207 and 617 (B) Squadrons
No. 555 Wing: Nos. 460 and 467 (B) Squadrons RAAF
No. 627 (Pathfinder) Squadron
No. 128 (Photo Reconnaissance/Meteorological) Squadron
Nos. 49 and 189 (B) Squadrons (in reserve).

No. 6 (RCAF) Group
Commander Designate: A/V/M C.M. McEwen, later A/V/M C.R. Slemon
Base Group Headquarters: No. 1 Maintenance Wing; No. 101 Equipment Park; No. 41 Repair Depot.
No. 661 Wing: Nos. 419 and 428 (B) Squadrons
No. 662 Wing: Nos. 431 and 434 (B) Squadrons
No. 663 Wing: Nos. 420 and 425 (B) Squadrons
No. 664 Wing: Nos. 405 and 408 (B) Squadrons
No. 6 Group (Long Range) Transport Wing: Nos. 422, 423 and 426 (T) Squadrons.

Plans and preparations were as follows:

1. The eight RCAF bomber squadrons were to be converted to Canadian-built Lancaster B.Mk.X aircraft in the United Kingdom, and ferry them back to Canada in May-June 1945. Following a 30-day leave, the bomber squadrons were to begin reorganization and training in Canada; Nos. 661 and 662 Wings on 10 August, and Nos. 663 and 664 Wings on 24 August.

2. The three transport squadrons were to be re-equipped with Liberator aircraft in the United Kingdom — No. 426 Squadron to be operational by July, No. 422 by August and No. 423 by December — and to ferry their aircraft to Cawnpore, India, the main supply base. Eventually, the Liberators were to be replaced by York aircraft.

3. The air echelon of the force headquarters was to arrive in the Pacific by 1 September 1945.

4. The first operational units — four Lancaster squadrons of No. 5 (RAF) Group, and the Mosquito-equipped pathfinder squadron — were to be operational in the Pacific by 1 December.

5. The second contingent — four Lancaster squadrons including Nos. 419 and 428 Squadrons of No. 662 Wing, plus the photo reconnaissance/meteorological squadron — were to be operational by 1 January 1946.

6. If the full force of 20 heavy bomber squadrons was approved, the build-up was to be at the rate of four per month. The air/sea rescue squadron was to be introduced in the first month.

7. The first eight Lancaster squadrons (including Nos. 419 and 428 Squadrons) were to be re-equipped with Lincoln aircraft in July 1946. In the meantime, the remaining six RCAF bomber squadrons would be converted to Lincolns at No. 76 (RCAF) Training Base in the United Kingdom. Plans were also under way to build Lincoln aircraft in Canada.

8. No. 5 (RCAF) Operational Training Unit at Boundary Bay, British Columbia was to train replacement crews; and was to be converted from Liberator to Lancaster aircraft and subsequently, as they became available, Lincolns. The surrender of Japan on 14 August 1945 obviated any need for the programme. All RCAF units intended for "Tiger Force" were ordered to cease functioning, and were disbanded by 6 September 1945.

Appendix 3

RCAF Fighter Pilots in Korea 1950-1953

Between November 1950 and July 1953, 22 RCAF pilots were attached to the United States Fifth Air Force and flew with Sabre-equipped fighter-interceptor squadrons.

The first of the twenty-two was F/L J.A.O. Levesque, who was on exchange duty with 334th Fighter Interceptor Squadron (334 FIS) of the 4th Fighter Interceptor Wing (4 FIW) in the United States when the unit was ordered to Korea in November 1950. As the wing considered him combat-ready, F/L Levesque was authorized by the Canadian government to fly combat sorties with the squadron.

The RCAF subsequently established that any pilot applying for combat duty in Korea must have 50 hours flying experience on jets, and his tour would consist of 50 combat missions or six months, whichever he completed first. The first two qualified volunteers, F/O's S.B. Fleming and G.W. Nixon, were sent to Korea in March 1952 and, thereafter, one pilot each month.

Nine Canadian pilots were credited with a total of 9 MiG-15's* destroyed, 2 probably destroyed and 10 damaged (9-2-10):

F/L E.A. Glover (3-0-3) Attached to 334 FIS from June to October 1952, flew 50 sorties (71 hours); credited with 2 damaged on 30 August, 1 destroyed and 1 damaged on 8 September, 1 destroyed on 9 September, 1 destroyed on 16 September; awarded both British and American DFC's.

S/L J.D. Lindsay (2-0-3) During Second World War, credited with 7 aircraft destroyed and 5 damaged over Europe, won DFC; in Korea, attached to 39 FIS, 51 FIW from July to December 1952, flew 70 sorties (73 hours); credited with 2 damaged on 5 September, 1 destroyed on 11 October (29th sortie), 1 damaged on 25 October, 1 destroyed on 26 November (49th sortie); awarded DFC by the United States.

S/L J. MacKay (1-0-0) During Second World War, credited with 11 aircraft destroyed and 6½ damaged over Europe, won DFC and Bar; in Korea, attached to 51 FIW from March to July 1953, flew 50 sorties (71 hours); scored on 30 June; awarded US Air Medal.

F/L J.C.A. LaFrance (1-0-0) Attached to 39 FIS, 51 FIW from May to October 1952, flew 49 sorties (67 hours); scored on 5 August (22nd sortie); awarded DFC by the United States.

F/L J.A.O. Levesque (1-0-0) During Second World war, credited with 4 aircraft destroyed over Europe; in Korea, on exchange duty with 334 FIS from November 1950 to May 1951, flew 71 sorties (48 hours); scored on 30 March; awarded DFC by the United States.

F/L L.E. Spurr (1-0-0) During Second World War, credited with 1 Me. 262 jet fighter damaged; in Korea, attached to 25 FIS, 51 FIW from April to September 1952, flew 50 sorties (76 hours); scored on 14 July; awarded DFC by the United States.

F/O S.B. Fleming (0-1-2) Attached to 336 FIS, 4 FIW from March to August 1952, flew the most sorties (82) and hours (113) of all RCAF pilots; credited with 1 probable on 13 May, 1 damaged on 1 April, and 1 damaged on 21 April; awarded DFC by the United States.

F/L G.H. Nichols (0-1-0) Attached to 16 FIS, 51 FIW from January to May 1953, flew 50 sorties (74 hours); scored on 7 April; awarded US Air Medal.

F/O A. Lambros (0-0-2) October 1952 to March 1953; credited with 1 MiG damaged on both 22 and 31 January 1953.

Thirteen other RCAF pilots flew in Korea but did not score:

F/O G.W. Nixon (March - July 1952; awarded US Air Medal).

F/O J.D. Donald (April-May 1952; returned to Canada before combat-ready).

G/C E.B. Hale (flew with 51 FIW April-September 1952; awarded DFC by the United States).

F/L R.E. Lowry (20 sorties, July-November 1952)

S/L E.G. Smith (August-December 1952; awarded US Air Medal).

W/C R.T.P. Davidson (during Second World War, credited with 2 Japanese, 2 Italian and 2 German aircraft, won DFC; served in Korea September-December 1952).

S/L A.R. Mackenzie (during Second World War credited with 8½ aircraft, won DFC; shot down and captured on his fifth Korean sortie, 8 December 1952; released on 5 December 1954).

F/L F.W. Evans (December 1952-May 1953).

F/O R.D. Carew (February-June 1953).

S/L W.H.F. Bliss (April-July 1953).

S/L Fox (May-July 1953).

F/O J.B. Mullin (June-July 1953; no combat missions).

S/L D. Warren (July 1953; no combat missions).

---

*The 15th variant of a Soviet fighter designed by Mikoyan and Gurevich.

Appendix 4

Unit Codes

Canada — 1939-1942

On 1 August 1939, two-letter unit codes were issued to the RCAF's 23 operational squadrons then on strength, and first appeared on aircraft in August 1940. As further squadrons were formed, new combinations were assigned.

To identify aircraft as belonging to a home based squadron, each letter on the side of the fuselage was underlined with a bar. A third letter, also with bar, was assigned to individual aircraft by the unit; for example, Blackburn Shark No. 545 of No. 4 (Bomber Reconnaissance) Squadron was coded and marked "FY (roundel)Y" on the side and, on the under wing surface, "545↑S↑S" (the arrow supplied to indicate direction of flight).

Following Japan's entry into the war the RCAF was authorized, early in 1942, to form an additional ten fighter and six bomber reconnaissance squadrons; and in May 1942 a revised list of two-letter unit codes was issued, with instructions that these be applied at once to all operational aircraft. On 16 October 1942, however, Air Force Headquarters sent a signal to both Eastern and Western Air Command ordering that, for security reasons, the use of unit codes on home-based operational aircraft be discontinued immediately; and accordingly, only the individual aircraft letter remained.

In the following list, unit codes shown as in use "39-42" were those originally issued on 1 August 1939 and used until May 1942; the "40-42" codes were issued during 1940 and as new units were formed, and used until May 1942; and "42" denotes from May to 16 October 1942.

Code	Unit	Code	Unit	Code	Unit
AE 42	130 Sqn	GK 40-42	162 Sqn	PQ 42	117 Sqn
AF 42	6 Sqn	GR 42	119 Sqn	QE 40-42	12 Sqn
AG 42	122 Sqn	GV 40-42	3 CAC Det.	QG 42	6 CAC Det
AN 40-42	13 Sqn	HA 42	129 Sqn	QN 39-42	5 Sqn
AN 42	160 Sqn	HJ 42	9 Sqn	RA 41-42	School of AC
AY 39-42	110 Sqn	JK 42	10 Sqn	RA 42	128 Sqn
AZ 42	3 OTU	JY 39-42	121 Sqn	RE 39-42	118 Sqn
BA 42	125 Sqn	KA 39-42	9 Sqn	RS 42	120 Sqn
BD 42	4 Sqn	KL 42	11 Sqn	SZ 42	147 Sqn
BF 40-42	14 Sqn	KO 39-42	2 Sqn	TF 42	127 Sqn
BK 39-42	115 Sqn	LM 42	113 Sqn	TM 39-42	111 Sqn
BT 39-42	113 Sqn	LR 42	31 OTU	TN 42	161 Sqn
BV 42	126 Sqn	LT 39-42	7 Sqn	TQ 40-42	2 CAC Det
DE 42	5 Sqn	LU 40-42	1 CAC Det	UV 42	115 Sqn
DM 39-42	119 Sqn	LV 42	5 CAC Det	UY 42	1 OTU
DZ 42	162 Sqn	LZ 42	111 Sqn	VD 42	123 Sqn
EA 42	145 Sqn	MA 42	114 Sqn	VW 42	118 Sqn
ED 42	4 OTU	MK 42	13 Sqn	XE 39-42	6 Sqn
EF 42	36 OTU	MR 42	4 CAC Det	XO 39-42	112 Sqn
EN 42	121 Sqn	MX 39-42	120 Sqn	XP 42	135 Sqn
ET 42	2 OTU	NA 39-42	1 Sqn	XY 42	1 CAC Det
EX 39-42	117 Sqn	NK 42	31 GRS	YA 42	14 Sqn
FD 39-42	114 Sqn	NO 42	116 Sqn	YH 42	1 GRS
FG 42	7 Sqn	OP 39-42	3 Sqn	YO 39-42	8 Sqn
FN 42	133 Sqn	OP 42	32 OTU	YZ 42	2 CAC Det
FY 39-42	4 Sqn	OY 39-42	11 Sqn	ZD 39-42	116 Sqn
FY 42	34 OTU	PB 39-42	10 Sqn	ZM 42	149 Sqn
GA 42	8 Sqn	PO 40-42	4 CAC Det	ZR 42	132 Sqn
				ZR 42	132 Sqn

In the summer of 1942, Dartmouth-based Hurricane aircraft of Nos. 126, 127 and 129 (Fighter) Squadron were maintained by a Central Maintenance Unit and the individual unit codes were replaced by the single digit "1" or "2".

Overseas — 1940-1946

For overseas units of the RCAF, the British Air Ministry assigned the following codes:

Code	Unit	Code	Unit	Code	Unit
AB	423 Sqn[1]	F3	438 Sqn	QY	1666 HCU[6]
AE	402 Sqn	6U	415 Sqn[3]	RA	410 Sqn
AL	429 Sqn	GX	415 Sqn[3]	RR	407 Sqn[2]
AN	417 Sqn	G9	441 Sqn	RU	414 Sqn
AU	421 Sqn	HU	406 Sqn	RV	1659 HCU[4]
AW	664 Sqn	I8	440 Sqn	SE	431 Sqn
BM	433 Sqn	9G	441 Sqn	SP	400 Sqn
BX	666 Sqn	KH	403 Sqn	SW	1679 HCU
2I	443 Sqn	KP	409 Sqn	TH	418 Sqn
CI	407 Sqn[2]	KW	425 Sqn	VR	419 Sqn
DB	411 Sqn	LQ	405 Sqn	VZ	412 Sqn
DG	422 Sqn	NA	428 Sqn	WL	434 Sqn
DH	1664 HCU[5]	ND	1666 HCU[6]	YI	423 Sqn[1]
DN	416 Sqn	NH	415 Sqn[3]	YO	401 Sqn
EE	404 Sqn	OW	426 Sqn	YZ	442 Sqn
EQ	408 Sqn	PT	420 Sqn	ZL	427 Sqn
5V	439 Sqn	QB	424 Sqn	ZU	1664 HCU[5]
FD	1659 HCU[4]	QL	413 Sqn	Z2	437 Sqn
		QO	432 Sqn		

For a time in 1943 and early 1944, RAF Coastal Command, as a security measure, discontinued using two-letter individual unit codes and had each of its stations identify their squadrons by a single digit number. Those known to be affected by this change were No. 404, No. 422 and No. 423 Squadron. Both 404 and 422 aircraft carried the digit "2", while 423 used "3".

On 8 September 1945, RAF Transport Command assigned to its units a four-letter unit code. The first letter, "O", denoted the command, the second letter indicated the type of aircraft — "A" Anson, "D" and "F" Dakota, "K" Lancaster, "L" Liberator, "Q" Sunderland, "R" Stirling, "S" Skymaster, "X" Warwick, "Y" York and "Z" miscellaneous civilian types — the third letter identified the squadron, and the fourth was an individual letter assigned by the squadron. The following three-letter codes were allotted to the four RCAF transport squadrons: No. 426, "OWL," No. 435 "ODM," No. 436, "ODN; No. 437, "ODO."

Post-War Codes — 1947-1958

On 9 May 1947, a five-letter "VC" registration was promulgated in Air Force Routine Order 250 in accordance with the new International Civil Aviation Organization system of registering individual service aircraft. For example, Sabre 19102 of No. 410 (Fighter) Squadron was registered as "VC-AMN," the "VC" denoting the RCAF and "AM" denoting No. 410 Squadron; and "N" was assigned by the squadron to that particular aircraft. The Royal Canadian Navy was "VG"; and civilian aircraft were "CF." Problems occurred under this system, for whenever an aircraft was transferred to another unit it was necessary to re-register it and apply new markings. Maintaining the VC code was found to be too costly and time-consuming, and since no other military organization had adopted the ICAO system it was discontinued on 19 November 1951. New two-letter codes were now assigned to all flying units of the RCAF, and the individual aircraft letter was replaced by the last three digits of the aircraft's serial number. This system remained in effect until early 1958, when the two- letter combination was replaced by "RCAF."

Using the example of Sabre 19102, and assuming that it remained continuously with No. 410 Squadron from 1947 to 1958, the following markings would have applied:

1947-1951: Lower wing surfaces VC↑AMN; upper wing AM↑N (arrows supplied to indicate direction of flight). Fuselage side AM [roundel] N;

1951-1958: Wing undersurface 102↑ZOL. Fuselage side AM [roundel] 102;

1958 onward: Wing undersurface as above. Fuselage side RCAF [roundel] 102.

In the following list, unit codes shown as in use prior to 1952 (e.g. "47-51") were part of the "VC" registration:

AA 47-51 400 Sqn	BA 47-58 424 Sqn	KE 53-58 440 Sqn
AB 47-51 401 Sqn	BB 52-58 427 Sqn	KH 51-58 411 Sqn
AC 47-51 402 Sqn	BC 47-51 426 Sqn	LP 54-58 409 Sqn
AD 48-51 403 Sqn	BH 51-58 430 Sqn	MN 51-58 408 Sqn
AF 51 404 Sqn	BN 47-51 435 Sqn	NQ 53-58 423 Sqn
AG 50-51 405 Sqn	BQ 47-58 438 Sqn	PF 51-58 443 Sqn
AH 47-51 406 Sqn	BR 52-58 434 Sqn	PR 51-58 403 Sqn
AK 49-51 408 Sqn	BT 51-58 441 Sqn	QP 51-58 406 Sqn
AM 48-56 410 Sqn	BU 47-58 442 Sqn	QT 51-58 401 Sqn
AN 50-51 411 Sqn	BV 53-58 444 Sqn	RX 52-58 407 Sqn[7]
AN 56-58 410 Sqn	DL 54-58 432 Sqn	SA 53-58 445 Sqn
AO 47-51 412 Sqn	FG 54-58 433 Sqn	SP 51-58 404 Sqn[7]
AP 47-58 413 Sqn	GW 51-58 400 Sqn	SV 51-58 402 Sqn
AQ 47-58 414 Sqn	HG 54-58 428 Sqn	TF 53-58 422 Sqn
AS 51-58 416 Sqn	HO 51-58 418 Sqn	UD 54-58 419 Sqn
AT 47-48 417 Sqn	IG 51-58 439 Sqn	VN 51-58 405 Sqn
AU 47-51 418 Sqn	JF 52-58 3 OTU	XK 51-58 406 Sqn
AW 47-51 420 Sqn	JW 52-58 3 OTU	XV 50-58 2 OTU
AX 49-58 421 Sqn		

---

[1] Aircraft were coded "AB" until 1943, then "YI".
[2] As a coastal strike squadron, to January 1943, "PR"; thereafter, as a general reconnaissance unit, "CI."
[3] As a torpedo bomber squadron, under Coastal Command, used "GX" on Beauforts, Hampdens and Wellingtons, and "NH" on Albacores, until July 1944 and, when transferred to Bomber Command, "6U."
[4] "RV" 1943-45; also, in 1944-45, "FD."
[5] "ZU" 1943-45; also, in 1944-45, "DH."
[6] Used both "ND" and "QY" 1943-45.
[7] When Lancaster and Neptune aircraft were transferred from No. 404 to 407 Squadron, they retained the "SP" code until repainted.

Appendix 5
RCAF Order of Battle

10 September 1939
Canada Declares War on Germany

Unit	Location	Aircraft Type*
EASTERN AIR COMMAND		
HQ Halifax, NS		
No. 10 (B) Sqn	Halifax, NS	Wapiti
No. 116 (F) Sqn (Aux)	Halifax, NS	nil
No. 5 (GR) Sqn	Dartmouth, NS	Stranraer
No. 8 (GP) Sqn	Sydney, NS	Delta
No. 2 (AC) Sqn	Saint John, NB	Atlas
No. 117 (CAC) Sqn (Aux)	Saint John, NB	nil
No. 1 (F) Sqn	St Hubert, Que	Hurricane I

Note: No. 1 Squadron was en route from Calgary to its "war station" at Dartmouth, arrived on 5 November.

WESTERN AIR COMMAND		
HQ Vancouver, BC		
No. 4 (GR) Sqn	Jericho Beach, BC	Stranraer
No. 6 (TB) Sqn	Jericho Beach, BC	Shark
No. 111 (CAC) Sqn (Aux)	Sea Island, BC	Avro 621, 626
No. 112 (AC) Sqn (Aux)	Winnipeg, Man	Avro 626
No. 113 (F) Sqn (Aux)	Calgary, Alta	nil
No. 120 (B) Sqn (Aux)	Regina, Sask	DH82A

AIR TRAINING COMMAND		
HQ Toronto, Ont		
No. 110 (AC) Sqn (Aux)	Toronto, Ont	Avro 626
No. 114 (B) Sqn (Aux)	London, Ont	nil
No. 119 (B) Sqn (Aux)	Hamilton, Ont	DH82A

*An oblique between two different types of aircraft indicates that the squadron was in the process of converting from one type of aircraft to another. A comma indicates that both types of aircraft were on squadron strength.

AIR FORCE HEADQUARTERS		
HQ Ottawa, Ont		
No. 7 (GP) Sqn	Rockcliffe, Ont	FC 71, Bellanca
No. 115 (F) Sqn (Aux)	Montreal, Que	Fawn
No. 118 (B) Sqn (Aux)	Montreal, Que	DH60
No. 121 (B) Sqn (Aux)	Quebec City, Que	nil

Notes:
1. The following were disbanded because either their organizations were incomplete or to provide needed reinforcements to other operational units: 30 September 1939 — No. 113 at Calgary and No. 121 at Quebec City; 3 October 1939 — No. 7 at Rockcliffe which was reduced to the AFHQ Communications Flight; 20 October 1939 — No. 114 at London; 28 October 1939 — No. 117 at Saint John; 2 November 1939 — No. 116 at Halifax; 16 December 1939 — No. 2 at Rockcliffe to provide reinforcements for Nos. 110 and 112 prior to their going overseas to England in early 1940; 26 May 1940 — No. 115 at Montreal which merged with No. 1 prior to its going overseas to England.
2. Following squadrons were redesignated Bomber Reconnaissance (BR) on 31 October 1939: Nos. 4, 5, 6, 8, 10, 111, 118, 119 and 120.
3. During May and June 1940, three squadrons were sent to England, Nos. 1 (F), 110 and 112 (AC).

1 January 1943
Formation of No. 6 (RCAF) Group in RAF Bomber Command

HOME WAR ESTABLISHMENT
EASTERN AIR COMMAND
HQ Halifax, NS

No. 10 (BR) Sqn	Dartmouth, NS	Digby
No. 11 (BR) Sqn	Dartmouth, NS	Hudson III
No. 116 (BR) Sqn	Dartmouth, NS	Catalina
No. 117 (BR) Sqn	Dartmouth, NS	Canso
No. 126 (F) Sqn	Dartmouth, NS	Hurricane I
No. 129 (F) Sqn	Dartmouth, NS	Hurricane I
No. 121 (K) Sqn	Dartmouth, NS	various
No. 113 (BR) Sqn	Yarmouth, NS	Hudson III
No. 162 (BR) Sqn	Yarmouth, NS	Canso A
No. 119 (BR) Sqn	Sydney, NS	Hudson III
No. 128 (F) Sqn	Sydney, NS	Hurricane I
No. 130 (F) Sqn	Bagotville, Que	Hurricane XII
No. 1 GROUP, HQ St John's, Nfld		
No. 5 (BR) Sqn	Gander, Nfld	Canso A
No. 127 (F) Sqn	Gander, Nfld	Hurricane I
No. 145 (BR) Sqn	Torbay, Nfld	Hudson III
No. 125 (F) Sqn	Torbay, Nfld	Hurricane I
No. 12 (OPERATIONAL TRAINING) GROUP, HQ Halifax, NS		
No. 31 (RAF) OTU	Debert, NS	ferry
No. 32 (RAF) OTU	Patricia Bay, BC	torpedo bomber
No. 34 (RAF) OTU	Pennfield Ridge, NB	light bomber
No. 1 (RCAF) OTU	Bagotville, Que	fighter
No. 3 (RCAF) OTU	Patricia Bay, BC	bomber recce

WESTERN AIR COMMAND
HQ Vancouver, BC

No. 2 GROUP, HQ Victoria, BC
Formed to cover the move of Western Air Command Headquarters from Victoria to Vancouver from 1 January to 15 March 1943

No. 147 (BR) Sqn	Sea Island, BC	Bolingbroke II
No. 14 (F) Sqn	Sea Island, BC	Kittyhawk I
No. 149 (TB) Sqn	Patricia Bay, BC	Beaufort
No. 135 (F) Sqn	Patricia Bay, BC	Hurricane XII
No. 122 (K) Sqn	Patricia Bay, BC	various
No. 4 (BR) Sqn	Ucluelet, BC	Canso A
No. 120 (BR) Sqn	Coal Harbour, BC	Stranraer
No. 132 (F) Sqn	Tofino, BC	Kittyhawk I
No. 133 (F) Sqn	Boundary Bay, BC	Hurricane XII
No. 4 GROUP, HQ Prince Rupert, BC		
No. 6 (BR) Sqn	Alliford Bay, BC	Stranraer
No. 7 (BR) Sqn	Prince Rupert, BC	Shark
No. 9 (BR) Sqn	Bella Bella, BC	Stranraer
RCAF WINGS, DETACHED OPERATIONS, ALASKA		
"X" WING, HQ Anchorage, Alaska		
No. 8 (BR) Sqn	Anchorage, Alaska	Bolingbroke IV
No. 111 (F) Sqn	Kodiak, Alaska	Kittyhawk I

Note: No. 8 (BR) Squadron transferred to Sea Island on 26 February 1943; replaced by No. 14 (F) Squadron with Kittyhawk I's on 3 March 1943.

"Y" WING, HQ Annette Island, Alaska

No. 115 (BR) Sqn	Annette Island, Alaska	Bolingbroke IV
No. 118 (F) Sqn	Annette Island, Alaska	Kittyhawk I

AIR FORCE HEADQUARTERS
Ottawa, Ont

No. 12 (Comm) Sqn	Rockcliffe, Ont	various
No. 124 (Ferry) Sqn	Rockcliffe, Ont	

BRITISH COMMONWEALTH AIR TRAINING PLAN
No. 1 Training Command, HQ Toronto, Ont
No. 2 Training Command, HQ Winnipeg, Man
No. 3 Training Command, HQ Montreal, Que
No. 4 Training Command, HQ Regina, Sask

OVERSEAS WAR ESTABLISHMENT
RAF BOMBER COMMAND

No. 4 GROUP
No. 429 (B) Sqn	East Moor, Yorks	Wellington III/X
No. 431 (B) Sqn	Burn, Yorks	Wellington III/X

No. 6 (RCAF) GROUP, HQ Allerton Park, Yorks
No. 408 (B) Sqn	Leeming, Yorks	Halifax II
No. 419 (B) Sqn	Middleton St. George, Durham	Halifax II
No. 420 (B) Sqn	Middleton St. George, Durham	Wellington III
No. 424 (B) Sqn	Topcliffe, Yorks	Wellington III
No. 425 (B) Sqn	Dishforth, Yorks	Wellington III
No. 426 (B) Sqn	Dishforth, Yorks	Wellington III
No. 427 (B) Sqn	Croft, Yorks	Wellington III
No. 428 (B) Sqn	Dalton, Yorks	Wellington III

Notes:
1. No. 429 Squadron joined No. 6 Group on 1 April and No. 431 on 15 July 1943.
2. No. 405 (B) Squadron was on loan to RAF Coastal Command.
3. From May to October 1943, No. 331 (Medium Bomber) Wing RCAF, composed of Nos. 420, 424 and 425 Squadrons, served in North Africa in support of the Allied invasions of Sicily and Italy.

RAF FIGHTER COMMAND

No. 10 GROUP
No. 400 (AC) Sqn	Middle Wallop, Hants	Mustang I
No. 406 (NF) Sqn	Middle Wallop, Hants	Beaufighter IV
No. 421 (F) Sqn	Angle, S. Wales	Spitfire VB

Note: No. 400 Squadron was on attachment from RAF Army Co-operation Command for training in fighter tactics and for daylight intruder operations over northern France.

No. 11 GROUP
No. 418 (I) Sqn	Bradwell Bay, Essex	Boston III

CANADIAN KENLEY WING
No. 402 (F) Sqn	Kenley, Surrey	Spitfire IX
No. 401 (F) Sqn	Kenley, Surrey	Spitfire VB
No. 412 (F) Sqn	Kenley, Surrey	Spitfire VB
No. 416 (F) Sqn	Redhill, Surrey	Spitfire VB

No. 12 GROUP
No. 409 (NF) Sqn	Coleby Grange, Lincs	Beaufighter VI

CANADIAN DIGBY WING
No. 411 (F) Sqn	Digby, Lincs	Spitfire VB

No. 13 GROUP
No. 403 (F) Sqn	Catterick, Yorks	Spitfire VB/IX
No. 410 (NF) Sqn	Acklington, Northumberland	Mosquito II

Notes:
1. Nos. 401 and 403 Squadrons exchanged stations on 23 January.
2. Nos. 421 and 412 Squadrons exchanged stations on 29 January.
3. No. 416 Squadron moved to Kenley on 1 February.

RAF COASTAL COMMAND

No. 15 GROUP
No. 422 (GR) Sqn	Oban, Scot	Sunderland III
No. 423 (GR) Sqn	Castle Archdale, N. Ire	Sunderland III

No. 16 GROUP
No. 407 (CS) Sqn	Docking, Norfolk	Hudson V
No. 415 (TB) Sqn	Thorney Island, Hants	Hampden I

No. 18 GROUP
No. 404 (CF) Sqn	Dyce, Scot	Beaufighter II
No. 405 (B) Sqn	Beaulieu, Hants	Halifax II

Note: No. 405 Squadron was on loan from RAF Bomber Command for anti-submarine patrols over the Bay of Biscay.

RAF ARMY CO-OPERATION COMMAND

No. 70 GROUP
No. 39 (RCAF) WING, HQ Leatherhead, Surrey
No. 414 (AC) Sqn	Dunsfold, Surrey	Mustang I
No. 430 (AC) Sqn	Hartford Bridge, Hants	Mustang I

Notes:
1. No. 400 (AC) Squadron was detached to RAF Fighter Command, No. 10 Group, for training in fighter tactics and for daylight intruder operations over northern France.
2. No. 430 Squadron moved to Dunsfold on 12 January.

RAF SOUTH EAST ASIA COMMAND

No. 222 GROUP
No. 413 (GR) Sqn	Koggala, Ceylon	Catalina

RAF MIDDLE EAST COMMAND

AHQ EGYPT
No. 417 (F) Sqn	Idku, Egypt	Spitfire VB

6 June 1944
Allied Invasion of Europe

HOME WAR ESTABLISHMENT
EASTERN AIR COMMAND
HQ Halifax, NS
No. 145 (BR) Sqn	Dartmouth, NS	Ventura GRV
No. 126 (F) Sqn	Dartmouth, NS	Hurricane XII
No. 121 (K) Sqn	Dartmouth, NS	various
No. 167 (Comm) Sqn	Dartmouth, NS	various
No. 113 (BR) Sqn	Sydney, NS	Ventura GRV
No. 160 (BR) Sqn	Yarmouth, NS	Canso A

No. 1 GROUP, HQ St. John's, Nfld
No. 5 (BR) Sqn	Torbay, Nfld	Canso A
No. 11 (BR) Sqn	Torbay, Nfld	Liberator III
No. 10 (BR) Sqn	Gander, Nfld	Liberator III
No. 116 (BR) Sqn	Gander, Nfld	Canso A
No. 127 (F) Sqn	Gander, Nfld	Hurricane XII
No. 129 (F) Sqn	Gander, Nfld	Hurricane XII

No. 5 (GULF) GROUP, HQ Gaspé, Que
No. 161 (BR) Sqn	Gaspé, Que	Canso A

Notes:
1. No. 162 (BR) Squadron was on loan to RAF Coastal Command, AHQ Iceland.
2. No. 116 Squadron moved to Sydney on 7 June.
3. No. 11 Squadron moved to Dartmouth on 18 June.
4. No. 113 Squadron moved to Torbay on 21 June, and was disbanded on 23 August.

No. 12 (OPERATIONAL TRAINING) GROUP, HQ Halifax, NS
No. 31 (RAF) OTU	Debert, NS	ferry
No. 36 (RAF) OTU	Greenwood, NS	fighter bomber
No. 1 (RCAF) OTU	Bagotville, Que	fighter
No. 3 (RCAF) OTU	Patricia Bay, BC	bomber recce
No. 5 (RCAF) OTU	Boundary Bay, BC	bomber (Liberator)
No. 6 (RCAF) OTU	Comox, BC	transport

WESTERN AIR COMMAND
HQ Vancouver, BC
No. 8 (BR) Sqn	Patricia Bay, BC	Ventura GRV
No. 135 (F) Sqn	Patricia Bay, BC	Kittyhawk I
No. 122 (K) Sqn	Patricia Bay, BC	various
No. 133 (F) Sqn	Sea Island, BC	Kittyhawk I
No. 166 (Comm) Sqn	Sea Island, BC	various
No. 4 (BR) Sqn	Ucluelet, BC	Canso A
No. 6 (BR) Sqn	Coal Harbour, BC	Canso A
No. 7 (BR) Sqn	Alliford Bay, BC	Canso A
No. 9 (BR) Sqn	Bella Bella, BC	Canso A
No. 115 (BR) Sqn	Tofino, BC	Ventura GRV
No. 132 (F) Sqn	Tofino, BC	Kittyhawk I

Note: No. 115 Squadron was disbanded on 23 August, and No. 132 on 30 September.

AIR FORCE HEADQUARTERS
Ottawa, Ont

DIRECTORATE OF AIR TRANSPORT COMMAND, HQ OTTAWA, ONT
No. 164 (T) Sqn	Moncton, NB	Dakota, Lodestar
No. 165 (T) Sqn	Sea Island, BC	Dakota, Lodestar
No. 168 (HT) Sqn	Rockcliffe, Ont	Fortress, Dakota
No. 12 (Comm) Sqn	Rockcliffe, Ont	various
No. 124 (Ferry) Sqn	St Hubert, Que	
No. 170 (Ferry) Sqn	Winnipeg, Man	

No. 7 (PHOTOGRAPHIC) WING, HQ Rockcliffe, Ont
No. 13 (P) Sqn	Rockcliffe, Ont	Mitchell, Canso A

BRITISH COMMONWEALTH AIR TRAINING PLAN
No. 1 Training Command, HQ Trenton, Ont
No. 2 Training Command, HQ Winnipeg, Man
No. 3 Training Command, HQ Montreal, Que
No. 4 Training Command, HQ Regina, Sask

NORTH WEST AIR COMMAND
HQ Edmonton, Alta
no operational flying units were assigned

OVERSEAS WAR ESTABLISHMENT
RAF BOMBER COMMAND
No. 6 (RCAF) GROUP, HQ Allerton Park, Yorks
No. 61 (TRAINING) BASE, HQ Topcliffe, Yorks

No. 1659 HCU	Topcliffe, Yorks	Halifax
No. 1664 HCU	Dishforth, Yorks	Halifax
No. 1666 HCU	Wombleton, Yorks	Halifax
No. 1695 Bomber (Defence) Training Flight	Dalton, Yorks	

Note: In November, Nos. 1659 and 1666 HCUs were converted to Lancaster Xs.

No. 62 "BEAVER" BASE, HQ Linton-on-Ouse, Yorks

No. 408 (B) Sqn	Linton-on-Ouse, Yorks	Lancaster II
No. 426 (B) Sqn	Linton-on-Ouse, Yorks	Halifax VII
No. 420 (B) Sqn	Tholthorpe, Yorks	Halifax III
No. 425 (B) Sqn	Tholthorpe, Yorks	Halifax III
No. 432 (B) Sqn	East Moor, Yorks	Halifax III

No. 63 BASE, HQ Leeming, Yorks

No. 427 (B) Sqn	Leeming, Yorks	Halifax III
No. 429 (B) Sqn	Leeming, Yorks	Halifax III
No. 424 (B) Sqn	Skipton-on-Swale, Yorks	Halifax III
No. 433 (B) Sqn	Skipton-on-Swale, Yorks	Halifax III

No. 64 BASE, HQ Middleton St. George, Durham

No. 419 (B) Sqn	Middleton St. George, Durham	Lancaster X
No. 428 (B) Sqn	Middleton St. George, Durham	Halifax II/Lancaster X
No. 431 (B) Sqn	Croft, Yorks	Halifax III
No. 434 (B) Sqn	Croft, Yorks	Halifax III

No. 8 (PATHFINDER) GROUP

No. 405 (B) Sqn	Gransden Lodge, Beds	Lancaster I

RAF COASTAL COMMAND
No. 15 GROUP

No. 422 (GR) Sqn	Castle Archdale, N. Ire	Sunderland III
No. 423 (GR) Sqn	Castle Archdale, N. Ire	Sunderland III

No. 16 GROUP

No. 415 (TB) Sqn	Bircham Newton, Norfolk	Wellington XII (L/L), Albacore I

No. 19 GROUP

No. 404 (CF) Sqn	Davidstow Moor, Cornwall	Beaufighter TFX
No. 407 (GR) Sqn	Chivenor, Devon	Wellington XIV (L/L)

AHQ ICELAND

No. 162 (BR) Sqn	Reykjavik, Iceland	Canso A

Notes:
1. No. 415 Squadron was redesignated Bomber and transferred to No. 6 (RCAF) Group on 12 July.
2. No. 162 Squadron was on loan from RCAF Eastern Air Command, and operations at this time were being carried out from Wick, Scotland.

ALLIED EXPEDITIONARY AIR FORCE
RAF AIR DEFENCE GREAT BRITAIN

Formerly RAF Fighter Command, it was redesignated ADGB on 15 November 1943 when Second Tactical Air Force was formed, and reverted to RAF Fighter Command on 15 October 1944.

No. 10 GROUP

No. 406 (NF) Sqn	Winkleigh, Devon	Beaufighter IV/Mosquito XII

No. 11 GROUP

No. 418 (I) Sqn	Holmsley South, Sussex	Mosquito II

SECOND TACTICAL AIR FORCE
No. 83 (COMPOSITE) GROUP
No. 17 (RCAF) SECTOR, HQ en route to France
No. 126 (F) WING, HQ Tangmere, Sussex

No. 401 (F) Sqn	Tangmere, Sussex	Spitfire IXB
No. 411 (F) Sqn	Tangmere, Sussex	Spitfire IXB
No. 412 (F) Sqn	Tangmere, Sussex	Spitfire IXB

No. 127 (F) WING, HQ Tangmere, Sussex

No. 403 (F) Sqn	Tangmere, Sussex	Spitfire IXB
No. 416 (F) Sqn	Tangmere, Sussex	Spitfire IXB
No. 421 (F) Sqn	Tangmere, Sussex	Spitfire IXB

No. 144 (F) WING, HQ en route to France

No. 441 (F) Sqn	Ford, Sussex	Spitfire IXB
No. 442 (F) Sqn	Ford, Sussex	Spitfire IXB
No. 443 (F) Sqn	Ford, Sussex	Spitfire IXB

No. 22 (RCAF) SECTOR, HQ en route to France
No. 143 (FB) WING, HQ Hurn, Hants

No. 438 (FB) Sqn	Hurn, Hants	Typhoon IB
No. 439 (FB) Sqn	Hurn, Hants	Typhoon IB
No. 440 (FB) Sqn	Hurn, Hants	Typhoon IB

No. 121 (RAF) WING
No. 124 (RAF) WING
No. 39 (RCAF) SECTOR, HQ en route to France
No. 128 (R) WING, HQ Odiham, Hants

No. 400 (FR) Sqn	Odiham, Hants	Spitfire XI
No. 414 (FR) Sqn	Odiham, Hants	Mustang I
No. 430 (FR) Sqn	Odiham, Hants	Mustang I

No. 15 (RAF) SECTOR
No. 129 (FB) WING RCAF, HQ Westhampnett, Sussex
no RCAF units assigned
No. 85 (BASE) GROUP
No. 142 (RAF) WING

No. 402 (F) Sqn	Horne, Yorks	Spitfire VC

No. 147 (RAF) WING

No. 409 (NF) Sqn	West Malling, Kent	Mosquito XIII

No. 149 (RAF) WING

No. 410 (NF) Sqn	Hunsdon, Herts	Mosquito XIII

Note: 2nd TAF was reorganized on 12 July. Disbanded were the Sector Headquarters, Nos. 144 and 129 (RCAF) Wings, and No. 39 (RCAF) Sector merged with No. 128 (RCAF) Wing to form No. 39 (R) Wing RCAF. The wing headquarters then reported directly to its respective group headquarters, and the wing strength was increased from three to four squadrons.

RAF SOUTH EAST ASIA COMMAND
No. 222 GROUP

No. 413 (GR) Sqn	Koggala, Ceylon	Catalina

ALLIED MEDITERRANEAN AIR COMMAND
RAF DESERT AIR FORCE
No. 211 GROUP
No. 244 (F) WING

No. 417 (F) Sqn	Cassino, Italy	Spitfire VIII

10 August 1945
Japan Surrenders

WAR HOME ESTABLISHMENT
EASTERN AIR COMMAND
HQ Halifax, NS

No. 10 (BR) Sqn	Torbay, Nfld	Liberator III/V
No. 121 (K) Sqn	Dartmouth, NS	various
No. 167 (Comm) Sqn	Dartmouth, NS	various

Note: No. 10 Squadron was disbanded on 15 August, No. 121 on 30 September, and No. 167 on 1 October.

No. 6 (RCAF) GROUP, Advance HQ Halifax, NS
No. 661 (HB) WING, HQ Yarmouth, NS

No. 419 (B) Sqn	Yarmouth, NS	Lancaster X
No. 428 (B) Sqn	Yarmouth, NS	Lancaster X

No. 662 (HB) WING, HQ Dartmouth, NS

No. 431 (B) Sqn	Dartmouth, NS	Lancaster X
No. 434 (B) Sqn	Dartmouth, NS	Lancaster X

No. 663 (HB) WING, HQ Debert, NS

No. 420 (B) Sqn	Debert, NS	Lancaster X
No. 425 (B) Sqn	Debert, NS	Lancaster X

No. 664 (HB) WING, HQ Greenwood, NS

No. 405 (B) Sqn	Greenwood, NS	Lancaster X
No. 408 (B) Sqn	Greenwood, NS	Lancaster X

No. 1 MAINTENANCE WING, HQ Scoudouc, NB

Note: No. 6 Group was in Canada for reorganization and training for its employment with RAF "Tiger Force" in the Pacific Theatre of Operations. The Group and its components were disbanded on 5 September.

WESTERN AIR COMMAND
HQ Vancouver, BC

No. 11 (BR) Sqn	Patricia Bay, BC	Liberator III/V
No. 133 (F) Sqn	Patricia Bay, BC	Mosquito FB26
No. 135 (F) Sqn	Patricia Bay, BC	Kittyhawk I
No. 122 (K) Sqn	Patricia Bay, BC	various
No. 166 (Comm) Sqn	Sea Island, BC	various

Note: Nos. 133 and 135 Squadrons were disbanded on 10 September, Nos. 11 and 122 on 15 September, and No. 166 on 31 October.

NORTH WEST AIR COMMAND
HQ Edmonton, Alta

no operational flying units assigned

No. 1 AIR COMMAND
HQ Trenton, Ont

no operational flying units assigned

No. 2 AIR COMMAND
HQ Winnipeg, Man

no operational flying units assigned

AIR FORCE HEADQUARTERS
Ottawa, Ont

No. 5 (RCAF) OTU	Boundary Bay, BC	Lancaster X
No. 6 (RCAF) OTU	Comox, BC	transport (Liberator)

Note: No. 5 OTU was disbanded on 31 October, and No. 6 on 31 March 1946.

No. 7 (PHOTOGRAPHIC) WING, HQ Rockcliffe, Ont

No. 13 (P) Sqn	Rockcliffe, Ont	Mitchell, Canso A

No. 9 (TRANSPORT) GROUP, HQ Rockcliffe, Ont

No. 164 (T) Sqn	Moncton, NB	Dakota
No. 165 (T) Sqn	Sea Island, BC	Dakota
No. 168 (HT) Sqn	Rockcliffe, Ont	Fortress, Liberator
No. 12 (Comm) Sqn	Rockcliffe, Ont	various
No. 124 (Ferry) Sqn	St Hubert, Que	
No. 170 (Ferry) Sqn	Winnipeg, Man	

Note: No. 170 Squadron was disbanded on 30 September, No. 165 on 31 October, No. 168 on 21 April 1946, and No. 124 on 30 September 1946.

OVERSEAS WAR ESTABLISHMENT
RAF BOMBER COMMAND

No. 1 GROUP
No. 63 (RCAF) BASE, HQ Leeming, Yorks

No. 424 (B) Sqn	Skipton-on-Swale, Yorks	Lancaster I
No. 433 (B) Sqn	Skipton-on-Swale, Yorks	Lancaster I
No. 427 (B) Sqn	Leeming, Yorks	Lancaster I
No. 429 (B) Sqn	Leeming, Yorks	Lancaster I

No. 6 (RCAF) GROUP, Rear HQ Allerton Park, Yorks
no flying units assigned

No. 7 (TRAINING) GROUP
No. 76 (RCAF) BASE, HQ Topcliffe, Yorks
RCAF Stations Dalton, Topcliffe and Wombleton

Note: No. 63 Base was disbanded on 30 August, No. 6 Group Rear Headquarters on 31 August, No. 76 Base on 1 September, Nos. 424 and 433 Squadrons on 15 October, Nos. 427 and 429 Squadrons on 31 May 1946.

RAF TRANSPORT COMMAND

No. 47 GROUP

No. 426 (T) Sqn	Tempford, Beds	Liberator

No. 111 (RAF) WING

No. 437 (T) Sqn	Melsbroek, Bel	Dakota

No. 301 (RAF) WING

No. 422 (T) Sqn	Bassingbourn, Hants	Liberator
No. 423 (T) Sqn	Bassingbourn, Hants	Liberator

Notes:
1. Nos. 422 and 423 Squadrons were disbanded on 4 September, and No. 426 on 31 December.
2. No. 120 (Transport) Wing RCAF was formed on 17 September, composed of Nos. 435, 436 and 437 (T) Squadrons with Dakota aircraft, and was disbanded on 30 June 1946.

BRITISH AIR FORCES OF OCCUPATION GERMANY
SECOND TACTICAL AIR FORCE

No. 83 GROUP
No. 126 (F) WING RCAF, HQ Utersen, Ger

No. 411 (F) Sqn	Utersen, Ger	Spitfire XIV
No. 412 (F) Sqn	Utersen, Ger	Spitfire XIV
No. 416 (F) Sqn	Utersen, Ger	Spitfire XIV
No. 443 (F) Sqn	Utersen, Ger	Spitfire XIV

No. 136 (RAF) WING

No. 418 (I) Sqn	Volkel, Ger	Mosquito VI

No. 84 GROUP
FIRST CANADIAN ARMY

No. 664 (AOP) Sqn	Apeldoorn, Neth	Auster IV, V
No. 666 (AOP) Sqn	Apeldoorn, Neth	Auster IV, V

Note: No. 418 Squadron was disbanded on 7 September, No. 666 on 31 October, No. 126 Wing and its four squadrons on 21 March 1946, and No. 664 on 31 March.

RAF SOUTH EAST ASIA COMMAND
No. 229 GROUP
No. 341 WING
No. 435 (T) Sqn   Tulihal, India   Dakota
No. 232 GROUP
No. 342 WING
No. 436 (T) Sqn   Ramree Island, Burma   Dakota

Note: Nos. 435 and 436 Squadrons were transferred to England in September where they joined No. 437 Squadron to form No. 120 (T) Wing RCAF in RAF Transport Command.

1 April 1958
CF-100s in Europe With No. 1 (RCAF) Air Division
Six Auxiliary Squadrons Flying Sabre 5s

AIR TRANSPORT COMMAND
HQ Lachine, Que

Unit	Base	Aircraft
No. 4 (T) OTU	Trenton, Ont	various
No. 408 (R) Sqn	Rockcliffe, Ont	Lancaster, Canso A, Dakota
No. 412 (T) Sqn	Uplands, Ont	Comet IA, C-5, North Star, Dakota
No. 426 (T) Sqn	Dorval, Que	North Star
No. 435 (T) Sqn	Namao, Alta	C-119, Dakota
No. 436 (T) Sqn	Downsview, Ont	C-119

TACTICAL AIR COMMAND
HQ Edmonton, Alta

No. 18 WING (AUX), HQ Edmonton, Alta
No. 418 Sqn (Aux)	Namao, Alta	Expeditor, Silver Star

No. 23 WING (AUX), HQ Saskatoon, Sask
No. 416 Sqn (Aux)	Saskatoon, Sask	Expeditor

TRAINING COMMAND
HQ Trenton, Ont

No. 14 (TRAINING) GROUP, HQ Winnipeg, Man
No. 17 WING (AUX), HQ Winnipeg, Man
No. 402 Sqn (Aux)	Winnipeg, Man	Expeditor

No. 30 WING (AUX), HQ Calgary, Alta
No. 403 Sqn (Aux)	Calgary, Alta	Expeditor

AIR MATERIEL COMMAND
HQ Rockcliffe, Ont
no flying units assigned

NORTH AMERICAN AIR DEFENCE COMMAND
RCAF AIR DEFENCE COMMAND
HQ St Hubert, Que

Unit	Base	Aircraft
No. 1 (F) OTU	Chatham, NB	Sabre 6
No. 3 AW(F) OTU	Cold Lake, Alta	CF-100 5
No. 410 AW (F) Sqn	Uplands, Ont	CF-100 5
No. 428 AW(F) Sqn	Uplands, Ont	CF-100 5
No. 414 AW(F) Sqn	North Bay, Ont	CF-100 5
No. 433 AW(F) Sqn	North Bay, Ont	CF-100 5
No. 413 AW(F) Sqn	Bagotville, Que	CF-100 5
No. 432 AW(F) Sqn	Bagotville, Que	CF-100 5
No. 416 AW(F) Sqn	St Hubert, Que	CF-100 5
No. 425 AW(F) Sqn	St Hubert, Que	CF-100 5

No. 11 WING (AUX), HQ Montreal, Que
No. 401 (F) Sqn (Aux)	St Hubert, Que	Sabre 5
No. 438 (F) Sqn (Aux)	St Hubert, Que	Sabre 5

No. 14 WING (AUX), HQ Toronto, Ont
No. 400 (F) Sqn (Aux)	Downsview, Ont	Sabre 5
No. 411 (F) Sqn (Aux)	Downsview, Ont	Sabre 5

No. 16 WING (AUX), HQ Hamilton, Ont
No. 424 Sqn (Aux)	Mount Hope, Ont	Expeditor

No. 5 AIR DIVISION, HQ Vancouver, BC
No. 409 AW(F) Sqn	Comox, BC	CF-100 5

No. 19 WING (AUX), HQ Vancouver, BC
No. 442 (F) Sqn (Aux)	Comox, BC	Sabre 5
No. 443 (F) Sqn (Aux)	Comox, BC	Sabre 5

NORTH ATLANTIC TREATY ORGANIZATION
SUPREME HEADQUARTERS ALLIED POWERS EUROPE

ALLIED COMMAND ATLANTIC
RCAF MARITIME AIR COMMAND
HQ Halifax, NS

Unit	Base	Aircraft
No. 2 (M) OTU	Summerside, PEI	Neptune
No. 404 (MR) Sqn	Greenwood, NS	Neptune
No. 405 (MR) Sqn	Greenwood, NS	Neptune/Argus

MARITIME AIR COMMAND PACIFIC, HQ Vancouver, BC
No. 407 (MR) Sqn	Comox, BC	Lancaster/Neptune

ALLIED COMMAND EUROPE
ALLIED AIR FORCES CENTRAL EUROPE
FOURTH ALLIED TACTICAL AIR FORCE
No. 1 (RCAF) AIR DIVISION EUROPE
HQ Metz, Fr

No. 1 (F) WING, HQ Marville, Fr
No. 439 (F) Sqn	Marville, Fr	Sabre 6
No. 441 (F) Sqn	Marville, Fr	Sabre 6
No. 445 AW(F) Sqn	Marville, Fr	CF-100 4B

No. 2 (F) WING, HQ Grostenquin, Fr
No. 421 (F) Sqn	Grostenquin, Fr	Sabre 6
No. 430 (F) Sqn	Grostenquin, Fr	Sabre 6
No. 423 AW(F) Sqn	Grostenquin, Fr	CF-100 4B

No. 3 (F) WING, HQ Zweibrucken, Ger
No. 427 (F) Sqn	Zweibrucken, Ger	Sabre 6
No. 434 (F) Sqn	Zweibrucken, Ger	Sabre 6
No. 440 AW(F) Sqn	Zweibrucken, Ger	CF-100 4B

No. 4 (F) WING, HQ Baden Soellingen, Ger
No. 422 (F) Sqn	Baden Soellingen, Ger	Sabre 6
No. 444 (F) Sqn	Baden Soellingen, Ger	Sabre 6
No. 419 AW(F) Sqn	Baden Soellingen, Ger	CF-100 4B

CANADIAN ARMED FORCES
Air Component
1 February 1968

CANADIAN FORCES TRANSPORT COMMAND
HQ Trenton, Ont

Unit	Base	Aircraft
No. 4 (T) OTU	Trenton, Ont	various
No. 437 (T) Sqn	Trenton, Ont	Yukon
No. 412 (T) Sqn	Uplands, Ont	Cosmopolitan, Falcon
No. 436 (T) Sqn	Uplands, Ont	C-130E
No. 435 (T) Sqn	Namao, Alta	C-130E

No. 11 WING (AUX), HQ Montreal, Que
No. 401 Sqn (Aux)	St Hubert, Que	Otter
No. 438 Sqn (Aux)	St Hubert, Que	Otter

No. 14 WING (AUX), HQ Toronto, Ont
| No. 400 Sqn (Aux) | Downsview, Ont | Otter |
| No. 411 Sqn (Aux) | Downsview, Ont | Otter |

No. 17 WING (AUX), HQ Winnipeg, Man
| No. 402 Sqn (Aux) | Winnipeg, Man | Otter |

No. 18 WING (AUX), HQ Edmonton, Alta
| No. 418 Sqn (Aux) | Namao, Alta | Otter |

CANADIAN FORCES MOBILE COMMAND
HQ St Hubert, Que

No. 434 (OT) Sqn	Cold Lake, Alta	Silver Star
No. 408 (TacS&R) Sqn	Rivers, Man	C-130B, Silver Star
No. 429 (TacT) Sqn	St Hubert, Que	Buffalo

CANADIAN FORCES MATERIEL COMMAND
HQ Rockcliffe, Ont

no flying units assigned

CANADIAN FORCES TRAINING COMMAND
HQ Winnipeg, Man

no operational flying units assigned

NORTH AMERICAN AIR DEFENCE COMMAND
CANADIAN FORCES AIR DEFENCE COMMAND
HQ North Bay, Ont

No. 417 (ST/R) Sqn	Cold Lake, Alta	CF-104D
No. 414 (EWOT) Sqn	St Hubert, Que	CF-100 5
No. 409 AW(F) Sqn	Comox, BC	CF-101B
No. 416 AW(F) Sqn	Chatham, NB	CF-101B
No. 425 AW(F) Sqn	Bagotville, Que	CF-101B
No. 445 (SAM) Sqn	North Bay, Ont	Bomarc
No. 447 (SAM) Sqn	La Macaza, Que	Bomarc

Note: No. 417 Squadron was formerly No. 6 (ST/R) OTU training CF-104 pilots for No. 1 Air Division Europe.

NORTH ATLANTIC TREATY ORGANIZATION
SUPREME HEADQUARTERS ALLIED POWERS EUROPE

ALLIED COMMAND ATLANTIC
CANADIAN FORCES MARITIME COMMAND
HQ Halifax, NS

No. 449 (MOT) Sqn	Summerside, PEI	Neptune, Argus
No. 415 (MP) Sqn	Summerside, PEI	Argus
No. 404 (MP) Sqn	Greenwood, NS	Argus
No. 405 (MP) Sqn	Greenwood, NS	Argus
VS880	Shearwater, NS	Tracker
HS50	Shearwater, NS	Sea King

Notes:
1. *No. 449 Squadron was formerly No. 2 (M) OTU.*
2. *VS880 and HS50 were squadrons with the Royal Canadian Navy.*
3. *H.M.C.S. Shearwater was formerly known as RCAF Station, Dartmouth*

CANADIAN FORCES MARITIME COMMAND PACIFIC,
HQ Esquimalt, BC

| No. 407 (MP) Sqn | Comox, BC | Neptune |

Note: No. 407 Squadron converted to Argus aircraft in June; the Neptunes were retired from operational service.

ALLIED COMMAND EUROPE
ALLIED AIR FORCES CENTRAL EUROPE
FOURTH ALLIED TACTICAL AIR FORCE
CANADIAN FORCES No. 1 AIR DIVISION
HQ Lahr, Ger

No. 1 WING, HQ Lahr, Ger
| No. 439 (ST/R) Sqn | Lahr, Ger | CF-104 |
| No. 441 (ST/R) Sqn | Lahr, Ger | CF-104 |

No. 3 WING, HQ Zweibrucken, Ger
| No. 427 (ST/A) Sqn | Zweibrucken, Ger | CF-104 |
| No. 430 (ST/A) Sqn | Zweibrucken, Ger | CF-104 |

No. 4 WING, HQ Baden Soellingen, Ger
| No. 421 (ST/A) Sqn | Baden Soellingen, Ger | CF-104 |
| No. 422 (ST/A) Sqn | Baden Soellingen, Ger | CF-104 |

INDEXES

Index 1
Squadron Commanders

The asterisk * denotes that the number so marked refers to a photograph.

A

Adamson, W/C D.R., 437 Sqn, 193
Adkins, W/C G.D., 401 Sqn, 152
Aitken, W/C D.F.R., 442 Sqn, 199
Alexander, W/C E.A., 442 Sqn, 199
Allan, W/C J.F., 414 Sqn, 171
Allatt, S/L L.D., 421 Sqn, 179
Allison, G/C C., 413 Sqn, 169; 4 (F) Wing, 217
Anderson, W/C A.L., 147 Sqn, 65
Anderson, A/V/M N.R., East Air Comd, 209
Annan, G/C D.B., 418 Sqn, 110; 424 Sqn, 183; 16 Wing (Aux), 219
Annis, A/V/M C.L., 10 (BR) Sqn, 31; Mat Comd, 208
Annis, W/C R.H., 421 Sqn, 179; 444 Sqn, 202
Archambault, G/C L.G.G.J., 423 Sqn, 118; 1 Gp (Aux), 212
Archer, S/L D.F., 444 Sqn, 202
Archer, W/C J.C. (RAF), 407 Sqn, 92
Archer, S/L P.L.I., 402 Sqn, 84
Ardeline, G/C P., 16 Wing (Aux), 219
Arnold, S/L E.L., 441 Sqn, 197
Arnold, S/L J.T., 120 Sqn, 55
Ashbury, W/C W.B., 415 Sqn, 173
Ashman, W/C R.A., 115 Sqn, 47; 407 Sqn, 92
Ashton, S/L A.J., 4 (GR) Sqn, 24
Austin, S/L C.C., 13 (OT) Sqn, 36
Austin, W/C G.S., 147 Sqn, 65
Avant, G/C A.F., 429 Sqn, 126; 1 (F) Wing, 216

B

Bailey, Captain W.I., 7, 8
Baillie, W/C J., 406 Sqn, 158
Ball, W/C F.W., 415 Sqn, 106
Ball, S/L G.E., 411 Sqn, 98
Bannock, W/C R., 406 Sqn, 91; 418 Sqn, 110
Barclay, S/L J.R., 167 Sqn, 72
Barker, W/C A., 418 Sqn, 110
Barker, G/C R.B., 443 Sqn, 200; 19 Wing (Aux), 219
Barker, W/C W.G., D/RCAF, 207
Barnett, W/C C.D., 430 Sqn, 188
Bartlett, W/C C.S., 434 Sqn, 131
Baskerville, S/L P.G., 11 (BR) Sqn, 33
Batty, S/L R.H., 145 Sqn, 64
Bauer, W/C A.J., 421 Sqn, 179; 430 Sqn, 188
Bean, W/C W.W., 415 Sqn, 106
Beardmore, S/L E.W., 4 (BR) Sqn, 24; 118 Sqn, 52
Beatty, S/L J.H., 439 Sqn, 137
Begg, W/C R.F., 414 Sqn, 104; 442 Sqn, 199
Beirnes, S/L J.R., 118 Sqn, 52; 438 Sqn, 135
Bell, W/C J.P., 439 Sqn, 195
Belleau, S/L K.J.H.M., 439 Sqn, 195
Bell-Irving, W/C A.D., 111 Sqn, 42; 100 Wing (Aux), 220
Bennell, W/C R.J., 418 Sqn, 110
Bennett, A/C W.E., North West Air Comd, 210; Tactical Air Comd, 210; 126 Wing, 221
Benson, S/L T., 7 (BR) Sqn, 28
Benton, S/L R.J.E., 120 Sqn, 55
Berg, G/C V.L., 3 (F) Wing, 217
Beveridge, W/C M.W., 409 Sqn, 95
Bing, S/L L.P.S., 423 Sqn, 182
Bishop, Lt.-Col. W.A., Commands CAF Section, 7
Bissky, S/L P., 438 Sqn, 135; 408 Sqn, 160
Black, W/C C.M., 426 Sqn, 122; 428 Sqn, 125
Black, S/L I.M., 113 Sqn, 46
Blackburn, W/C A.P., 434 Sqn, 131; 413 Sqn, 169
Blagrave, W/C R.D.P., 427 Sqn, 119, 124
Blanchard, W/C S.S., 8 (BR) Sqn, 29; 116 Sqn, 49; 426 Sqn, 122
Bland, S/L E.A., 124 Sqn, 58
Blane, W/C J.D., 424 Sqn, 119
Blatchford, W/C H.P., (Can/RAF), Cdn Digby Wing, 215
Bliss, W/C W.H.F., 422 Sqn, 181

Blyth, Major D.W., 664 Sqn, 143
Bodien, S/L H.E., 410 Sqn, 163
Bolduc, W/C R.L., 429 Sqn, 126
Boomer, S/L K.A., 111 Sqn, 42; 132 Sqn, 62
Borden, W/C J.W., 412 Sqn, 167
Boulton, S/L F.H., 416 Sqn, 107
Bourque, S/L C.D.A., 421 Sqn, 179
Bowie, W/C T., 401 Sqn, 152
Boyd, S/L M.D., 412 Sqn, 100
Bradley, S/L C.W., 10 (BR) Sqn, 31
Bradshaw, A/V/M D.A.R., 420 Sqn, 113; 1 Air Div, 211
Braham, W/C J.R.D., 432 Sqn, 189
Brannagan, S/L T.A., 441 Sqn, 139
Breadner, A/M L.S., D/RCAF, 207, CAS, 207; AOC-in-C, O/S HQ, 207
Breckon, W/C W.B., 402 Sqn, 153
Bretz, W/C N.H., 402 Sqn, 84; 411 Sqn, 98; Cdn Digby Wing, 215
Bricker, S/L C.D., 430 Sqn, 127; 443 Sqn, 142; 439 Sqn, 195
Bridges, W/C H.E., 409 Sqn, 162
Briese, S/L E.C., 128 Sqn, 60
Briese, W/C R.G., 13 (OT) Sqn, 36; 413 Sqn, 102
Brookes, A/V/M G.E., No. 1 Trg Comd, 19; 5 (GR) Sqn, 25; 6 (B) Gp., 212
Brooks, S/L W.T., 133 Sqn, 62
Brown, W/C A.C. (Can/RAF), 407 Sqn, 92
Brown, W/C J.A., 402 Sqn, 153
Brown, W/C J.C.R., 427 Sqn, 124
Brown, W/C M., 127 Wing, 222
Brown, S/L M.G., 400 Sqn, 80
Brown, W/C O.C., 419 Sqn, 177
Brown, W/C W.W., 8 (BR) Sqn, 29
Browne, S/L J.D., 421 Sqn, 115; 441 Sqn, 139
Bryan, W/C M.G., 412 Sqn, 167
Bryans, A/V/M J.G., Trg Comd, 210; 14 (Trg) Gp., 213; 63 (O) Base, 214
Bryson, W/C E.M., 427 Sqn, 124; 431 Sqn, 128
Buchanan, Captain W.K., 665 Sqn, 143
Buckham, W/C R.A., 403 Sqn, 86; 127 Wing, 222
Bullock, S/L D.J., 439 Sqn, 195
Burgess, A/C C.W., 426 Sqn, 122; Trg Comd, 210
Burke, S/L D.K., 427 Sqn, 186

Burns, W/C G.M., 420 Sqn, 178
Burnside, W/C D.H. (RAF), 427 Sqn, 124
Butler, S/L F., 167 Sqn, 72
Buzza, W/C W.J., 416 Sqn, 173; 422 Sqn, 181; 423 Sqn, 182

C

Cameron, W/C A.M., 10 (BR) Sqn, 31
Cameron, G/C D.C., 442 Sqn, 199; 19 Wing (Aux), 219
Cameron, W/C L.M., 401 Sqn, 82; 402 Sqn, 153
Campbell, S/L C.N.S. (RAF), 403 Sqn, 86
Campbell, A/M H.L., CAS, 207; North West Air Comd., 210; 1 Air Div., 211
Campbell, W/C J.D.W., 423 Sqn, 182
Campbell, G/C J.K., 418 Sqn, 176; 18 Wing (Aux), 219
Cannon, S/L A.E.L., 128 Sqn, 60; 130 Sqn, 61; 132 Sqn, 62
Cannon, S/L R.Y., 439 Sqn, 195
Carling-Kelly, W/C F.C., 426 Sqn, 122; 1 Gp (Aux), 212
Carpenter, A/C F.S., 9 (BR) Sqn, 31; 117 Sqn, 50; 160 Sqn, 67; ATC, 209; Maritime Air Comd, 210
Carr, W/C W.K., 412 Sqn, 167
Carr-Harris, F/L B.G., 3 (B) Sqn, 18n, 24; 1 (F) Sqn, 21
Carruthers, W/C R.E., 427 Sqn, 186
Carscallen, A/V/M H.M., 5 (Br) Sqn, 25; 10 (BR) Sqn, 31; 424 Sqn, 119; ATC, 209; Trg Comd, 210
Carson, W/C C.V.B., 440 Sqn, 196
Carswell, W/C H.A., 404 Sqn, 156
Chadburn, W/C L.V., 402 Sqn, 84; 416 Sqn, 107; Cdn Digby Wing, 215; 127 Wing, 222; 222*
Chapman, W/C C.G.W., 162 Sqn, 68; 426 Sqn, 185
Cherrington, S/L G.E., 14 (P) Sqn, 39
Chesters, S/L F.H., 430 Sqn, 127
Chevrier, S/L J.A.J., 130 Sqn 61, 61n
Christie, G/C R.G., 1 (F) Wing, 216
Christie, S/L T.H., 115 Sqn, 48
Christmas, G/C B.E., 133 Sqn, 62; 4 (F) Wing, 217
Clarke, S/L J.W., 161 Sqn, 67
Clayton, W/C A.C.P. (Can/RAF), 405 Sqn, 89; 408 Sqn, 94
Clement, W/C R.J., 402 Sqn, 153
Clements, A/C W.I., Maritime Air Comd, 210
Cleveland, W/C H.D., 418 Sqn, 110
Cliff, S/L M.F., 411 Sqn, 165
Coffey, S/L R.E., 440 Sqn, 138
Coggins, W/C J.P., 406 Sqn, 158
Colborne, W/C F.C., 160 Sqn, 67
Coleman, A/C S.W., 5 (BR) Sqn, 25; Tactical Air Comd, 210
Collier, W/C J.D.D. (RAF), 420 Sqn, 113
Collier, S/L J.E., 403 Sqn, 86
Collis, F/L R., 15
Comar, W/C J., 429 Sqn, 126
Connell, S/L W.C., 133 Sqn, 62; 135 Sqn, 63
Connolly, F/L D.W.P., 163 Sqn, 69

Conrad, S/L W.A.G., 403 Sqn, 86; 421 Sqn, 115
Cook, A/V/M C.A., Mat Comd, 208
Cook, W/C G.M., 11 (BR) Sqn, 33
Corbett, S/L V.B., 402 Sqn, 84
Cornish, W/C O.W., 418 Sqn, 176
Corrigan, W/C J.F., 440 Sqn, 196
Cosco, S/L J.E., 166 Sqn, 71
Costello, A/C M., Maritime Air Comd, 209; Tactical Air Comd, 210
Coumans, W/C O.B., 15 (TT) Wing (Aux), 218
Coverdale, W/C J. (RAF), 431 Sqn, 128
Cowan, W/C S.F., 408 Sqn, 160; 440 Sqn, 196
Cowley, A/V/M A.T.N., No. 1 Trg Comd, 19; 4 Trg. Comd. 19
Cox, S/L P.B., 120 Sqn, 55
Crabb, S/L H.P., 112 Sqn, 45
Creeper, W/C J.E., 405 Sqn, 157
Creighton, S/L G.E., 116 Sqn, 49
Cribb, W/C C.J. (RAF), 427 Sqn, 124
Croft, W/C L.H., 407 Sqn, 159
Croil, A/V/M G.M., SAR, 207; CAS, 75; 207
Crooks, W/C L. (RAF), 426 Sqn, 122
Crosby, S/L R.G., 439 Sqn, 137
Cruickshank, W/C A.R., 403 Sqn, 155
Cuffe, A/V/M A.A.L., 2 (Ops) Sqn, 15; 1 Trg Comd, 19; 4 (FB) Sqn, 24; D/RCAF, 207; Air Trg Comd, 208; Eastern Air Comd, 209
Curtis, A/M W.A., 110 Sqn, 40; CAS, 207; 101 Wing (Aux), 221
Curtis, S/L W.A. Jr., 400 Sqn, 150
Cuthbertson, S/L D.R., 441 Sqn, 197

D

Darragh, W/C H.F., 444 Sqn, 202
Darrow, W/C C., 411 Sqn, 165
Davenport, W/C R.F., 431 Sqn, 128
Davidson, W/C R.T.P. (RAF), 421 Sqn, 179; 143 Wing, 223, 224*
Davis, W/C H.B., 400 Sqn, 150
Davis, F/L R.C., 7 (GP) Sqn, 28
Davoud, G/C P.Y., 409 Sqn, 95; 410 Sqn, 96; 418 Sqn, 110; 22 Sector, 214; 143 Wing, 223
Day, S/L C.J., 439 Sqn, 195
De Courcy, S/L T.J., 443 Sqn, 142
de Bombasle, Col. G.C., 7
Deere, S/L A.C. (RAF), 403 Sqn, 86
Delaney, W/C P.S., 113 Sqn, 46; 145 Sqn, 64
Delhaye, W/C R.A., 120 Sqn, 55
Denison, W/C R.L., 436 Sqn, 133
de Niverville, A/V/M J.L.E.A., 3 Trg Comd, 19
Dennis, W/C R.R., 149 Sqn, 66; 415 Sqn, 106
Dewan, S/L D.J., 412 Sqn, 100
Diamond, A/C G.G., 12 (Comm) Sqn, 34; 122 Sqn, 57; ATC, 209
Doak, S/L J.B., 133 Sqn, 62
Dobbin, S/L K.C.M., 436 Sqn, 192
Dobson, S/L J.M., 114 Sqn, 47
Dobson, S/L R., 7 (BR) Sqn, 28
Dodd, S/L W.G., 402 Sqn, 84

Doherty, W/C W.M., 5 (BR) Sqn, 25
Dooher, W/C M.J., 425 Sqn, 184
Douglas, S/L A.G. (RAF), 401 Sqn, 82; 403 Sqn, 86
Douglas, S/L P.H., 119 Sqn, 53
Dover, S/L D.H., 412 Sqn, 100
Dow, W/C H.R., 431 Sqn, 128
Dowding, S/L H.J., 442 Sqn, 141
Doyle, W/C M.F., 410 Sqn, 163; 428 Sqn, 187
Doyle, W/C M.G., 6 (BR) Sqn, 27; 116 Sqn, 49
Drake, W/C J.F., 405 Sqn, 157
Drake, S/L W.L., 416 Sqn, 173
Draper, G/C J.W.P., 411 Sqn, 165; 14 Wing (Aux), 218
Dubuc, S/L M.C., 118 Sqn, 51
Dunlap, A/M C.R., CAS, 207; ADC, 208; North West Air Comd, 210; 64 (O) Base, 214; 331 Wing, 224
Dunlop, W/C J.F., 427 Sqn, 186; 434 Sqn, 197
Dupuis, S/L L.P.J., 425 Sqn, 120
Du Temple, S/L G.W., 111 Sqn, 42
Dyer, W/C H.R.F., 419 Sqn, 112

E

Earle, W/C A. (RAF), 428 Sqn, 125
Easton, A/C J.A., 12 Gp, 211
Easton, W/C J.F., 408 Sqn, 94
Edser, S/L W.E., 124 Sqn, 58
Edwards, W/C A.J., 401 Sqn, 152
Edwards, W/C G.J.J., 420 Sqn, 113; 435 Sqn, 191
Edwards, A/M H., 5 (GR) Sqn, 25; APC RCAF O/S, 207
Edwards, W/C J.F., 430 Sqn, 188; 127 Wing, 222
Edwards, W/C J.L.T., 442 Sqn, 199
Edwards, W/C R.M., 441 Sqn, 197
Egan, S/L W.G., 116 Sqn, 49
Elliott, S/L G.J., 111 Sqn, 42; 132 Sqn, 62
Ellis, W/C R.A., 400 Sqn, 80; 430 Sqn, 127
Elms, W/C G.H., 410 Sqn, 96
Ely, Major D.R., 664 Sqn, 665 Sqn & 666 Sqn, 143
Etienne, W/C P.E., 419 Sqn, 177
Ettles, S/L C.I.M., 400 Sqn, 150
Evans, W/C E.H., 110 Sqn, 40; 429 Sqn, 126; 12 Gp. 211
Evans, W/C G.H.D., 415 Sqn, 106
Evans, W/C T.J., 409 Sqn, 162
Everard, G/C H.J., 401 Sqn, 82, 152; 11 Wing (Aux), 218
Ewart, W/C F.J., 5 (BR) Sqn, 25; 120 Sqn, 55; 5 (Gulf) Gp., 212

F

Fallis, W/C E.K., 448 Sqn, 205
Farrell, W/C C.M.G., 4 (BR) Sqn, 24; North West Staging Route, 210
Fauquier, A/C J.E., 405 Sqn, 89; 62 (O) Base, 214
Fee, W/C J.C., Canadian Kenley Wing, 76, 215; 412 Sqn, 100
Fellows, W/C F.G., 446 Sqn, 204

Fenwick-Wilson, W/C R.M. (Can/RAF), 405 Sqn, 89
Ferguson, W/C M.E., 419 Sqn, 112
Ferris, W/C W.D.S., 408 Sqn, 94
Fiander, S/L J.W., 434 Sqn, 190
Finley, S/L K.J., 443 Sqn, 142
Fiset, S/L K.J., 439 Sqn, 137
Fisher, S/L A.W., 410 Sqn, 163; 439 Sqn, 195; 441 Sqn, 197
Fisher, G/C J.D., 438 Sqn, 194; 443 Sqn, 200; 11 Wing (Aux), 218
Fitzsimmons, W/C J.W., 429 Sqn, 188
Fleming, W/C A., 5(BR) Sqn, 25; 116 Sqn, 49
Fleming, W/C A.E., 403 Sqn, 86; 400 Sqn, 150
Fleming, W/C M.M. (Can/RAF), 419 Sqn, 112
Flint, G/C J.M., 418 Sqn, 176; 18 Wing (Aux), 219
Ford, W/C L.S., 403 Sqn, 86; Cdn Digby Wing, 215
Forsyth, W/C C.H., 424 Sqn, 183
Foss, G/C R.H., 115 Sqn, 47; 4 Gp., 212
Foster, S/L F.B., 417 Sqn, 109
Foster, W/C W.D., 407 Sqn, 159
Fowlow, S/L N.R., 403 Sqn, 86; 411 Sqn, 98
France, W/C K.A., 432 Sqn, 129
Fraser, W/C L.G.D., 168 Sqn, 73; 405 Sqn, 89
Frazer, S/L J.L., 439 Sqn, 195
Freeman, W/C D.B., 403 Sqn, 155
French, W/C D.T., 428 Sqn, 125; 405 Sqn, 157
Frizzle, W/C J.R., 168 Sqn, 73; 422 Sqn, 116
Frostad, G/C G.C., 424 Sqn, 183; 16 Wing (Aux), 219
Fullerton, S/L E.G., 1 (F) Sqn, 21; 7 (GP) Sqn, 28
Fulton, W/C J. (Can/RAF), 419 Sqn, 112
Fumerton, W/C R.C., 406 Sqn, 91

G

Gagne, W/C J.L.Y., 438 Sqn, 194
Gagnon, S/L D.J., 416 Sqn, 173
Gagnon, F/L J.A.M.G., 121 Sqn, 56
Gall, W/C M.W., 428 Sqn, 125
Galloway, W/C D.D., 404 Sqn, 156
Gamble, W/C A.E., 438 Sqn, 194
Ganderton, W/C V.F., 427 Sqn, 124
Gardner, S/L E.F., 121 Sqn, 56
Gardiner, S/L W.G., 122 Sqn, 57
Garrett, S/L E.W., 410 Sqn, 163
Garry, S/L E., 444 Sqn, 202
Gatheral, W/C G.H. (RAF), 418 Sqn, 110
Gatwood, W/C A.K. (RAF), 404 Sqn, 88
Gauthier, G/C J.E.M.M., 438 Sqn, 194; 11 Wing (Aux), 218
Georgas, W/C G.M., 400 Sqn, 150
Gervais, S/L C.L.V., 427 Sqn, 186
Gibbs, S/L P.L., 430 Sqn, 188
Gilbertson, S/L F.S., 414 Sqn, 104
Gilbertson, S/L P.A., 126 Sqn, 59; 127 Sqn, 60; 128 Sqn, 61
Gilchrist, W/C P.A. (Can/RAF), 405 Sqn, 89
Gill, S/L W.T.H., 441 Sqn, 197

Gilroy, W/C G.E., 400 Sqn, 150
Gledhill, S/L J.W., 122 Sqn, 57
Glen, Captain J.A., 7; 8
Gobeil, S/L F.M., Comd 242 (Cdn) Sqn, 22n
Godefroy, W/C H.C., 403 Sqn, 86; 127 Wing, 222
Godfrey, A/C A.E., 14, Air Trg Comd, 208; Western Air Comd, 210
Godfrey, W/C J.M., 414 Sqn, 104; 39 (R) Wing, 220: 128 Wing, 222
Godwin, A/V/M H.B., Mat Comd, 208; 1 Air Div, 211; 64 (O) Base, 214
Goldberg, G/C D., 417 Sqn, 109; 16 Wing (Aux), 219
Goldsmith, W/C J.E., 425 Sqn, 184
Gooderham, G/C G.W., 400 Sqn, 150; 14 Wing (Aux), 218
Gooding, S/L H.O., 440 Sqn, 138
Goodwin, W/C R.W., 164 Sqn, 70; 168 Sqn, 73
Gordon, S/L D.C., 402 Sqn, 84
Gordon, G/C J.L., D/CAF, 207
Gordon, A/C R.A., 436 Sqn, 133; Maritime Air Comd, 210
Gordon, A/C R.C., 10 (BR) Sqn, 31; Maritime Air Comd, 209; North West Air Comd, 210; 12 AD Gp., 211; Gp., 212
Gordon, S/L W.I., 413 Sqn, 169
Grandy, S/L R.S., 8 (BR) Sqn, 29
Grant, S/L F.E., 403 Sqn, 86; 416 Sqn, 107; 438 Sqn, 135
Grant, W/C F.G., 118 Sqn, 52; 401 Sqn, 152; 143 Wing, 223
Grant, S/L P.G., 7 (BR) Sqn, 28
Grant, W/C P.J., 423 Sqn, 118; 435 Sqn, 191
Gray, S/L C.E. (RAF), 403 Sqn, 86
Gray, G/C D.M., 402 Sqn, 153; 17 Wing (Aux), 219
Gray, S/L E.R.B., 434 Sqn, 190
Gray, W/C R.G., 406 Sqn, 91
Gray, W/C R.J., 420 Sqn, 113
Green, S/L F.E., 416 Sqn, 107; 421 Sqn, 115
Green, W/C J.F., 145 Sqn, 64; 426 Sqn, 122
Greenway, A/V/M C.H., Trg Comd, 210
Greenway, S/L P.F., 428 Sqn, 187
Grisdale, W/C J.M., 401 Sqn, 152
Guest, G/C F.T., 418 Sqn, 176; 18 Wing (Aux), 219
Gulyus, W/C S.P., 439 Sqn, 195
Guthrie, A/V/M K.M., 2 Trg Comd, 19; North West Air Comd, 210; 2 Air Comd, 211

H

Hagerman, W/C D.C., 419 Sqn, 112
Hale, G/C E.B., 12 (Comm) Sqn, 34; 161 Sqn, 67; 165 Sqn, 71; 1 (F) Wing, 216
Haley, S/L F.E., 409 Sqn, 162
Halkett, W/C A.M., 404 Sqn, 156
Hall, G/C D.P., 1 (F) Wing, 216
Hall, S/L E.O.W., 124 Sqn, 58
Hall, S/L J.D., 421 Sqn, 115; 128 Wing, 222

Hall, S/L L.A., 410 Sqn, 163
Hamber, W/C E.C., 426 Sqn, 122
Hammond, W/C A.B., 432 Sqn, 189
Handley, W/C K.B., 423 Sqn, 182; 432 Sqn, 189
Hanna, S/L W.F., 112 Sqn, 45; 402 Sqn, 84
Harding, S/L D.A., 7 (Gp) Sqn, 28
Hare, W/C C.D.L., 414 Sqn, 171
Harling, S/L L.A., 6(BR) Sqn, 27
Harris, W/C C.E. (Can/RAF), 434 Sqn, 131
Harrop, F/L B.N., 4(ops) Sqn, 16; 6 (BR) Sqn, 27
Hartnett, W/C T.P., 435 Sqn, 132
Harvey, S/L W.S., 430 Sqn, 188
Hatton, W/C R.F., 409 Sqn, 95; 445 Sqn, 203
Hay, S/L W.B., 417 Sqn, 109
Haylett, G/C A.D., 420 Sqn, 178; 22 Wing (Aux), 220
Hayward, S/L R.K., 411 Sqn, 98
Heakes, A/V/M F.V., O/S RCAF, 207; Western Air Comd, 210 ; 1 Gp (A), 211
Heard, S/L S.F., 113 (F) Sqn, 46
Hebert, W/C R.J.C., 438 Sqn, 194
Heggtveit, S/L E.R., 444 Sqn, 202
Henderson, S/L E., 121 Sqn, 56
Hendrick, A/V/M M.M., Air Def Comd, 208
Henry, W/C J.C., 416 Sqn, 173
Henry, W/C R.J., 406 Sqn, 158
Herbert, W/C R.F., 440 Sqn, 196
Hetherington, W/C R.O., 443 Sqn, 200
Heybroek, W/C E.P., 410 Sqn, 96
Higgins, S/L F.C., 5 (FB) Sqn, 25
Higgs, W/C P.J.S., 422 Sqn, 181; 427 Sqn, 186
Hill, S/L G.U., 441 Sqn, 139
Hill, S/L L.J., 411 Sqn, 197
Hillock, W/C F.W., 410 Sqn, 96; 425 Sqn, 184; 143 Wing, 223
Hiltz, G/C G.A., 410 Sqn, 96; 2 Gp (Aux), 212
Hobbs, S/L A.G., 115 Sqn, 48
Hodson, W/C K.L.B., 401 Sqn, 82; Cdn Kenley Wing 215; 126 Wing, 221
Hogg, S/L J.E., 438 Sqn, 135
Holmes, W/C A.R., 437 Sqn, 134
Hooper, S/L A.W., 124 Sqn, 58
Houle, S/L A.U., 417 Sqn, 109
Houser, W/C W.M., 405 Sqn, 157
Hovey, W/C J.C., 433 Sqn, 190
Howard, G/C B.A., 400 Sqn, 150; 14 Wing (Aux), 218
Howard, F/L J.P., 1 (F) Sqn, 22
Howsam, A/V/M G.R., 4 Trg Comd, 19; 2 (AC) Sqn, 23; 7 (GP) Sqn, 28
Hoyt, W/C C.W., 164 Sqn, 70
Huggard, S/L J.C., 1 (F) Sqn, 22
Hull, A/C A.C., 428 Sqn, 125; ATC, 209; 3 (F) Wing, 217
Hull, S/L A.H., 3 (B) Sqn, 18n, 23n, 24; 6 (BR) Sqn, 27
Hulburt, W/C E.E., 435 Sqn, 191
Hurley, A/C J.L., 6 (B) Gp., 212; 61 (Trg) Base, 214; 62 (O) Base, 214; 63 (O) Base, 214
Husch, W/C R.G., 437 Sqn, 193

Hussey, W/C L.W., 436 Sqn, 192
Huston, W/C W.H., 403 Sqn, 155
Hutchinson, S/L R.T., 414 Sqn, 104

I

Inglis, W/C G., 409 Sqn, 162
Innes, S/L B.E., 411 Sqn, 98
Ireland, W/C A.J., 404 Sqn, 156
Ireland, W/C E.G., 409 Sqn, 162; 419 Sqn, 177; 445 Sqn, 203

J

Jackson, W/C D.J.G., 408 Sqn, 160; 413 Sqn, 169
Jacobi, S/L G.W., 120 Sqn, 55
Jacobs, W/C D.S., 408 Sqn, 94
Jacox, G/C D.R., 418 Sqn, 176; 18 Wing (Aux), 219
James, A/V/M A.L., Air Def Comd, 208
James, W/C C.A., 400 Sqn, 150
Jasieniuk, W/C T., 406 Sqn, 158
Jellison, S/L J.E., 7 (BR) Sqn, 28; 120 Sqn, 55
Johnson, A/C B.F., 61 (Trg) Base, 214
Johnson, A/V/M G.O., 1 (Ops) Wing, 13; 1 Trg Coms, 19; (SA), 207; AOC-in-C, O/S HQ, 207; Eastern Air Comd, 209; Western Air Comd, 210
Johnson, W/C G.W., 424 Sqn, 183
Johnson, W/C J.E. (RAF), Cdn Kenley Wing, 215; 127 Wig, 222; 144 Wing, 224
Johnson, S/L J.R.F., 421 Sqn, 179
Johnston, G/C E.R., 2 (F) Wing, 217
Johnston, S/L M., 442 Sqn, 141
Johnstone, S/L H.T., 403 Sqn, 155
Johnstone, S/L N.R., 126 Sqn, 59; 128 Sqn, 60; 130 Sqn, 61; 401 Sqn, 82
Jordon, G/C J.J., 4 (F) Wing, 217
Jowsey, S/L M.E., 442 Sqn, 141
Judd, W/C M.T. (RAF), 143 Wing, 223

K

Kallio, S/L O.C., 417 Sqn, 109
Kaufman, W/C F.J., 422 Sqn, 181
Keefer, W/C G.C., 412 Sqn, 100; 126 Wing, 221
Kelly, G/C D.J., 23 Wing (Aux), 220
Kelly, W/C E.D., 413 Sqn, 169; 416 Sqn, 173
Kelly, S/L F.W., 126 Sqn, 59; 412 Sqn, 100; 421 Sqn, 115
Kelly, G/C G.M., 403 Sqn, 155; 30 Wing (Aux), 220
Kennedy, W/C H.M., 12 (Comm) Sqn, 34
Kennedy, S/L I.F., 401 Sqn, 82
Kenyon, W/C A.G., 113 Sqn, 46; 405 Sqn, 157
Ker, S/L E.A., 401 Sqn, 82
Kerby, W/C H.W., 400 Sqn, 80; 432 Sqn, 129
Kerr, A/V/M J.G., Tactical Air comd, 210; Trg Comd, 210; 6 (B) Gp., 212; 213N; 61 (Trg) Base, 214; 62 (O) Base, 214; 661 Wing, 225N

Kerwin, S/L J.W., 111 Sqn, 42
King, W/C C.F. (RAF), 407 Sqn, 92
Kipp, S/L R.A., 410 Sqn, 163
Klassen, W/C W.C., 435 Sqn, 191
Klersy, S/L W.T., 401 Sqn, 82
Klesko, W/C R., 418 Sqn, 176
Knight, W/C H.R., 427 Sqn, 186; 430 Sqn, 188
Knight, W/C W.R., 427 Sqn, 186
Krug, S/L K.E., 5 (BR) Sqn, 25

L

Laflamme, W/C J.E.A., 447 Sqn, 205
Laidler, W/C D., 427 Sqn, 186
Lambert, S/L J.F., 421 Sqn, 115
Lane, W/C E.M., 411 Sqn, 165
Lane, A/V/M R.J., 405 Sqn, 89; ATC, 209; 1 Air Div, 211; 120 (T) Wing, 221
Lapp, S/L E.G., 411 Sqn, 98
Laubman, G/C D.C., 402 Sqn, 84; 416 Sqn, 173; 3 (F) Wing, 217
Laut, W/C A., 10 (BR) Sqn, 31; 113 Sqn, 46
Lawlor, W/C R.J., 423 Sqn, 182
Lawrence, W/C A.G., 432 Sqn, 189; 440 Sqn, 196; 446 Sqn, 204
Lawrence, W/C J.B., 402 Sqn, 84; 421 Sqn, 179; 444 Sqn,202
Lawrence, A/V/M T.A., 2 (AC) Sqn, 18N, 23; 2 Trg Comd, 19; North West Air Comd, 210
Lawson, Captain W.B., 8, 8*
Lay, S/L H.M., 8 (BR) Sqn, 29
Leach, Captain J.O., 8
Leckie, A/M R., 7, 10n, 207
Lecompte, G/C J.H.L., 415 Sqn, 106; 425 Sqn, 120; 423 Sqn, 182; 432 Sqn, 189; 663 Wing, 225
Ledoux, W/C H.C., 425 Sqn, 120
Lee, W/C C.C., 422 Sqn, 181
Lee, W/C K.C., 435 Sqn, 191
Lee-Knight, S/L R.A. (RAF), 403 Sqn, 86
Lee, S/L R.L., 145 Sqn, 64
Leigh, G/C Z.L., 13 (OT) Sqn, 36; 9 (T) Gp, 209; ATC, 209; 12 Gp, 211; 2 Gp., (Aux), 212
Lehn, W/C C.A., 405 Sqn, 157
Lett, S/L K.C., 416 Sqn, 173
Lewington, W/C A.J., 433 Sqn, 130
Lewis, W/C A., 3 (B) Sqn, 24; 10 (BR) Sqn, 31; 11 (BR) Sqn, 33
Lewis, W/C C.E., 433 Sqn, 130
Lewis, G/C K.E., 434 Sqn, 190; 3 (F) Wing, 217
Liggett, S/L L.J., 414 Sqn, 171
Lindsay, S/L J.D., 413 Sqn, 169
Lipton, W/C M., 129 Sqn, 61; 410 Sqn, 96
Lisson, S/L H.S., 133 Sqn, 62
Lister, A/V/M M.D., Air Def Comd, 208
Little, W/C J.H. (RAF), 418 Sqn, 110
Little, S/L R.H., 112 Sqn, 45
Lloyd, W/C W.R. 436 Sqn, 192
Lowe, G/C A.D.R., 432 Sqn, 129; 418 Sqn, 176; 18 Wing (Aux), 219
Lowe, W/C K.F., 414 Sqn, 171
Lowry, S/L R.H., 4 (BR) Sqn, 24
Lupton, W/C H.W., 426 Sqn, 185

M

MacBride, G/C R.E., 419 Sqn, 177; 2 (F) Wing, 217
MacBrien, A/V/M W.R., RCAF Station Kenley, 79N; Air Def Comd, 208; 17 Sector, 213; 127 Wing, 222
MacDonald, W/C D.C., 418 Sqn, 110; 442 Sqn, 199; 129 Wing, 223
MacDonald, S/L F.K., 4 (BR) Sqn, 24
MacDonald, G/C J.K.F., 432 Sqn, 129; 426 Sqn, 185; 2 (F) Wing, 217; 3 (F) Wing, 217
MacDonald, G/C J.R., 662 Wing, 225
MacDonald, W/C K.W., 410 Sqn, 163; 445 Sqn, 203
MacDonnell, W/C F.W.H., 408 Sqn, 160
MacFarlane, W/C I.A.H., 404 Sqn, 156
MacGregor, W/C N.S., 119 Sqn, 53
MacKay, W/C J., 416 Sqn, 173; 439 Sqn, 195; 444 Sqn,202
MacKenzie, S/L A.R., 135 Sqn, 63; 441 Sqn, 197
MacKenzie, W/C D.R., 412 Sqn, 167
Mackie, W/C A.J., 426 Sq, 185
Macklin, S/L D.I., 1 (F) Sqn, 22
MacLaren, Major D.R., 7
MacLean, A/C L.L., 12 (OT) Gp., 213
MacLeod, W/C E.L., 4 (GR) Sqn, 24
MacWilliam, W/C D.L.S., 425 Sqn, 184
McBratney, W/C V.H.A. (RAF), 413 Sqn, 102
McBurney, A/V/M R.E., Mat Comd, 208; 61 (Trg) Base, 214; 64 (O) Base, 214
McCarthy, W/C F.S., 420 Sqn, 113
McCarthy, W/C J.C., 4-7 Sqn, 159
McCarthy, W/C J.P., 424 Sqn, 119
McClure, W/C J.E., 410 Sqn, 163
McCutcheon, W/C J.T., 436 Sqn, 192
McElroy, S/L J.F., 416 Sqn, 107
McEwen, A/V/M C.M., 3 Trg Comd, 19; 1 Gp (A), 211; 6 (B) Gp., 212; 61 (Trg) Base, 214; 62 (O) Base, 214; Tiger Force, 213
McFarlane, S/L J.D., 411 Sqn, 98
McFarlane, W/C W.J., 1 (F) Sqn, 22; 111 Sqn, 42; North West Staging Route, 210
McGill, A/V/M F.S., No. 1 Trg Comd, 19; 15 & 115 Sqn, 47
McGregor, S/L D.U., 119 Sqn, 53; 120 Sqn, 55
McGregor, G/C G.R., 1 (F) Sqn, 21; 402 Sqn, 84; Digby Wing, 215; 126 Wing, 221; "X" Wing, 225; 226n
McHardy, W/C E.H. (NZ/RAF), 404 Sqn., 88
McIntosh, W/C D. (Can/RAF), 420 Sqn, 113
McKay, W/C H.A., 408 Sqn, 160; 414 Sqn, 171
McKay, W/C R.M., 110 Sqn, 40; 163 Sqn, 69; 40 Sqn, 80
McKay, W/C W.A., 432 Sqn, 129
McKeever, Major A.E., 6n; 8
McKenna, W/C A.G. 420 Sqn, 113
McKernan, P/O F.P., 62
McKinnon, W/C W.F., 431 Sqn, 128
McLachlan, W/C H., 441 Sqn, 197
McLaws, W/C J.G., 403 Sqn, 155

McLeish, G/C W.A.G., 428 Sqn, 125; 440 Sqn, 196; 664 Wing, 225
McLeod, A/C E.L., 2 Gp., 212
McLeod, S/L H.W., 127 Sqn, 60; 443 Sqn, 142, 224
McLeod, W/C W., 407 Sqn, 159
McLernon, W/C A.R., 408 Sqn, 94; 425 Sqn, 120
McMillan, W/C S.R., 117 Sqn, 50; 413 Sqn, 102; 423 Sqn, 118
McMurdy, W/C G.A., 419 Sqn, 112
McNab, G/C E.A., 1 (F) Sqn, 21; 118 Sqn, 52; 19 Wing (Aux), 219
McNair, G/C R.W., 416 Sqn, 107; 421 Sqn, 115; 4(F) Wing, 217; 126 Wing, 221
McNee, S/L J.W., 9 (BR) Sqn, 31
McNeill, W/C C.W., 8 (BR) Sqn, 29; 407 Sqn, 159
McNeill, W/C J.G., 415 Sqn, 106
McRae, S/L R.W., 4 (BR) Sqn, 24; 9 (BR) Sqn, 31
McVeigh, W/C C.N., 435 Sqn, 132
Madden, W/C H.O., 124 Sqn, 58; 165 Sqn, 71
Magee, S/L C.C., 422 Sqn, 181
Magee, S/L K.L., 417 Sqn, 109
Magwood, S/L C.M., 403 Sqn, 86; 421 Sqn, 115
Mair, W/C A.C., 408 Sqn, 94
Mair, S/L R.C.S, 4 (BR) Sqn, 24; 5 (BR) Sqn, 25
Maitland, W/C J.O., 426 Sqn, 185; 437 Sqn, 193
Malfroy, S/L C.E. (RAF), 417 Sqn, 109
Malloy, S/L D.G., 402 Sqn, 84
Maloney, W/C W.B., 406 Sqn, 158
Mann, W/C E.H., 15 (TT) Wing (Aux), 218
Margetts, S/L V.A., 6 (BR) Sqn, 27
Marples, W/C R. (RAF), 143 Wing, 223
Marsh, W/C W.J., 425 Sqn, 184
Marshall, W/C C.C.W., 424 Sqn, 119; 435 Sqn, 191
Marshall, G/C A.N., 424 Sqn, 183; 16 Wing (Aux), 219
Martin, W/C A.N., 424 Sqn, 119
Martyn, W/C M.P., 10 (BR) Sqn, 31; 11 (BR) Sqn, 33
Mawdesley, S/L F.J., 4 (BR) Sqn, 24
Merritt, S/L H.J.L., 130 Sqn, 61
Meyers, W/C A.A., 406 Sqn, 158
Michalski, S/L A.L., 121 Sqn, 56
Michalski, W/C W.J., 164 Sqn, 70; 167 Sqn, 72
Middlemiss, S/L R.G., 421 Sqn, 179; 427 Sqn, 186
Middleton, A/V/M E.E., Central Air Comd, 209; Trg Comd, 210; 1 Air Comd, 211
Middleton, W/C R.B., 164 Sqn, 70; 168 Sqn, 73
Middleton, W/C W.M., 425 Sqn, 184
Miles, A/C H.T., 64 (O) Base, 214
Millar, W/C W.B.M., 414 Sqn, 171
Miller, W/C D.R., 426 Sqn, 122
Miller, A/C/M F.R., CDS 148; Mat Comd, 208; 61 (Trg) Base, 214; 63 (O) Base, 214
Miller, W/C J.S., 436 Sqn, 192

Miller, W/C S.A., 416 Sqn, 173
Miller, W/C W.G.S., 412 Sqn, 167
Mills, W/C F.J., 411 Sqn, 165
Millward, A/V/M J.B., Mat Comd, 208
Milne, S/L R.F., 14 (P) Sqn, 39; 414 Sqn, 171
Miners, S/L E.L., 122 Sqn, 57
Minhinnick, S/L S.H., 432 Sqn, 129
Miscampbell, S/L G.V., 13 (P) Sqn, 37
Mitchell, S/L A.W., 9 (BR) Sqn, 31
Mitchell, W/C E.M., 431 Sqn, 128
Mitchell, W/C J.F., 408 Sqn, 160
Mitchell, W/C S.S., 415 Sqn, 173
Mitchener, W/C J.D., 416 Sqn, 107; 417 Sqn, 175; 434 Sqn, 190
Moffit, W/C B.H., 404 Sqn, 156; 412 Sqn, 167
Molson, S/L H. deM., 118 Sqn, 52; 126 Sqn, 59
Moncrieff, G/C E.H.G., 430 Sqn, 127; 39 (R) Wing, 220; 129 Wing, 223
Monnon, S/L H.F., 161 Sqn, 67
Monson, S/L A.E., 440 Sqn, 138
Moore, W/C K.O., 407 Sqn, 159
Moore, S/L L.A., 402 Sqn, 84
Moran, S/L C.C., 128 Sqn, 60; 29 Sqn, 61
Morfee, A/V/M A.L., Eastern Air Comd, 209; 1 Gp., (A), 211
Morris, S/L B.G. (RAF), 403 Sqn, 86
Morris, S/L R.H., No. 7 (BR) Sqn, 28; 122 Sqn, 57
Morrison, W/C H.A., 405 Sqn, 89; 412 Sqn, 167
Morrison, S/L J.D., 412 Sqn, 100
Morrissette, W/C A.R., 438 Sqn, 194
Morrow, W/C R.E.E., 402 Sqn, 84; 'X' Wing, 225
Morton, S/L J.A., 400 Sqn, 80
Mostyn-Brown, G/C W.A., 403 Sqn, 155; 30 Wing (Aux), 220
Mullen, W/C J.T., 430 Sqn, 188
Mulock, Col. R.H., 7
Mulvihill, S/L J.C., 160 Sqn, 67; 434 Sqn, 131; 405 Sqn, 157
Murray, W/C R.B., 423 Sqn, 182
Murray, S/L R.G., 422 Sqn, 181
Mussells, W/C C.H., 426 Sqn, 185
Myers, G/C A.A., 23 Wing (Aux), 220

N

Napier, S/L W.F., 129 Sqn, 61
Neal, S/L E.L., 130 Sqn, 61; 401 Sqn, 82
Neale, W/C A.C., 6 (BR) Sqn, 27; 7 (BR) Sqn, 28
Neroutsos, W/C R.R., 411 Sqn, 165
Nesbitt, W/C A.D., 111 Sqn, 42; 401 Sqn, 82; 143 Wing, 223; 144 Wing, 224
Newell, S/L J.N., 411 Sqn, 98
Newson, G/C W.F.M., 405 Sqn, 89; 431 Sqn, 128; 2 (F) Wing, 217
Newton, S/L R.B. (RAF), 411 Sqn, 98
Nichols, W/C G.H., 425 Sqn, 184; 433 Sqn, 190
Nickerson, W/C G.E., 445 Sqn, 203
Nixon, W/C J.A. (RAF), 404 Sqn, 88
Norris, W/C D.G. (RAF), 406 Sqn, 91
Norris, W/C H.R., 432 Sqn, 189

Norris, W/C R.W., 125 Sqn, 59; 424 Sqn, 119
Norsworthy, S/L H.H., 439 Sqn, 137
Northcott, G/C G.W., 402 Sqn, 84; 442 Sqn, 199; 19 Wing (Aux), 219; 126 Wing, 221

O

O'Brian, S/L G.S., 110 Sqn, 40; 114 Sqn, 47
O'Callaghan, W/C J.P., 435 Sqn, 141
Olmstead, S/L W.A., 442 Sqn, 141
Olsen, S/L C.S., 413 Sqn, 169
Olsson, W/C C.L., 408 Sqn, 160
Ormston, S/L I.C., 133 Sqn, 62; 411 Sqn, 98
Orpen, W/C R.G., 408 Sqn, 160
Orr, A/C W.A., 5 Air Div, 211; 5 (Gulf) Gp., 212
Owens, W/C J.A.P., 429 Sqn, 126

P

Paisley, W/C W.G., 430 Sqn, 188
Palmer, W C.W., 405 Sqn, 89
Parks, W/C W.F., 435 Sqn, 191
Pate, W/C W.G., 111 Sqn, 49
Patterson, W/C G.W., 409 Sqn, 162
Patterson W/C J.D., 429 Sqn, 126
Patterson, W/C J.T., 402 Sqn, 153
Pattison, W/C J.D., 145 Sqn, 64
Pentland, S/L W.H., 440 Sqn, 138
Peters, S/L H.P., 414 Sqn, 104
Petersen, W/C N.B., 409 Sqn, 95
Petriarche, G/C V.H., North West Staging Route, 210; 2 Gp (Aux), 212
Phelan, W/C P.J., 167 Sqn, 72
Phelan, W/C W.G., 420 Sqn, 113
Philip, W/C O.B., 434 Sqn, 190
Piddington, W/C J.A. (Can/RAF), 429 Sqn, 126
Pierce, W/C E.W., 404 Sqn, 88
Pierpoint, W/C J.V., 415 Sqn, 173
Pike, W/C H.I., 425 Sqn, 184
Pitcher, S/L P.B., 1 (F) Sqn, 21; 401 Sqn, 82; 411 Sqn, 98; 417 Sqn, 109; "X" Wing, 226n
Plant, A/V/M J.L., 12 Sqn, 34; 413 Sqn, 102; Mat Comd, 208; 9 (T) Gp, 209; Western Air Comd, 210; 12 Gp., 211
Pleasance, W/C W.P., 419 Sqn, 112; 405 Sqn, 157
Poag, W/C W.F., 162 Sqn, 68
Pollard, A/V/M M.E., 435 Sqn, 191; Air Def Comd, 208; 2 (F) Wing, 217
Prendergast, S/L J.B., 414 Sqn, 104
Prest, S/L W.A., 421 Sqn, 115
Preston, S/L G.D., 122 Sqn, 57; 166 Sqn, 71
Price, S/L L.C., 441 Sqn, 197

R

Ramsay, S/L D.L., 111 Sqn, 42; 163 Sqn, 69
Randall, W/C L.H., 413 Sqn, 102
Rankin, S/L L.C., 123 Sqn, 58
Rathwell, W/C D.W., 402 Sqn, 153

Raymond, A/V/M A., 3 Trg Comd, 19; 118 Sqn, 51; 1 Air Comd, 211
Redpath, W/C R.F., D/CAF, 207
Reid, W/C J.M., 402 Sqn, 153
Reid, W/C J.W., 409 Sqn, 95; 401 Sqn, 152; 442 Sqn, 199
Reid, S/L R.F., 438 Sqn, 135
Reilander, Major N.W., 665 Sqn, 143; 444 Sqn, 201
Reyno, S/L E.M., 115 Sqn, 47; 135 Sqn, 63
Richards, A/C H.G., Tactical Air Comd, 210
Richardson, G/C G.T., 12 (OT) Gp., 213
Richer, G/C J.A.D.B., 425 Sqn, 120; 1 Gp (Aux), 212
Riddell, S/L W.I., 4 (GR) Sqn, 24
Ripley, A/V/M R.C., Mat Comd, 208; ATC, 209
Ritch, S/L J.R., 414 Sqn, 171
Ritzel, S/L D.F., 166 Sqn, 71
Roberts, W/C J.H., 117 Sqn, 50
Roberts, F/L W.P., 127 Sqn, 60
Robertson, S/L G.D., 411 Sqn, 98
Robinson, S/L F.V., 124 Sqn, 58
Rogers, S/L W.W., 117 Sqn, 50
Rohmer, W/C R.H., 400 Sqn, 150; 411 Sqn, 165
Ross, A/C A.D., 5 (BR) Sqn, 25; ATC, 209; Maritime Air Comd, 209; 5 Air Div, 211; 62 (O) Base, 214
Ross, S/L W.W.S., 123 Sqn, 58
Roussell, W/C J.L.A., 416 Sqn, 173; 444 Sqn, 202; 447 Sqn, 205
Roy, W/C G.A., 424 Sqn, 119
Roy, W/C P.J.P., 447 Sqn, 205
Rump, W/C F.J., 423 Sqn, 118
Russel, S/L A.H.K., 110 Sqn, 40
Russel, W/C B.D., 14 (F) Sqn, 39; 411 Sqn, 98; 442 Sqn, 141; 126 Wing, 221
Russell, W/C D.W., 170 Sqn, 74
Russell, S/L H.W., 430 Sqn, 127
Rutledge, A/C H.H.C., 8 (BR) Sqn, 29; 14 (Trg) Gp., 213; 120 (T) Wing, 221
Ruttan, W/C C.G., 415 Sqn, 106

S
Sager, S/L H., 443 Sqn, 142
St. Louis, W/C P.B., 414 Sqn, 171; 427 Sqn, 186; 439 Sqn, 195
St. Pierre, W/C J.M.W., 425 Sqn, 120
Sampson, W/C F.A., 5 (BR) Sqn, 25
Sanderson, S/L J.H., 170 Sqn, 74
Sargenia, G/C G.E., 30 Wing (Aux), 220
Saunders, W/C A.E. (RAF), 418 Sqn, 110
Savard, W/C J.L., 429 Sqn, 126
Schultz, W/C J.R.D., 413 Sqn, 169; 425 Sqn, 184
Schwartz, W/C J.A.E., 7 (P) Wing, 218
Scott, W/C D.R., 402 Sqn, 153
Scott, G/C H.N., 17 Wing (Aux), 219
Scott, W/C J.C., 413 Sqn, 102
Scott, W/C J.S., D/CAF & D/RCAF, 207
Scott, W/C R.K., 422 Sqn, 181
Searle, G/C A.B., 2 (F) Wing, 217
Sellers, G/C G.H., 17 Wing (Aux), 219; 39 (R) Wing, 220
Semple, S/L G.C., 411 Sqn, 98

Seward, A/C W.J., 12 (OT) Gp., 213
Sharp, W/C F.R., 408 Sqn, 94; 661 Wing, 225
Shaw, S/L R.O., 119 Sqn, 53
Shearer, A/V/M A.B., 2 Trg Comd, 19; 4 (FB) Sqn, 24
Sheppard, S/L J.E., 133 Sqn, 62; 135 Sqn, 63; 412 Sqn, 100
Showler, W/C J.G., 408 Sqn, 160
Sinclair, S/L A.L., 434 Sqn, 190
Sinton, W/C C.B., 433 Sqn, 130
Skey, W/C L.W. (Can/RAF), 422 Sqn, 116
Slemon, A/M C.R., 8 (BR) Sqn, 29; CAS, 207; Tr Comd, 210; "Tiger Force," 231
Sloat, W/C R.D., 423 Sqn, 182
Smale, W/C H.E., 407 Sqn, 159
Small, S/L N.E., 113 Sqn, 46; 162 Sqn, 68
Small, W/C R.C., 424 Sqn, 183
Smiley, W/C D.A.B., 441 Sqn, 197
Smith, W/C C.O.P., 443 Sqn, 200
Smith, S/L D.J., 135 Sqn, 63
Smith, G/C D.M., 414 Sqn, 104; 39 (R) Wing, 220
Smith, W/C D.W.M. (Can/RAF), 428 Sqn, 125
Smith, W/C E.G., 413 Sqn, 169; 414 Sqn, 171
Smith, W/C E.J., 405 Sqn, 157
Smith, W/C E.W., 428 Sqn, 187
Smith, W/C G.M., 442 Sqn, 199
Smith, W/C H.M., 408 Sqn, 160
Smith, W/C R.I.A., 401 Sqn, 82; 411 Sqn, 165
Smith, W/C R.V., 444 Sqn, 202
Smith, S/L W.M., 439 Sqn, 137
Somerville, G/C J.D., 409 Sqn, 95; 1 (F) Wing, 216
Sorenson, S/L P.E., 9 (BR) Sqn, 31
Souchen, W/C D.W., 404 Sqn, 156
Speed, W/C A.W., 418 Sqn, 176
Springall, F/L E.A. (RAF), 6 (TB) Sqn, 27
Sproule, W/C J.A. (Can/RAF), 437 Sqn, 134
Stacey, W/C W.J., 422 Sqn, 181
Stephenson, W/C I.R. (RAF), 406 Sqn, 91
Stevenson, A/V/M L.F., 4 Trg Comd, 19; AOC RCAF O/S, 207; Western Air Comd., 210
Stewart, Major A.B., 666 Sqn, 143
Stewart, W/C H.C., 413 Sqn, 169; 434 Sqn, 190
Stover, S/L C.H., 414 Sqn, 104
Stowe, W/C W.N., 400 Sqn, 150
Straub, F/L S.H., 416 Sqn, 107
Stroud, S/L J.M., 147 Sqn, 69
Styles, W/C H.M. (RAF), 407 Sqn, 92
Suggitt, W/C W.R., 428 Sqn, 125
Sully, W/C J.A., 112 Sqn, 45; 102 Wing (Aux), 221
Sully, W/C J.K., 162 Sqn, 68
Sumner, W/C J.R., 422 Sqn, 116
Sutherland, W/C G., 445 Sqn, 203
Sutherland, W/C J.M., 440 Sqn, 196
Swetman, W/C W.H., 11 (BR) Sqn, 33; 12 (K) Sqn, 34; 426 Sqn, 122; 412 Sqn, 167

T
Tambling, W/C G.A., 433 Sqn, 130
Tew, W/C W.R., 427 Sqn, 186
Thomas, W/C R.I., 120 Sqn, 55; 7 (P) Wing, 218
Thompson, S/L J.A., 132 Sqn, 62
Thorneycroft, W/C K.J., 444 Sqn, 202
Timerman, A/C N.W. (Can/RAF), 408 Sqn, 94; 61 (Trg) Base, 214
Torontow, W/C C., 405 Sqn, 157
Trainer, S/L H.C., 401 Sqn, 82
Trecarten, F/L C.L., 6 (TB) Sqn, 18n, 27
Trevena, S/L C.W., 125 Sqn, 59; 412 Sqn, 100
Trickett, W/C R.I., 412 Sqn, 167
Trotter, W/C E.J., 445 Sqn, 203
Trumley, S/L R.K., 436 Sqn, 192
Truscott, A/C G.G., 404 Sqn, 88; 5 Air Div, 211
Turnbull, G/C R.S., 427 Sqn, 124; 4(F) Wing, 217
Turner, S/L D.E., 162 Sqn, 68
Turner, G/C P.S. (Can/RAF), 411 Sqn, 98; 417 Sqn, 109; 127 Wing, 222
Turner, G/C W.R.D., 17 Wing (Aux), 219
Twigg, W/C J.D., 408 Sqn, 94; 413 Sqn, 102
Tylee, A/C (Lt.-Col.) A.K., AOC CAF, 10n, 207

U
Upson, S/L G.C., 6 (BR) Sqn, 27
Ursulak, S/L J., 434 Sqn, 190
Ussher, W/C T.H., 411 Sqn, 165

V
Valois, W/C J.A.G.P.F., 438 Sqn, 194
Vadboncoeur, F/L G., 118 Sqn, 51
Van Adel, W/C R., 421 Sqn, 179
Van Camp, W/C W.C., 11 (BR) Sqn, 33
Vanhee, S/L A., 160 Sqn, 67
Van Slyck, G/C E.B., 23 Wing (Aux), 220
Van Vliet, W/C W.D., 2 (AC) Sqn, 23; 110 Sqn, 40
Varnam, G/C G.S., 17 Wing (Aux), 219
Viau, S/L J.M., 5 (BR) Sqn, 25
Villeneuve, W/C J.A.G.F., 434 Sqn, 190
Vincent, W/C W.H., 409 Sqn, 162
Vinnicombe, W/C H.C., 160 Sqn, 67
Virr, S/L R.V., 430 Sqn, 188
Volk, W/C S., 7 (P) Wing, 218

W
Waddell, G/C R.C.A., 400 Sqn, 80; 14 Wing (Aux), 218; 39 (R) Wing, 220
Wait, A/C F.G., 2 (AC) Sqn, 23; Maritime Air Comd, 209; 1 Gp (Aux), 212; 61 (Trg) Base, 214
Walker, S/L B.R., 14 Sqn, 39; 442 Sqn, 141
Walker, W/C J.E., 126 Wing, 221; 144 Wing, 224
Walker, S/L R.H., 416 Sqn, 107; 441 Sqn, 139

Walker, F/L W.H., 123 Sqn, 58
Walsh, W/C A.P. (Can/RAF), 419 Sqn, 112
Walsh, A/C G.V., 3 Trg Comd, 19; O/S RCAF, 207
Warren, S/L D., 410 Sqn, 163
Watkins, W/C F.H., 434 Sqn, 131
Watts, S/L J., 430 Sqn, 127
Webb, S/L P.C., 416 Sqn, 107
Weiser, G/C W., 2 (F) Wing, 217
Wenz, W/C H.F., 434 Sqn, 190
West, W/C R.B., 416 Sqn, 173
Weston, A/C R.C., 126 Sqn, 59; 412 Sqn, 100; 5 Aid Div., 211
White, S/L C.W., 430 Sqn, 188
Whiteley, S/L J.W., 410 Sqn, 163; 430 Sqn, 188
Wickett, W/C J.C., 418 Sqn, 110
Wigle, W/C D.H., 119 Sqn, 53; 424 Sqn, 183
Wilkinson, S/L P., 11 (BR) Sqn, 33
Williams, G/C D.J., 406 Sqn, 91; 1 (F) Wing, 216
Williams, S/L E.M., 145 Sqn, 64
Willis, S/L F.C., 421 Sqn, 115
Willis, W/C C.A., 8 (BR) Sqn, 29; 164 Sqn, 70; 404 Sqn, 88; 426 Sqn, 185
Wills, W/C R.A. (RAF), 406 Sqn, 91
Wilson, W/C A.H., 111 Sqn, 42
Wilson, W/C E., 414 Sqn, 171; 433 Sqn, 190
Wilson, S/L J.T., 149 Sqn, 66; 163 Sqn, 69
Wilson, S/L K.C., 11 (BR) Sqn, 33; 407 Sqn, 92
Wilson, S/L P., 438 Sqn, 135
Winny, S/L H.J., 6 (BR) Sqn, 27
Wiseman, S/L J.A., 13 (P) Sqn, 37; 413 Sqn, 169
Wonnacott, S/L G., 414 Sqn, 104
Wood, S/L E.P., 403 Sqn, 86; 413 Sqn, 169; 421 Sqn, 179
Wood, W/C S.E.W.J., 418 Sqn, 176
Woodruff, W/C H.P. (Can/RAF), 414 Sqn, 88
Woods, S/L W.B., 400 Sqn, 80
Woolfenden, S/L J., 117 Sqn, 50
Wray, A/V/M L.E., 6 (BR) Sqn, 27; 7 (GP) Sqn, 28; Air Def Comd, 208; Air Transport Comd, 209; 1 Air Div. 211
Wright, W/C G.G., 422 Sqn, 181
Wurtele, W/C E.L., 415 Sqn, 106
Wylie, S/L R.B., 8 (BR) Sqn, 29
Wynn, G/C G.K., 18 Wing (Aux), 219

Y
Young, W/C J.M., 10 (BR) Sqn, 31
Yuile, S/L A.M., 118 Sqn, 52; 126 Sqn, 59

Z
Zary, S/L H.P.M., 403 Sqn, 86

Index 2

"Squadron Aces"

A
Audet, F/L R.J., 411 Sqn, 99

B
Banks, F/L W.J., 412 Sqn, 101
Bannock, W/C R., 418 Sqn, 111
Bishop, Lt.-Col. W.A., 5
Boak, F/L W.A., 406 Sqn, 92
Bodard, F/L G.P.A., 410 Sqn, 97
Boulton, S/L F.H., 416 Sqn, 108
Bouskill, F/O R.R., 401 Sqn, 83
Boyle, F/L J.J., 411 Sqn, 99
Brannagan, F/L T.A., 441 Sqn, 140
Britten, F/L R.I.E., 409 Sqn

C
Cameron, F/O G.D., 401 Sqn, 83
Cameron, S/L L.M., 401 Sqn, 83
Charron, F/O P.M., 412 Sqn, 101
Christie, F/O J.S. (RAF), 410 Sqn, 97
Cleveland, S/L H.D., 418 Sqn, 111
Collishaw, Major R., 5
Cotterill, F/L S.H.R., 418 Sqn, 111

D
Dack, F/O D.B., 401 Sqn, 83

E
Edinger, F/L C.E., 410 Sqn, 97
Evans, F/L C.J., 418 Sqn, 111

F
Ford, S/L L.S., 403 Sqn, 87
Forsyth, F/L D.E., 418 Sqn, 111
Francis, F/O J.P.W., 401 Sqn, 83

G
Godefroy, S/L H.C., 403 Sqn, 87
Gordon, F/L D.C., 442 Sqn, 141
Graham, F/O M.G., 411 Sqn, 99
Gray, S/L R., 418 Sqn, 111

H
Hall, F/L D.I., 414 Sqn, 104
Harrington, Lt. A.A. (USAAF), 410 Sqn, 97
Houle, S/L A.U., 417 Sqn, 110

J
Jamieson, F/O D.R.C., 412 Sqn, 101
Jasper, S/L C.M., 418 Sqn, 111
Johnson, F/L G.W., 401 Sqn, 83
Johnson, F/L P.G., 421 Sqn, 115

K
Kerr, S/L J.B., 418 Sqn, 111
Kimball, F/L D.H., 441 Sqn, 140
Kirkpatrick, F/O C.J., 406 Sqn, 92
Klersy, S/L W.T., 401 Sqn, 83

L
Lapp, S/L E.G., 411 Sqn, 99
Laubman, F/L D.C., 412 Sqn, 101
Leggat, F/L P.S., 418 Sqn, 111
Lindsay, F/L J.D., 403 Sqn, 87

M
MacDonald, F/L H.D., 403 Sqn, 87
MacKay, F/L J., 401 Sqn, 83
McLeod, S/L H.W., 443 Sqn, 142
McNair, S/L R.W., 421 Sqn, 115
Miller, F/L H.E., 418 Sqn, 111
Mitchner, P/O I.D., 402 Sqn, 85
Morrison, F/L D.R., 401 Sqn, 83
Mott, F/L G.E., 441 Sqn, 140

N
Noonan, F/L D.E., 416 Sqn, 108
Northcott, S/L G.W., 402 Sqn, 85

R
Reid, F/O S.P., 418 Sqn, 111
Robinson, F/O G.D., 410 Sqn, 97

S
Schultz, F/L R.D., 410 Sqn, 97
Smith, F/L R.I.A., 412 Sqn, 101
Somerville, S/L J.D., 410 Sqn, 97

T
Tonque, F/O D.G. (RAF), 410 Sqn, 97
Trainer, F/L H.C., 411 Sqn, 99

V
Vaessen, F/O C.L., 410 Sqn, 97

W
Williams, W/C D.J., 406 Sqn, 92
Williams, F/L V.A., 410 Sqn, 97

Index 3
Aircraft Citations and Photographs

The asterisk * denotes that the number so marked refers to a photograph.

A
Airspeed Oxford, 22
Armstrong Whitworth Atlas Mk.I, 23, 42, 43*, 51, 52*
Armstrong Whitworth Atlas Mk.IAC, 23*
Armstrong Whitworth Siskin III, 11, 15*, 18, 18n; Mk.IIIA, 21, 22*, 24
Auster A.O.P. Mk.IV, 143; Mk.V, 143; Mk.VI, 201, 201*
Avro 504K, 8
—— 504N, 15
—— 552A Viper, 10n, 13, 13*, 14*, 15, 16*
—— 621 Tutor, 40, 41*, 42, 43*, 45, 45*
—— 626, 40, 41*, 42, 43*, 45
—— 652 Anson Mk.I, 56, 72; Mk.V, 34, 36*, 39*, 57, 71; Mk.VP, 39, 40*
—— 683 Lancaster Mk.I, 90, 119, 120*, 124, 126, 127*, 130, 132; Mk.II, 94, 95*, 122, 123*, 129, 130*; Mk.III, 90, 91*, 119, 120*, 124, 126, 130, 132; Mk.X, 78, 90, 91*, 94, 112, 113*, 114, 114*, 121, 121*, 125, 126*, 129, 129*, 131, 156, 156*, 157, 157*, 159, 159*, 160, 160*, 169, 169*; Mk.XP, 37
—— 694 Lincoln Mk.XV, 225*
—— CF-100 Canuck, 144-6; Mk.3, 162*, 174*; Mk.3B, 182, 182*, 196, 196*, 203, 203*; Mk.4A, 162, 162*, 177, 177**, 184, 187, 187*, 189; Mk.4B, 162, 162*, 177, 182, 182*, 184, 187, 187*, 189, 189*, 190, 190*, 196, 197*, 203, 203*; Mk.5, 162, 163*, 164, 164*, 169, 170*, 171, 172**, 174, 174*, 182, 184, 185*, 187, 187*, 189, 189*, 190, 190*, 197*
—— CF-105 Arrow Mk.I, 147

B
Baddeck No. 1, 1, 3*
Barkley Crow, 34, 35*
Breechcraft Expeditor, 147; Mk.II, 34, 34*;Mk.3, 150, 151*, 165, 153*, 154, 154*, 155, 158, 158*, 165, 166*, 167, 176, 177*, 183, 184*, 194, 199, 200; Mk.3T, 71, 72, 217*
Bellanca Pacemaker, 11, 28, 28*, 30, 30*
Blackburn Shark, 18, 18n; Mk.II, 27, 27*, 42, 43, 57, 57*: Mk.III, 24, 25*, 27, 28, 28*, 51, 52*, 57
Boeing 247D, 34, 56, 56*
—— Fortress Mk.IIA, 73, 73*
Boulton Paul Defiant Mk.I, 95; Mk.IF, 97, 97*
Bristol F.2B, 7*, 8
—— Beaufighter, 76; Mk.IIF, 88, 91, 92*, 95, 97; Mk.VIF, 91, 95; Mk.XC, 88, 89*; Mk.XIC, 88
—— Beaufort Mk.I, 66, 66*, 106
—— Blenheim Mk.I, 91; Mk.IV, 88, 89*, 91, 93
—— Bolingbroke, 20, 22; Mk.I, 30, 48*, 53, 54*, 65; Mk.III, 26*; Mk.IV, 30, 30*, 36, 37*, 48, 48*, 54, 65, 65*, 69; Mk.IVT, 56; Mk.IVW, 54
Burgess Dunne, 1, 3*

C
Canadair CL-11 (RCAF C-5), 167, 168*
—— CP-107 Argus Mk.I, 156, 157, 157*, 173, 173*; Mk.2, 156*, 157
—— CC-109 Cosmopolitan, 167, 168*
—— CL-2 North Star Mk.M1, 167, 167*, 185, 185*
—— Sabre, 144, 146; Mk.2, 163, 164, 169, 170*, 174, 174*, 179, 180*, 181, 186, 186*, 189, 189*, 190, 191*, 195, 196*, 197, 216; Mk.4, 171, 181, 181*, 202, 202*; Mk.5, 147, 150, 151*, 152, 153*, 163, 165, 166*, 169, 170*, 171, 174, 174*, 179, 180*, 181, 181*, 186, 189, 189*, 190, 194, 195*, 196, 196*, 198, 198*, 199, 199*, 200, 201*, 202; Mk.6, 169, 171, 172*, 174, 179, 180*, 181, 181*, 186, 186*, 189, 189*, 190, 191*, 196, 196*, 198, 198*, 202, 202*
—— Silver Star, 147; Mk.3, 150, 151*, 183*, 152, 152*, 153, 154*, 155, 158, 158*, 160, 161*, 165, 165*, 176, 178, 183, 183*, 194, 194*, 205, 205*, 217*
—— CF-104 Starfighter, 144, 179, 180*, 181, 186, 189, 190, 196, 198, 198*, 202, 205, 206*, 218*
—— CC-106 Yukon, 167, 168*, 185n, 193, 193*
Canadian Vickers Vancouver Mk.II, 24, 25*, 36
—— Varuna, 13, 15; Mk.I, 16*; Mk.II, 14*, 16, 16*
—— Vedette Mk.I, 13, 15, 24, 25*, 27, 30, 36; Mk.II, 13*, 16*
Cessna Crane Mk.I, 36, 71, 71*
Consolidated Canso A, 24, 25*, 26, 26*, 27, 28, 29*, 37, 38*, 49, 49*, 50, 50*, 55, 67, 67*, 68, 68*, 160, 160*, 169, 170*
—— Catalina, 20, 77; Mk.I, 26, 26*, 31, 49, 49*, 50, 51*, 103, 103*; Mk.IB, 24, 27, 31, 31*, 49, 50, 103, 103*, 116; Mk.III, 116; Mk.IIIA, 27; Mk.IV, 24, 28, 29*, 31, 103; Mk.IVA, 50, 51*, 55; Mk.VB, 116
—— Liberator, 20; Mk.III, 31, 33, Mk.V, 31, 32*, 33, 33*; C.Mk.VI, 116, 118, 122; G.R.Mk.VI, 31, 32*, 33, 33*; G.R.Mk.VIT, 73, 74*, 166*; C.Mk.VII, 118; C.Mk.VIII, 116, 122, 123*
Curtiss HS-2L, 5, 5*, 10n, 10*, 14, 14*, 15, 15*, 16, 16*
—— JN-4 "Canuck," 1, 3*, 4*
—— Kittyhawk, 20, 20n; Mk.I, 39, 40*, 42, 44*, 52, 53*, 61, 62, 63*, 69; Mk.IA, 61*, 62; Mk.III, 62, 69, 69*; Mk.IV, 42, 43n, 63, 64*
—— Tomahawk Mk.I, 81, 81*, 86, 87*, 104, 105*, 128; Mk.II, 128; Mk.IIA, 81, 86, 87*, 104; Mk.IIB, 81
—— P40K-1 Warhawk, 39
—— P40N Warhawk, 43n

D
de Havilland D.H.4, 10n, 14*, 15
—— D.H.9a, 8, 8*, 10*
—— CC-115 Buffalo, 148, 188, 188*
—— Chipmunk Mk.I, 201, 201*
—— Comet Mk.1A, 167, 168*
—— Mosquito Mk.II, 110; N.F.Mk.II, 97, 97*, ; F.B.Mk.VI, 91, 97, 110, 111*; P.R.Mk.VIC, 88; Mk.XII, 91; N.F.Mk.XIII, 96, 96*, 97; P.R.Mk.XVI, 81, 82*; B.Mk.25, 37; F.B.Mk.26, 62, 63*; F.B.Mk.XXX, 91, 92*; N.F.Mk.XXX, 97
—— Moths, 18n
—— DH-60 Moths, 40, 41*, 42, 45, 45*, 47, 51, 52*, 53
—— Otter, 147, 150, 151*, 152, 154, 154*, 155, 158, 161*, 165, 166*, 176, 177*, 183, 184*, 194, 195*, 199, 200*, 200
—— DH-82 Tiger Moth, 40, 41*
—— DH-82A Tiger Moth, 42, 43*, 53, 53*, 55, 55*
—— Vampire, 145, 147; Mk.III, 150, 151*, 152, 152*, 153, 154*, 163, 164*, 165, 165*, 169, 179, 179*, 194, 194*, 197, 198*, 199, 199*; Mk.5, 179, 179*
Douglas Boston Mk.III, 110, 111*
—— Dakota Mk.I, 70, 70*, 71, 73; Mk.III, 34, 35*, 39, 58, 73, 74*, 132, 133, 134*, 134, 135*, 160, 161*, 167, 167*, 171, 172*; Mk.IV, 39, 73, 132, 133, 133*, 134, 160, 167, 169, 171, 185, 191, 205, 206*
—— DB-1 Digby, 31, 32*, 67, 67*, 72,72*

F
Fairchild 51, 34, 34*, 52*
—— 71, 11, 24, 25*, 26, 26*, 28, 30, 30*, 34, 36, 37*, 42
—— 71B, 44*
—— C-119G Flying Boxcar, 160, 161*, 191, 192, 192*
Fairey Albacore Mk.I, 106, 107*
—— Battle Mk.I, 18, 22, 42, 43*, 47, 52*
—— Transatlantic, 9*
Felixstowe F-3, 10, 10*, 12*
Fleet Fawn Mk.I, 40, 41*, 47, 48*, 53, 54*; Mk.II, 34, 35*, 47
Fokker D.VII, 7*, 8, 8*

G
Gloster Meteor T.Mk&, 179, 180*
Grumman Goblin, 20n, 52, 53*, 58
—— Goose Mk.II, 34, 35*, 36, 37*, 56, 56*, 57, 57*, 71, 72, 166*, 167, 43*

H
Handley Page Halifax Mk.II, 94; B.Mk.II, 90, 90*, 112, 113*, 125, 126, 127*; B.Mk.III, 94, 106, 114, 114*, 119, 120*, 121, 121*, 122, 124, 124*, 126,

128, 129, 130, 131; Mk.V, 94; B.Mk.V, 124, 125, 128, 131; A.Mk.VII, 94, 95*; B.Mk.VII, 106, 122, 123*, 129
—— Hampden Mk. I, 94, 95*, 106, 107*, 114
Hawker Hurricane Mk.I, 18, 20, 20n, 21, 22*, 36, 58, 58*, 59, 60, 83, 84*, 85; Mk.IIA, 84; Mk.IIB, 85, 85*, 109; Mk.IIC, 109; Mk.IV, 136, 136*, 137, 138; Mk.XII, 58, 59, 60, 60*, 61, 62, 63, 64*, 69, 69*; Mk.XIIA, 59, 59*, 60, 61
—— Tomtit, 34, 36*
—— Typhoon Mk.IB, 136, 136*, 137, 137*, 138, 138*, 139*, 224*
HS-2L (Flying Boats, U.S.), 6n

L
Lockheed 10-A, 34*, 57, 71
—— 10-B, 36, 37*, 38*
—— 12-A, 34*
—— 212, 36*
—— CIM-10B Bomarc, 146, 204, 204*, 205
—— Lockheed Electra, 34*, 57, 71
—— C-130B Hercules, 160, 161*, 191
—— C-130E Hercules, 191, 192, 192*, 193*
—— Hudson Mk.I, 20, 33n, 33, 33*, 34, 36, 37*, 55, 55*, 64, 65*; Mk.II, Mk.III, 33, 46, 54, 56, 57, 72, 72*, 93, 93n, 93*; Mk.V, 93, 93*
—— Lodestar, 35*, 70, 70*, 71, 71*, 73, 73*
—— 26 Neptune, 156, 156*, 157, 157*, 159, 159*
—— Vega Ventura G.R.Mk.V, 30, 30*, 46, 48, 48*, 57, 64, 65*, 66, 66*

M
Maurice Farman, No. 731, 6n
McDonnell CF-101 Voodoo, 144, 146, 148
—— CF-101B Voodoo, 162, 163*, 164, 164*, 171, 172*, 174, 175*, 184
—— CF-101F Voodoo, 184, 185*

N
Noorduyn Norseman Mk.III, 57; Mk.IV, 27, 27*, 36, 37, 37*, 38*, 56, 56*, 57, 71; Mk.IVWA, 71; Mk.VI, 34, 36*, 71, 72, 72*, 160, 160*, 169, 170*
North American Harvard Mk.I, 22, 39, 47, 48*, 147; Mk.II, 69, 150, 151*, 152, 153, 154*, 155, 158, 165, 165*, 175, 176, 176*, 178, 183, 183*, 194, 194*, 199, 200, 200* ; Mk.IIB, 34, 35*, 58, 71, 152*, 175*, 178*, 199*
—— Mitchell Mk.II, 36-37, 38*, 169, 170*, 176, 176*; Mk.III, 147, 158, 158*, 167, 167*
—— Mustang Mk. I, 81, 82*, 104, 105*, 128, 145; Mk.III, 140, 140*, 141; Mk.IV, 153, 154*, 155, 155*, 174, 174*, 175, 175*, 178, 178*, 183, 183*, 199, 200, 201*
—— Yale, 22

Northrop Delta Mk.I, 36; Mk.II, 22*, 30, 30*, 34, 34*, 36, 38*, 43*, 53, 54*, 55, 55*

S
Saro Lerwick Mk.I, 116, 117*
S.E. 5a, 5*, 6n, 8, 8*
Short Sunderland Mk.III, 116, 117*, 118, 118*
Sikorsky H-34A Helicopter, 205, 205*
Silver Dart, 1, 2*, 5n
Sopwith Dolphins, 5*, 6n, 8
—— Pup, 8
—— Snipe, 7*, 8
Stinson 105, 34*
Supermarine Spitfire Mk.I, 86; Mk.IA, 98; Mk.IIA, 83, 86, 98, 99*, 101, 101*, 108, 109; Mk.IIB, 108, 109; P.R.Mk.V, 36, 37n, 38*; Mk.VA, 115; Mk.VB, 83, 84*, 85, 86, 98, 101, 101*, 108, 109, 115, 140, 141, 142; Mk.VC, 85, 86, 108, 109; Mk.VIII, 109, 110*; IX, 83, 85, 85*, 86, 87*, 108, 115; Mk.IXB, 83, 84*, 85, 86, 98, 99*, 101, 102*, 108, 108*, 109, 115, 116*, 140, 140*, 141, 142; F.Mk.IX, 140; H.F.Mk.IX, 150; L.F.Mk.IX, 98, 99*, 104, 105*, 140; Mk.IXE, 98, 100*, 101, 102*, 141; P.R.Mk.XI, 81, 82*; Mk.XIV, 98, 142; Mk.XIVE, 83, 85, 85*, 101, 108, 142; F.R.Mk.XIV, 104, 105*, 128; Mk.XVI, 83, 85, 86, 87*, 98, 101, 108, 108*, 115, 116*; Mk.XVIE, 142*
—— Stranraer, 24, 25*, 26, 26*, 27, 27*, 31, 36, 50, 50*, 55, 71

V
Vickers Viking, 13; Mk.IV, 13*, 15, 15*
—— Wellington Mk.IC, 112; Mk.II, 90, 90*; Mk.III, 112, 113*, 122, 123*; B.Mk.III, 114, 119, 121, 124, 125, 126; B.Mk.X, 77, 114, 119, 121, 122, 124, 125, 126, 126*, 128, 129; Mk.XI, 93; Mk.XII (L/L), 93, 93*; Mk.XIII (L/L), 106; Mk.XIV (L/L), 93, 93*

W
Westland Lysander Mk.II, 23, 23*, 40, 41*, 42, 43*, 44, 45, 45*, 51, 52*, 58; Mk.IITT, 56, 57, 57*; Mk.III, 40, 81, 81*, 104
—— Wapiti, 18, 18n; Mk.IIA, 24, 24*, 31, 32*

Index 4
General Index

The asterisk * denotes that the number so marked refers to a photograph

A
Admiralty, British, 4, 20
Adshead, F/O W.W., 32
Aerial Experimental Association, 5n, 6n
Aero Club of America, 6n
Aeronautical Engineering Division, 12n, 17
Air Battalion, See: Royal Flying Corps
Air Board, 9, 10
Air Board Act, 9
Air Council, 147
Air Defence Great Britain, See: Fighter Command, RAF
"airdrome of democracy," 18
Air Mail, 11, 73
Air Ministry, British, 2, 4, 75, 79n
Alaska, 20
Aldwinkle, F/L R.M., 32
Allerton Park, Yorks, 213*
Aleutian Islands, 19
Allied Air Forces Central Europe (AIRCENT) 146, (NATO)
Allied Command Atlantic, (NATO), 145, 146
Allied Command Europe, (NATO) , 146
Allied Tactical Air Forces, (NATO), 146
Second Allied Tactical Air Force, (NATO), 146
Fourth Allied Tactical Air Force, 146, 149n, (NATO)
Anchorage, Alaska, 20
Anderson, P/O C.M., 15
Anderson, F/O C.W., 128
Annette Island, 20
Annis, S/L C.L., 32n
Anthony, P/O K.H., 86, 87
Anti-Smuggling Patrols, 10n
Anti-submarine sweeps, 20
Archer, F/L.P.L.I., 84
Armour Heights, Ont., 2, 3*
Army Council (British), 2
Arnhem, Holland, 78
Arctic flying, 9
"Article 15," 75,78
Asia, South East, 77
Athenia, S.S., 3*
Athens, Greece, 78
Aitken, F/L G.P., 64
Atlantic Ocean, 19, 20, 145-6
Audet, F/L R.J., 98-99, 100*, 99n
Australia, 75, 78, 79n
Australian Flying Corps, 6n

B
Baddeck Bay, N.S. 1
Baden-Soellingen, Germany, 146
Baffin Island, 9
Baldwin, R.W., 5n, 6n
Bannock, S/L.R. 110
Banks, F/L W.J. 101
Bannock, W/C R. credited with 19 V-Is, 76; His aircraft "Hairless Joe," 111*

Barker, Major W.G., V.C.; 5; total score in Italy, 6n
"Basterpiece," 33n, 33*
Bateman, F/O L.J. 93
Battle of Britain, 75-6
Battle of Malta, 78
Bedford Basin, Halifax, N.S. 4, 20
Behan, F/O J.A.T, 62
Belgium, 146, 149n
Bell, Dr. Alexander Graham, 5n
Bell, F/O J.R., 67
Bellis, F/O J.W. 117n
Berchtesgaden, Germany, 90, 141n
Bill C-90, 148
Bill C-243, 148
Bing, Sgt. L.P.S. 79n, 92
Birks, P/O P.F., 47n
Birchall, S/L L.J. 77, 102; Saviour of Ceylon, 103n
Bishop, Lt. Col. W.A., 4, 5, 7, 6n
Bissett, Man. 106
Blakeslee, P/O D.J.M. (later Lt. Col. USAAF), 99n
Bliss, S/L W.H.F. 232
Blitzkrieg, 77
Bomber Offensive, German, Changed from daylight to night, 76
Bomber Operational Base system, 77
Boomer, S/L K.A. 20, 42, 43n, 225, 226n
Boulton, S/L F.H. 108n
Bradford, Yorks, Adopts 429 Sqn. 126
Brice, S/L L.B.B., 94n
British Army, Eighth, 77; Fourteenth, 77; Second, 77-8, 79n; 21st Army Group, 77
British Columbia, Government of, and Civil Operations Branch, 10n
British Commonwealth Air Training Plan, 18, 19, 75, 79n, 229-30
Bruce, F/O R.R. (RAF), 76
Brussels, Treaty of, 146
Burgess-Dunne Company, 1, 6n
Burma, 77
Butler, F/L S.W. (RAF), 117
"Buzz-bombs," 76
Bylot Island, 9
Byrd, Lt. Richard E., 5

C

Calgary, Alta, Adopts 403 (F) Sqn., 86
Cameron, F/O G.D. 83, 83n
Campbell, F/L D.F. 98
Camp Borden, Ont, RFC Training at, 2; training programme of CAF at, 9; 9*
Camp Taliferro, Texas, 2
Campbell, F/O W.E. 46
Canada, 1, 75, 79n, 144-6, 148n, 149n
Canadian Aerodrome Company, 6n
Canadian Air Force Association; Formed 31 August 1920, 9
Canadian Air Force (1920-1923), 9-11
Canadian Air Force (England) 2, 4
Canadian Air Force Packing Section, 4
Canadian Armed Forces (CAF): 144, 145, 147, 148, 149n
Canadian Army, 148; First Canadian Army, 76-78, 79n, 143; Corps, Canadian, 2; Divisions: 1st, 75; 4th, 58n; 7th, 58n; 1st Field Regiment, 143; 1st Parachute Battalion, 71n; Royal Canadian Artillery, 75,143
Canadian Aviation Corps 1, 6n
Canadian Expeditionary Force: 1
Canadian Forces Reorganization Act, 148
"Canadianize," 79n
Canadian National Exhibition; 164n
Cape Canso; 24n
Capp, Al, 176n
Carew, F/O R.D. 232
Carling, F/L H.E. 133
Carlson, F/O A.T. 81
Carpenter, Sgt. T. (RAF), 96
Carscallen, S/L H.M., 31
Carter, Capt. A.D. 8n
Casualties, 2, 5, 6n, 30n, 58, 79
Ceylon, 77, 102, "Saviour of Ceylon", 103n
Chairman of the Chiefs of Staff Committee, 148
Chapman, W/C C.G.W. 68
Chief of the Air Staff, and memo to MND re overseas service, 75
Chief of the Defence Staff, 148
Chief of the General Staff, 17
Chief of Staff, 11, 17
Christofferson Aircraft Company, 6n
Churchill, Winston Spencer, At Quebec Conference, 61n
Cirko, F/L A. 70
Civil Aviation, 12n, 18n
Civil Aviation Branch (Air Board), 9, 17
Civil Government Air Operations, 12n
Civil Operations Branch (Air Board), 9, 10n
Clark, F/L J.M., 33
Clarke, F/L N.S., 128
Clements, S/L W.I., 75
Codes, unit, 233-34
Comox, B.C., 145
Convoy escort, 20
Convoy systems, 4, 20
Cooke, F/O T.C., 68
Cornwall, F/O H.G., 49
Craig, F/O D.L., 33
Craig, F/O W.A., 122
Creed, P/O D., 97
Curry, F/L F.B., 48
Curtiss, Glen, 5n
Curtiss School of Aviation, Toronto, 3*, 6n
Cybulski, P/O M.A., 79n, 97n
Czechoslovakia, 146

D

Daniels, F/L D., 70n
Dartmouth, N.S., 10n, 12*
Davenport, F/L R.M., 83n
Davidson, W/C R.T.P., 232
Davoud, W/C P.Y., 96
Defence Committee, 148
Defence Research Board, 148
Department of the Interior and Arctic flying, 9
Denmark, 146, 149n
Department of Militia and Defence and flight demonstrations at Petawawa, 2 Aug. 1909, 1
Department of Transport, Formed 1 November 1936, 17; and the BCATP, 19
Deputy Minister of National Defence, 12n, 18n
Destroyers, German, 6 June 44, 404 Sqn. part of strike force that sinks three, 88
CAF Directorate of Air Services, 4, 7
Directorate of Air Transport Command, 145, 209
Directorate of Civil Government Air Operations, Formed 1 July 1927, 11-2; Consolidated with the RCAF, 17; other refs: 24n
Distant Early Warning Line (DEW), 148n
Distinguished Flying Cross, First RCAF members to be decorated, 76
Doan, F/S J.E., 30n
Donald, F/O J.D., 232
Donald, Captain R.A., 143
Drake, F/L J.F., 169n
Drumheller, HMCS, 118
Dunver, HMCS, 118

E

East Alburg, Vt., 6n
Eastern Passage, N.S., 4, 6n
"Easy does it" (Aircraft), 429 Sqn., 127*
Edmonton, Alta., Adopts 418 Sqn. 110
Edwards, F/O R.L., 22
Ellesmere Island, 9
Ely, Major D.R., 143
Empringham, F/O L.G., 26
English Channel; and Flying Bombs, 76
Establishment, 11, 144
Evans, F/L F.W., 232
Everard, F/L J.H., 83n

F

"FB," "Flying Boat," "Fighter Bomber," 136n
Farr, S/Sgt. H.A., 1
Farren, F/O J.N., 118
Fédération Aeronautique Internationale, 1
Fellows, F/L F.G., 118
Ferguson, P/O R.R., 97
First French Tactical Air Corps, 149n
Fisher, F/L R.F., 32
Fleming, F/O J.W., 140
Fleming, F/O S.B., 232
Flying Training School at Trenton, 18n
Forest Fire Patrols, 10n
Fort Worth, Texas, 2
Fox, S/L, 232
Foymount, Ont., 208*
France, 146, 149n
Fraser, F. M.P., 55
"French-Canadian," 425 Sqn. 120
Fulton, W/C John "Moose," 419 Sqn. 112n
Fumerton, F/O R.C., 79n, 92
Furious, HMS, 33
Fursman, F/L R.W., 103n

G

Gee, F/L A. 65
George V, H.M. King, Approves granting title "Royal," 12n

Glover, F/L E.A., 232
Gobeil, S/L F.M., 22n, 75, 79n
Godefroy, S/L H.C., 215, 222
Gohl, F/O J.G., 226n
Goldsmith, F/O J.E., 169n
Gooding, F/O H.O., 226n
Goose Bay, Labrador, 144, 209*
Gordon, W/C J.L., 12n
Government, Canadian, 4, 5, 9, 144, 147-8, 232
Grant, F/O D.M., 81
Greece, 149n
Greenburgh, F/L L., 122
Greenland, 146
Greenwood, N.S., 145
Grostenquin, France, 146
Gunnery Training, 4*
"Gutsy Girty" bomber, 124*
Gwatkin, Maj-Gen Sir Willoughby (Air Vice-Marshal), 10n

H
Haight, FS B.M., 97
Hale, G/C E.B., 232
Halifax, N.S., 4, 9, 20, 146
Hamilton, Ont., Adopts 424 Sqn, 119
Hamm, F/L D., 96
Hanton, F/L F.E.W., 81, 223
Hardin, Sgt. G.W.C. (RAF), 418 Sqn, 111
Haskell, Sgt. H., 418 Sqn, 111
"Hawk," Operation, 426 Sqn, 185
Henderson, Captain G.M., 143
Hermanson, F/O E., 96
Heron, F/O P.W., 93
Hespeler, HMCS, 118
High Commissioner, Canadian, 2
High River, Alta., Civil Operations Branch operates a station at, 10n, Air Station at, 12*
Hill, F/O G.O.L., 67
Hill, S/L G.U., 140
Hillman, Sgt. A.B., 169n
Hills, F/O H.H., 104
Hiroshima, Japan, 78
Hoare, Lt-Col. C.G., 6n
Home Defence, and BCATP, require RCAF to retain greater part of its strength in Canada, 75
Hornell, F/L D.E., Awarded the Victoria Cross, 68, 79
Hudson's Bay Company, Winnipeg, Adopts 437 Sqn, 134
Hughes, Sir Sam, 1
Hunt, F/L G.R.M., 65
Hunter, S/L R.E., 117n

I
Iceland, 146
Imperial Order of the Daughters of the Empire, Toronto, Adopt 428 Sqn, 125
India, 77
Irving, Sgt. H.J.H., 418 Sqn, 111
Italy, 77, 146, 149n

J
Janney, Captain E.L., 1, 6n
Japan, Enters War, 19, 20; acceptance of the Allied terms of surrender, 78
Japanese Fire Balloons, 27; 133 Sqn., 62
Japanese Forces, 20
Jericho Beach, B.C., 11*
Johnson, W/C J.E., 140
Jordan, F/O M.P., 93

K
Kamloops, B.C., Adopts 419 Sqn, 112
Keene, P/O N.A. 215
Kelly Field, Texas, 6n
Kenny, Captain W.R., 5*
Kerwin, F/O J.W., 22
Kingcombe, W/C C.B., 76
King's Regulations and Orders for the Royal Canadian Air Force 1924, Promulgated, 12n
Kingsville, Ontario, Adopts 408 Sqn., 94
Kipling, Sgt. T. (RAF), 97
Kiska, Island of, 20
Klersy, S.L. W.T., 83
Korea, South, 146
Korean War, 145, 232

L
Ladbrook, P/O H.H. (RAF), 79n, 97n
LaFrance, F/L J.C.A., 232
Lagan, HMS, 118
Lambros, F/O A., 232
La Presse Newspaper Auxiliary, Montreal, Adopt 425 Sqn, 120
Laubman, S/L D.C., 83n, 101n
Lavery, F/L W.R., 70n73
Lawrence, F/L T.A., 16
Laye, LAC R.S., 143
"Leapfrog I," Operation, 146
Leaside, Ontario, 2; Adopts 432 Sqn, 129
Leckie, A/M R., 7, 10n, 207
Leguerrier, FS J.H.G., 417 Sqn, 109
Lethbridge, Alta., City Council of, Adopts 429 Sqn, 126
Levesque, F/L J.A.O., 232
Lewis, S/L.A., 33
"Li' Abner," 176n
Lincoln Park, 155n
Lindsay, S/L. J.D, 87, 232
Linpert, WO L.W.C., 117n
Little, F/O T.B., 22
"Loco busting," 79n
Logan, S/L R.A., 9
Long Beach, Ont., 2
Lowry, F R.E., 232
Lucas, P/O A. (RAF), 418 Sqn, 111
Luftwaffe, 77
Luxembourg, 146
Lynch, F/O R., 226n

Mac
MacBride, F/L R.E., 68
MacDonald, Sgt. E.N., 215
MacDonald, F/L H.D., 222
Macdonell, Sgt. E.N., 101
MacKay, S/L J., 83, 83n, 232
MacKenzie, S/L A.R., 198n, 232
MacNeil, Captain R.R., 143
MacPherson, F/L S., 26n

M and Mc
Magnificent, HMCS, 146, 216*
Mail, To Canadian servicemen overseas, 73
"Malton Mile," (Lancaster), 90n
Manuel, F/L L.W., 93
Marblehead, Mass., 6n
Marriot, F/L L.T., 141
Marshall, F/O W.O., 68
Marville, France, 146
Matheson, F/O D.R., 215
Maxwell, P/O E.E., 62
McClusky, Sgt. G., 85
McColl-Frontenac Oil Company of Canada, Adopts 421 Sqn, 115
McCurdy, J.A.D., 1, 5n, 6n
McDonald, Sgt. (Later P/O C.E.), Awarded the Military Medal, 87n
McEwen, Lt. C.M., 6n
McFarlane, F/S I.A.H., 55
McFarlane, F/L W.J., 42
McGregor, W/C G.R., 22, 76, and command of Digby Wing, 76, 85
McIntosh, F/L W.H., 73
McLeod, 2/Lt. A.A. V.C., 5
McLeod, S/L H.W., 142
McNab, S/L E.A., 22, 22n, 76
McNair, S/L R.W., 99, 215, 221, 222
McNee, F/S J.W., 24
Metro-Goldwyn Studios, Adopts 427 Sqn, 124
Michalski, F/L W.J., 46
Mid-Canada Line, 148n
Middleton, W/C R.B., 73
Miller, FS K.S., 88
Milliken, Captain W.G., 143
Mills, F/O R.J., 68
Minister of National Defence, RCAF becomes an independent arm under, 18, and Formation of overseas units, 75, and Unification of the forces, 148
Miskiman, F/O H.E., 122
Moffit, S/L B.H., 26
Mohawk, Ontario, 2
Monnon, F/O H.F., 42
Montreal, P.Q., Adopts 438 Sqn, 135
Morris, S/L B.G. (RAF), 87
Morton, WO2 W.F., 117
Mossman, Peter, Illustration by, 100
Moul, FS J., 108
Mullin, F/O J.B., 232
Musgrave, F/L J., 118
Mutual Aid Plan, 147, 149n
Myers, Sgt. C.D., 115
Myers, F/O W.J., 140
Mynarski, P/O A.C., and Victoria Cross, 68n, 79, 112

N
Nagasaki, Japan, 78
Naples, Italy, 78
National Defence Act, passed 28 June 1922, 10n, amended, 148
Neal, F/O E.L., 83
Netherlands, 146, 149n
New Zealand, and Article 15 of the BCATP agreement, 75, and "Tiger

Force", 78; allotted Nos. 485-490 for its sqns, 79n
Nichols, F/L G.H, 232
Nixon, F/O G.W., 232
"Noball" (V-1 flying bomb), 80, 104; See also V-1 flying bomb, 76
NORAD, See: North American Air Defence Command, 145
North Africa, Nos. 420, 424 & 425 detached and on loan to, 77
North American Air Defence Command (NORAD), 145, 148, 148n, Regional HQ at North Bay, Ont., 146
North Atlantic Treaty, 146
North Atlantic Treaty Organization, 145, 146, 148
North Bay, Ont., 146
North Devon Island, 9
North Luffenham, Eng., 146
North Sydney, N.S., 6n
North West Staging Route, 71n
Norway, 146, 149n

O
O'Brien, F/L T.L., 98
O'Donnell, P/O E.M., 93
Omand, Sgt. J.A., 115
Ontario, Government of and Civil Operations Branch, 10n
Ortona, Italy, 143
Oshawa, City of, Adopts 416 Sqn., 107
Oslo, Norway, 78
Ottawa, Ontario, 2, 9, 10n, 12*; City of adopts 440 Sqn, 138
Overseas Military Forces of Canada, 2

P
Pacific Ocean, 20
Paradrop, first, 70n
Parkdale Lions Club, Toronto, Ont, Adopt 411 Sqn, 98; adopt 412 Sqn (13 May 1943), 100
Pathfinder, 405 Sqn., 90
Patten, P/O J.G., 62
Pear Harbour, 20
Perley, Sir George, 6n
Permanent Active Air Force, 11
Petawawa, Ont, and flights of the Silver Dart and Baddeck No. 1, 1
Pethick, F/O T.M., 128
Pinetree Line, 146, 147, 148n, 208*
Plummer, F/O L.A., 140
"Polco," Operation, 169n
No. 310 (Polish) Squadron, 83
Porcupine district of Northern Ontario, Adopts 433 Sqn, 130
Port Mouton, 24n
Portugal, Crew from 425 Sqn interned in, 121n, 146, 149n
Prince Rupert, B.C., 20
Prince Helene (Ferry), 67
Privy Council (Canadian), 2, 4, 9, 11

Q
Quebec City, 6n; Quebec City Detachment, 129 Sqn, 61n; Adopts 425 Sqn, 120
Quebec Conference, 61n

Quebec, Government of and Civil Operations Branch, 10n

R
Radar, Boon to night fighting, 76; Site, 208*
Ramillies, HMS, 54
Randolph, Sgt. W.S., 418 Sqn, 111
Rathbun, Ont, 2
Raymes, F/L D.F., 31
Recruiting, suspended in June 1944, 19
Regina, Sask., Adopts 426 Sqn, 122
Remembrance, Book of, 5
Renfrew, Scotland, 146
Rennie, LAC D.A., 30n
Renown, HMS, 32
Repulse, HMS, 33
Reyno, F/O E.M., 2
Rhine River, 78
"Rhubarbs," 79n
Robb, F/O J.M., 233
Roberts, F/L J.H., 50
Roberts, F/O W.P., 52n
Robertson, G.D. 402 Sqn as Sgt., 85, 215; 421 Sqn as F/L, 115
Roberval, P.Q., Civil Operations Branch operates station at, 10n
Robillard, F/O R.J., 6, 141n
Robinson, F/O E.L., 64
Rogers, F/O A.W., 48
Roosevelt, President Franklin Delano, 18, 61n
Rotary Club of Halifax, N.S., Adopts 434 Sqn, 131
Rouse, Captain D.G., 143
"Royal," Granting of title, 11, 12n
Royal Aero Club, 1
Royal Air Force, 4-5, 6n, 7-8, 18, 68, 75-9, 83-5, 87*, 89, 94n, 98, 99n, 103n, 106n, 111, 133n, 134n, 141n, 143, 164n, 196n, 214-5, 216n, 220-1, 223
Royal Canadian Air Cadets, 144
Royal Flying Corps, 1-2, 3*, 5, 6n; Units, 2, 6n
Royal Canadian Mounted Police, 25, 26n
Royal Canadian Naval Air Service, 4-5
Royal Canadian Navy, 24n, 145, 148
Royal Military Colleges, 148
Royal Naval Air Service, 1, 5, 6n
Royal Navy, 77
"Ruhr Express", 90n, 91*
Rumbold, F/L P.A.S., 103n
Russel, S/L B.D., 22, 76, 141, 215, 221
Russell, F/O A.H., 118

S
SAGE (Semi-automatic ground environmental), 146
Sager, F/L A.H., 222
Saint John, NB, 96
San Antonio Mines, 106
Sargent F/O P.T., 117n
Sarnia, Ont., 104
Scharf, P/O W.K., 58
Schnorkel "breathing" device, 20
School of Aerial Fighting, 2
School of Aerial Gunnery, 2
School of Armament, 2
School of Army Co-operation, 58, 137

Scott, W/C J.C., 103n
Scott, Lt.-Col. J.S., 10n
Selfridge, Lt. Thomas, 5n
Semple, S/L G.C., 221
Senior Air Officer, 17
Shank, F/L W.V., 141n
Shannon, Sgt. R.P., 111
Sharpe, Lt. W.F.N., 1, 6n
Sherman, F/O L., 68
Shoreham-by-Sea, Sussex, 4, 4*
Sicily, 77
Simcoe, Ont., 128
Simpson, F/O H.A., 215
Sims, F/S I.R., 88
Skey, W/C L.W., 117
Small, S/L N.E., 46, 68
Smith, S/L E.G., 232
Smith, S/L R.I.A., 83n
South Africa, 78
Sparling, F/O R.H., 140
"Spartan", Exercise, 79n
Spurr, F/L L.E., 232
Steadman, W/C E.W., 12n
Stevenson, P/O R.R., 32n
St. Hubert, PQ, 145
St. Lawrence River, 20
St. Pierre, F/O J.W., 51
Strength, 2, 5, 17, 19-20, 75, 78, 144
Styles, W/C H.M. (RAF), 93
Submarines (German), 6n, 20; See also under U-BOATS
Sudbury, Ont., 127
Summerside, PEI, 145
Supreme Allied Commander Europe, 146
Supreme Headquarters Allied Powers in Europe, 146
Surrender, German, 78
Sutherland, F/O H.B., 67
Sydney, N.S., 4

T
Taylor, S/L C.I.W., 93
Terrace, B.C., 20
Texas, 2
Thomas, F/L J.F., 119n
"Tiger Force", 78, 225, 225*, 231
Toronto, Ont., 80, 88
Torpedoes, Acoustic, 20
'Train busting', 79n
Trans-Canada Flight, 9
Transport Work, 10n
'Treaty money' flights, 10n, 11
Tudhope, S/L J.H., 14
Turkey, 149n

U
U-Boats, 19-20, U-263, 90; U-283, 93; U-300, 68; U-311, 118; U-341, 32; U-342, 68; U-420, 32; U-456, 118; U-477, 68; U-478, 68; U-484, 118; U-489, 118; U-520, 31; U-610, 118; U-625, 117: U-630, 26; U-658, 64; U-669, 93; U-715, 68; U-754, 46; U-772, 93; U-846, 93; U-980, 68; U-1225, 68
Unification, 148
United Kingdom, 146, 149n
United Nations Organization, 145-6, 148

United States of America, military forces, 2, 5, 6n, 19-20, 99n, 145-6, 148n, 232
Uplands, Ont., 146
Upper Heyford, 4
U.S.S.R., 145-6

V
V-1 Flying-bombs, 76, 85n
Vancouver, B.C., 9, 10n, 89
Victoria Beach, Man., 12*
Victoria, B.C., 95
Victoria Cross, 5, 68, 68n, 79
Vienna, Austria, 78

W
Waddell, W/C R.C.A., 128
Wallace, F/O C.A.B., 83
War Cabinet, Canadian, 143
War Establishment, Home, 19-20, 144
War Establishment, Overseas, 18
Warner, F/L K.W., 122
War Office, British, 6n
Warren, S/L D., 232
Watch, Captain B.R.H., 143
Webster, Clifford, 6n
Weeks, P/O W.R., 141
Wells, W/C E.P. (NZ/RAF), 224
Wesel, Germany, 78
Western Union Defence Organization, 146
West Germany, 149n
Westmount, PQ, 137
White Paper, 148
Wilhelmshaven, Germany, 94n
Wilson, Mr. J.A., 12n
Wilson, F/O L.H., 141n
Windsor, Ont., 109
"Wing Commander Flying", 79n
Winnipeg, Man., 10n, 45n, 45, 84
Women's Air Force Auxiliary, 45, 84, 91, 113
Woodruff, W/C H.P.J. (Can/RAF), 88
Wright, LAC M.L., 143
Wright, Orville, 6n
Wright, Wilbur, 6n

Y
Young, P/O F.B., 141
Young, Sgt. J. (RAF), 88

Z
Zary, F/L H.P.M., 115
Zweibrucken, Germany, 146

PHOTOGRAPH CREDITS

The photographs in this volume, as numbered below, were generously provided by the following collections:

The Aeroplane, 349
D. Anderson, 46, 52, 80
D. Arnold, 270
J. Beilby, 93
The Bell Telephone Co., Ltd., 2
D. Binns, 250
W. Breadman, 105
W. Brown, 136, 137
Canadair Co., Ltd., 306
Canadian Aeroplanes Ltd., 6
Canadian Armed Forces, 86, 87, 88, 89, 90, 92, 117, 182, 197, 210, 257, 285, 320, 321, 323, 352, 353, 363, 366, 377, 379, 385, 399, 403, 406, 408, 414, 423, 427, 460
Canadian War Museum, 228
C. Catalano, 17
G. Chivers, 344, 429
D.R. Dunsmore, 356
J. Ellis, 59, 238
G. Fuller, 291, 292, 339, 415, 417, 418, 445
B. Gibbens, 331, 361, 373
J.A. Griffin, 287, 288, 319, 333, 369, 370, 413, 452
S. Harper, 44
Headquarters, Allied Air Force Central Europe, 381
Imperial War Museum, 199, 202, 204, 223, 227, 230, 231, 239, 240, 251, 252, 258, 266, 279, 455
D. Kennedy, 101

L. Milberry, 364, 412
J. McNulty, 47, 53, 58, 63, 68, 72, 82, 83, 84, 85, 102, 104, 110, 112, 114, 122, 128, 129, 135, 140, 144, 145, 146, 147, 158, 165, 175, 177, 297, 299, 322, 334, 362, 387, 388, 390, 391, 419, 438, 444, 450
K.M. Molson, 5, 7, 10, 11, 14, 15, 31, 42, 66, 134, 153, 174, 286, 293, 294, 298, 300, 302, 307, 308, 309, 310, 313, 314, 315, 316, 318, 329, 337, 338, 351, 354, 358, 365, 367, 368, 371, 372, 384, 386, 392, 396, 397, 401, 402, 404, 405, 407, 430, 433, 435, 436, 441, 449
National Museum of Science and Technology, 36
G. Neal, 8
D.F. Parrott, 150
Public Archives of Canada, 1, 3, 4, 9, 12, 13, 16, 18, 19, 20, 21, 22, 23, 24, 25, 26, 27, 28, 29, 30, 32, 33, 34, 35, 37, 38, 39, 40, 41, 43, 45, 48, 49, 50, 51, 54, 55, 56, 57, 61, 62, 67, 69, 70, 71, 73, 74, 76, 77, 78, 91, 95, 96, 97, 98, 99, 100, 103, 107, 108, 109, 113, 115, 118, 119, 120, 121, 123, 124, 125, 130, 131, 132, 133, 138, 139, 141, 142, 143, 148, 149, 151, 154, 155, 156, 159, 160, 161, 612, 164, 166, 167, 168, 169, 170, 171, 172, 173, 176, 178, 179, 180, 181, 183, 184, 185, 186, 187, 188, 189, 190, 191, 192, 194, 195, 196, 198, 200, 201, 203, 205, 206, 207, 208, 209, 211, 212, 213, 214, 215, 216, 217, 218, 219, 220, 221, 222, 224, 225, 226, 232, 233, 234, 235, 236, 241, 243, 244, 245, 246, 247, 248, 249, 253, 254, 255, 256, 259, 260, 261, 262, 263, 264, 265, 267, 268, 271, 273, 274, 275, 276, 277, 278, 280, 281, 282, 284, 289, 296, 301, 303, 304, 311, 312, 317, 324, 325, 326, 327, 328, 332, 335, 336, 340, 341, 342, 345, 346, 347, 348, 355, 359, 360, 374, 375, 376, 378, 380, 393, 394, 395, 398, 400, 409, 410, 416, 420, 421, 422, 424, 425, 428, 432, 440, 442, 443, 447, 448, 451, 453, 454, 456, 457, 458, 459
R. Richardson, 437
G. Rowe, 269
W. Sawchuck, 446
M.J. Smedley, 193, 295, 343, 350, 382, 383
T. Stachiw, 431
G.E. Stewart, 389
H. Stone, 94
J.M. Taillefer, 163
C. Vincent, 64, 65, 75, 79, 157, 229
A. Walker, 272
W. Wheeler, 111, 305
I. Wikene, 434
G. Wohl, 242
G. Wragg, 106, 290
R. Wylie, 60, 81, 126, 127, 152, 237, 283, 411, 426
No. 4 Wing Album, 439
No. 400 Squadron Album, 116
No. 411 Squadron Album, 330

ERRATA

Page 49, line 24: *for* W/C M.C. Doyle *read* W/C M.G. Doyle

Page 70, line 19: *for* W/C W.U. Michalski *read* W/C W.J. Michalski

Page 72, line 8: *for* W/C W.J. Michalski, AFC 15 Aug 43-12 Jul 43 *read* Jul 44

Page 81, line 22 from bottom: *for* F/O F.E. Hanton *read* F/O F.E.W. Hanton

Page 110, line 25: *for* W/C D.C.S. MacDonald *read* W/C D.C. MacDonald

Page 119, line 34: *for* W/C C.W. Marshall *read* W/C C.C.W. Marshall

Page 199, caption 432: *for* North American Sabre Mk. 5 *read* Canadian Sabre Mk. 5

Page 212, line 19: *for* G/C F.R.C. Carling-Kelly *read* G/C F.C. Carling-Kelly

Page 219, line 3 from bottom: *for* G/C K.M. Flint *read* G/C J.M. Flint

Page 220, line 11 from bottom: *for* W/C J.H. Godfrey *read* W/C J.M. Godfrey

Page 223, line 26: *for* W/C D.C.S. MacDonald *read* W/C D.C. MacDonald